CALCULUS

for Business, Economics, and the Social and Life Sciences

Also Available from McGraw-Hill

Schaum's Outline Series in Mathematics & Statistics

Most outlines include basic theory, definitions, and hundreds of
solved problems and supplementary problems with answers.

Titles on the Current List Include:

Advanced Calculus
Advanced Mathematics
Analytic Geometry
Beginning Calculus
Boolean Algebra
Calculus, 3d edition
Calculus for Business, Economics,
    & The Social Sciences
Calculus of Finite Differences
    & Difference Equations
College Algebra
Complex Variables
Descriptive Geometry
Differential Equations
Differential Geometry
Discrete Mathematics

Elementary Algebra
Essential Computer Mathematics
Finite Mathematics
First Year College Mathematics
Fourier Analysis
General Topology
Geometry, 2d edition
Group Theory
Laplace Transforms
Linear Algebra, 2d edition
Mathematical Handbook of
    Formulas & Tables
Matrices
Matrix Operations
Modern Algebra
Modern Elementary Algebra

Modern Introductory Differential
    Equations
Numerical Analysis, 2d edition
Partial Differential Equations
Probability
Probability & Statistics
Projective Geometry
Real Variables
Review of Elementary
    Mathematics
Set Theory & Related Topics
Statistics, 2d edition
Technical Mathematics
Tensor Calculus
Trigonometry, 2d edition
Vector Analysis

Schaum's Solved Problems Books

Each title in this series is a complete and expert source of solved
problems containing thousands of problems with worked out solutions.

Titles on the Current List Include:

3000 Solved Problems in Calculus
2500 Solved Problems in College Algebra and
    Trigonometry
2500 Solved Problems in Differential Equations

2000 Solved Problems in Discrete Mathematics
3000 Solved Problems in Linear Algebra
2000 Solved Problems in Numerical Analysis
3000 Solved Problems in Precalculus

Available at your College Bookstore. A complete list of Schaum titles may be obtained by writing to:

Schaum Division
McGraw-Hill, Inc.
Princeton Road S-1
Hightstown, NJ 08520

CALCULUS

For Business, Economics, and the Social and Life Sciences

FIFTH EDITION

LAURENCE D. HOFFMANN

Prudential Securities

GERALD L. BRADLEY

Department of Mathematics

Claremont McKenna College

McGRAW-HILL, INC.

NEW YORK ST. LOUIS SAN FRANCISCO AUCKLAND BOGOTÁ CARACAS LISBON

LONDON MADRID MEXICO MILAN MONTREAL NEW DELHI PARIS

SAN JUAN SINGAPORE SYDNEY TOKYO TORONTO

CALCULUS for Business, Economics, and the Social and Life Sciences

2 3 4 5 6 7 8 9 0 VNH VNH 9 0 9 8 7 6 5 4 3 2

ISBN 0-07-029344-9

This book was set in Times Roman by Better Graphics, Inc.
The editors were Richard Wallis and Jack Maisel;
the designer was Hermann Strohbach;
the production supervisor was Friederich W. Schulte.
Von Hoffmann Press, Inc., was printer and binder.
Cover photo: © Eric Schweikardt for Sports Illustrated.

Photo Credits for Chapter-Opening Photos

1 Terje Rakke/The Image Bank	7 Daniel Brody/Stock, Boston
2 Globus/The Stock Market	8 Berenice Abbott/Photo Researchers
3 Comstock	9 Roy Bishop/Stock, Boston
4 R. Michael Stuckey/Comstock	10 George Holton/Photo Researchers
5 Hugh Rogers/Monkmeyer	11 Joel Gordon
6 Hugh Rogers/Monkmeyer	12 Catherine Ursello/Photo Researchers

Library of Congress Cataloging-in-Publication Data

Hoffmann, Laurence D., (date).
   Calculus for business, economics, and the social and life sciences
/ Laurence D. Hoffmann, Gerald L. Bradley. —5th ed.
     p.   cm.
   Includes index.
   ISBN 0-07-029344-9
   1. Calculus.   I. Bradley, Gerald L., (date).   II. Title
QA303.H569   1992
515—dc20   91-38983

About the Authors

Laurence D. Hoffmann is a former professor at Claremont McKenna College, where he taught mathematics from 1970 to 1985. He is a three-time recipient of the College's Huntoon Award for Excellence in Teaching, an award for "best teacher" voted by the students. He received his B.A. from Brown University and his Ph.D. from the University of Wisconsin. He is also the coauthor of *Mathematics with Applications* and *Finite Mathematics with Applications* (both McGraw-Hill). Since 1985, Hoffmann has been a financial advisor with Prudential Securities, Inc., where he is currently Vice President for Investments.

Gerald L. Bradley has taught at Claremont McKenna College since 1966. He received his B.S. degree from Harvey Mudd College and his Ph.D. from the California Institute of Technology. Bradley's major field of interest is matrix theory. He is the author of the text *A Primer of Linear Algebra* (Prentice Hall). His personal interests include archaeology, history, and astronomy, and he is an ardent bridge player.

Contents

Preface

Overview

If you are preparing for a career in business, economics, psychology, sociology, architecture, or biology, and if you have taken high school algebra, then this book was written for you. Its primary goal is to teach you the techniques of differential and integral calculus that you are likely to encounter in undergraduate courses in your major and in your subsequent professional activities. The exposition is designed to give you a sound, intuitive understanding of the basic concepts without sacrificing mathematical accuracy. Thus, the main results are stated carefully and completely, and whenever possible, explanations are intuitive or geometric.

Applications The text is applications-oriented. Each new concept you learn is applied to a variety of practical situations. The techniques and strategies you will need to solve applied problems are stressed. The applications are drawn from the social, managerial, and life sciences, with special emphasis on business and economics.

Problems You learn mathematics by doing it. Each section in this text is followed by an extensive set of problems. Many involve routine computation and are designed to help you master new techniques. Others ask you to apply the new techniques to practical situations. There is a set of review problems at the end of each chapter. At the back of the book, you will find the answers to the odd numbered problems and to all the review problems.

Algebra Review If you need to brush up on your high school algebra, there is an extensive algebra review in the appendix that includes worked examples and practical problems for you to do. You will be advised throughout the text when it might be appropriate to consult this material.

Features New to this Edition

The fifth edition retains the straightforward style, intuitive approach, and applications orientation of the first four editions, while improving the focus of certain sections and adding examples and exercises. Many new applications have been added, such as: the effect of taxation on a monopoly; Pareto's law; currency replacement; optimum sail design; Zeno's race course paradox; and the Gompertz curve. The design of the text has been improved by using color for emphasis and placing certain figures in the margins. Here is a list of the major changes:

- Section 1.2 now includes intersections of graphs and a discussion of how to graph parabolas of the general form $y = Ax^2 + Bx + C$.
- There is a more traditional treatment of rectilinear motion and other rates of change in Section 2.3. The material on marginal analysis is now in Section 2.4, as part of a general discussion of approximation by increments and differentials.
- In Chapter 4, there is new material on compound interest and graphing exponential and logarithm functions. The half-life of a radioactive substance, carbon dating, and logarithmic differentiation are explained in more detail.
- Some elementary applications involving integration have been transferred from Chapter 6 to Chapter 5, leaving only those applications that use definite integration in some essential way in Chapter 6. Examples and problems involving rectilinear motion have been added to Chapter 5. Chapter 6 has been streamlined by combining the old Sections 6.4 and 6.5 into a single section emphasizing the use of the definite integral in applications.
- Chapter 7 now includes a section on first-order linear differential equations and features examples that illustrate how non-separable linear differential equations can arise in practical problems.
- The material on level curves, isoquants, and indifference curves is now in Section 9.1. Chapter 9 also includes new material on computer-generated figures.
- Newton's method of approximating roots has been moved from Chapter 11 to Chapter 3.

Supplementary Materials

Instructor's Manual: This manual by Henri Feiner (West Los Angeles College and California Academy of Mathematics and Science) contains the answers to the even-numbered problems. (Answers to odd-numbered problems and all review problems are in the back of the book.) Complimentary copies of this manual are available to adopters.

Student's Solutions Manual: This manual by Stanley M. Lukawecki (Clemson University) and Henri Feiner contains complete step-by-step solutions of all the odd-numbered problems and all review problems. Students may purchase this manual through their local bookstore.

Matrix Algebra Supplement: This textual supplement is available for student purchase. Contact your McGraw-Hill sales representative for details.

Print Test Bank: Instructors can choose multiple-choice questions, open-ended problems, and short conceptual test questions from each chapter to make their own tests and quizzes.

Computerized Testing System: Instructors can create and print tests by entering their own additional questions or selecting from items in the printed test bank. This testing system includes editing capabilities and an algorithmic generator. (Available for IBM and Macintosh.)

Computerized Study Guide: This tutorial provides additional coverage and support for all sections of the text. Students can choose different types of questions to work by chapter. The study guide indicates if the student has the correct answer; if not, an explanation of the correct answer will be given and the guide will refer the student to the appropriate section in the text for additional information. Students may purchase this guide through their local bookstore.

Full-Color Overhead Transparencies: An extensive collection of full-color overhead transparencies depicting selected calculus concepts, both single and multivariable. Contact your McGraw-Hill sales representative for details.

Computer Software: IBM and Macintosh packages for numerical computation and graphical displays are available. Contact your McGraw-Hill sales representative for details.

Acknowledgments

Many people helped with the preparation of this edition, but we are again especially grateful for Henri Feiner's insightful comments and thorough editing work.

The following reviewers provided thoughtful comments and suggestions for improvement on the manuscript for this edition: Joel W. Irish, University of Southern Maine; Donald R. LaTorre, Clemson University; James E. McClure, University of Kentucky; Ann B. Megaw, University of Texas at Austin; and Robert E. Sharpton, Miami-Dade Community College. Special thanks also go to John Gresser at Bowling Green State University and to Randall C. Hicks, Mary S. Hopper, Ashok Kumar, and Sudhir Kumar Goel at Valdosta State College.

Reviewers of previous editions include: Dan Anderson, University of Iowa; Randall Brian, Vincennes University; James F. Brooks, Eastern Kentucky University; Laura Cameron, University of New Mexico; Gerald R. Chachere, Howard University; Charles C. Clever, South Dakota State University; Raúl Curto, University of Iowa; Bruce Edwards, University of Florida; Ronnie Goolsby, Winthrop College; Erica Jen, University of Southern California; Melvin Lax, California State University at Long Beach; James Osterburg, University of Cincinnati; Richard Randell, University of Iowa; Dolores Schaffner, University of South Dakota; Anthony Shersin, Florida International University; Jane E. Sieberth, Franklin University; Joseph F. Stokes, Western Kentucky University; Keith Stroyan, University of Iowa; Martin Tangora, University of Illinois at Chicago; Lee Topham, North Harris County College; Charles Votaw, Fort Hays State University; and Jonathan Weston-Dawkes, University of North Carolina.

The information and advice provided by Susan Feigenbaum during the preparation of the business and economics material of this text were invaluable.

Finally, special thanks go to Richard H. Wallis and Jack Maisel, our McGraw-Hill editors, for their dedicated efforts and support.

Laurence D. Hoffmann

Gerald L. Bradley

CALCULUS

for Business, Economics, and the Social and Life Sciences

CHAPTER 1
FUNCTIONS, GRAPHS, AND LIMITS

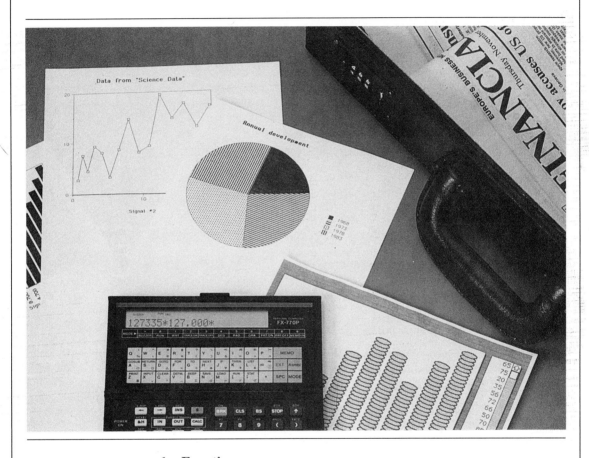

1   Functions

In many practical situations, the value of one quantity may depend on the value of a second. For example, the consumer demand for beef may depend on its current market price; the amount of air pollution in a metropolitan area may depend on the number of cars on the road; the value of a bottle of wine may depend on its age. Such relationships can often be represented mathematically as **functions.**

> **Function**
>
> A function is a rule that assigns to each object in a set A one and only one object in a set B.

This definition is illustrated in Figure 1.1.

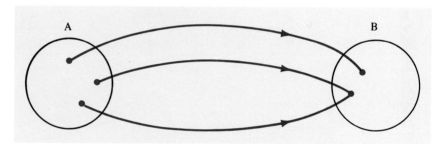

Figure 1.1 A visual representation of a function.

For most of the functions in this book, the sets A and B will be collections of real numbers. You can think of such a function as a rule that assigns "new" numbers to "old" numbers. To be called a function, the rule must have the property that it assigns one and only one "new" number to each "old" number. Here is an example.

EXAMPLE 1.1

According to a certain function, the "new" number is obtained by adding 4 to the square of the "old" number. What number does this function assign to 3?

Solution

The number assigned to 3 is $3^2 + 4$, or 13.

Variables   Often you can write a function compactly by using a mathematical formula. It is traditional to let x denote the old number and y the new number and write an equation relating x and y. For instance, you can express the function in Example 1.1 by the equation

$$y = x^2 + 4$$

The letters x and y that appear in such an equation are called **variables.** The numerical value of the variable y is determined by that of the variable x. For this reason, y is sometimes referred to as the **dependent variable** and x as the **independent variable.**

Functional Notation   There is an alternative notation for functions that is widely used and somewhat more versatile. A letter such as f is chosen to stand for the function itself, and the value that the function assigns to x is denoted by $f(x)$ instead of y. The symbol $f(x)$ is read "f of x." Using this **functional notation,** you can rewrite Example 1.1 as follows.

EXAMPLE 1.2

Find $f(3)$ if $f(x) = x^2 + 4$.

Solution

$$f(3) = 3^2 + 4 = 13$$

Observe the convenience and simplicity of this notation. In Example 1.2, the compact formula $f(x) = x^2 + 4$ completely defines the function and the simple equation $f(3) = 13$ indicates that 13 is the number that the function assigns to 3.

The use of functional notation is illustrated further in the following examples. Notice that in Example 1.3, letters other than f and x are used to denote the function and its independent variable.

EXAMPLE 1.3

If $g(t) = (t - 2)^{1/2}$, find (if possible) $g(27)$, $g(5)$, $g(2)$, and $g(1)$.

Solution

Rewrite the function as $g(t) = \sqrt{t - 2}$. (If you need to brush up on fractional powers, you can consult the discussion of exponential notation in the Algebra Review at the back of the book.) Then,

$$g(27) = \sqrt{27 - 2} = \sqrt{25} = 5$$

$$g(5) = \sqrt{5 - 2} = \sqrt{3} \approx 1.7321$$

and

$$g(2) = \sqrt{2 - 2} = \sqrt{0} = 0$$

However, $g(1)$ is undefined since

$$g(1) = \sqrt{1 - 2} = \sqrt{-1}$$

and negative numbers do not have real square roots.

In the next example, two formulas are needed to define the function.

EXAMPLE 1.4

Find $f\left(-\dfrac{1}{2}\right)$, $f(1)$, and $f(2)$ if

$$f(x) = \begin{cases} \dfrac{1}{x - 1} & \text{if } x < 1 \\ 3x^2 + 1 & \text{if } x \geq 1 \end{cases}$$

Solution

From the first formula,

$$f\left(-\frac{1}{2}\right) = \frac{1}{-1/2 - 1} = \frac{1}{-3/2} = -\frac{2}{3}$$

and from the second formula,

$$f(1) = 3(1)^2 + 1 = 4 \qquad \text{and} \qquad f(2) = 3(2)^2 + 1 = 13$$

The next example illustrates how functional notation is used in a practical situation. Notice that to make the algebraic formula easier to interpret, letters suggesting the relevant practical quantities are used for the function and its independent variable. (In this example, the letter C stands for "cost" and q stands for "quantity" manufactured.)

EXAMPLE 1.5

Suppose the total cost in dollars of manufacturing q units of a certain commodity is given by the function $C(q) = q^3 - 30q^2 + 500q + 200$.

(a) Compute the cost of manufacturing 10 units of the commodity.
(b) Compute the cost of manufacturing the 10th unit of the commodity.

Solution

(a) The cost of manufacturing 10 units is the value of the total cost function when $q = 10$. That is,

$$\text{Cost of 10 units} = C(10)$$
$$= (10)^3 - 30(10)^2 + 500(10) + 200$$
$$= \$3,200$$

(b) The cost of manufacturing the 10th unit is the difference between the cost of manufacturing 10 units and the cost of manufacturing 9 units. That is,

$$\text{Cost of 10th unit} = C(10) - C(9) = 3,200 - 2,999 = \$201$$

The Domain of a Function

The set of values of the independent variable for which a function can be evaluated is called the **domain** of the function. For instance, the function $f(x) = x^2 + 4$ in Example 1.2 can be evaluated for any real number x. Thus, the domain of this function is the set of all real numbers. The domain of the function $C(q) = q^3 - 30q^2 + 500q + 200$ in Example 1.5 is also the set of all real numbers [although $C(q)$ represents total cost only for nonnegative values of q]. In the next example are two functions whose domains are restricted for algebraic reasons.

EXAMPLE 1.6

Find the domain of each of the following functions:

$$\text{(a) } f(x) = \frac{1}{x - 3}$$
$$\text{(b) } g(x) = \sqrt{x - 2}$$

Solution

(a) Since division by any real number except zero is possible, the only value of x for which $f(x) = \dfrac{1}{x - 3}$ cannot be evaluated is $x = 3$, the value that makes the denominator of f equal to zero. Hence, the domain of f consists of all real numbers except 3.

(b) Since negative numbers do not have real square roots, the only values of x for which $g(x) = \sqrt{x - 2}$ can be evaluated are those for which $x - 2$ is nonnegative, that is, for which

$$x - 2 \geq 0 \quad \text{or} \quad x \geq 2$$

That is, the domain of g consists of all real numbers that are greater than or equal to 2.

Composition of Functions

There are many situations in which a quantity is given as a function of one variable which, in turn, can be written as a function of a second variable. By combining the functions in an appropriate way, you can express the original quantity as a function of the second variable. This process is known as the **composition of functions.**

Composition of Functions

The composite function $g[h(x)]$ is the function formed from the two functions $g(u)$ and $h(x)$ by substituting $h(x)$ for u in the formula for $g(u)$.

The situation is illustrated in Figure 1.2.

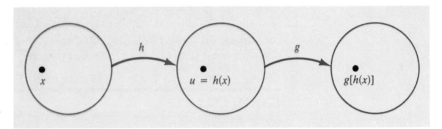

Figure 1.2 The composition of functions.

EXAMPLE 1.7

Find the composite function $g[h(x)]$ if $g(u) = u^2 + 3u + 1$ and $h(x) = x + 1$.

Solution

Replace u by $x + 1$ in the formula for g to get

$$\begin{aligned} g[h(x)] &= (x + 1)^2 + 3(x + 1) + 1 \\ &= (x^2 + 2x + 1) + (3x + 3) + 1 \\ &= x^2 + 5x + 5 \end{aligned}$$

Example 1.7 could have been worded more compactly as follows: Find the composite function $g(x + 1)$ where $g(u) = u^2 + 3u + 1$. The use of this compact notation is illustrated further in the next example.

EXAMPLE 1.8

Find $f(x - 1)$ if $f(x) = 3x^2 + \dfrac{1}{x} + 5$.

Solution

At first glance, this problem may look confusing because the letter x appears both as the independent variable in the formula defining f and as part of the expression $x - 1$. Because of this, you may find it helpful to begin by writing the formula for f in more neutral terms, say as

$$f(\square) = 3(\square)^2 + \frac{1}{\square} + 5$$

To find $f(x - 1)$, you simply insert the expression $x - 1$ inside each box, getting

$$f(x - 1) = 3(x - 1)^2 + \frac{1}{x - 1} + 5$$

Occasionally, you will have to be able to "take apart" a given composite function $g[h(x)]$ and identify the functions $g(u)$ and $h(x)$ from which it was formed. The procedure is illustrated in the next example.

EXAMPLE 1.9

If $f(x) = \dfrac{5}{x - 2} + 4(x - 2)^3$, find functions $g(u)$ and $h(x)$ such that $f(x) = g[h(x)]$.

Solution

The form of the given function is

$$f(x) = \frac{5}{\square} + 4(\square)^3$$

where each box contains the expression $x - 2$. Thus $f(x) = g[h(x)]$, where

$$g(u) = \frac{5}{u} + 4u^3 \quad \text{and} \quad h(x) = x - 2$$

Actually, in Example 1.9, there are infinitely many pairs of functions $g(u)$ and $h(x)$ that combine to give $g[h(x)] = f(x)$. [For example, $g(u) = \dfrac{5}{u + 1} + 4(u + 1)^3$ and $h(x) = x - 3$.] The particular pair selected in the solution to this example is the most natural one and reflects most clearly the structure of the original function $f(x)$.

An Application of Composite Functions   The next example illustrates how a composite function may arise in a practical problem.

EXAMPLE 1.10

An environmental study of a certain community suggests that the average daily level of carbon monoxide in the air will be $c(p) = 0.5p + 1$ parts per million when the population is p thousand. It is estimated that t years from now the population of the community will be $p(t) = 10 + 0.1t^2$ thousand.

(a) Express the level of carbon monoxide in the air as a function of time.
(b) When will the carbon monoxide level reach 6.8 parts per million?

Solution

(a) Since the level of carbon monoxide is related to the variable p by the equation
$$c(p) = 0.5p + 1$$
and the variable p is related to the variable t by the equation
$$p(t) = 10 + 0.1t^2$$
it follows that the composite function
$$c[p(t)] = c(10 + 0.1t^2) = 0.5(10 + 0.1t^2) + 1 = 6 + 0.05t^2$$
expresses the level of carbon monoxide in the air as a function of the variable t.

(b) Set $c[p(t)]$ equal to 6.8 and solve for t to get
$$6 + 0.05t^2 = 6.8$$
$$0.05t^2 = 0.8$$
$$t^2 = \frac{0.8}{0.05} = 16$$
$$t = \sqrt{16} = 4$$

That is, 4 years from now the level of carbon monoxide will be 6.8 parts per million.

Problems 1.1

In Problems 1 through 11, compute the indicated values of the given function.

1. $f(x) = 3x^2 + 5x - 2; f(1), f(0), f(-2)$

2. $h(t) = (2t + 1)^3; h(-1), h(0), h(1)$

3. $g(x) = x + \dfrac{1}{x}; g(-1), g(1), g(2)$

4. $f(x) = \dfrac{x}{x^2 + 1}; f(2), f(0), f(-1)$

5. $h(t) = \sqrt{t^2 + 2t + 4}; h(2), h(0), h(-4)$

6. $g(u) = (u + 1)^{3/2}; g(0), g(-1), g(8)$

7. $f(t) = (2t - 1)^{-3/2}; f(1), f(5), f(13)$

8. $g(x) = 4 + |x|; g(-2), g(0), g(2)$

9. $f(x) = x - |x - 2|; f(1), f(2), f(3)$

10. $h(x) = \begin{cases} -2x + 4 & \text{if } x \le 1 \\ x^2 + 1 & \text{if } x > 1 \end{cases}; h(3), h(1), h(0), h(-3)$

11. $f(t) = \begin{cases} 3 & \text{if } t < -5 \\ t + 1 & \text{if } -5 \le t \le 5; f(-6), f(-5), f(16) \\ \sqrt{t} & \text{if } t > 5 \end{cases}$

In Problems 12 through 24, specify the domain of the given function.

12. $f(x) = x^3 - 3x^2 + 2x + 5$

13. $g(x) = \dfrac{x^2 + 5}{x + 2}$

14. $f(t) = \dfrac{t + 1}{t^2 - t - 2}$

15. $y = \sqrt{x - 5}$

16. $f(x) = \sqrt{2x - 6}$

17. $g(t) = \sqrt{t^2 + 9}$

18. $h(u) = \sqrt{u^2 - 4}$

19. $f(t) = (2t - 4)^{3/2}$

20. $y = \dfrac{x - 1}{\sqrt{x^2 + 2}}$

21. $f(x) = (x^2 - 9)^{-1/2}$

22. $h(t) = \dfrac{\sqrt{t^2 - 4}}{\sqrt{t - 4}}$

23. $g(t) = \dfrac{1}{|t - 1|}$

24. $h(x) = \sqrt{|x - 3|}$

Manufacturing Cost   25. Suppose the total cost in dollars of manufacturing q units of a certain commodity is given by the function $C(q) = q^3 - 30q^2 + 400q + 500$.

(a) Compute the cost of manufacturing 20 units.

(b) Compute the cost of manufacturing the 20th unit.

Worker Efficiency   26. An efficiency study of the morning shift at a certain factory indicates that an average worker who arrives on the job at 8:00 A.M. will have assembled $f(x) = -x^3 + 6x^2 + 15x$ transistor radios x hours later.

(a) How many radios will such a worker have assembled by 10:00 A.M.? (*Hint*: At 10:00 A.M., $x = 2$.)

(b) How many radios will such a worker assemble between 9:00 and 10:00 A.M.?

Temperature Change   27. Suppose that t hours past midnight, the temperature in Miami was $C(t) = -\frac{1}{6}t^2 + 4t + 10$ degrees Celsius.

(a) What was the temperature at 2:00 P.M.?

(b) By how much did the temperature increase or decrease between 6:00 and 9:00 P.M.?

Population Growth   28. It is estimated that t years from now, the population of a certain suburban community will be $P(t) = 20 - \dfrac{6}{t + 1}$ thousand.

(a) What will the population of the community be 9 years from now?

(b) By how much will the population increase during the 9th year?

(c) What will happen to the size of the population in the long run?

Experimental   29. To study the rate at which animals learn, a psychology student
Psychology   performed an experiment in which a rat was sent repeatedly through a laboratory maze. Suppose that the time required for the rat to traverse the maze on the nth trial was approximately $f(n) = 3 + \dfrac{12}{n}$ minutes.

(a) What is the domain of the function f?

(b) For what values of n does $f(n)$ have meaning in the context of the psychology experiment?

(c) How long did it take the rat to traverse the maze on the 3rd trial?

(d) On which trial did the rat first traverse the maze in 4 minutes or less?

(e) According to the function f, what will happen to the time required for the rat to traverse the maze as the number of trials increases? Will the rat ever be able to traverse the maze in less than 3 minutes?

Poiseuille's Law   30. Biologists have found that the speed of blood in an artery is a function of the distance of the blood from the artery's central axis. According to **Poiseuille's law,** the speed (in centimeters per second) of blood that is r centimeters from the central axis of an artery is given by the function $S(r) = C(R^2 - r^2)$, where C is a constant and R is the radius of the artery. Suppose that for a certain artery, $C = 1.76 \times 10^5$ and $R = 1.2 \times 10^{-2}$ centimeters.

(a) Compute the speed of the blood at the central axis of this artery.

(b) Compute the speed of the blood midway between the artery's wall and central axis.

Distribution Cost 31. Suppose that the number of worker-hours required to distribute new telephone books to x percent of the households in a certain rural community is given by the function $f(x) = \dfrac{600x}{300 - x}$.

(a) What is the domain of the function f?

(b) For what values of x does $f(x)$ have a practical interpretation in this context?

(c) How many worker-hours were required to distribute new telephone books to the first 50 percent of the households?

(d) How many worker-hours were required to distribute new telephone books to the entire community?

(e) What percentage of the households in the community had received new telephone books by the time 150 worker-hours had been expended?

Immunization 32. Suppose that during a nationwide program to immunize the population against a certain form of influenza, public health officials found that the cost of inoculating x percent of the population was approximately $f(x) = \dfrac{150x}{200 - x}$ million dollars.

(a) What is the domain of the function f?

(b) For what values of x does $f(x)$ have a practical interpretation in this context?

(c) What was the cost of inoculating the first 50 percent of the population?

(d) What was the cost of inoculating the second 50 percent of the population?

(e) What percentage of the population had been inoculated by the time 37.5 million dollars had been spent?

Speed of a Moving Object 33. A ball has been dropped from the top of a building. Its height (in feet) after t seconds is given by the function $H(t) = -16t^2 + 256$.

(a) How high will the ball be after 2 seconds?

(b) How far will the ball travel during the 3rd second?

(c) How tall is the building?

(d) When will the ball hit the ground?

In Problems 34 through 41, find the composite function $g[h(x)]$.

34. $g(u) = u^2 + 4$, $h(x) = x - 1$

35. $g(u) = 3u^2 + 2u - 6$, $h(x) = x + 2$

36. $g(u) = (2u + 10)^2$, $h(x) = x - 5$

37. $g(u) = (u - 1)^3 + 2u^2$, $h(x) = x + 1$

38. $g(u) = \dfrac{1}{u}$, $h(x) = x^2 + x - 2$

39. $g(u) = \dfrac{1}{u^2}$, $h(x) = x - 1$

40. $g(u) = u^2$, $h(x) = \dfrac{1}{x - 1}$

41. $g(u) = \sqrt{u + 1}$, $h(x) = x^2 - 1$

In Problems 42 through 49, find the indicated composite function.

42. $f(x + 1)$ where $f(x) = x^2 + 5$

43. $f(x - 2)$ where $f(x) = 2x^2 - 3x + 1$

44. $f(x + 3)$ where $f(x) = (2x - 6)^2$

45. $f(x - 1)$ where $f(x) = (x + 1)^5 - 3x^2$

46. $f\left(\dfrac{1}{x}\right)$ where $f(x) = 3x + \dfrac{2}{x}$

47. $f(x^2 + 3x - 1)$ where $f(x) = \sqrt{x}$

48. $f(x^2 - 2x + 9)$ where $f(x) = 2x - 20$

49. $f(x + 1)$ where $f(x) = \dfrac{x - 1}{x}$

In Problems 50 through 55, find functions $h(x)$ and $g(u)$ such that $f(x) = g[h(x)]$.

50. $f(x) = (x^5 - 3x^2 + 12)^3$

51. $f(x) = \sqrt{3x - 5}$

52. $f(x) = (x - 1)^2 + 2(x - 1) + 3$

53. $f(x) = \dfrac{1}{x^2 + 1}$

54. $f(x) = \sqrt{x + 4} - \dfrac{1}{(x + 4)^3}$

55. $f(x) = \sqrt{x + 3} - \dfrac{1}{(x + 4)^3}$

Air Pollution   56. An environmental study of a certain suburban community suggests that the average daily level of carbon monoxide in the air will be

$c(p) = 0.4p + 1$ parts per million when the population is p thousand. It is estimated that t years from now the population of the community will be $p(t) = 8 + 0.2t^2$ thousand.

(a) Express the level of carbon monoxide in the air as a function of time.

(b) What will the carbon monoxide level be 2 years from now?

(c) When will the carbon monoxide level reach 6.2 parts per million?

Manufacturing Cost   57. At a certain factory, the total cost of manufacturing q units during the daily production run is $C(q) = q^2 + q + 900$ dollars. On a typical workday, $q(t) = 25t$ units are manufactured during the first t hours of a production run.

(a) Express the total manufacturing cost as a function of t.

(b) How much will have been spent on production by the end of the 3rd hour?

(c) When will the total manufacturing cost reach $11,000?

Consumer Demand   58. An importer of Brazilian coffee estimates that local consumers will buy approximately $Q(p) = \dfrac{4,374}{p^2}$ kilograms of the coffee per week when the price is p dollars per kilogram. It is estimated that t weeks from now the price of this coffee will be $p(t) = 0.04t^2 + 0.2t + 12$ dollars per kilogram.

(a) Express the weekly consumer demand for the coffee as a function of t.

(b) How many kilograms of the coffee will consumers be buying from the importer 10 weeks from now?

(c) When will the demand for the coffee be 30.375 kilograms?

2   The Graph of a Function

Graphs have visual impact. They also reveal information that may not be evident from verbal or algebraic descriptions. Two graphs depicting practical relationships are shown in Figure 1.3.

The graph in Figure 1.3a describes the variation in total industrial production in a certain country over a 4-year period of time. Notice that the highest point on the graph occurs near the end of the 3rd year, indicating that production was greatest at that time.

The graph in Figure 1.3b represents population growth when environmental factors impose an upper bound on the possible size of the population. It indicates that the rate of population growth increases at first and then decreases as the size of the population gets closer and closer to the upper bound.

To represent a function $y = f(x)$ geometrically as a graph, it is common

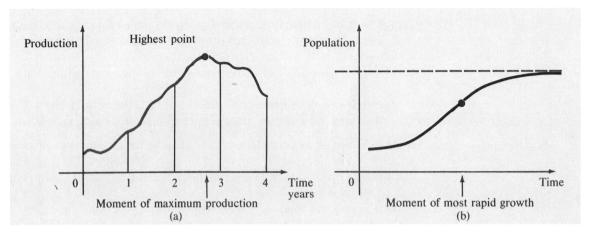

Figure 1.3 (a) A production function. (b) Bounded population growth.

practice to use a rectangular coordinate system on which units for the independent variable x are marked on the horizontal axis and those for the dependent variable y are marked on the vertical axis.

The Graph of a Function

The graph of a function f consists of all points (x, y) where x is in the domain of f and $y = f(x)$.

In Chapter 3, you will see efficient techniques involving calculus that can be used to draw accurate graphs of functions. For many functions, however, you can make a fairly good sketch by the elementary method of **plotting points,** which may be summarized as follows.

How to Sketch the Graph of a Function f by Plotting Points

1. Choose a representative collection of numbers x from the domain of f and construct a table of function values $y = f(x)$ for those numbers.
2. Plot the corresponding points (x, y).
3. Connect the plotted points with a smooth curve.

EXAMPLE 2.1

Graph the function $f(x) = x^2$.

Solution

Begin by constructing the table

x	-3	-2	-1	$-\frac{1}{2}$	0	$\frac{1}{2}$	1	2	3
$y = x$	9	4	1	$\frac{1}{4}$	0	$\frac{1}{4}$	1	4	9

Then plot the points (x, y) and connect them with the smooth curve shown in Figure 1.4a.

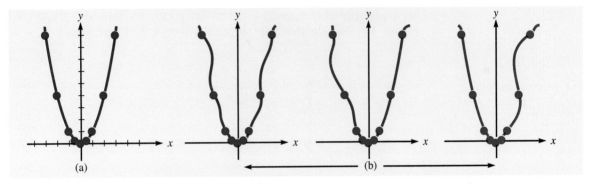

Figure 1.4 (a) The graph of $y = x^2$. (b) Other graphs through the points in Example 2.1.

Comment: Many different curves pass through the points in Example 2.1. Several of these curves are shown in Figure 1.4b. There is no way to guarantee that the curve we pass through the plotted points is the actual graph of f. However, in general, the more points that are plotted, the more likely the graph is to be reasonably accurate.

The next example illustrates how to sketch the graph of a function defined by more than one formula.

EXAMPLE 2.2

Graph the function

$$f(x) = \begin{cases} 2x & \text{if } 0 \le x < 1 \\ \dfrac{2}{x} & \text{if } 1 \le x < 4 \\ 3 & \text{if } x \ge 4 \end{cases}$$

Solution

When making a table of values for this function, remember to use the formula that is appropriate for each particular value of x. Using the formula $f(x) = 2x$ when $0 \leq x < 1$, the formula $f(x) = \dfrac{2}{x}$ when $1 \leq x < 4$, and the formula $f(x) = 3$ when $x \geq 4$, you can compile the following table:

x	0	$\dfrac{1}{2}$	1	2	3	4	5	6
$f(x)$	0	1	2	1	$\dfrac{2}{3}$	3	3	3

Now plot the corresponding points $(x, f(x))$ and draw the graph as in Figure 1.5. Notice that the pieces for $0 \leq x < 1$ and $1 \leq x < 4$ are

Figure 1.5 The graph for Example 2.2.

connected to each other at $(1, 2)$ but that the piece for $x > 4$ is separated from the rest of the graph. [The "open dot" at $\left(4, \dfrac{1}{2}\right)$ indicates that the graph approaches this point but that the point is not actually on the graph.]

The x and y Intercepts The points (if any) at which a graph crosses the x and y axes are called the x **and** y **intercepts,** respectively. The y intercept is the point on the graph with x coordinate zero, and the x intercepts are the points on the graph with y coordinate zero.

How to Find the x and y Intercepts

To find any y intercept of $y = f(x)$, set x equal to 0 and compute y.

To find any x intercepts of $y = f(x)$, set y equal to 0 and solve for x.

EXAMPLE 2.3

Graph the function $f(x) = -x^2 + x + 2$. Include all x and y intercepts.

Solution

The y intercept is $f(0) = 2$. To find the x intercepts, solve the equation $f(x) = 0$. Factoring, we find that

$$-x^2 + x + 2 = 0$$
$$-(x + 1)(x - 2) = 0$$
$$x = -1, x = 2$$

Thus, the x intercepts are $(-1, 0)$ and $(2, 0)$.

Next, make a table of values and plot the corresponding points $(x, f(x))$.

x	-3	-2	-1	0	1	2	3	4
$f(x)$	-10	-4	0	2	2	0	-4	-10

Figure 1.6 The graph of $f(x) = -x^2 + x + 2$.

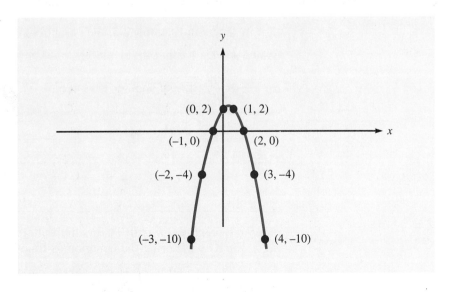

The graph of f is shown in Figure 1.6.

Comment: The graphs in Figures 1.4 and 1.6 are called **parabolas.** In general, the graph of $y = Ax^2 + Bx + C$ is a parabola as long as $A \neq 0$. All parabolas have a U shape, and $y = Ax^2 + Bx + C$ opens either up (if $A > 0$) or down (if $A < 0$). The "peak" or "valley" of the parabola is called its **vertex,** and in either case, the x coordinate of the vertex is $x = \dfrac{-B}{2A}$ (see Figure 1.7 and Problem 60). These features of the parabola

Figure 1.7 The graph of the parabola $y = Ax^2 + Bx + C$. (a) If $A > 0$, the parabola opens up. (b) If $A < 0$, the parabola opens down.

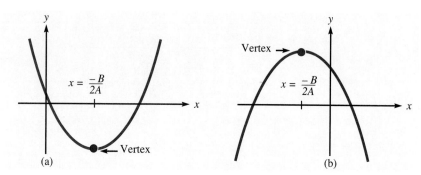

are derived by the methods of calculus discussed in Chapter 3. Note that to get a reasonable sketch of the parabola $y = Ax^2 + Bx + C$, you need only determine

1. The location of the vertex $\left(\text{where } x = \dfrac{-B}{2A}\right)$
2. Whether the parabola opens up $(A > 0)$ or down $(A < 0)$
3. Any intercepts

For instance, in Example 2.3, the parabola $y = -x^2 + x + 2$ turns down (since $A = -1$ is negative) and has its vertex (high point) at $x = \dfrac{-B}{2A} = \dfrac{-1}{2(-1)} = \dfrac{1}{2}$. Here is another example.

EXAMPLE 2.4

Graph the function $f(x) = x^3 - x^2 - 6x$. Include all the x and y intercepts.

Solution

The y intercept is $f(0) = 0$. To find the x intercepts, set $f(x)$ equal to 0 and solve for x by factoring the equation as follows:

$$0 = x^3 - x^2 - 6x = x(x^2 - x - 6)$$
$$= x(x - 3)(x + 2)$$

so that $\qquad\qquad x = 0, x = 3, \text{ or } x = -2$

It follows that the x intercepts are $(0, 0)$, $(3, 0)$, and $(-2, 0)$.

Next, make a table of values (including the intercepts) and plot the corresponding points $(x, f(x))$:

x	-3	-2	-1	0	1	2	3	4
$f(x)$	-18	0	4	0	-6	-8	0	24

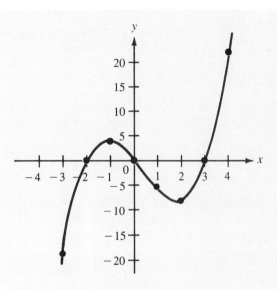

Figure 1.8 The graph of the function $y = x^3 - x^2 - 6x$.

The graph is shown in Figure 1.8.

Discontinuities   A function whose graph is unbroken is said to be **continuous,** while one whose graph has a gap or hole is **discontinuous.** For the functions you will encounter in this text, discontinuities may arise in one of the following two ways.

> **Two Common Types of Discontinuities**
>
> A function defined in several pieces will have discontinuities if the graphs of the individual pieces are not connected to each other.
>
> A function defined as a quotient will have a discontinuity whenever the denominator is zero.

The first kind of discontinuity was illustrated in Figure 1.5 (note the "gap" at $x = 4$). The second kind of discontinuity is illustrated in the following example. Continuity is discussed in more detail in Section 1.5.

EXAMPLE 2.5

Graph the function $f(x) = \dfrac{x^2 + x - 2}{x - 2}$.

Solution

Since the denominator is zero when $x = 2$, the function is undefined and will have a discontinuity for this value of x. To remind yourself that there is no point on the graph whose x coordinate is 2, you might begin your sketch by drawing a broken vertical line at $x = 2$. (See Figure 1.9.) Your graph should not cross this line.

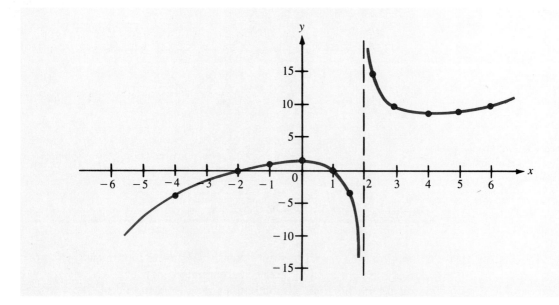

Figure 1.9 The graph of $y = \dfrac{x^2 + x - 2}{x - 2}$.

The y intercept is $f(0) = \dfrac{-2}{-2} = 1$. To find the x intercepts, put the function in factored form.

$$f(x) = \frac{(x + 2)(x - 1)}{x - 2}$$

From the factored numerator you see that $f(x)$ is zero when $x = -2$ and $x = 1$. Hence the x intercepts are $(-2, 0)$ and $(1, 0)$.

Next, make a table of values. To find out what the graph looks like near the discontinuity at $x = 2$, you should include some values of x that are close to 2. The following is a representative table.

x	-50	-4	-2	-1	0	1	1.5	2	2.5	3	4	5	6	50
$f(x)$	-47.1	-1.7	0	0.7	1	0	-3.5	✕	13.5	10	9	9.3	10	53.1

Finally, plot the corresponding points $(x, f(x))$ and pass a smooth curve through the points on the left of the vertical line $x = 2$ and another smooth curve through the points on the right of the line $x = 2$. The graph is shown in Figure 1.9.

Intersections of Graphs   Sometimes it is necessary to determine when two functions are equal. For instance, an economist may wish to compute the market price at which the consumer demand for a commodity will be equal to supply. Or a political analyst may wish to predict how long it will take for the popularity of a certain challenger to reach that of the incumbent. We shall examine some of these applications in Section 4.

In geometric terms, the values of x for which two functions $f(x)$ and $g(x)$ are equal are the x coordinates of the points at which their graphs intersect. In Figure 1.10, the graph of $y = f(x)$ intersects that of $y = g(x)$ at two points, labeled P and Q.

To find the points of intersection algebraically, set $f(x)$ equal to $g(x)$ and solve for x. This procedure is illustrated in Example 2.6.

Figure 1.10 The graphs of $y = f(x)$ and $y = g(x)$ intersect at P and Q.

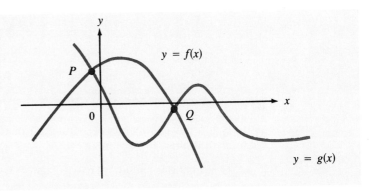

EXAMPLE 2.6

Find all points of intersection of the graphs of $f(x) = 3x + 2$ and $g(x) = x^2$.

Solution

You must solve the equation $x^2 = 3x + 2$. Rewrite the equation as $x^2 - 3x - 2 = 0$ and apply the quadratic formula to obtain

$$x = \frac{-(-3) \pm \sqrt{(-3)^2 - 4(1)(-2)}}{2(1)} = \frac{3 \pm \sqrt{17}}{2}$$

The solutions are

$$x = \frac{3 + \sqrt{17}}{2} \approx 3.56 \quad \text{and} \quad x = \frac{3 - \sqrt{17}}{2} \approx -0.56$$

(The computations were done with a calculator, and the results were rounded off to two decimal places.)

Computing the corresponding y coordinates from the equation $y = x^2$, you find that the points of intersection are approximately (3.56, 12.67) and (−0.56, 0.31). (As a result of round-off errors, you will get slightly different values for the y coordinates if you substitute into the equation $y = 3x + 2$.) The graphs and the intersection points are shown in Figure 1.11.

Figure 1.11 The intersection of the graphs of $f(x) = 3x + 2$ and $g(x) = x^2$.

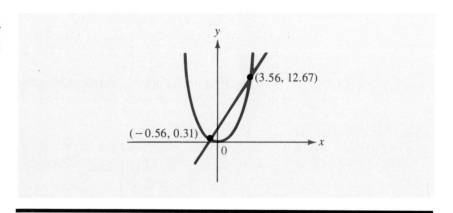

Power Functions   A **power function** is a function of the form $f(x) = x^n$, where n is a real number. For example $f(x) = x^2$, $f(x) = x^{-3}$, and $f(x) = x^{1/2}$ are all power functions. So are $f(x) = \dfrac{1}{x^2}$ and $f(x) = \sqrt[3]{x}$ since they can be rewritten as $f(x) = x^{-2}$ and $f(x) = x^{1/3}$, respectively.

Polynomials   A **polynomial** is a function of the form

$$f(x) = a_0 + a_1 x + a_2 x^2 + \cdots + a_n x^n$$

where n is a nonnegative integer and $a_0, a_1, \ldots, a_n$ are constants. If $a_n \neq 0$, the integer n is called the **degree** of the polynomial. For example, the function $f(x) = 3x^5 - 6x^2 + 7$ is a polynomial of degree 5. Polynomials are continuous functions. It can be shown that the graph of a polynomial of degree n is an unbroken curve that crosses the x axis no more than n times. To illustrate some of the possibilities, the graphs of three polynomials of degree 3 are shown in Figure 1.12.

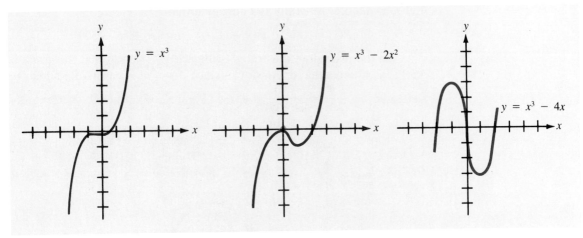

Figure 1.12 Three polynomials of degree 3.

Polynomial functions play a useful role in business and economics. In the following example, a profit function is described by a quadratic polynomial $Ax^2 + Bx + C$, and the point where maximum profit occurs is then found by locating the vertex of the parabola $y = Ax^2 + Bx + C$ (recall Figure 1.7). In Chapter 3, you will learn how to use calculus to solve such optimization problems.

EXAMPLE 2.7

A manufacturer can produce radios at a cost of $10 apiece and estimates that if they are sold for x dollars apiece, consumers will buy approximately $80 - x$ radios each month. Express the manufacturer's monthly profit as a function of the price x, graph this function, and determine the price at which the manufacturer's profit will be greatest.

Solution

Begin by stating the desired relationship in words:

$$\text{Profit} = (\text{number of radios sold})(\text{profit per radio})$$

Now replace the words by algebraic expressions. You know that

$$\text{Number of radios sold} = 80 - x$$

and since the radios are produced at a cost of $10 apiece and sold for x dollars apiece, it follows that

$$\text{Profit per radio} = x - 10$$

Let $P(x)$ denote the profit and conclude that

$$P(x) = (80 - x)(x - 10) = -x^2 + 90x - 800$$

Using the procedure illustrated in Figure 1.7, we see that the graph of $P(x)$ is a parabola that opens down (since $A = -1$ is negative) and has its vertex at $x = \dfrac{-B}{2A} = \dfrac{-90}{2(-1)} = 45$ (see Figure 1.13). Since the vertex is the

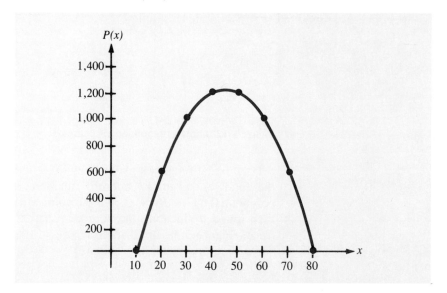

Figure 1.13 The profit function $P(x) = (80 - x)(x - 10)$.

highest point on the graph, it follows that the manufacturer's profit will be greatest when the price is $45. [Incidentally, notice from the factored form of the profit function that the x intercepts of the profit function are $(10, 0)$ and $(80, 0)$. Can you explain these x intercepts in economic terms?]

Rational Functions   The quotient of two polynomials is called a **rational function.** For example, the function $f(x) = \dfrac{x^2 + x - 2}{x - 2}$ in Example 2.5 is rational. So is the function $f(x) = 1 + \dfrac{1}{x}$ since it can be rewritten as $f(x) = \dfrac{x + 1}{x}$. Since division by zero is impossible, a rational function has a discontinuity whenever its denominator is zero. Many cost functions in economics are rational. Here is an example.

EXAMPLE 2.8

For each shipment of raw materials, a manufacturer must pay an ordering fee to cover handling and transportation. After they are received, the raw materials must be stored until needed, and storage costs result. If each shipment of raw materials is large, ordering costs will be low because few shipments are required, but storage costs will be high. If each shipment is small, ordering costs will be high because many shipments will be required, but storage costs will drop. A manufacturer estimates that if each shipment contains x units, the total cost of obtaining and storing the year's supply of raw materials will be $C(x) = x + \dfrac{160,000}{x}$ dollars. Sketch the relevant portion of this cost function and estimate the optimal shipment size.

Solution

$C(x)$ is a rational function with a discontinuity when $x = 0$ and represents cost for nonnegative values of x. Compile a table for some representative nonnegative values of x and plot the corresponding points to get the graph shown in Figure 1.14.

x	100	200	300	400	500	600	700	800
$C(x)$	1,700	1,000	833	800	820	867	929	1,000

Figure 1.14 The cost function $C(x) = x + \dfrac{160,000}{x}$.

The graph indicates that total cost will be high if shipments are very small or very large and that the optimal shipment size is approximately 400 units.

Problems 1.2

In Problems 1 through 24, sketch the graph of the given function. Include all x and y intercepts. (You may use a calculator to help with the computations.)

1. $f(x) = x$

2. $f(x) = x^2$

3. $f(x) = x^3$

4. $f(x) = x^4$

5. $f(x) = \dfrac{1}{x}$

6. $f(x) = \dfrac{1}{x^2}$

7. $f(x) = \dfrac{1}{x^3}$

8. $f(x) = \sqrt{x}$

9. $f(x) = 2x - 1$

10. $f(x) = 2 - 3x$

11. $f(x) = -x^3 + 1$

12. $f(x) = -\dfrac{1}{x^2}$

13. $f(x) = \begin{cases} x - 1 & \text{if } x \le 0 \\ x + 1 & \text{if } x > 0 \end{cases}$

14. $f(x) = \begin{cases} x^2 - 1 & \text{if } x \le 2 \\ 3 & \text{if } x > 2 \end{cases}$

15. $f(x) = (x - 1)(x + 2)$

16. $f(x) = (x + 2)(x + 1)$

17. $f(x) = x^2 - x - 6$

18. $f(x) = x^3 - 4x$

19. $f(x) = \dfrac{1}{x - 2}$

20. $f(x) = \dfrac{1}{(x + 2)^2}$

21. $f(x) = \dfrac{x^2 - 2x}{x + 1}$

22. $f(x) = \dfrac{x^2 - 2x - 3}{x - 1}$

23. $f(x) = x + \dfrac{1}{x}$

24. $f(x) = x - \dfrac{1}{x}$

Manufacturing Cost   25. A manufacturer can produce cassette tape recorders at a cost of $20 apiece. It is estimated that if the tape recorders are sold for x dollars apiece, consumers will buy $120 - x$ of them a month. Express the manufacturer's monthly profit as a function of price, graph this function, and use the graph to estimate the optimal selling price.

Retail Sales   26. A bookstore can obtain an atlas from the publisher at a cost of $5 per copy and estimates that if it sells the atlas for x dollars per copy, approximately $20(22 - x)$ copies will be sold each month. Express the bookstore's monthly profit from the sale of the atlas as a function of price, graph this function, and use the graph to estimate the optimal selling price.

Consumer Expenditure   27. The consumer demand for a certain commodity is $D(p) = -200p +$

12,000 units per month when the market price is p dollars per unit.

(a) Graph this demand function.

(b) Express consumers' total monthly expenditure for the commodity as a function of p. (The total monthly expenditure is the total amount of money consumers spend each month on the commodity.)

(c) Graph the total monthly expenditure function.

(d) Discuss the economic significance of the p intercepts of the expenditure function.

(e) Use the graph in part (c) to estimate the market price that generates the greatest consumer expenditure.

Speed of a Moving Object
28. If an object is thrown vertically upward from the ground with an initial speed of 160 feet per second, its height (in feet) t seconds later is given by the function $H(t) = -16t^2 + 160t$.

(a) Graph the function $H(t)$.

(b) Use the graph in part (a) to determine when the object will hit the ground.

(c) Use the graph in part (a) to estimate how high the object will rise.

Distribution Cost
29. Suppose the number of worker-hours required to distribute new telephone books to x percent of the households in a certain rural community is given by the function $f(x) = \dfrac{600x}{300 - x}$. Sketch this function and specify what portion of the graph is relevant to the practical situation under consideration.

Immunization
30. Suppose that during a nationwide program to immunize the population against a certain form of influenza, public health officials found that the cost of inoculating x percent of the population was approximately $f(x) = \dfrac{150x}{200 - x}$ million dollars. Sketch this function and specify what portion of the graph is relevant to the practical situation under consideration.

Experimental Psychology
31. To study the rate at which animals learn, a psychology student performed an experiment in which a rat was sent repeatedly through a laboratory maze. Suppose the time required for the rat to traverse the maze on the nth trial was approximately $f(n) = 3 + \dfrac{12}{n}$ minutes.

(a) Graph the function $f(n)$.

(b) What portion of the graph is relevant to the practical situation under consideration?

(c) What happens to the graph as n increases without bound? Interpret your answer in practical terms.

Inventory Cost
32. A manufacturer estimates that if each shipment of raw materials

contains x units, the total cost of obtaining and storing the year's supply of raw materials will be $C(x) = 2x + \dfrac{80{,}000}{x}$ dollars. Sketch the relevant portion of the graph of this cost function and estimate the optimal shipment size.

Production Cost   33. A manufacturer estimates that if x machines are used, the cost of a production run will be $C(x) = 20x + \dfrac{2{,}000}{x}$ dollars. Sketch the relevant portion of this cost function and estimate how many machines the manufacturer should use to minimize cost.

Average Cost   34. Suppose the total cost of manufacturing x units of a certain commodity is $C(x) = \dfrac{1}{6}x^3 + 2x + 5$ dollars. Express the average cost per unit as a function of the number of units produced and, on the same set of axes, sketch the total cost and average cost functions. (*Hint:* Average cost is total cost divided by the number of units produced.)

Average Cost   35. Suppose the total cost (in dollars) of manufacturing x units is given by the function $C(x) = x^2 + 6x + 19$. Express the average cost per unit as a function of the number of units produced and, on the same set of axes, sketch the total cost and average cost functions.

In Exercises 36 through 51, find the points of intersection (if any) of the given pair of curves and draw the graphs.

36. $y = 3x + 5$ and $y = -x + 3$

37. $y = 5x - 14$ and $y = 4 - x$

38. $y = 3x + 8$ and $y = 3x - 2$

39. $y = x^2$ and $y = 6 - x$

40. $y = x^2 - x$ and $y = x - 1$

41. $y = x^3 - 6x^2$ and $y = -x^2$

42. $y = x^3$ and $y = x^2$

43. $y = x^2$ and $y = 2x + 2$

44. $y = x^2 + 2$ and $y = x$

45. $3y - 2x = 5$ and $y + 3x = 9$

46. $2x - 3y = -8$ and $3x - 5y = -13$

47. $y = \dfrac{1}{x}$ and $y = x^2$

48. $y = \dfrac{1}{x^2}$ and $y = 4$

49. $y = \dfrac{1}{x}$ and $y = \dfrac{1}{x^2}$

50. $y = \dfrac{1}{x^2}$ and $y = -x^2$

51. $y = x^2 - 2x$ and $y = x - 1$

52. (a) Graph the functions $y = x^2$ and $y = x^2 + 3$. How are the graphs related?
    (b) Without further computation, graph the function $y = x^2 - 5$.
    (c) Suppose $g(x) = f(x) + c$, where c is a constant. How are the graphs of f and g related? Explain.

53. (a) Graph the functions $y = x^2$ and $y = -x^2$. How are the graphs related?
    (b) Suppose $g(x) = -f(x)$. How are the graphs of f and g related? Explain.

54. Graph the function $y = \dfrac{1}{x}$. Then, without further computation, graph the function $y = 3 - \dfrac{1}{x}$. (*Hint:* Combine the rules you discovered in Problems 52 and 53.)

55. (a) Graph the functions $y = x^2$ and $y = (x - 2)^2$. How are the graphs related?
    (b) Without further computation, graph the function $y = (x + 1)^2$.
    (c) Suppose $g(x) = f(x - c)$, where c is a constant. How are the graphs of f and g related? Explain.

56. Graph the function $y = \dfrac{1}{x}$. Then, without further computation, graph the function $y = 3 - \dfrac{1}{x + 2}$.

57. Show that the distance d between the two points (x_1, y_1) and (x_2, y_2) is given by the formula

$$d = \sqrt{(x_2 - x_1)^2 + (y_2 - y_1)^2}$$

(*Hint:* Apply the pythagorean theorem to the right triangle whose hypotenuse is the line segment joining the two given points.)

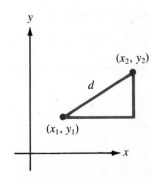

58. Use the distance formula in Problem 57 to find the distance between each of the following points:
    (a) $(5, -1)$ and $(2, 3)$
    (b) $(2, 6)$ and $(2, -1)$

59. Use the distance formula in Problem 57 to show that the circle

centered at (a, b) with radius R has the equation

$$(x - a)^2 + (y - b)^2 = R^2$$

60. Show that the vertex of the parabola $y = Ax^2 + Bx + C$ $(A \neq 0)$ occurs at the point where $x = -B/(2A)$. [*Hint:* First verify that

$$Ax^2 + Bx + C = A\left[\left(x + \frac{B}{2A}\right)^2 + \left(\frac{C}{A} - \frac{B^2}{4A^2}\right)\right]$$

Then note that the largest or smallest value of $f(x) = Ax^2 + Bx + C$ must occur where $x = \dfrac{-B}{2A}$.]

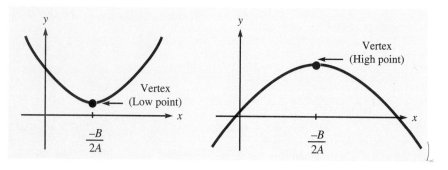

In Problems 61 through 64, the graph of the given function is a parabola. For each graph, find the intercepts and the vertex (using Problem 60). Then sketch the graph.

61. $f(x) = -x^2 - 2x + 15$ 62. $f(x) = x^2 + 2x - 8$

63. $f(x) = 6x^2 + 13x - 5$ 64. $f(x) = -26x^2 + 10x + 21$

3 Linear Functions

In many practical situations, the rate at which one quantity changes with respect to another is constant. Here is a simple example from economics.

EXAMPLE 3.1

A manufacturer's total cost consists of a fixed overhead of $200 plus production costs of $50 per unit. Express the total cost as a function of the number of units produced and draw the graph.

Solution

Let x denote the number of units produced and $C(x)$ the corresponding

total cost. Then,

$$\text{Total cost} = (\text{cost per unit})(\text{number of units}) + \text{overhead}$$

where

$$\text{Cost per unit} = 50$$

$$\text{Number of units} = x$$

$$\text{Overhead} = 200$$

Hence,

$$C(x) = 50x + 200$$

Figure 1.15 The cost function $C(x) = 50x + 200$.

The graph of this cost function is sketched in Figure 1.15.

The total cost in Example 3.1 increases at a constant rate of \$50 per unit. As a result, its graph in Figure 1.15 is a straight line that increases in height by 50 units for each 1-unit increase in x.

In general, a function whose value changes at a constant rate with respect to its independent variable is said to be a **linear function.** This is because the graph of such a function is a straight line. In algebraic terms, a linear function is a function of the form

$$f(x) = a_0 + a_1 x$$

where a_0 and a_1 are constants. For example, the functions $f(x) = \frac{3}{2} + 2x$, $f(x) = -5x$, and $f(x) = 12$ are all linear. Linear functions are traditionally written in the form

$$y = mx + b$$

where m and b are constants. This standard notation will be used in the discussion that follows.

> **Linear Functions**
>
> A linear function is a function that changes at a constant rate with respect to its independent variable.
>
> The graph of a linear function is a straight line.
>
> The equation of a linear function can be written in the form
>
> $$y = mx + b$$
>
> where m and b are constants.

To work with linear functions, you will need to know the following things about straight lines.

The Slope of a Line   The **slope** of a line is the amount by which the y coordinate of a point on the line changes when the x coordinate is increased by 1. You can compute the slope of a nonvertical line if you know any two of its points. Suppose (x_1, y_1) and (x_2, y_2) lie on a line as indicated in Figure 1.16.

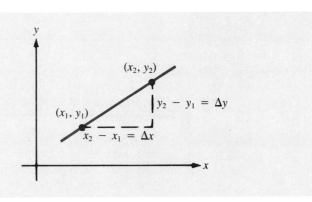

Figure 1.16

Slope $= \dfrac{y_2 - y_1}{x_2 - x_1} = \dfrac{\Delta y}{\Delta x}$.

Between these points, x changes by the amount $x_2 - x_1$ and y by the amount $y_2 - y_1$. The slope is the ratio

$$\text{Slope} = \frac{\text{change in } y}{\text{change in } x} = \frac{y_2 - y_1}{x_2 - x_1}$$

It is sometimes convenient to use the symbol Δy instead of $y_2 - y_1$ to denote the change in y. The symbol Δy is read "delta y." Similarly, the symbol Δx is used to denote the change $x_2 - x_1$.

$\dfrac{\Delta y}{\Delta x}$

The Slope of a Line

The slope of the nonvertical line passing through the points (x_1, y_1) and (x_2, y_2) is given by the formula

$$\text{Slope} = \frac{\Delta y}{\Delta x} = \frac{y_2 - y_1}{x_2 - x_1}$$

The use of this formula is illustrated in the following example.

EXAMPLE 3.2

Find the slope of the line joining the points $(-2, 5)$ and $(3, -1)$.

Solution

$$\text{Slope} = \frac{\Delta y}{\Delta x} = \frac{-1 - 5}{3 - (-2)} = \frac{-6}{5}$$

$\dfrac{5 + 1}{-2 - 3} = \dfrac{-6}{5}$

Figure 1.17 The line joining $(-2, 5)$ and $(3, -1)$.

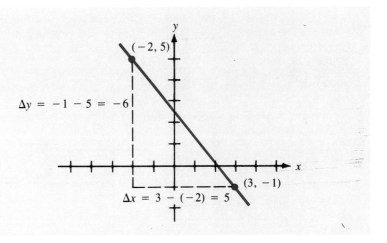

The situation is illustrated in Figure 1.17.

The sign and magnitude of the slope of a line indicate the line's direction and steepness, respectively. The slope is positive if the height of the line increases as x increases and is negative if the height decreases as x increases. The absolute value of the slope is large if the slant of the line is severe and small if the slant of the line is gradual. The situation is illustrated in Figure 1.18.

Figure 1.18 The direction and steepness of a line.

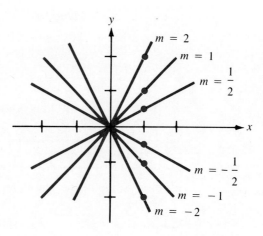

Horizontal and Vertical Lines

Horizontal and vertical lines (Figure 1.19a and 1.19b) have particularly simple equations. The y coordinates of all the points on a horizontal line are the same. Hence, a horizontal line is the graph of a linear function of the form $y = b$, where b is a constant. The slope of a horizontal line is zero, since changes in x produce no changes in y.

Figure 1.19 Horizontal and vertical lines.

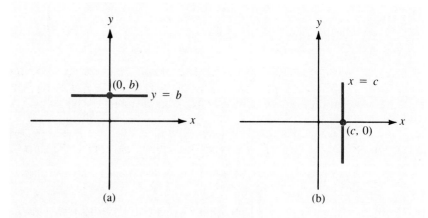

The x coordinates of all the points on a vertical line are equal. Hence, vertical lines are characterized by equations of the form $x = c$, where c is a constant. The slope of a vertical line is undefined. This is because only the y coordinates of points on a vertical line can change, and so the denominator of the quotient $\dfrac{\text{change in } y}{\text{change in } x}$ is zero.

The Slope-Intercept Form of the Equation of a Line   The constants m and b in the equation $y = mx + b$ of a nonvertical line have geometric interpretations. The coefficient m is the slope of the line. To see this, suppose that (x_1, y_1) and (x_2, y_2) are two points on the line $y = mx + b$. Then, $y_1 = mx_1 + b$ and $y_2 = mx_2 + b$, and so

$$\text{Slope} = \frac{y_2 - y_1}{x_2 - x_1} = \frac{(mx_2 + b) - (mx_1 + b)}{x_2 - x_1}$$

$$= \frac{mx_2 - mx_1}{x_2 - x_1} = \frac{m(x_2 - x_1)}{x_2 - x_1} = m$$

The constant b in the equation $y = mx + b$ is the value of y corresponding to $x = 0$. Hence, b is the height at which the line $y = mx + b$ crosses the y axis, and the corresponding point $(0, b)$ is the y intercept of the line. The situation is illustrated in Figure 1.20.

Figure 1.20 The slope and y intercept of the line $y = mx + b$.

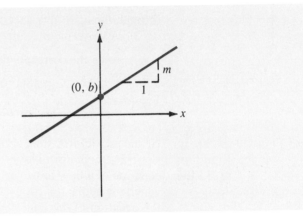

slope intercept form $y = mx + b$

Because the constants m and b in the equation $y = mx + b$ correspond to the slope and y intercept, respectively, this form of the equation of a line is known as the **slope-intercept form.**

The Slope-Intercept Form of the Equation of a Line

The equation
$$y = mx + b$$
is the equation of the line whose slope is m and whose y intercept is $(0, b)$.

The slope-intercept form of the equation of a line is particularly useful when geometric information about a line (such as its slope or y intercept) is to be determined from the line's algebraic representation. Here is a typical example.

$y = -\frac{2}{3}x + 2$

EXAMPLE 3.3

Find the slope and y intercept of the line $3y + 2x = 6$ and draw the graph.

Solution

The first step is to put the equation $3y + 2x = 6$ in slope-intercept form $y = mx + b$. To do this, solve for y to get

$$3y = -2x + 6 \quad \text{or} \quad y = -\frac{2}{3}x + 2$$

It follows that the slope is $-\frac{2}{3}$ and the y intercept is $(0, 2)$.

To graph a linear function, plot two of its points and draw a straight line through them. In this case, you already know one point, the y intercept $(0, 2)$. A convenient choice for the x coordinate of the second point is $x = 3$, since the corresponding y coordinate is $y = -\frac{2}{3}(3) + 2 = 0$. Draw a line through the points $(0, 2)$ and $(3, 0)$ to obtain the graph shown in Figure 1.21.

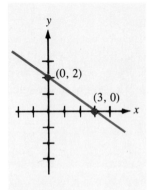

Figure 1.21 The line $3y + 2x = 6$.

The Point-Slope Form of the Equation of a Line

Geometric information about a line can be obtained readily from the slope-intercept formula $y = mx + b$. There is another form of the equation of a line, however, that is usually more efficient for problems in which the geometric properties of a line are known and the goal is to find the equation of the line.

> **The Point-Slope Form of the Equation of a Line**
>
> The equation $\qquad y - y_0 = m(x - x_0)$
>
> is an equation of the line that passes through the point (x_0, y_0) and that has slope equal to m.

The point-slope form of the equation of a line is simply the formula for slope in disguise. To see this, suppose that the point (x, y) lies on the line that passes through a given point (x_0, y_0) and that has slope m. Using the points (x, y) and (x_0, y_0) to compute the slope, you get

$$\frac{y - y_0}{x - x_0} = m$$

which you can put in point-slope form

$$y - y_0 = m(x - x_0)$$

by simply multiplying both sides by $x - x_0$.

    The use of the point-slope form of the equation of a line is illustrated in the next two examples.

EXAMPLE 3.4

Find an equation of the line that passes through the point (5, 1) and whose slope is equal to $\frac{1}{2}$.

Solution

Use the formula $y - y_0 = m(x - x_0)$ with $(x_0, y_0) = (5, 1)$ and $m = \frac{1}{2}$ to get

$$y - 1 = \frac{1}{2}(x - 5)$$

which you can rewrite as

$$y = \frac{1}{2}x - \frac{3}{2}$$

Figure 1.22 The line $y = \frac{1}{2}x - \frac{3}{2}$.

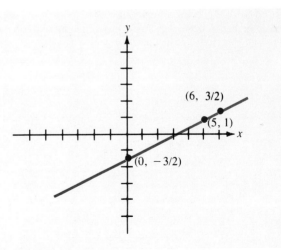

The graph is shown in Figure 1.22.

Instead of the point-slope form, the slope-intercept form could have been used to solve the problem in Example 3.4. For practice, solve the problem this way. Notice that the solution based on the point-slope formula is more efficient.

The next example illustrates how you can use the point-slope form to find an equation of a line that passes through two given points.

EXAMPLE 3.5

Find an equation of the line that passes through the points $(3, -2)$ and $(1, 6)$.

Solution

First compute the slope

$$m = \frac{6 - (-2)}{1 - 3} = \frac{8}{-2} = -4$$

Then use the point-slope formula with $(1, 6)$ as the given point (x_0, y_0) to get
$$y - 6 = -4(x - 1) \quad \text{or} \quad y = -4x + 10$$

Convince yourself that the resulting equation would have been the same if you had chosen $(3, -2)$ to be the given point (x_0, y_0).

Figure 1.23 The line $y = -4x + 10$.

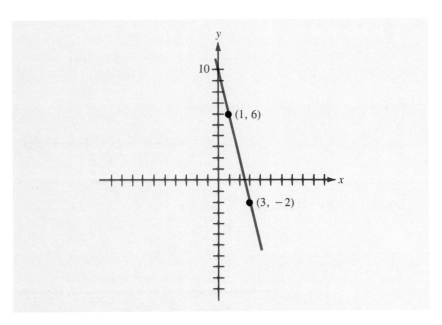

The graph is shown in Figure 1.23.

Practical Applications   If the rate of change of one quantity with respect to a second quantity is constant, the function relating the quantities must be linear. The constant rate of change is the slope of the corresponding line. The next two examples illustrate techniques you can use to find the appropriate linear functions in such situations.

EXAMPLE 3.6

Since the beginning of the year, the price of whole-wheat bread at a local discount supermarket has been rising at a constant rate of 2 cents per month. By November first, the price had reached $1.06 per loaf. Express the price of the bread as a function of time and determine the price at the beginning of the year.

Solution

Let x denote the number of months that have elapsed since the first of the year and y the price of a loaf of bread (in cents). Since y changes at a constant rate with respect to x, the function relating y to x must be linear and its graph is a straight line. Since the price y increases by 2 each time x increases by 1, the slope of the line must be 2. The fact that the price was 106 cents ($1.06) on November first, 10 months after the first of the year, implies that the line passes through the point (10, 106). To write an equation defining y as a function of x, use the point-slope formula

$$y - y_0 = m(x - x_0)$$

with $$m = 2, x_0 = 10, y_0 = 106$$

to get $$y - 106 = 2(x - 10) \quad \text{or} \quad y = 2x + 86$$

The corresponding line is shown in Figure 1.24. Notice that the y

Figure 1.24 The rising price of bread: $y = 2x + 86$.

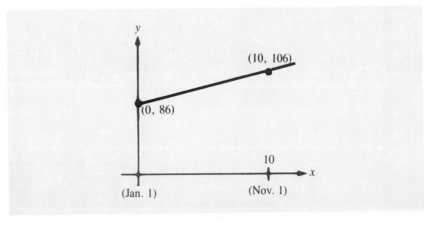

intercept is (0, 86), which implies that the price of bread at the beginning of the year was 86 cents per loaf.

EXAMPLE 3.7

The average scores of incoming students at an eastern liberal arts college in the SAT mathematics examination have been declining at a constant rate in recent years. In 1986, the average SAT score was 575, while in 1991 it was 545.

(a) Express the average SAT score as a function of time.
(b) If the trend continues, what will the average SAT score of incoming students be in 1996?
(c) If the trend continues, when will the average SAT score be 527?

Solution

(a) Let x denote the number of years since 1986 and y the average SAT score of incoming students. Since y changes at a constant rate with respect to x, the function relating y to x must be linear. Since $y = 575$ when $x = 0$ and $y = 545$ when $x = 5$ (in 1991), the corresponding straight line must pass through the points (0, 575) and (5, 545). The slope of this line is

$$m = \frac{575 - 545}{0 - 5} = -6$$

Since one of the given points happens to be the y intercept (0, 575), you can use the slope-intercept form to conclude immediately that

$$y = -6x + 575$$

The corresponding line is shown in Figure 1.25.

Figure 1.25 Declining SAT scores: $y = -6x + 575$.

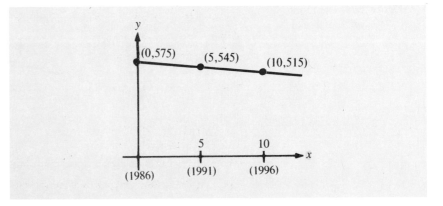

wcwm
sports info
student pahol
coach
soccer
clinic

(b) To predict the average SAT score in 1996, compute y when $x = 10$:

$$y = -6(10) + 575 = 515$$

(c) Set y equal to 527 and solve for x to get

$$527 = -6x + 575 \qquad 6x = 48 \qquad \text{or} \qquad x = 8$$

Thus the SAT score will be 527 in 1994.

Problems 1.3

In Problems 1 through 6, find the slope (if possible) of the line that passes through the given pair of points.

1. $(2, -3)$ and $(0, 4)$

2. $(-1, 2)$ and $(2, 5)$

3. $(2, 0)$ and $(0, 2)$

4. $(5, -1)$ and $(-2, -1)$

5. $(2, 6)$ and $(2, -4)$

6. $\left(\dfrac{2}{3}, -\dfrac{1}{5}\right)$ and $\left(-\dfrac{1}{7}, \dfrac{1}{8}\right)$

In Problems 7 through 18, find the slope and y intercept (if they exist) of the given line and draw a graph.

7. $y = 3x$

8. $y = 5x + 2$

9. $y = 3x - 6$

10. $x + y = 2$

11. $3x + 2y = 6$

12. $2x - 4y = 12$

13. $5y - 3x = 4$

14. $4x = 2y + 6$

15. $\dfrac{x}{2} + \dfrac{y}{5} = 1$

16. $y = 2$

17. $x = -3$

18. $\dfrac{x + 3}{-5} + \dfrac{y - 1}{2} = 1$

In Problems 19 through 30, write an equation for the line with the given properties.

19. Through $(2, 0)$ with slope 1

20. Through $(-1, 2)$ with slope $\dfrac{2}{3}$

21. Through $(5, -2)$ with slope $-\dfrac{1}{2}$

22. Through $(0, 0)$ with slope 5

23. Through $(2, 5)$ and parallel to the x axis

24. Through (2, 5) and parallel to the y axis

25. Through (1, 0) and (0, 1)

26. Through (2, 5) and (1, −2)

27. Through $\left(-\frac{1}{5}, 1\right)$ and $\left(\frac{2}{3}, \frac{1}{4}\right)$

28. Through (−2, 3) and (0, 5)

29. Through (1, 5) and (3, 5)

30. Through (1, 5) and (1, −4)

Manufacturing Cost   31. A manufacturer's total cost consists of a fixed overhead of $5,000 plus production costs of $60 per unit. Express the total cost as a function of the number of units produced and draw the graph.

Car Rental   32. A certain car rental agency charges $25 per day plus 30 cents per mile.
(a) Express the cost of renting a car from this agency for 1 day as a function of the number of miles driven and draw the graph.
(b) How much does it cost to rent a car for a 1-day trip of 50 miles?
(c) How many miles were driven if the daily rental cost was $45.20?

Course Registration   33. Students at a state college may preregister for their fall classes by mail during the summer. Those who do not preregister must register in person in September. The registrar can process 35 students per hour during the September registration period. Suppose that after 4 hours in September, a total of 360 students (including those who preregistered) have been registered.
(a) Express the number of students registered as a function of time and draw the graph.
(b) How many students were registered after 3 hours?
(c) How many students preregistered during the summer?

Membership Fees   34. Membership in a swimming club costs $150 for the 12-week summer season. If a member joins after the start of the season, the fee is prorated; that is, it is reduced linearly.
(a) Express the membership fee as a function of the number of weeks that have elapsed by the time the membership is purchased and draw the graph.
(b) Compute the cost of a membership that is purchased 5 weeks after the start of the season.

Linear Depreciation   35. A doctor owns $1,500 worth of medical books which, for tax purposes, are assumed to depreciate linearly to zero over a 10-year period. That is, the value of the books decreases at a constant rate so that it is equal to zero at the end of 10 years. Express the value of the books as a function of time and draw the graph.

Linear Depreciation 36. A manufacturer buys $20,000 worth of machinery that depreciates linearly so that its trade-in value after 10 years will be $1,000.
(a) Express the value of the machinery as a function of its age and draw the graph.
(b) Compute the value of the machinery after 4 years.

Water Consumption 37. Since the beginning of the month, a local reservoir has been losing water at a constant rate. On the 12th of the month the reservoir held 200 million gallons of water, and on the 21st it held only 164 million gallons.
(a) Express the amount of water in the reservoir as a function of time and draw the graph.
(b) How much water was in the reservoir on the 8th of the month?

Car Pooling 38. To encourage motorists to form car pools, the transit authority in a certain metropolitan area has been offering a special reduced rate at toll bridges for vehicles containing four or more persons. When the program began 30 days ago, 157 vehicles qualified for the reduced rate during the morning rush hour. Since then, the number of vehicles qualifying has been increasing at a constant rate, and today 247 vehicles qualified.
(a) Express the number of vehicles qualifying each morning for the reduced rate as a function of time and draw the graph.
(b) If the trend continues, how many vehicles will qualify during the morning rush hour 14 days from now?

Metric Conversion 39. (a) Temperature measured in degrees Fahrenheit is a linear function of temperature measured in degrees Celsius. Use the fact that 0° Celsius is equal to 32° Fahrenheit and 100° Celsius is equal to 212° Fahrenheit to write an equation for this linear function.
(b) Use the function you obtained in part (a) to convert 15° Celsius to Fahrenheit.
(c) Convert 68° Fahrenheit to Celsius.

Appreciation of Assets 40. The value of a certain rare book doubles every 10 years. The book was originally worth $3.
(a) How much is the book worth when it is 30 years old? When it is 40 years old?
(b) Is the relationship between the value of the book and its age linear? Explain.

Retail Sales 41. A manufacturer can produce radios at a cost of $2 apiece. The radios have been selling for $5 apiece, and at this price, consumers have been buying 4,000 radios a month. The manufacturer is planning to raise the price of the radios and estimates that for each $1 increase in the price, 400 fewer radios will be sold each month. Express the number of radios sold as a function of the manufacturer's selling price. [*Hint:* Note that the number of radios sold y is a linear function of the selling price x and that its graph passes through the point (5,4000). What is its slope?]

Retail Sales 42. A retailer can obtain cameras from the manufacturer at a cost of $50 apiece. The retailer has been selling the cameras at the price of $80 apiece, and at this price, consumers have been buying 40 cameras a month. The retailer is planning to lower the price to stimulate sales and estimates that for each $1 reduction in the price, 2 more cameras will be sold. Express the number of cameras sold as a function of the selling price.

Parallel Lines 43. What is the relationship between the slopes of parallel lines? Explain your answer. Using this relationship, write equations for the lines with the following properties:
(a) Through (1, 3) and parallel to the line $4x + 2y = 7$
(b) Through (0, 2) and parallel to the line $2y - 3x = 5$
(c) Through $(-2, 5)$ and parallel to the line through the points (1, 2) and $(6, -1)$

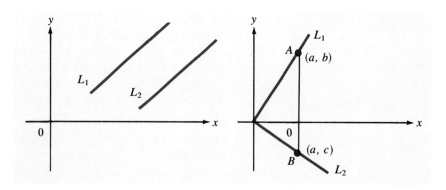

Problem 43   Problem 44

Perpendicular Lines 44. Show that if a line L_1 with slope m_1 is perpendicular to a line L_2 with slope m_2, then $m_1 = -\dfrac{1}{m_2}$. (*Hint:* Find expressions for the slopes of the perpendicular lines L_1 and L_2 in the accompanying figure. Then apply the pythagorean theorem, together with the distance formula from Section 2, Problem 57, to the right triangle OAB to obtain the desired relationship between the slopes.)

Using this result, write equations for the lines with the following properties:
(a) Through $(-1, 3)$ and perpendicular to the line $4x + 2y = 7$

(b) Through (0, 0) and perpendicular to the line $2y - 3x = 5$

(c) Through (2, 1) and perpendicular to the line joining (0, 3) and (2, -1)

4 Functional Models

A mathematical representation of a practical situation is called a **mathematical model.** In preceding sections, you saw models representing such quantities as manufacturing cost, air pollution levels, population size, supply, and demand. In this section, you will see examples illustrating some of the techniques you can use to build mathematical models of your own.

A Profit Function   In the following example, profit is expressed as a function of the price at which a product is sold.

EXAMPLE 4.1

A manufacturer can produce radios at a cost of $2 apiece. The radios have been selling for $5 apiece, and at this price, consumers have been buying 4,000 radios a month. The manufacturer is planning to raise the price of the radios and estimates that for each $1 increase in the price, 400 fewer radios will be sold each month. Express the manufacturer's monthly profit as a function of the price at which the radios are sold.

Solution

Begin by stating the desired relationship in words.

Profit = (number of radios sold)(profit per radio)

Since the goal is to express profit as a function of price, the independent variable is price and the dependent variable is profit. Let x denote the price at which the radios will be sold and $P(x)$ the corresponding profit.

Next, express the number of radios sold in terms of the variable x. You know that 4,000 radios are sold each month when the price is $5 and that 400 fewer will be sold each month for each $1 increase in the price. Thus,

Number of radios sold = 4,000 − 400(number of $1 increases)

The number of $1 increases in the price is the difference $x - 5$ between the new and old selling prices. Hence,

Number of radios sold = 4,000 − 400(x − 5)

$$= 400[10 - (x - 5)]$$
$$= 400(15 - x)$$

The profit per radio is simply the difference between the selling price x and the cost \$2. That is,

$$\text{Profit per radio} = x - 2$$

If you now substitute the algebraic expressions for the number of radios sold and the profit per radio into the verbal equation with which you began, you will get

$$P(x) = 400(15 - x)(x - 2)$$

The graph of this factored polynomial is sketched in Figure 1.26.

Figure 1.26 The profit function $P(x) =$ $400(15 - x)(x - 2)$.

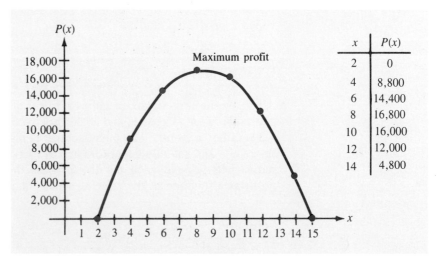

x	$P(x)$
2	0
4	8,800
6	14,400
8	16,800
10	16,000
12	12,000
14	4,800

Actually, only the portion of the graph for $x \geq 5$ is relevant to the original problem as stated. (Can you give a practical interpretation of the portion between $x = 2$ and $x = 5$?) Notice that the profit function reaches a maximum for some value of x near $x = 8$. In Chapter 3, you will learn how to use calculus to find this optimal selling price.

Comment: The number of radios sold is a linear function of the selling price x. The equation for this linear function can be obtained from the point-slope formula with $(5,4000)$ as the point and -400 as the slope. See Problem 41 of Section 3.

Elimination of Variables   In the next two examples, the quantity you are seeking is expressed most naturally in terms of two variables. You will have to eliminate one of these variables before you can write the quantity as a function of a single variable.

EXAMPLE 4.2

The highway department is planning to build a picnic area for motorists along a major highway. It is to be rectangular with an area of 5,000 square yards and is to be fenced off on the three sides not adjacent to the highway. Express the number of yards of fencing required as a function of the length of the unfenced side.

Solution

It is natural to start by introducing two variables, say x and y, to denote the length of the sides of the picnic area (Figure 1.27) and to express the

Figure 1.27 Rectangular picnic area.

number of yards F of required fencing in terms of these two variables:

$$F = x + 2y$$

Since the goal is to express the number of yards of fencing as a function of x alone, you must find a way to express y in terms of x. To do this, use the fact that the area is to be 5,000 square yards and write

$$\underbrace{xy}_{\text{area}} = 5{,}000$$

Solve this equation for y

$$y = \frac{5{,}000}{x}$$

and substitute the resulting expression for y into the formula for F to get

$$F(x) = x + 2\left(\frac{5{,}000}{x}\right) = x + \frac{10{,}000}{x}$$

A graph of the relevant portion of this rational function is sketched in Figure 1.28. Notice that there is some length x for which the amount of required fencing is minimal. In Chapter 3 you will compute this optimal value of x using calculus.

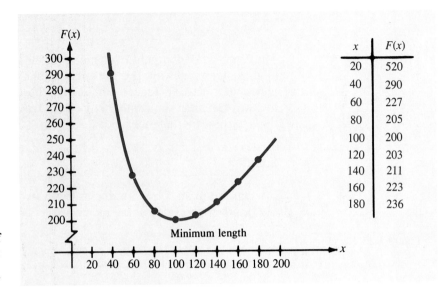

x	F(x)
20	520
40	290
60	227
80	205
100	200
120	203
140	211
160	223
180	236

Figure 1.28 The length of fencing:
$$F(x) = x + \frac{10,000}{x}.$$

EXAMPLE 4.3

A cylindrical can is to have capacity (volume) of 24π cubic inches. The cost of the material used for the top and bottom of the can is 3 cents per square inch, and the cost of the material used for the curved side is 2 cents per square inch. Express the cost of constructing the can as a function of its radius.

Solution

Let r denote the radius of the circular top and bottom, h the height of the can, and C the cost (in cents) of constructing the can. Then,

$$C = \text{cost of top} + \text{cost of bottom} + \text{cost of side}$$

where, for each component of cost,

$$\text{Cost} = (\text{cost per square inch})(\text{number of square inches})$$

$$= (\text{cost per square inch})(\text{area})$$

The area of the circular top (or bottom) is πr^2, and the cost per square inch of the top (or bottom) is 3 cents. Hence,

$$\text{Cost of top} = 3\pi r^2 \qquad \text{and} \qquad \text{Cost of bottom} = 3\pi r^2$$

To find the area of the curved side, imagine the top and bottom of the can removed and the side cut and spread out to form a rectangle as shown in Figure 1.29.

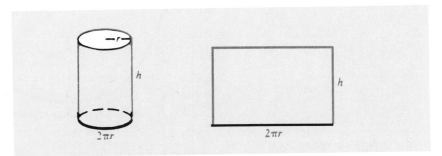

Figure 1.29 Cylindrical can for Example 4.3.

The height of the rectangle is the height h of the can. The length of the rectangle is the circumference $2\pi r$ of the circular top (or bottom) of the can. Hence, the area of the rectangle (or curved side) is $2\pi rh$ square inches. Since the cost of the side is 2 cents per square inch, it follows that

$$\text{Cost of side} = 2(2\pi rh) = 4\pi rh$$

Putting it all together,

$$C = 3\pi r^2 + 3\pi r^2 + 4\pi rh = 6\pi r^2 + 4\pi rh$$

Since the goal is to express the cost as a function of the radius alone, you must find a way to express the height h in terms of r. To do this, use the fact that the volume $V = \pi r^2 h$ is to be 24π. That is, set $\pi r^2 h$ equal to 24π and solve for h to get

$$\pi r^2 h = 24\pi \qquad \text{or} \qquad h = \frac{24}{r^2}$$

Now substitute this expression for h into the formula for C:

$$C(r) = 6\pi r^2 + 4\pi r\left(\frac{24}{r^2}\right)$$

or

$$C(r) = 6\pi r^2 + \frac{96\pi}{r}$$

A graph of the relevant portion of this cost function is sketched in Figure 1.30. Notice that there is some radius r for which the cost is minimal. In Chapter 3 you will learn how to find this optimal radius using calculus.

Functions Involving Multiple Formulas

In the next example, you will need three formulas to define the desired function.

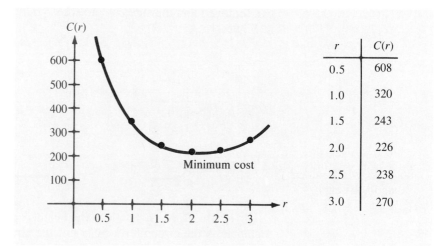

r	$C(r)$
0.5	608
1.0	320
1.5	243
2.0	226
2.5	238
3.0	270

Figure 1.30 The cost function:
$$C(r) = 6\pi r^2 + \frac{96\pi}{r}.$$

EXAMPLE 4.4

During a drought, residents of Marin County, California, were faced with a severe water shortage. To discourage excessive use of water, the county water district initiated drastic rate increases. The monthly rate for a family of four was $1.22 per 100 cubic feet of water for the first 1,200 cubic feet, $10 per 100 cubic feet for the next 1,200 cubic feet, and $50 per 100 cubic feet thereafter. Express the monthly water bill for a family of four as a function of the amount of water used.

Solution

Let x denote the number of hundred-cubic-feet units of water used by the family during the month and $C(x)$ the corresponding cost in dollars. If $0 \leq x \leq 12$, the cost is simply the cost per unit times the number of units used:
$$C(x) = 1.22x$$

If $12 < x \leq 24$, each of the first 12 units costs $1.22, and so the total cost of these 12 units is $1.22(12) = 14.64$ dollars. Each of the remaining $x - 12$ units costs $10, and hence the total cost of these units is $10(x - 12)$ dollars. The cost of all x units is the sum
$$C(x) = 14.64 + 10(x - 12) = 10x - 105.36$$

If $x > 24$, the cost of the first 12 units is $1.22(12) = 14.64$ dollars, the cost of the next 12 units is $10(12) = 120$ dollars, and the cost of the remaining $x - 24$ units is $50(x - 24)$ dollars. The cost of all x units is the sum

$$C(x) = 14.64 + 120 + 50(x - 24) = 50x - 1{,}065.36$$

Combining these three formulas, you get

$$C(x) = \begin{cases} 1.22x & \text{if } 0 \leq x \leq 12 \\ 10x - 105.36 & \text{if } 12 < x \leq 24 \\ 50x - 1{,}065.36 & \text{if } x > 24 \end{cases}$$

The graph of this function is shown in Figure 1.31. Notice that the graph

Figure 1.31 The cost of water in Marin County.

x	$C(x)$
0	0
12	14.64
24	134.64
30	434.64

consists of three line segments, each one steeper than the preceding one. What aspect of the practical situation is reflected by the increasing steepness of the lines?

Proportionality   The following concepts of proportionality are used frequently in the construction of mathematical models.

Direct Proportionality

To say that Q is proportional (or directly proportional) to x means that there is a constant k for which $Q = kx$.

Inverse Proportionality

To say that Q is inversely proportional to x means that there is a constant k for which $Q = \dfrac{k}{x}$.

> **Joint Proportionality**
>
> To say that Q is jointly proportional to x and y means that Q is directly proportional to the product of x and y; that is, there is a constant k for which $Q = kxy$.

Here is an example from biology.

EXAMPLE 4.5

When environmental factors impose an upper bound on its size, population grows at a rate that is jointly proportional to its current size and the difference between its current size and the upper bound. Express the rate of population growth as a function of the size of the population.

Solution

Let p denote the size of the population, $R(p)$ the corresponding rate of population growth, and b the upper bound placed on the population by the environment. Then,

$$\text{Difference between population and bound} = b - p$$

and so
$$R(p) = kp(b - p)$$

where k is the constant of proportionality.

A graph of this factored polynomial is sketched in Figure 1.32. In

Figure 1.32 The rate of bounded population growth: $R(p) = kp(b - p)$.

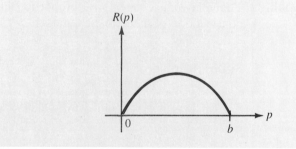

Chapter 3, you will use calculus to compute the population size for which the rate of population growth is greatest.

Break-Even Analysis   Intersections of graphs arise in business in the context of **break-even analysis.** In a typical situation, a manufacturer wishes to determine how many units of a certain commodity have to be sold for total revenue to equal total cost. Suppose x denotes the number of units manufactured and sold, and let $C(x)$ and $R(x)$ be the corresponding total cost and total revenue, respectively. A pair of cost and revenue curves is sketched in Figure 1.33.

Figure 1.33 Cost and revenue curves, with a break-even point at P.

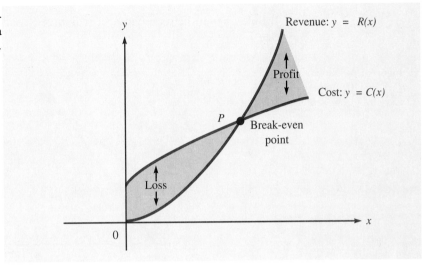

Because of fixed overhead costs, the total cost curve is initially higher than the total revenue curve. Hence, at low levels of production, the manufacturer suffers a loss. At higher levels of production, however, the total revenue curve is the higher one and the manufacturer realizes a profit. The point at which the two curves cross is called the **break-even point,** because when total revenue equals total cost, the manufacturer breaks even, experiencing neither a profit nor a loss. Here is an example.

EXAMPLE 4.6

A manufacturer can sell a certain product for $110 per unit. Total cost consists of a fixed overhead of $7,500 plus production costs of $60 per unit.

(a) How many units must the manufacturer sell to break even?
(b) What is the manufacturer's profit or loss if 100 units are sold?
(c) How many units must be sold for the manufacturer to realize a profit of $1,250?

Solution

If x is the number of units manufactured and sold, the total revenue is given by $R(x) = 110x$ and the total cost by $C(x) = 7,500 + 60x$.

(a) To find the break-even point, set $R(x)$ equal to $C(x)$ and solve:

$$110x = 7,500 + 60x$$
$$50x = 7,500$$

so that $x = 150$

It follows that the manufacturer will have to sell 150 units to break even. See Figure 1.34.

Figure 1.34 Revenue $R(x) = 110x$ and cost $C(x) = 7,500 + 60x$.

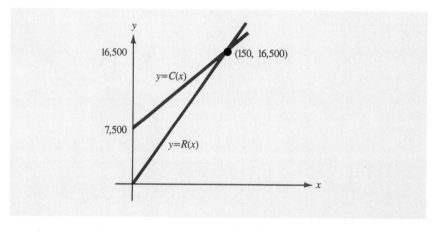

(b) The profit $P(x)$ is revenue minus cost. Hence,

$$P(x) = R(x) - C(x) = 110x - (7,500 + 60x) = 50x - 7,500$$

The profit from the sale of 100 units is

$$P(100) = 5,000 - 7,500 = -2,500$$

The minus sign indicates a negative profit (or loss), which was expected since 100 units is less than the break-even level of 150 units. It follows that the manufacturer will lose $2,500 if 100 units are sold.

(c) To determine the number of units that must be sold to generate a profit of $1,250, set the formula for profit $P(x)$ equal to 1,250 and solve for x. You get

$$50x - 7,500 = 1,250$$
$$50x = 8,750$$
$$x = \frac{8,750}{50} = 175$$

from which you can conclude that 175 units must be sold to generate the desired profit.

The next example illustrates how break-even analysis can be used as a tool for decision making.

EXAMPLE 4.7

A certain car rental agency charges $25 plus 60 cents per mile. A second agency charges $30 plus 50 cents per mile. Which agency offers the better deal?

Solution

The answer depends on the number of miles the car is driven. For short trips, the first agency charges less than the second, but for long trips, the second agency charges less. Break-even analysis can be used to find the number of miles for which the two agencies charge the same.

Suppose a car is to be driven x miles. Then the first agency will charge $C_1(x) = 25 + 0.60x$ dollars and the second will charge $C_2(x) = 30 + 0.50x$ dollars. If you set these expressions equal to each other and solve, you get

$$25 + 0.60x = 30 + 0.50x$$

so that $0.1x = 5$ or $x = 50$

This implies that the two agencies charge the same amount if the car is driven 50 miles. For shorter distances, the first agency offers the better deal, and for longer distances, the second agency is better. The situation is illustrated in Figure 1.35.

Figure 1.35 Car rental costs at competing agencies.

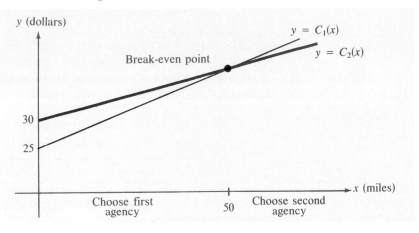

Market Equilibrium   An important economic application involving intersections of graphs arises in connection with the **law of supply and demand.** In this context, we think of the market price p of a commodity as determining the number of units of the commodity that manufacturers are willing to supply as well as the number of units that consumers are willing to buy. In most cases, manufacturers' supply $S(p)$ increases and consumers' demand $D(p)$ decreases as the market price p increases. A pair of supply and demand curves is sketched in Figure 1.36. (The letter q used to label the vertical axis stands for "quantity.")

(Actually, an economist's graph of these functions would not look quite like the one in Figure 1.36. For technical reasons, when dealing with supply and demand curves, economists usually break with mathematical tradition and use the horizontal axis for the dependent variable q and the vertical axis for the independent variable p.)

Figure 1.36 Market equilibrium: the intersection of supply and demand.

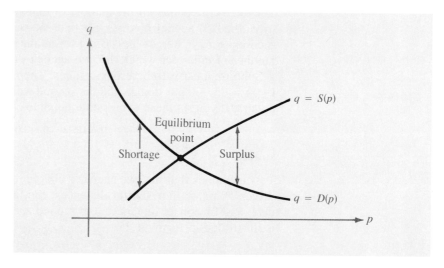

The point of intersection of the supply and demand curves is called the point of **market equilibrium.** The p coordinate of this point (the **equilibrium price**) is the market price at which supply equals demand, that is, the market price at which there will be neither a surplus nor a shortage of the commodity.

The law of supply and demand asserts that in a situation of pure competition, a commodity will tend to be sold at its equilibrium price. If the commodity is sold for more than the equilibrium price, there will be an unsold surplus on the market and retailers will tend to lower their prices. On the other hand, if the commodity is sold for less than the equilibrium price, the demand will exceed the supply and retailers will be inclined to raise their prices.

Here is an example.

EXAMPLE 4.8

Find the equilibrium price and the corresponding number of units supplied and demanded if the supply function for a certain commodity is $S(p) = p^2 + 3p - 70$ and the demand function is $D(p) = 410 - p$.

Solution

Set $S(p)$ equal to $D(p)$ and solve for p to get

$$p^2 + 3p - 70 = 410 - p$$
$$p^2 + 4p - 480 = 0$$
$$(p - 20)(p + 24) = 0$$
$$p = 20 \quad \text{or} \quad p = -24$$

Since only positive values of p are meaningful in this practical problem, you can conclude that the equilibrium price is $20. Since the corresponding supply and demand are equal, use the simpler demand equation to compute this quantity to get

$$D(20) = 410 - 20 = 390$$

Hence, 390 units are supplied and demanded when the market is in equilibrium.

Figure 1.37 The supply and demand curves for Example 4.8.

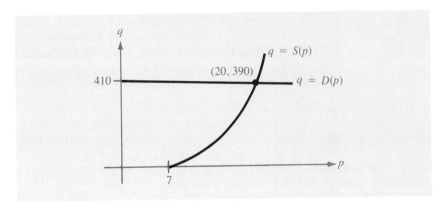

The supply and demand curves are sketched in Figure 1.37. Notice that the supply curve crosses the p axis when $p = 7$. (Verify this.) What is the economic interpretation of this fact?

Problems 1.4

Fencing

1. A farmer wishes to fence off a rectangular field with 1,000 feet of fencing. If one side of the field is along a stream (and does not require fencing), express the area of the field as a function of its width.

Landscaping

2. A landscaper wishes to make a rectangular flower garden that is twice as long as it is wide. Express the area of the garden as a function of its width.

Algebra   3. The sum of two numbers is 18. Express the product of the numbers as a function of the smaller number.

Algebra   4. The product of two numbers is 318. Express the sum of the numbers as a function of the smaller number.

Sales Revenue   5. Each unit of a certain commodity costs $p = 35x + 15$ cents when x units of the commodity are produced. If all x units are sold at this price, express the revenue derived from the sales as a function of x.

Retail Sales   6. A bookstore can obtain a certain gift book from the publisher at a cost of $3 per book. The bookstore has been offering the book at the price of $15 per copy and, at this price, has been selling 200 copies a month. The bookstore is planning to lower its price to stimulate sales and estimates that for each $1 reduction in the price, 20 more books will be sold each month. Express the bookstore's monthly profit from the sale of this book as a function of the selling price, draw the graph, and estimate the optimal selling price.

Retail Sales   7. A manufacturer has been selling lamps at the price of $6 per lamp, and at this price, consumers have been buying 3,000 lamps a month. The manufacturer wishes to raise the price and estimates that for each $1 increase in the price, 1,000 fewer lamps will be sold each month. The manufacturer can produce the lamps at a cost of $4 per lamp. Express the manufacturer's monthly profit as a function of the price that the lamps are sold, draw the graph, and estimate the optimal selling price.

Transportation Costs   8. A bus company is willing to charter buses only to groups of 35 or more people. If a group contains exactly 35 people, each person pays $60. In larger groups, everybody's fare is reduced by 50 cents for each person in excess of 35. Express the bus company's revenue as a function of the size of the group, draw the graph, and estimate the size of the group that will maximize the revenue.

Agricultural Yield   9. A Florida citrus grower estimates that if 60 orange trees are planted, the average yield per tree will be 400 oranges. The average yield will decrease by 4 oranges per tree for each additional tree planted on the same acreage. Express the grower's total yield as a function of the number of additional trees planted, draw the graph, and estimate the total number of trees the grower should plant to maximize yield.

Harvesting   10. Farmers can get $2 per bushel for their potatoes on July first, and after that, the price drops by 2 cents per bushel per day. On July first, a farmer has 80 bushels of potatoes in the field and estimates that the crop is increasing at the rate of 1 bushel per day. Express the farmer's revenue from the sale of the potatoes as a function of the time at which the crop is harvested, draw the graph, and estimate when the farmer should harvest the potatoes to maximize revenue.

Recycling 11. To raise money, a service club has been collecting used bottles that it plans to deliver to a local glass company for recycling. Since the project began 80 days ago, the club has collected 24,000 pounds of glass for which the glass company currently offers 1 cent per pound. However, since bottles are accumulating faster than they can be recycled, the company plans to reduce by 1 cent each day the price it will pay for 100 pounds of used glass. Assuming that the club can continue to collect bottles at the same rate and that transportation costs make more than one trip to the glass company unfeasible, express the club's revenue from its recycling project as a function of the number of additional days the project runs. Draw the graph and estimate when the club should conclude the project and deliver the bottles to maximize its revenue.

Fencing 12. A city recreation department plans to build a rectangular playground 3,600 square meters in area. The playground is to be surrounded by a fence. Express the length of the fencing as a function of the length of one of the sides of the playground, draw the graph, and estimate the dimensions of the playground requiring the least amount of fencing.

Area 13. Express the area of a rectangular field whose perimeter is 320 meters as a function of the length of one of its sides. Draw the graph and estimate the dimensions of the field of maximum area.

Surface Area 14. A closed box with a square base is to have volume 1,500 cubic inches. Express its surface area as a function of the length of its base.

Volume 15. A closed box with a square base has surface area 4,000 square centimeters. Express its volume as a function of the length of its base.

Construction Cost 16. A closed box with a square base is to have a volume of 250 cubic meters. The material for the top and bottom of the box costs $2 per square meter, and the material for the sides costs $1 per square meter. Express the construction cost of the box as a function of the length of its base.

Construction Cost 17. An open box with a square base is to be built for $48. The sides of the box will cost $3 per square meter, and the base will cost $4 per square meter. Express the volume of the box as a function of the length of its base.

Volume 18. An open box is to be made from a square piece of cardboard, 18 inches by 18 inches, by removing a small square from each corner and folding up the flaps to form the sides. Express the volume of the resulting box as a function of the length x of a side of the removed squares. Draw the graph and estimate the value of x for which the volume of the resulting box is greatest.

In Problems 19 through 23, you will need to know that a cylinder of radius r and height h has volume $V = \pi r^2 h$. Also, recall that a circle of radius r has area $A = \pi r^2$ and circumference $C = 2\pi r$.

Packaging   19. A soda can holds 12 fluid ounces (approximately 6.89π cubic inches). Express the surface area of the can as a function of its radius.

Volume   20. A closed cylindrical can has surface area 120π square inches. Express the volume of the can as a function of its radius.

Volume   21. A closed cylindrical can has radius r and height h.
(a) If the surface area S of the can is a constant, express the volume V of the can in terms of S and r.
(b) If the volume V of the can is a constant, express the surface area S in terms of V and r.

Packaging   22. A cylindrical can is to hold 4π cubic inches of frozen orange juice. The cost per square inch of constructing the metal top and bottom is twice the cost per square inch of constructing the cardboard side. Express the cost of constructing the can as a function of its radius if the cost of the side is 0.02 cent per square inch.

Volume   23. A cylindrical can with no top has been made from 27π square inches of metal. Express the volume of the can as a function of its radius.

Admission Fees   24. A local natural history museum charges admission to groups according to the following policy: Groups of fewer than 50 people are charged a rate of $1.50 per person, while groups of 50 people or more are charged a reduced rate of $1 per person.
(a) Express the amount a group will be charged for admission as a function of its size and draw the graph.
(b) How much money will a group of 49 people save in admission costs if it can recruit 1 additional member?

Discounts 25. A record club offers the following special sale: If 5 compact disks are bought at the full price of $10 apiece, additional disks can then be bought at half price. There is a limit of 9 records per customer. Express the cost of the records as a function of the number bought and draw the graph.

Income Tax 26. The following data are taken from the 1990 federal income tax rate schedule for single taxpayers:

If the Taxable Income Is:		The Income Tax Is:	
Over	But Not Over		Of the Amount Over
$0	$19,450	15%	$0
$19,450	$47,050	$2,917.50 + 28%	$19,450
$47,050	$97,620	$10,645.50 + 33%	$47,050

(a) Express an individual's income tax as a function of the taxable income x for $0 \leq x \leq 97, 620$ and draw the graph.
(b) The graph in part (a) should consist of three line segments. Compute the slope of each segment. What happens to these slopes as the taxable income increases? Explain the behavior of the slopes in practical terms.

Transportation Cost 27. A bus company has adopted the following pricing policy for groups that wish to charter its buses: Groups containing no more than 40 people will be charged a fixed amount of $2,400 (40 times $60). In groups containing between 40 and 80 people, everyone will pay $60 minus 50 cents for each person in excess of 40. The company's lowest fare of $40 per person will be offered to groups that have 80 members or more. Express the bus company's revenue as a function of the size of the group. Draw the graph.

Population Growth 28. In the absence of environmental constraints, population grows at a rate proportional to its size. Express the rate of population growth as a function of the size of the population.

Radioactive Decay 29. A sample of radium decays at a rate proportional to the amount of radium remaining. Express the rate of decay of the sample as a function of the amount remaining.

Temperature Change 30. The rate at which the temperature of an object changes is proportional to the difference between its own temperature and the temperature of the surrounding medium. Express this rate as a function of the temperature of the object.

The Spread of an Epidemic 31. The rate at which an epidemic spreads through a community is jointly proportional to the number of people who have caught the disease and the number who have not. Express this rate as a function of the number of people who have caught the disease.

Political Corruption   32. The rate at which people are implicated in a government scandal is jointly proportional to the number of people already implicated and the number of people involved who have not yet been implicated. Express this rate as a function of the number of people who have been implicated.

Production Cost   33. At a certain factory, setup cost is directly proportional to the number of machines used and operating cost is inversely proportional to the number of machines used. Express the total cost as a function of the number of machines used.

Transportation Cost   34. A truck is hired to transport goods from a factory to a warehouse. The driver's wages are figured by the hour and so are inversely proportional to the speed at which the truck is driven. The cost of gasoline is directly proportional to the speed. Express the total cost of operating the truck as a function of the speed at which it is driven.

Distance   35. A car traveling east at 80 kilometers per hour and a truck traveling south at 60 kilometers per hour start at the same intersection. Express the distance between them as a function of time. (*Hint:* Use the pythagorean theorem.)

Distance   36. A truck is 300 miles due east of a car and is traveling west at the constant speed of 30 miles per hour. Meanwhile, the car is going north at the constant speed of 60 miles per hour. Express the distance between the car and truck as a function of time.

Installation Cost 37. A cable is to be run from a power plant on one side of a river 900 meters wide to a factory on the other side, 3,000 meters downstream. The cable will be run in a straight line from the power plant to some point P on the opposite bank and then along the bank to the factory. The cost of running the cable across the water is $5 per meter, while the cost over land is $4 per meter. Let x be the distance from P to the point directly across the river from the power plant and express the cost of installing the cable as a function of x.

Poster Design 38. A rectangular poster contains 25 square centimeters of print surrounded by margins of 2 centimeters on each side and 4 centimeters on the top and bottom. Express the total area of the poster (printing plus margins) as a function of the width of the printed portion.

Production Cost 39. A company has received an order from the city recreation department to manufacture 8,000 Styrofoam kickboards for its summer swimming program. The company owns several machines, each of which can produce 30 kickboards an hour. The cost of setting up the machines to produce these particular kickboards is $20 per machine. Once the machines have been set up, the operation is fully automated and can be overseen by a single production supervisor earning $4.80 per hour. Express the cost of producing the 8,000 kickboards as a function of the number of machines used, draw the graph, and estimate the number of machines the company should use to minimize cost.

Break-Even Analysis 40. A furniture manufacturer can sell dining-room tables for $70 apiece. The manufacturer's total cost consists of a fixed overhead of $8,000 plus production costs of $30 per table.
(a) How many tables must the manufacturer sell to break even?
(b) How many tables must the manufacturer sell to make a profit of $6,000?
(c) What will be the manufacturer's profit or loss if 150 tables are sold?
(d) On the same set of axes, graph the manufacturer's total revenue and total cost functions. Explain how the overhead can be read from the graph.

Break-Even Analysis 41. During the summer, a group of students builds kayaks in a converted garage. The rental for the garage is $600 for the summer, and the materials needed to build a kayak cost $25. The kayaks can be sold for $175 apiece.
(a) How many kayaks must the students sell to break even?
(b) How many kayaks must the students sell to make a profit of $450?

Checking Accounts 42. The charge for maintaining a checking account at a certain bank is $4 per month plus 10 cents for each check that is written. A competing bank charges $3 per month plus 14 cents per check. Find a criterion for deciding which bank offers the better deal.

Membership Fees   43. Membership in a private tennis club costs $1,000 per year and entitles the member to use the courts for a fee of $3 per hour. At a competing club, membership costs $800 per year and the charge for the use of the courts is $4 per hour. If only financial considerations are to be taken into account, how should a tennis player choose which club to join?

Property Tax   44. Under the provisions of a proposed property tax bill, a homeowner will pay $100 plus 8 percent of the assessed value of the house. Under the provisions of a competing bill, the homeowner will pay $1,900 plus 2 percent of the assessed value. If only financial considerations are taken into account, how should a homeowner decide which bill to support?

Supply and Demand   45. The supply and demand functions for a certain commodity are $S(p) = 4p + 200$ and $D(p) = -3p + 480$, respectively. Find the equilibrium price and the corresponding number of units supplied and demanded, and draw the supply and demand curves on the same set of axes.

Supply and Demand   46. When electric blenders are sold for p dollars apiece, manufacturers will supply $\frac{p^2}{10}$ blenders to local retailers while the local demand will be $60 - p$ blenders. At what market price will the manufacturers' supply of electric blenders be equal to the consumers' demand for the blenders? How many blenders will be sold at this price?

Supply and Demand   47. The supply and demand functions for a certain commodity are $S(p) = p - 10$ and $D(p) = \dfrac{5,600}{p}$, respectively.

(a) Find the equilibrium price and the corresponding number of units supplied and demanded.
(b) Draw the supply and demand curve on the same set of axes.
(c) Where does the supply curve cross the p axis? What is the economic significance of this point?

Spy Story   48. The hero of a popular spy story has escaped from the headquarters of an international diamond-smuggling ring in the tiny Mediterranean country of Azusa. Our hero, driving a stolen milk truck at 72 kilometers per hour, has a 40-minute head start on his pursuers, who are chasing him in a Ferrari going 168 kilometers per hour. The distance from the smugglers' headquarters to the border, and freedom, is 83.8 kilometers. Will our hero make it?

Air Travel   49. Two jets bound for Los Angeles leave New York 30 minutes apart. The first travels 550 miles per hour, while the second goes 650 miles per hour. At what time will the second plane pass the first?

Supply and Demand   50. Suppose the supply and demand functions for a certain commodity are $S(p) = ap + b$ and $D(p) = cp + d$, respectively.

(a) What can you say about the signs of the coefficients a, b, c, and d if the supply and demand curves are oriented as shown in the accompanying figure?

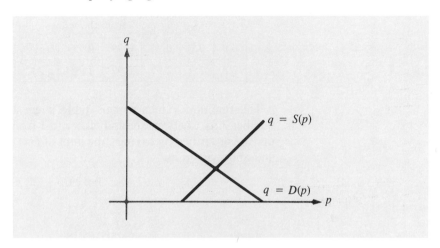

(b) Express the equilibrium price in terms of the coefficients a, b, c, and d.

(c) Use your answer in part (b) to determine what happens to the equilibrium price as a increases.

(d) Use your answer in part (b) to determine what happens to the equilibrium price as d increases.

5 Limits and Continuity

As you will see in subsequent chapters, **calculus** is an enormously powerful branch of mathematics with a wide range of applications, including curve sketching, optimization of functions, analysis of rates of change, and computation of area. What gives calculus its power and distinguishes it from algebra is the concept of **limit.** The purpose of this section is to give you an introduction to this important concept. The approach will be intuitive rather than formal. The ideas outlined here form the basis for a more rigorous development of the laws and procedures of calculus and lie at the heart of much of modern mathematics.

The Limit Concept Limits describe what happens to a function $f(x)$ as its variable x approaches a particular number c. To illustrate this concept, suppose you want to know what happens to the function $f(x) = \dfrac{x^2 + x - 2}{x - 1}$ as x approaches 1. Although $f(x)$ is not defined *at $x = 1$*, you can get a feel for the situation by evaluating $f(x)$ using values of x that get closer and closer to 1 from both the left and the right. The following table summarizes the behavior of $f(x)$ for x near 1.

			x approaches 1 from the left →				← x approaches 1 from the right			
x	0.8	0.9	0.95	0.99	0.999	1	1.001	1.01	1.05	1.1
$f(x)$	2.8	2.9	2.95	2.99	2.999		3.001	3.01	3.05	3.10

The function values in this table suggest that $f(x)$ approaches the number 3 as x gets closer and closer to 1 from either side. This behavior may be described by saying "the limit of $f(x)$ as x approaches 1 equals 3" and abbreviated as

$$\lim_{x \to 1} f(x) = 3$$

Limit

If $f(x)$ gets closer and closer to a number L as x gets closer and closer to c from either side, then L is the limit of $f(x)$ as x approaches c. This behavior is expressed by writing

$$\lim_{x \to c} f(x) = L$$

Geometric Interpretation of Limits

Geometrically, the limit statement $\lim_{x \to c} f(x) = L$ means that the height of the graph of $y = f(x)$ approaches L as x approaches c (Figure 1.38). For instance, the graph of the function $f(x) = \dfrac{x^2 + x - 2}{x - 1}$ is a line with

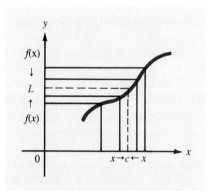

Figure 1.38 If $\lim_{x \to c} f(x) = L$, the height of the graph of $f(x)$ approaches L as x approaches c.

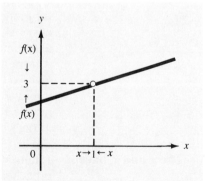

Figure 1.39 Geometric interpretation of the limit statement

$$\lim_{x \to 1} \frac{x^2 + x - 2}{x - 1} = 3$$

a "hole" at (1, 3), and the points (x, y) on the graph approach this hole as x approaches 1 from either side (Figure 1.39).

It is important to remember that limits describe the behavior of a function *near* a particular point, not necessarily *at* the point itself. This is illustrated in Figure 1.40. For all three functions graphed, the limit of $f(x)$ as x approaches c is equal to L. Yet the functions behave quite differently at $x = c$ itself. In Figure 1.40a, $f(c)$ is equal to the limit L; in Figure 1.40b, $f(c)$ is different from L; and in Figure 1.40c, $f(c)$ is not defined at all.

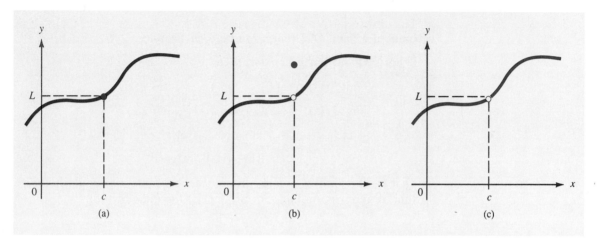

Figure 1.40 Three functions for which $\lim\limits_{x \to c} f(x) = L$.

Figure 1.41 shows the graphs of two functions that do not have a limit as x approaches c. The limit does not exist in Figure 1.41a because $f(x)$ approaches 5 as x approaches c from the right and approaches a different value, 3, as x approaches c from the left. The function in Figure 1.41b has no limit as x approaches c because the values of $f(x)$ increase without bound as x gets close to c and do not approach any (finite) number.

Figure 1.41 Two functions for which $\lim\limits_{x \to c} f(x)$ does not exist.

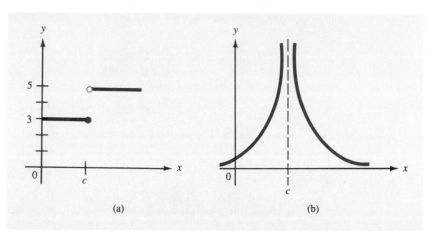

Functions like the one in Figure 1.41b are sometimes said to have an **infinite limit** as x approaches c.

Properties of Limits

Limits obey certain algebraic rules that may be used in computations. These rules, which should seem plausible on the basis of our informal definition of limit, are proved formally in more theoretical courses. They are important because they simplify the calculation of limits of algebraic functions.

Limit of a Sum, Difference, Multiple, or Product

If $\lim\limits_{x \to c} f(x)$ and $\lim\limits_{x \to c} g(x)$ exist, then

$$\lim_{x \to c} [f(x) + g(x)] = \lim_{x \to c} f(x) + \lim_{x \to c} g(x)$$

$$\lim_{x \to c} [f(x) - g(x)] = \lim_{x \to c} f(x) - \lim_{x \to c} g(x)$$

$$\lim_{x \to c} [k\, f(x)] = k \lim_{x \to c} f(x) \qquad \text{for any constant } k$$

$$\lim_{x \to c} [f(x)g(x)] = [\lim_{x \to c} f(x)][\lim_{x \to c} g(x)]$$

That is, the limit of a sum, difference, multiple, or product is the sum, difference, multiple, or product of the individual limits.

Limit of a Quotient

If $\lim\limits_{x \to c} f(x)$ and $\lim\limits_{x \to c} g(x)$ exist and $\lim\limits_{x \to c} g(x) \neq 0$, then

$$\lim_{x \to c} \frac{f(x)}{g(x)} = \frac{\lim\limits_{x \to c} f(x)}{\lim\limits_{x \to c} g(x)}$$

That is, if the limit of the denominator is not 0, the limit of a quotient is the quotient of the individual limits.

Limit of a Power

If $\lim\limits_{x \to c} f(x) = L$ and p is a real number for which L is defined, then

$$\lim_{x \to c} [f(x)]^p = [\lim_{x \to c} f(x)]^p = L^p$$

That is, the limit of a power is the power of the limit.

The next two properties deal with the limits of two elementary linear functions, which can be used to build all other algebraic functions.

Limits of Two Linear Functions

For any number c and constant k

$$\lim_{x \to c} k = k$$

and

$$\lim_{x \to c} x = c$$

That is, the limit of a constant is the constant itself, and the limit of $f(x) = x$ as x approaches c is c.

In geometric terms, the limit statement $\lim_{x \to c} k = k$ says that the height of the graph of the constant function $f(x) = k$ approaches k as x approaches c. Similarly, $\lim_{x \to c} x = c$ says that the height of the linear function $f(x) = x$ approaches c as x approaches c. These statements are illustrated in Figure 1.42.

Figure 1.42 Limits of two linear functions.

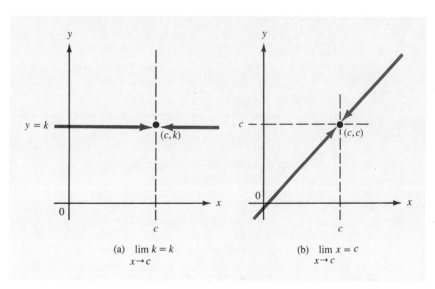

(a) $\lim_{x \to c} k = k$ (b) $\lim_{x \to c} x = c$

Computation of Limits

The following examples illustrate how the properties of limits can be used to calculate limits of algebraic functions. In the first example, you will see how to find the limit of a polynomial.

EXAMPLE 5.1

Find $\lim_{x \to -1} (3x^3 - 4x + 8)$.

Solution

Apply the properties of limits to obtain

$$\lim_{x \to -1} (3x^3 - 4x + 8) = 3\Big(\lim_{x \to -1} x\Big)^3 - 4\Big(\lim_{x \to -1} x\Big) + \lim_{x \to -1} 8$$
$$= 3(-1)^3 - 4(-1) + 8 = 9$$

In the next example, you will see how to find the limit of a rational function whose denominator does not approach zero.

EXAMPLE 5.2

Find $\displaystyle\lim_{x \to 0} \frac{3x^3 - 8}{x - 2}$.

Solution

Since $\displaystyle\lim_{x \to 0} (x - 2) \neq 0$, you can use the quotient rule for limits to get

$$\lim_{x \to 0}\frac{3x^3 - 8}{x - 2} = \frac{\displaystyle\lim_{x \to 0} (3x^3 - 8)}{\displaystyle\lim_{x \to 0} (x - 2)} = \frac{3 \displaystyle\lim_{x \to 0} x^3 - \lim_{x \to 0} 8}{\displaystyle\lim_{x \to 0} x - \lim_{x \to 0} 2} = \frac{-8}{-2} = 4$$

In general, you can use the properties of limits to obtain the following formulas, which may be used to evaluate many limits that occur in practical problems.

Limits of Polynomials and Rational Functions

If $p(x)$ and $q(x)$ are polynomials, then

$$\lim_{x \to c} p(x) = p(c)$$

and

$$\lim_{x \to c} \frac{p(x)}{q(x)} = \frac{p(c)}{q(c)} \qquad \text{if } q(c) \neq 0$$

In the next example, the denominator of the given rational function approaches zero, while the numerator does not. When this happens, you can conclude that the limit does not exist. The absolute value of such a quotient increases without bound and hence does not approach any finite number.

EXAMPLE 5.3

Find $\lim\limits_{x \to 2} \dfrac{x + 1}{x - 2}$.

Solution

The quotient rule for limits does not apply in this case since the limit of the denominator is

$$\lim_{x \to 2} (x - 2) = 0$$

Since the limit of the numerator is $\lim\limits_{x \to 2} (x + 1) = 3$, which is not equal to zero, you can conclude that the limit of the quotient does not exist.

The graph of the function $f(x) = \dfrac{x + 1}{x - 2}$ in Figure 1.43 gives you a

Figure 1.43 The graph of $f(x) = \dfrac{x + 1}{x - 2}$.

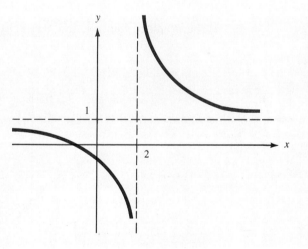

better idea of what is actually happening in this example. Note that $f(x)$ increases without bound as x approaches 2 from the right and decreases without bound as x approaches 2 from the left.

In the next example, both the numerator and the denominator of the given rational function approach zero. When this happens, you must simplify the function algebraically in order to find the desired limit.

EXAMPLE 5.4

Find $\lim\limits_{x \to 1} \dfrac{x^2 - 1}{x^2 - 3x + 2}$.

Solution

As x approaches 1, both the numerator and the denominator approach zero, and you can draw no conclusion about the size of the quotient. To proceed, observe that the given function is not defined when $x = 1$ but that for all other values of x, you can divide the numerator and denominator by $x - 1$ to obtain

$$\frac{x^2 - 1}{x^2 - 3x + 2} = \frac{(x - 1)(x + 1)}{(x - 1)(x - 2)} = \frac{x + 1}{x - 2}$$

(Since $x \neq 1$, you are not dividing by zero.) Now take the limit as x approaches (but is not equal to) 1 to get

$$\lim_{x \to 1} \frac{x^2 - 1}{x^2 - 3x + 2} = \frac{\lim\limits_{x \to 1} (x + 1)}{\lim\limits_{x \to 1} (x - 2)} = \frac{2}{-1} = -2$$

The graph of the function $f(x) = \dfrac{x^2 - 1}{x^2 - 3x + 2}$ is given in Figure 1.44.

Figure 1.44 The graph of $f(x) = \dfrac{x^2 - 1}{x^2 - 3x + 2}$.

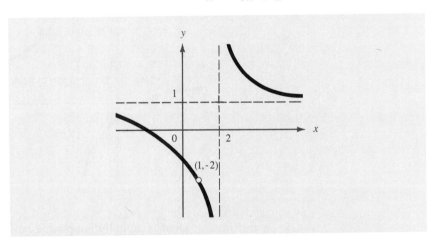

Note that it is like the graph in Figure 1.43 with a hole at the point $(1, -2)$.

In general, when both the numerator and denominator of a quotient approach zero as x approaches c, your strategy will be to simplify the quotient algebraically (as in Example 5.4 by canceling $x - 1$). In most cases, the simplified form of the quotient will be valid for all values of x except $x = c$. Since you are interested in the behavior of the quotient *near* $x = c$ and not at $x = c$, you may use the simplified form of the quotient to calculate the limit. Here is another example illustrating this technique.

EXAMPLE 5.5

Find $\lim\limits_{x \to 1} \dfrac{\sqrt{x} - 1}{x - 1}$.

Solution

Both the numerator and denominator approach zero as x approaches 1. To simplify the quotient, multiply the numerator and denominator by $\sqrt{x} + 1$ to get

$$\frac{\sqrt{x} - 1}{x - 1} = \frac{(\sqrt{x} - 1)(\sqrt{x} + 1)}{(x - 1)(\sqrt{x} + 1)} = \frac{x - 1}{(x - 1)(\sqrt{x} + 1)} = \frac{1}{\sqrt{x} + 1}$$

and then take the limit to obtain

$$\lim_{x \to 1} \frac{\sqrt{x} - 1}{x - 1} = \lim_{x \to 1} \frac{1}{\sqrt{x} + 1} = \frac{1}{2}$$

Continuity   In Section 2, we defined a continuous function as one whose graph is an unbroken curve with no holes or gaps. Alternatively, continuity may be thought of as the quality of having parts that are in immediate connection with one another. Not all functions have this property, but those which do have special features that make them extremely important in the development of calculus. Continuity was discussed intuitively in Examples 2.4 and 2.5 of Section 2, and it can be defined more precisely in terms of limits.

Continuity

A function f is continuous at c if

(a) $f(c)$ is defined

(b) $\lim\limits_{x \to c} f(x)$ exists

(c) $\lim\limits_{x \to c} f(x) = f(c)$

If $f(x)$ is not continuous at c, it is said to have a **discontinuity** there.

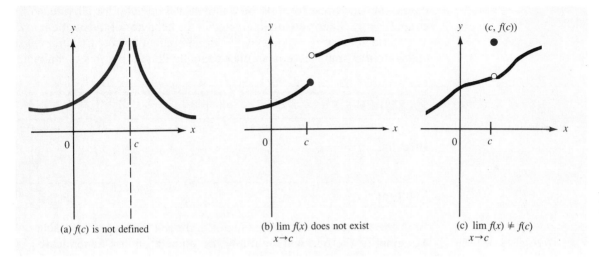

(a) $f(c)$ is not defined

(b) $\lim\limits_{x \to c} f(x)$ does not exist

(c) $\lim\limits_{x \to c} f(x) \neq f(c)$

Figure 1.45 Three functions with discontinuities at $x = c$.

Figure 1.45 shows the graphs of three functions that have discontinuities at $x = c$. In Figure 1.45a, the function is not continuous at $x = c$ because $f(c)$ is not defined. In Figure 1.45b, the function is not continuous at $x = c$ because $\lim\limits_{x \to c} f(c)$ does not exist. In Figure 1.45c, the function is not continuous at $x = c$ because $\lim\limits_{x \to c} f(x) \neq f(c)$ even though $f(c)$ is defined and the limit exists.

Continuity of Polynomials and Rational Functions   Recall that if $p(x)$ and $q(x)$ are polynomials, then

$$\lim_{x \to c} p(x) = p(c)$$

and

$$\lim_{x \to c} \frac{p(x)}{q(x)} = \frac{p(c)}{q(c)} \quad \text{if } q(c) \neq 0$$

These limit formulas may be interpreted as saying that **a polynomial or a rational function is continuous wherever it is defined.** This is illustrated in the following examples.

EXAMPLE 5.6

Show that the polynomial $p(x) = 3x^3 - x + 5$ is continuous at $x = 1$.

Solution

Verify that the three criteria for continuity are satisfied. Clearly $p(1)$ is defined; in fact, $p(1) = 7$. Moreover, $\lim\limits_{x \to 1} p(x)$ exists and

$$\lim_{x \to 1} p(x) = p(1)$$

as required for $p(x)$ to be continuous at $x = 1$.

EXAMPLE 5.7

Show that the rational function $f(x) = \dfrac{x + 1}{x - 2}$ is continuous at $x = 3$.

Solution

Note that $f(3) = \dfrac{3 + 1}{3 - 2} = 4$. Since $\lim\limits_{x \to 3} x - 2 \neq 0$, you find that

$$\lim_{x \to 3} f(x) = \lim_{x \to 3} \frac{x + 1}{x - 2} = \frac{\lim\limits_{x \to 3} (x + 1)}{\lim\limits_{x \to 3} (x - 2)} = \frac{4}{1} = 4 = f(3)$$

as required for $f(x)$ to be continuous at $x = 3$.

EXAMPLE 5.8

Discuss the continuity of each of the following functions:

(a) $f(x) = \dfrac{1}{x}$ (b) $g(x) = \dfrac{x^2 - 1}{x + 1}$

(c) $h(x) = \begin{cases} x + 1 & \text{if } x < 1 \\ 2 - x & \text{if } x \geq 1 \end{cases}$

Solution

The functions in parts (a) and (b) are rational and are therefore continuous wherever they are defined (that is, wherever their denominators are not zero).

(a) $f(x) = \dfrac{1}{x}$ is defined everywhere except at $x = 0$, and thus it is continuous for all $x \neq 0$ (Figure 1.46a).

(b) Since $x = -1$ is the only value of x for which $g(x)$ is undefined, $g(x)$ is continuous except at $x = -1$ (Figure 1.46b).

(c) This function is defined in two pieces. First check for continuity at $x = 1$, the value of x that separates the two pieces. You find that

(a) Continuous for $x \neq 0$ (b) Continuous for $x \neq -1$ (c) Continuous for $x \neq 1$

Figure 1.46 Functions for Example 5.8.

$\lim\limits_{x \to 1} h(x)$ does not exist, since $h(x)$ approaches 2 from the left and 1 from the right. Thus, $h(x)$ is not continuous at 1 (Figure 1.46c). However, since the polynomials $x + 1$ and $2 - x$ are each continuous for every value of x, it follows that $h(x)$ is continuous at every number x other than 1.

EXAMPLE 5.9

For what value of the constant A is the following function continuous for all real x?

$$f(x) = \begin{cases} Ax + 5 & \text{if } x < 1 \\ x^2 - 3x + 4 & \text{if } x \geq 1 \end{cases}$$

Solution

Since $Ax + 5$ and $x^2 - 3x + 4$ are both polynomials, it follows that $f(x)$ will be continuous everywhere except possibly at $x = 1$. Moreover, $f(x)$ approaches $A + 5$ as x approaches 1 from the left and approaches 2 as x approaches 1 from the right. Thus, for $\lim\limits_{x \to 1} f(x)$ to exist, we must have $A + 5 = 2$, which means $f(x)$ is continuous at $x = 1$ only when $A = -3$.

Continuity on an Interval   For many applications of calculus, it is useful to have the following definitions of continuity on open and closed intervals.

Continuity on an Interval

A function $f(x)$ is said to be continuous on an open interval $a < x < b$ if it is continuous at each x in the interval. It is said to be continuous on a closed interval $a \leq x \leq b$ if $f(x)$ is continuous on the open interval $a < x < b$ and $f(x)$ approaches $f(a)$ as x approaches a from the right (for $a < x$) and $f(x)$ approaches $f(b)$ as x approaches b from the left (for $x < b$).

EXAMPLE 5.10

Discuss the continuity of the function $f(x) = \dfrac{x + 2}{x - 3}$ on the open interval $-2 < x < 3$ and on the closed interval $-2 \leq x \leq 3$.

Solution

The rational function $f(x)$ is continuous for all x except $x = 3$. Therefore, it is continuous on the open interval $-2 < x < 3$ but not on the closed interval $-2 \leq x \leq 3$ since it is discontinuous at the endpoint 3 (where its denominator is zero).

Problems 1.5

In Problems 1 through 6, find $\lim\limits_{x \to a} f(x)$ if it exists.

4.

5.

6.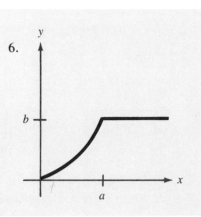

In Problems 7 through 28, find the indicated limit if it exists.

7. $\lim\limits_{x \to 2} (3x^2 - 5x + 2)$

8. $\lim\limits_{x \to -1} (x^3 - 2x^2 + x - 3)$

9. $\lim\limits_{x \to 0} (x^5 - 6x^4 + 7)$

10. $\lim\limits_{x \to 0} (1 - 5x^3)$

11. $\lim\limits_{x \to 3} (x - 1)^2(x + 1)$

12. $\lim\limits_{x \to -1} (x^2 + 1)(1 - 2x)^2$

13. $\lim\limits_{x \to 2} \dfrac{x + 1}{x + 2}$

14. $\lim\limits_{x \to 1} \dfrac{2x + 3}{x + 1}$

15. $\lim\limits_{x \to 5} \dfrac{x + 3}{5 - x}$

16. $\lim\limits_{x \to 3} \dfrac{2x + 3}{x - 3}$

17. $\lim\limits_{x \to 1} \dfrac{x^2 - 1}{x - 1}$

18. $\lim\limits_{x \to 3} \dfrac{9 - x^2}{x - 3}$

19. $\lim\limits_{x \to 5} \dfrac{x^2 - 3x - 10}{x - 5}$

20. $\lim\limits_{x \to 2} \dfrac{x^2 + x - 6}{x - 2}$

21. $\lim\limits_{x \to 4} \dfrac{(x + 1)(x - 4)}{(x - 1)(x - 4)}$

22. $\lim\limits_{x \to 0} \dfrac{x(x^2 - 1)}{x^2}$

23. $\lim\limits_{x \to -2} \dfrac{x^2 - x - 6}{x^2 + 3x + 2}$

24. $\lim\limits_{x \to 1} \dfrac{x^2 + 4x - 5}{x^2 - 1}$

25. $\lim\limits_{x \to 4} \dfrac{\sqrt{x} - 2}{x - 4}$

26. $\lim\limits_{x \to 9} \dfrac{\sqrt{x} - 3}{x - 9}$

27. $\lim\limits_{x \to 1} \dfrac{x - 1}{\sqrt{x} - 1}$

28. $\lim\limits_{x \to 9} \dfrac{x - 9}{\sqrt{x} - 3}$

In Problems 29 through 41, decide if the given function is continuous at the specified value of x.

29. $f(x) = 5x^2 - 6x + 1; x = 2$

30. $f(x) = x^3 - 2x^2 + x - 5; x = 0$

31. $f(x) = \dfrac{x + 2}{x + 1}; x = 1$

32. $f(x) = \dfrac{2x - 4}{3x - 2}; x = 2$

33. $f(x) = \dfrac{x + 1}{x - 1}; x = 1$

34. $f(x) = \dfrac{2x + 1}{3x - 6}; x = 2$

35. $f(x) = \dfrac{\sqrt{x} - 2}{x - 4}; x = 4$

36. $f(x) = \dfrac{\sqrt{x} - 2}{x - 4}; x = 2$

37. $f(x) = \begin{cases} x + 1 & \text{if } x \le 2 \\ 2 & \text{if } x > 2 \end{cases} \quad ; \quad x = 2$

38. $f(x) = \begin{cases} 0 & \text{if } x < 1 \\ x - 1 & \text{if } x \ge 1 \end{cases} \quad ; \quad x = 1$

39. $f(x) = \begin{cases} x + 1 & \text{if } x < 0 \\ x - 1 & \text{if } x \ge 0 \end{cases} \quad ; \quad x = 0$

40. $f(x) = \begin{cases} x^2 + 1 & \text{if } x \le 3 \\ 2x + 4 & \text{if } x > 3 \end{cases} \quad ; \quad x = 3$

41. $f(x) = \begin{cases} \dfrac{x^2 - 1}{x + 1} & \text{if } x < -1 \\ x^2 - 3 & \text{if } x \ge -1 \end{cases} \quad ; \quad x = -1$

In Problems 42 through 54, list all the values of x for which the given function is *not* continuous.

42. $f(x) = 3x^2 - 6x + 9$

43. $f(x) = x^5 - x^3$

44. $f(x) = \dfrac{x + 1}{x - 2}$

45. $f(x) = \dfrac{3x - 1}{2x - 6}$

46. $f(x) = \dfrac{3x + 3}{x + 1}$

47. $f(x) = \dfrac{x^2 - 1}{x + 1}$

48. $f(x) = \dfrac{3x - 2}{(x + 3)(x - 6)}$

49. $f(x) = \dfrac{x}{(x + 5)(x - 1)}$

50. $f(x) = \dfrac{x}{x^2 - x}$

51. $f(x) = \dfrac{x^2 - 2x + 1}{x^2 - x - 2}$

52. $f(x) = \begin{cases} 2x + 3 & \text{if } x \le 1 \\ 6x - 1 & \text{if } x > 1 \end{cases}$

53. $f(x) = \begin{cases} x^2 & \text{if } x \le 2 \\ 9 & \text{if } x > 2 \end{cases}$

54. $f(x) = \begin{cases} x - 1 & \text{if } x < 1 \\ 1 & \text{if } x = 1 \\ 1 - x & \text{if } x > 1 \end{cases}$

In Problems 55 and 56, find the values of the constant A such that the function $f(x)$ will be continuous for all x.

55. $f(x) = \begin{cases} Ax - 3 & \text{if } x < 2 \\ 3 - x + 2x^2 & \text{if } x \ge 2 \end{cases}$

56. $f(x) = \begin{cases} 1 - 3x & \text{if } x < 4 \\ Ax^2 + 2x - 3 & \text{if } x \ge 4 \end{cases}$

57. Find A and B such that the function defined by

$$f(x) = \begin{cases} Ax^2 + 5x - 9 & \text{if } x < 1 \\ B & \text{if } x = 1 \\ (3 - x)(A - 2x) & \text{if } x > 1 \end{cases}$$

is continuous for all x.

58. Discuss the continuity of the function $f(x) = \dfrac{x - 3}{4 + x}$ on the open interval $-4 < x < 1$ and on the closed interval $-4 \le x \le 1$.

59. Discuss the continuity of the function $f(x) = x\left(1 + \dfrac{1}{x}\right)$ on the open interval $0 < x < 1$ and on the closed interval $0 \le x \le 1$.

60. Discuss the continuity of the function

$$f(x) = \begin{cases} x^2 - 3x & \text{if } x < 2 \\ 4 + 2x & \text{if } x \ge 2 \end{cases}$$

on the open interval $0 < x < 2$ and on the closed interval $0 \le x \le 2$.

61. Discuss the continuity of the function.

$$f(x) = \begin{cases} x(x - 1) & \text{if } x < 3 \\ \dfrac{x^2 - 9}{x - 3} & \text{if } x \ge 3 \end{cases}$$

on the open interval $0 < x < 3$ and on the closed interval $0 \le x \le 3$.

Problems 62 through 65 are based on the following property of continuous functions:

The Intermediate Value Property

If $f(x)$ is continuous on the closed interval $a \le x \le b$ and L is a number between $f(a)$ and $f(b)$, then there is at least one number c between a and b $(a < c < b)$ such that $f(c) = L$.

62. Use the intermediate value property to show that if f is continuous on the interval $a \le x \le b$ and if $f(a)$ and $f(b)$ have opposite signs (one is positive and the other negative), then $f(c) = 0$ for at least one number c between a and b.

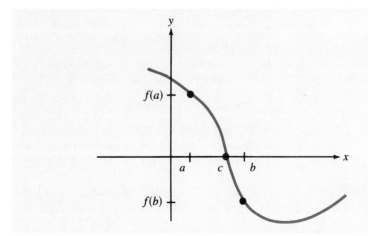

63. Tom weighs 150 pounds at age 20. Use the intermediate value property to show that at some time he weighed exactly 100 pounds.

64. The supply and demand functions for a certain commodity are $S(p)$ and $D(p)$, respectively, where $S(p)$ and $D(p)$ are continuous functions of the price p on the interval $p_1 \le p \le p_2$. If $D(p_1) > S(p_1)$ and $D(p_2) < S(p_2)$, explain why market equilibrium must occur at some price p_e between p_1 and p_2.

65. Show that the equation $\sqrt[3]{x} = x^2 + 2x - 1$ must have at least one solution on the interval $0 \le x \le 1$. (*Hint:* Use the result of Problem 62.)

Chapter Summary and Review Problems

Important Terms, Symbols, and Formulas

Function
Independent and dependent variables
Functional notation: $f(x)$
Domain of a function
Composition of functions: $g[h(x)]$
Graph of a function: the points $(x, f(x))$
x and y intercepts
Intersection of graphs
Limit of a function
Continuous function
Discontinuity
Polynomial
Rational function
Linear function; constant rate of change
Slope: $m = \dfrac{\Delta y}{\Delta x} = \dfrac{y_2 - y_1}{x_2 - x_1}$
Slope-intercept formula: $y = mx + b$
Point-slope formula: $y - y_0 = m(x - x_0)$
Break-even analysis
Market equilibrium; law of supply and demand
Direct proportionality: $Q = kx$
Inverse proportionality: $Q = \dfrac{k}{x}$
Joint proportionality: $Q = kxy$
Continuity of polynomials and rational functions
The intermediate value property of continuous functions

Review Problems

1. Specify the domain of each of the following functions:
   (a) $f(x) = x^2 - 2x + 6$
   (b) $f(x) = \dfrac{x - 3}{x^2 + x - 2}$
   (c) $f(x) = \sqrt{x^2 - 9}$

2. As advances in technology result in the production of increasingly powerful and compact calculators, the price of calculators currently on the market drops. Suppose that x months from now, the price of a certain model will be $P(x) = 40 + \dfrac{30}{x + 1}$ dollars.

   (a) What will the price be 5 months from now?
   (b) By how much will the price drop during the 5th month?
   (c) When will the price be $43?
   (d) What will happen to the price in the long run?

3. Find the composite function $g[h(x)]$.

   (a) $g(u) = u^2 + 2u + 1$, $h(x) = 1 - x$

   (b) $g(u) = \dfrac{1}{2u + 1}$, $h(x) = x + 2$

   (c) $g(u) = \sqrt{1 - u}$, $h(x) = 2x + 4$

4. (a) Find $f(x - 2)$ if $f(x) = x^2 - x + 4$.

   (b) Find $f(x^2 + 1)$ if $f(x) = \sqrt{x} + \dfrac{2}{x - 1}$.

   (c) Find $f(x + 1) - f(x)$ if $f(x) = x^2$.

5. Find functions $h(x)$ and $g(u)$ such that $f(x) = g[h(x)]$

   (a) $f(x) = (x^2 + 3x + 4)^5$

   (b) $f(x) = (3x + 1)^2 + \dfrac{5}{2(3x + 2)^3}$

6. An environmental study of a certain community suggests that the average daily level of smog in the air will be $Q(p) = \sqrt{0.5p + 19.4}$ units when the population is p thousand. It is estimated that t years from now, the population will be $p(t) = 8 + 0.2t^2$ thousand.

   (a) Express the level of smog in the air as a function of time.
   (b) What will the smog level be 3 years from now?
   (c) When will the smog level reach 5 units?

7. Find the value of c for which the curve $y = 3x^2 - 2x + c$ passes through the point $(2, 4)$.

8. Graph the following functions:

   (a) $f(x) = x^2 + 2x - 8$

   (b) $f(x) = \begin{cases} \dfrac{1}{x^2} & \text{if } x < 0 \\ x & \text{if } 0 \leq x < 3 \\ 4 & \text{if } x \geq 3 \end{cases}$

   (c) $f(x) = \dfrac{-2}{x - 3}$

   (d) $f(x) = x + \dfrac{2}{x}$

   (e) $f(x) = \dfrac{x^2 - 1}{x - 3}$

9. The consumer demand for a commodity is $D(p) = -50p + 800$ units per month when the market price is p dollars per unit.

(a) Graph this demand function.

(b) Express consumers' total monthly expenditure for the commodity as a function of p and draw the graph.

(c) Use the graph in part (b) to estimate the market price at which the total expenditure for the commodity is greatest.

10. A private college in the southwest has launched a fund-raising campaign. Suppose that college officials estimate that it will take $f(x) = \dfrac{10x}{150 - x}$ weeks to reach x percent of their goal.

(a) Sketch the relevant portion of the graph of this function.

(b) How long will it take to reach 50 percent of the campaign's goal?

(c) How long will it take to reach 100 percent of the goal?

11. Find the slope and y intercept of the given line and draw the graph.

(a) $y = 3x + 2$ (b) $5x - 4y = 20$

(c) $2y + 3x = 0$ (d) $\dfrac{x}{3} + \dfrac{y}{2} = 4$

12. Find the equation of the line with slope 5 and y intercept $(0, -4)$.

13. Find the equation of the line that passes through $(1, 3)$ and has slope -2.

14. Find the equation of the line through the points $(2, 4)$ and $(1, -3)$.

15. Since the beginning of the year, the price of unleaded gasoline has been increasing at a constant rate of 2 cents per gallon per month. By June first, the price had reached $1.03 per gallon.

(a) Express the price of unleaded gasoline as a function of time and draw the graph.

(b) What was the price at the beginning of the year?

(c) What will the price be on October first?

16. The circulation of a newspaper is increasing at a constant rate. Three months ago the circulation was 3,200. Today it is 4,400.

(a) Express the circulation as a function of time and draw the graph.

(b) What will the circulation be 2 months from now?

17. Find the points of intersection (if any) of the given pair of curves and draw the graphs.

(a) $y = -3x + 5$ and $y = 2x - 10$ (d) $y = x^2$ and $y = 15 - 2x$

(b) $y = x + 7$ and $y = -2 + x$

(c) $y = x^2 - 1$ and $y = 1 - x^2$ (e) $y = \dfrac{24}{x^2}$ and $y = 3x$

18. A manufacturer can produce bookcases at a cost of $10 apiece. Sales figures indicate that if the bookcases are sold for x dollars apiece, approximately $50 - x$ will be sold each month. Express the manufacturer's monthly profit as a function of the selling price x, draw the graph, and estimate the optimal selling price.

19. A retailer can obtain cameras from the manufacturer at a cost of $50 apiece. The retailer has been selling the cameras at the price of $80 apiece, and at this price, consumers have been buying 40 cameras a month. The retailer is planning to lower the price to stimulate sales and estimates that for each $5 reduction in the price, 10 more cameras will be sold each month. Express the retailer's monthly profit from the sale of the cameras as a function of the selling price. Draw the graph and estimate the optimal selling price.

20. A cylindrical can with no top is to be constructed for 80 cents. The cost of the material used for the bottom is 3 cents per square inch, and the cost of the material used for the curved side is 2 cents per square inch. Express the volume of the can as a function of its radius.

21. A manufacturing firm has received an order to make 400,000 souvenir silver medals commemorating the 25th anniversary of the landing of Apollo 11 on the moon. The firm owns several machines, each of which can produce 200 medals per hour. The cost of setting up the machines to produce the medals is $80 per machine, and the total operating cost is $5.76 per hour. Express the cost of producing the 400,000 medals as a function of the number of machines used. Draw the graph and estimate the number of machines the firm should use to minimize cost.

22. One plumber charges $25 plus $16 per half hour. A second charges $31 plus $14 per half hour. Find a criterion for deciding which plumber to call if only financial considerations are to be taken into account.

23. A manufacturer can sell a certain product for $80 per unit. Total cost consists of a fixed overhead of $4,500 plus production costs of $50 per unit.
    (a) How many units must the manufacturer sell to break even?
    (b) What is the manufacturer's profit or loss if 200 units are sold?
    (c) How many units must the manufacturer sell to realize a profit of $900?

24. Suppose it is known that the supply and demand for a certain commodity are functions of the form $S(p) = p^2 + Ap - 3$ and $D(p) = Cp + 32$ where p is the price per unit and A and C are constants. It is also known that no units will be supplied until the price is at least $3 and that when the price exceeds $8 per unit, no items will be demanded. How many units are supplied and demanded at equilibrium? What is the excess of supply over demand when the price is $6?

25. An author must decide between two publishers who are vying for the rights to her new book. Publisher A offers royalties of 12 percent of net proceeds on the first 30,000 copies sold and 17 percent on all

copies in excess of that figure and expects to net $5 on each copy sold. Publisher B will pay no royalties on the first 4,000 copies sold but will pay 15 percent on the net proceeds of all copies sold in excess of 4,000 and expects to net $6 on each copy sold. Which publisher offers the better deal?

26. Some psychologists believe that when a person is asked to recall a set of facts, the rate at which the facts are recalled is proportional to the number of relevant facts in the subject's memory that have not yet been recalled. Express the recall rate as a function of the number of facts that have been recalled.

27. Evaluate each of the following limits:

(a) $\lim_{x \to 2} \dfrac{x^2 - 3x}{x + 1}$

(b) $\lim_{x \to 1} \dfrac{x^2 + x - 2}{x^2 - 1}$

(c) $\lim_{x \to 1} \left(\dfrac{1}{x^2} - \dfrac{1}{x} \right)$

(d) $\lim_{x \to 2} \dfrac{x^3 - 8}{2 - x}$

28. In each of the following cases, list all values of x for which the given function is not continuous.

(a) $f(x) = 5x^3 - 3x + \sqrt{x}$

(b) $f(x) = \dfrac{x^2 - 1}{x + 3}$

(c) $g(x) = \dfrac{x^3 + 5x}{(x - 2)(2x + 3)}$

(d) $h(x) = \begin{cases} x^3 + 2x - 33 & \text{if } x \le 3 \\ \dfrac{x^2 - 6x + 9}{x - 3} & \text{if } x > 3 \end{cases}$

29. In each of the following cases, find the value of the constant A that makes the given function $f(x)$ continuous for all x.

(a)
$$f(x) = \begin{cases} 2x + 3 & \text{if } x < 1 \\ Ax - 1 & \text{if } x \ge 1 \end{cases}$$

(b)
$$f(x) = \begin{cases} \dfrac{x^2 - 1}{x + 1} & \text{if } x < -1 \\ Ax^2 + x - 3 & \text{if } x \ge -1 \end{cases}$$

30. The population (in thousands) of a colony of bacteria t minutes after the introduction of a toxin is given by the function

$$f(t) = \begin{cases} t^2 + 7 & \text{if } t < 5 \\ -8t + 72 & \text{if } t \ge 5 \end{cases}$$

(a) When does the colony die out?

(b) Explain why the population must be 10,000 some time between $t = 1$ and $t = 7$. (*Hint:* Use the intermediate value property stated in the problems in Section 5.)

CHAPTER 2
DIFFERENTIATION: BASIC CONCEPTS

1   The Derivative

Differentiation is a mathematical technique of exceptional power and versatility. It is one of the two central concepts in the branch of mathematics called **calculus** and has a variety of applications, including curve sketching, the optimization of functions, and the analysis of rates of change. To set the stage, here are brief introductory discussions of two of the types of practical problems that can be solved using calculus.

Optimization Problems   A typical problem to which calculus can be applied is the profit maximization problem you saw in Example 4.1 of Chapter 1. Recall that in that problem, a manufacturer's monthly profit from the sale of radios was $P(x) = 400(15 - x)(x - 2)$ dollars when the radios were sold for x dollars apiece. The graph of this profit function, which is reproduced in Figure 2.1, suggests that there is an optimal selling price x at which the manufacturer's profit will be greatest. In geometric terms, the optimal price is the x coordinate of the peak of the graph.

Figure 2.1 The profit function $P(x) = 400(15 - x)(x - 2)$.

In this relatively simple example, the peak can be characterized in terms of lines that are **tangent** to the graph. In particular, the peak is the only point on the graph at which the tangent line is horizontal, that is, at which the slope of the tangent is zero. To the left of the peak, the slope of the tangent is positive. To the right of the peak, the slope is negative. But just at the peak itself, the curve "levels off" and the slope of its tangent is zero. (The complete solution of this optimization problem using these ideas will be given a little later in this section.)

Rate of Change   Calculus is also one of the techniques used to find the rate of change of a function. Recall from Chapter 1, Section 3, that the rate of change of a linear function with respect to its independent variable is equal to the steepness or slope of its straight-line graph. Moreover, this steepness or rate of change is constant.

If the function under consideration does not happen to be linear, its rate of change with respect to its independent variable is still the steepness of its graph, measured in this case by the slope of the line that is tangent to the graph at the point in question. Since the graph is not a straight line, its steepness, or rate of change, is not constant but varies from point to point. (These ideas are of fundamental importance and will be developed much more carefully in Section 3.) The situation is illustrated in Figure 2.2.

Figure 2.2 Rate of change = slope (of tangent).

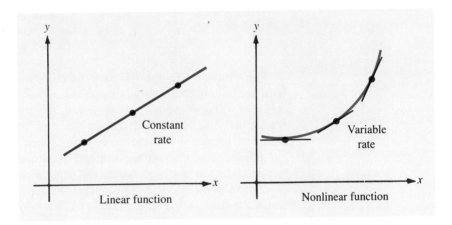

The Slope of a Tangent

The preceding discussion suggests that you could solve optimization problems and compute rates of change if you had a procedure for finding the slope of the tangent to a curve at a given point. Such a procedure will now be developed. Throughout the development, you may rely on your intuitive understanding that the tangent to a curve at a point is the line through the point that indicates the direction of the curve.

The goal is to solve the following general problem: Given a point $(x, f(x))$ on the graph of a function f, find the slope of the line that is tangent to the graph at this point. The situation is illustrated in Figure 2.3.

In Chapter 1, Section 3, you saw that the slope of the line passing through two points (x_1, y_1) and (x_2, y_2) is given by the formula

$$\text{Slope} = \frac{\Delta y}{\Delta x} = \frac{y_2 - y_1}{x_2 - x_1}$$

Unfortunately, in the present situation, you know only one point on the tangent line, namely, the point of tangency $(x, f(x))$. Hence, direct computation of the slope is impossible and you are forced to try an indirect approach.

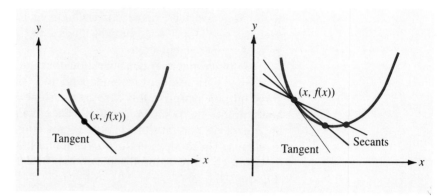

Figure 2.3 A tangent to the curve $y = f(x)$.

Figure 2.4 Secants approximating a tangent.

The strategy is to approximate the tangent by other lines whose slopes can be computed directly. In particular, consider lines joining the given point $(x, f(x))$ to neighboring points on the graph of f. These lines, shown in Figure 2.4, are called **secants** and are good approximations to the tangent, provided that the neighboring point is close to the given point.

In fact, you can make the slope of the secant as close as you like to the slope of the tangent by choosing the neighboring point sufficiently close to the given point $(x, f(x))$. This suggests that you should be able to determine the slope of the tangent itself by first computing the slopes of related secants and then studying the behavior of these slopes as the neighboring points get closer and closer to the given point.

To compute the slope of a secant, first label the coordinates of the neighboring point as indicated in Figure 2.5. In particular, let Δx denote the change in the x coordinate between the given point $(x, f(x))$ and the neighboring point. The x coordinate of the neighboring point is $x + \Delta x$, and since the point lies on the graph of f, its y coordinate is $f(x + \Delta x)$.

Figure 2.5 A secant through the curve $y = f(x)$.

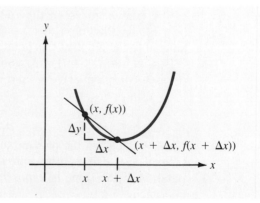

Since the change in the y coordinate is $\Delta y = f(x + \Delta x) - f(x)$, it follows that

$$\text{Slope of secant} = \frac{\Delta y}{\Delta x} = \frac{f(x + \Delta x) - f(x)}{x + \Delta x - x} = \frac{f(x + \Delta x) - f(x)}{\Delta x}$$

Remember that this quotient is not the slope of the tangent but only an approximation to it. If Δx is small, however, the neighboring point $(x + \Delta x, f(x + \Delta x))$ is close to the given point $(x, f(x))$ and the approximation is a good one. In fact, the slope of the actual tangent is the number that this quotient approaches as Δx approaches zero.

A calculation illustrating this procedure is performed in the following example.

EXAMPLE 1.1

Find the slope of the line that is tangent to the graph of the function $f(x) = x^2$ at the point (2, 4).

Solution

A sketch of f showing the given point (2, 4) and a related secant is drawn in Figure 2.6.

Since the x coordinate of the given point is 2, it follows that the x coordinate of the neighboring point is $2 + \Delta x$ and the y coordinate of this point is $(2 + \Delta x)^2$. Hence,

$$\text{Slope of secant} = \frac{(2 + \Delta x)^2 - 4}{\Delta x}$$

Figure 2.6 The curve $y = x^2$ and a secant through (2, 4).

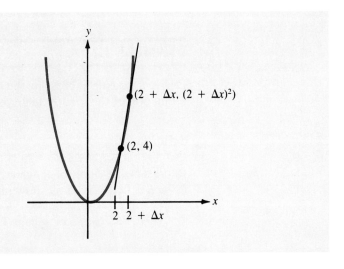

Your goal is to find the number that this quotient approaches as Δx approaches zero. Before you can do this, you must rewrite the quotient in simpler form. (Do you see what would happen if you let Δx become zero in the numerator and denominator of the quotient in its present form?) To simplify the quotient, expand the term $(2 + \Delta x)^2$, rewrite the numerator, and then divide numerator and denominator by Δx as follows:

$$\text{Slope of secant} = \frac{(2 + \Delta x)^2 - 4}{\Delta x}$$

$$= \frac{4 + 4\Delta x + (\Delta x)^2 - 4}{\Delta x}$$

$$= \frac{4\Delta x + (\Delta x)^2}{\Delta x}$$

$$= \frac{\Delta x(4 + \Delta x)}{\Delta x}$$

$$= 4 + \Delta x$$

Now you can let Δx approach zero. Since the slope $4 + \Delta x$ of the secant approaches 4 as Δx approaches zero, it follows that at the given point (2, 4), the slope of the tangent must be 4.

In the preceding example, you found the slope of the tangent to the curve $y = x^2$ at a particular point (2, 4). In the next example, you will perform the same calculation again, this time representing the given point algebraically as (x, x^2). The result will be a formula into which you can substitute any value of x to calculate the slope of the tangent to the curve at the point (x, x^2).

EXAMPLE 1.2

Derive a formula expressing the slope of the tangent to the curve $y = x^2$ as a function of the x coordinate of the point of tangency.

Solution

Represent the point of tangency as (x, x^2) and the neighboring point as $(x + \Delta x, (x + \Delta x)^2)$ as shown in Figure 2.7. Then,

$$\text{Slope of secant} = \frac{(x + \Delta x)^2 - x^2}{\Delta x}$$

$$= \frac{x^2 + 2x \Delta x + (\Delta x)^2 - x^2}{\Delta x}$$

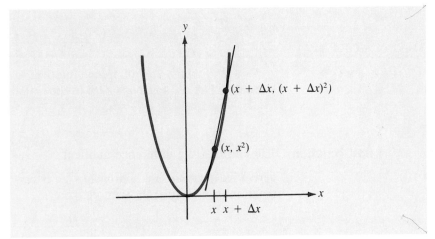

Figure 2.7 The curve $y = x^2$ and a secant through (x, x^2).

$$= \frac{2x\,\Delta x + (\Delta x)^2}{\Delta x}$$

$$= \frac{\Delta x(2x + \Delta x)}{\Delta x}$$

$$= 2x + \Delta x$$

Since the slope $2x + \Delta x$ of the secant approaches $2x$ as Δx approaches zero, it follows that at the point (x, x^2), the slope of the tangent is $2x$.

For example, at the point $(2, 4)$, $x = 2$, and so the slope of the tangent is $2(2) = 4$, as you found in Example 1.1.

The Derivative   In Example 1.2, you started with a function f and derived a related function that expressed the slope of its tangent in terms of the x coordinate of the point of tangency. This derived function is known as the **derivative** of f and is frequently denoted by the symbol f', which is read "f prime." In Example 1.2, you discovered that the derivative of x^2 is $2x$; that is, you found that if $f(x) = x^2$, then $f'(x) = 2x$.

Here is a summary of the situation.

Geometric Interpretation of the Derivative

The derivative $f'(x)$ expresses the slope of the tangent to the curve $y = f(x)$ as a function of the x coordinate of the point of tangency.

How to Compute the Derivative of $f(x)$

Step 1. Form the difference quotient (the slope of a secant).

$$\frac{f(x + \Delta x) - f(x)}{\Delta x}$$

Step 2. Simplify the difference quotient algebraically.
Step 3. Let Δx approach zero in the simplified difference quotient.

Limit Notation   The fact that the difference quotient $\dfrac{f(x + \Delta x) - f(x)}{\Delta x}$ approaches the derivative $f'(x)$ as Δx approaches zero can be expressed in the limit notation (see Chapter 1, Section 5) as

$$f'(x) = \lim_{\Delta x \to 0} \frac{f(x + \Delta x) - f(x)}{\Delta x}$$

The use of limit notation in the computation of a derivative is illustrated in the next example. Also illustrated is the geometric interpretation of the derivative as the slope of the tangent line.

EXAMPLE 1.3

(a) Find the derivative of the function $f(x) = \dfrac{1}{x}$.

(b) Find the equation of the line that is tangent to the graph of the function $f(x) = \dfrac{1}{x}$ when $x = 2$.

Solution

(a) $f'(x) = \displaystyle\lim_{\Delta x \to 0} \frac{f(x + \Delta x) - f(x)}{\Delta x}$ definition of derivative

$\qquad = \displaystyle\lim_{\Delta x \to 0} \frac{1/(x + \Delta x) - 1/x}{\Delta x}$ since $f(x) = \dfrac{1}{x}$

$\qquad = \displaystyle\lim_{\Delta x \to 0} \frac{1/(x + \Delta x) - 1/x}{\Delta x} \cdot \frac{x(x + \Delta x)}{x(x + \Delta x)}$

$\qquad = \displaystyle\lim_{\Delta x \to 0} \frac{x - (x + \Delta x)}{\Delta x[x(x + \Delta x)]}$

$\qquad = \displaystyle\lim_{\Delta x \to 0} \frac{-\Delta x}{\Delta x[x(x + \Delta x)]}$

$\qquad = \displaystyle\lim_{\Delta x \to 0} \frac{-1}{x(x + \Delta x)}$

$$= -\frac{1}{x^2} \qquad \text{after } \Delta x \to 0$$

Notice the algebraic technique used in the third line of the computation that led to the simplification of the difference quotient. Notice also the form of the computation, in particular that the expression $\lim\limits_{\Delta x \to 0}$ is written on each line until after the limit has actually been performed in the last line.

(b) To find the slope of the tangent when $x = 2$, compute $f'(2)$.

$$\text{Slope of tangent} = f'(2) = -\frac{1}{4}$$

To find the y coordinate of the point of tangency, compute $f(2)$.

$$y = f(2) = \frac{1}{2}$$

Now use the point-slope formula $y - y_0 = m(x - x_0)$ with $m = -\frac{1}{4}$

and $(x_0, y_0) = \left(2, \frac{1}{2}\right)$ to conclude that the equation of the tangent is

$$y - \frac{1}{2} = -\frac{1}{4}(x - 2) \qquad \text{or} \qquad y = -\frac{1}{4}x + 1$$

Figure 2.8 The curve $y = \frac{1}{x}$ and the tangent when $x = 2$.

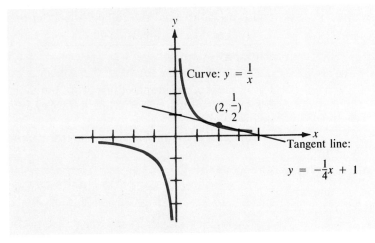

The situation is illustrated in Figure 2.8.

Derivative Notation   Symbols other than f' are often used to denote the derivative. For example, if y rather than $f(x)$ is used to denote the function itself, the symbol $\frac{dy}{dx}$ $\left(\text{suggesting slope } \frac{\Delta y}{\Delta x}\right)$ is frequently used instead of $f'(x)$. Hence, instead of the statement   If $f(x) = x^2$, then $f'(x) = 2x$

you could write   If $y = x^2$, then $\dfrac{dy}{dx} = 2x$

(The symbol $\dfrac{dy}{dx}$ is read "the derivative of y with respect to x.") By omitting reference to y, you can condense these statements and write

$$\frac{d}{dx}(x^2) = 2x$$

which is read "the derivative of x^2 with respect to x is $2x$."

The Maximization of Profit   To illustrate how the derivative can be applied to practical situations, here, as promised, is the solution to the profit maximization problem first discussed in Example 4.1 of Section 1.4 and then again at the beginning of this section. A more thorough treatment of the use of calculus to solve optimization problems will be given in Chapter 3.

EXAMPLE 1.4

Suppose a manufacturer's profit from the sale of radios is given by the function $P(x) = 400(15 - x)(x - 2)$, where x is the price at which the radios are sold. Find the selling price that maximizes profit.

Solution

For reference, the graph of this profit function is sketched again in Figure 2.9. (See also Figure 1.26 of Section 1.4 and Figure 2.1 at the beginning of this section.)

Your goal is to find the value of x that maximizes the profit $P(x)$. This is

Figure 2.9 The profit function $P(x) = 400(15 - x)(x - 2)$.

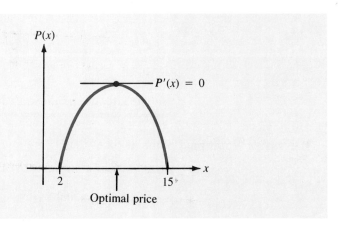

the value of x for which the slope of the tangent line is zero. Since the slope of the tangent line is given by the derivative, begin by computing $P'(x)$. For simplicity, apply the definition of the derivative to the unfactored form of the profit function.

$$P(x) = -400x^2 + 6,800x - 12,000$$

You find that

$$P'(x) = \lim_{\Delta x \to 0} \frac{P(x + \Delta x) - P(x)}{\Delta x}$$

$$= \lim_{\Delta x \to 0} \frac{-400(x + \Delta x)^2 + 6,800(x + \Delta x) - 12,000 - (-400x^2 + 6,800x - 12,000)}{\Delta x}$$

$$= \lim_{\Delta x \to 0} \frac{-400(\Delta x)^2 - 800x\,\Delta x + 6,800\Delta x}{\Delta x}$$

$$= \lim_{\Delta x \to 0} (-400\Delta x - 800x + 6,800)$$

$$= -800x + 6,800$$

To find the value of x for which the slope of the tangent is zero, set the derivative equal to zero and solve the resulting equation for x as follows:

$$P'(x) = 0$$

$$-800x + 6,800 = 0$$

$$800x = 6,800$$

$$x = \frac{6,800}{800} = 8.5$$

It follows that $x = 8.5$ is the x coordinate of the peak of the graph and that the optimal selling price is $8.50 per radio.

Differentiability and Continuity

Not all functions have a derivative for every value of x. Three functions that do not have derivatives when $x = 0$ are sketched in Figure 2.10. A function is said to be **differentiable** at $x = c$ if it has a derivative when

Figure 2.10 Three functions that are not differentiable at $x = 0$. (a) Corner: no tangent at $x = 0$. (b) Cusp: vertical tangent at $x = 0$. (c) The function is undefined at $x = 0$.

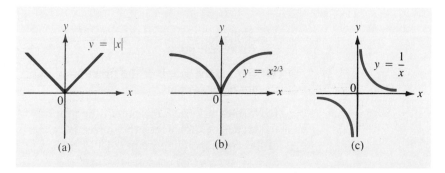

$x = c$. A function that is differentiable at all values of x in its domain is called a **differentiable function.** The graph of a differentiable function must be "well-behaved" in the sense that there is a nonvertical tangent line at each point on the graph. The graph of a differentiable function cannot have a sharp "corner" (Figure 2.10a), a "cusp" (Figure 2.10b), or a break (Figure 2.10c).

Most functions you will encounter in this text will be differentiable at almost all points. For example, it can be shown that polynomials are differentiable everywhere and that rational functions fail to be differentiable only at those values of x where their denominators are zero.

Recall from Chapter 1, Section 5, that a function is **continuous** if its graph is an unbroken curve. It can be shown that every differentiable function is continuous, but as the function $f(x) = |x|$ in Figure 2.10a demonstrates, not every continuous function is differentiable.

Problems 2.1

In Problems 1 through 8, compute the derivative of the given function and find the slope of the line that is tangent to its graph for the specified value of x.

1. $f(x) = 5x - 3; x = 2$

2. $f(x) = x^2 - 1; x = -1$

3. $f(x) = 2x^2 - 3x + 5; x = 0$

4. $f(x) = x^3 - 1; x = 2$

5. $f(x) = \dfrac{2}{x}; x = \dfrac{1}{2}$

6. $f(x) = \dfrac{1}{x^2}; x = 2$

7. $f(x) = \sqrt{x}; x = 9$

8. $f(x) = \dfrac{1}{\sqrt{x}}; x = 4$

In Problems 9 through 12, compute the derivative of the given function and find the equation of the line that is tangent to its graph for the specified value of x.

9. $f(x) = x^2 + x + 1; x = 2$

10. $f(x) = x^3 - x; x = -2$

11. $f(x) = \dfrac{3}{x^2}; x = \dfrac{1}{2}$

12. $f(x) = 2\sqrt{x}; x = 4$

13. Sketch the graph of the function $y = x^2 - 3x$ and use calculus to find its lowest point.

14. Sketch the graph of the function $y = 1 - x^2$ and use calculus to find its highest point.

In Problems 15 and 16, sketch the graph of the function $f(x)$. Determine the values of x for which the derivative is zero. What happens to the graph at the corresponding points?

15. $f(x) = x^3 + 3x^2$ 16. $f(x) = x^3 - x^2$

Maximization of Profit

17. A manufacturer can produce tape recorders at a cost of \$20 apiece. It is estimated that if the tape recorders are sold for x dollars apiece, consumers will buy $120 - x$ of them each month. Use calculus to determine the price at which the manufacturer's profit will be the greatest.

Minimal Fencing

18. A rectangular field has one side along a stream and is to have area 7,200 square yards. Use calculus to find the least amount of fencing required to do the job.

19. Suppose $f(x) = x^2$.
    (a) Compute the slope of the secant line joining the points on the graph of f whose x coordinates are $x = -2$ and $x = -1.9$.
    (b) Use calculus to compute the slope of the line that is tangent to the graph when $x = -2$ and compare this slope with your answer in part (a).

20. Suppose $f(x) = x^3$.
    (a) Compute the slope of the secant line joining the points on the graph of f whose x coordinates are $x = 1$ and $x = 1.1$.
    (b) Use calculus to compute the slope of the line that is tangent to the graph when $x = 1$ and compare this slope with your answer in part (a).

21. (a) Find the derivative of the linear function $f(x) = 3x - 2$.
    (b) Find the equation of the tangent line to the graph of this function at the point where $x = -1$.
    (c) Explain how the answers to parts (a) and (b) could have been obtained from geometric considerations with no calculation whatsoever.

22. (a) Find the derivatives of the functions $y = x^2$ and $y = x^2 - 3$ and account geometrically for their similarity.
    (b) Without further computation, find the derivative of the function $y = x^2 + 5$.

23. (a) Find the derivative of the function $y = x^2 + 3x$.
    (b) Find the derivatives of the functions $y = x^2$ and $y = 3x$ separately.
    (c) How is the derivative in part (a) related to those in part (b)?
    (d) In general, if $f(x) = g(x) + h(x)$, what would you guess is the relationship between the derivative of f and those of g and h?

24. (a) Compute the derivatives of the functions $y = x^2$ and $y = x^3$.
    (b) Examine your answers in part (a). Can you detect a pattern? What do you think is the derivative of $y = x^4$? How about the derivative of $y = x^{27}$?

25. Explain why the graph of a function $f(x)$ is rising over an interval $a \le x \le b$ if $f'(x) > 0$ throughout the interval. What can you say about the graph if $f'(x) < 0$ throughout the interval $a \le x \le b$?

In Problems 26 and 27, sketch the graph of a function f that has all the given properties. You may need to refer to the result of Problem 25.

26. (a) $f'(x) > 0$ when $x < 1$ and when $x > 5$
    (b) $f'(x) < 0$ when $1 < x < 5$
    (c) $f'(1) = 0$ and $f'(5) = 0$

27. (a) $f'(x) > 0$ when $x < -2$ and when $-2 < x < 3$
    (b) $f'(x) < 0$ when $x > 3$
    (c) $f'(-2) = 0$ and $f'(3) = 0$

2   Techniques of Differentiation

In Section 1, you learned how to find the derivative of a function by letting Δx approach zero in the expression for the slope of a secant. For even the simplest functions, this process is tedious and time-consuming. In this section, you will see some rules that simplify the process. Justification of some of these rules will be given at the end of the section, after you have had a chance to practice using them.

The Derivative of a Power Function   Recall from Section 1.1 that a **power function** is a function of the form $f(x) = x^n$, where n is a real number. Here is a simple rule you can use to find the derivative of any power function. The general proof of this rule will be given in Chapter 4.

The Power Rule

For any number n,

$$\frac{d}{dx}(x^n) = nx^{n-1}$$

That is, to find the derivative of x^n, reduce the power of x by 1 and multiply by the original power.

According to the power rule, the derivative of x^2 is $2x^1 = 2x$, which agrees with the result in Example 1.2. Similarly, the derivative of x^3 is $3x^2$, which should agree with the result you obtained in Problem 24 of the preceding section. You can use the power rule to differentiate radicals and reciprocals by first converting them to power functions with fractional and negative exponents, respectively. Here are a few more calculations that illustrate the use of the power rule.

EXAMPLE 2.1

Differentiate (find the derivative of) each of the following functions:

(a) $y = x^{27}$ (b) $y = \dfrac{1}{x^{27}}$ (c) $y = \sqrt{x}$ (d) $y = \dfrac{1}{\sqrt{x}}$

Solution

In each case, use exponential notation to write the function as a power function and then apply the general rule. (You can find a review of exponential notation in the appendix at the back of the book.)

(a) $\dfrac{d}{dx}(x^{27}) = 27x^{27-1} = 27x^{26}$

(b) $\dfrac{d}{dx}\left(\dfrac{1}{x^{27}}\right) = \dfrac{d}{dx}(x^{-27}) = -27x^{-27-1} = -27x^{-28} = -\dfrac{27}{x^{28}}$

(c) $\dfrac{d}{dx}(\sqrt{x}) = \dfrac{d}{dx}(x^{1/2}) = \dfrac{1}{2}x^{1/2-1} = \dfrac{1}{2}x^{-1/2} = \dfrac{1}{2\sqrt{x}}$

(d) $\dfrac{d}{dx}\left(\dfrac{1}{\sqrt{x}}\right) = \dfrac{d}{dx}(x^{-1/2}) = -\dfrac{1}{2}x^{-1/2-1} = -\dfrac{1}{2}x^{-3/2} = -\dfrac{1}{2\sqrt{x^3}}$

The Derivative of a Constant   The derivative of any constant function is zero. This is because the graph of a constant function $y = c$ is a horizontal line and the slope is zero. Thus, for example, if $f(x) = 5$, then $f'(x) = 0$.

> **The Derivative of a Constant**
>
> For any constant c,
>
> $$\frac{d}{dx}(c) = 0$$
>
> That is, the derivative of a constant is zero.

The Derivative of a Constant Times a Function   The next rule expresses the fact that the curve $y = cf(x)$ is c times as steep as the curve $y = f(x)$.

> **The Constant Multiple Rule**
>
> For any constant c,
>
> $$\frac{d}{dx}(cf) = c\frac{df}{dx}$$

That is, the derivative of a constant times a function is equal to the constant times the derivative of the function.

EXAMPLE 2.2

Differentiate the function $y = 3x^5$.

Solution

You already know that $\frac{d}{dx}(x^5) = 5x^4$. Combining this with the constant multiple rule, you get

$$\frac{d}{dx}(3x^5) = 3\frac{d}{dx}(x^5) = 3(5x^4) = 15x^4$$

The Derivative of a Sum   The next rule states that a sum can be differentiated term by term.

The Sum Rule

$$\frac{d}{dx}(f + g) = \frac{df}{dx} + \frac{dg}{dx}$$

That is, the derivative of a sum is the sum of the individual derivatives.

EXAMPLE 2.3

Differentiate the function $y = x^2 + 3x^5$.

Solution

You know that $\frac{d}{dx}(x^2) = 2x$ and that $\frac{d}{dx}(3x^5) = 15x^4$. According to the sum rule, you simply add these derivatives to get the derivative of the sum $x^2 + 3x^5$. That is,

$$\frac{d}{dx}(x^2 + 3x^5) = \frac{d}{dx}(x^2) + \frac{d}{dx}(3x^5) = 2x + 15x^4$$

By combining the sum rule with the power and constant multiple rules, you can differentiate any polynomial. Here is an example.

EXAMPLE 2.4

Differentiate the polynomial $y = 5x^3 - 4x^2 + 12x - 8$.

Solution

Differentiate this sum term by term to get

$$\frac{dy}{dx} = \frac{d}{dx}(5x^3) + \frac{d}{dx}(-4x^2) + \frac{d}{dx}(12x) + \frac{d}{dx}(-8)$$

$$= 15x^2 - 8x^1 + 12x^0 + 0$$

$$= 15x^2 - 8x + 12$$

The Derivative of a Product

Suppose you wanted to differentiate the product $y = x^2(3x + 1)$. You might be tempted to differentiate the factors x^2 and $3x + 1$ separately and then multiply your answers. That is, since $\frac{d}{dx}(x^2) = 2x$ and $\frac{d}{dx}(3x + 1) = 3$, you might suspect that $\frac{dy}{dx} = 6x$. However, this answer is wrong. To see this, rewrite the function as $y = 3x^3 + x^2$ and observe that the derivative is $9x^2 + 2x$ and not $6x$. The derivative of a product is *not* the product of the individual derivatives. Here is the correct formula for the derivative of a product.

> **The Product Rule**
>
> $$\frac{d}{dx}(fg) = f\frac{dg}{dx} + g\frac{df}{dx}$$
>
> That is, the derivative of a product is the first factor times the derivative of the second plus the second factor times the derivative of the first.

The use of this rule is illustrated in the next example.

EXAMPLE 2.5

Differentiate the function $y = x^2(3x + 1)$.

Solution

According to the product rule,

$$\frac{d}{dx}[x^2(3x + 1)] = x^2\frac{d}{dx}(3x + 1) + (3x + 1)\frac{d}{dx}(x^2)$$

$$= x^2(3) + (3x + 1)(2x)$$

$$= 9x^2 + 2x$$

which is precisely the result that was obtained when the product was multiplied out and differentiated as a sum.

The Derivative of a Quotient

The derivative of a quotient is not the quotient of the individual derivatives. Here is the correct rule.

The Derivative of a Quotient

$$\frac{d}{dx}\left(\frac{f}{g}\right) = \frac{g\dfrac{df}{dx} - f\dfrac{dg}{dx}}{g^2}$$

The quotient rule is probably the most complicated formula you have had to learn so far in this book. Here is one way to remember it. The numerator resembles the product rule except that it contains a minus sign, which makes the order in which the terms are written important. Begin by squaring the denominator (since this is easy to do) and then, while still thinking of the original denominator, copy it in the numerator. This gets you started with the proper term in the numerator, and you can easily write down the rest while thinking of the product rule. Don't forget to insert the minus sign, without which the rule would not have been so hard to remember in the first place!

Using the quotient rule, you can now differentiate any rational function. Here is an example.

EXAMPLE 2.6

Differentiate the rational function $y = \dfrac{x^2 + 2x - 21}{x - 3}$.

Solution

According to the quotient rule,

$$\frac{dy}{dx} = \frac{(x - 3)\frac{d}{dx}(x^2 + 2x - 21) - (x^2 + 2x - 21)\frac{d}{dx}(x - 3)}{(x - 3)^2}$$

$$= \frac{(x - 3)(2x + 2) - (x^2 + 2x - 21)(1)}{(x - 3)^2}$$

$$= \frac{2x^2 - 4x - 6 - x^2 - 2x + 21}{(x - 3)^2}$$

$$= \frac{x^2 - 6x + 15}{(x - 3)^2}$$

A Word of Advice   The quotient rule is somewhat cumbersome, so don't use it unnecessarily. Consider the following example.

EXAMPLE 2.7

Differentiate the function $y = \dfrac{2}{3x^2} - \dfrac{x}{3} + \dfrac{4}{5} + \dfrac{x + 1}{x}$.

Solution

Don't use the quotient rule! Instead, rewrite the function as

$$y = \frac{2}{3}x^{-2} - \frac{1}{3}x + \frac{4}{5} + 1 + x^{-1}$$

and then apply the power rule term by term to get

$$\frac{dy}{dx} = \frac{2}{3}(-2x^{-3}) - \frac{1}{3} + 0 + 0 + (-1)x^{-2}$$

$$= -\frac{4}{3}x^{-3} - \frac{1}{3} - x^{-2}$$

$$= -\frac{4}{3x^3} - \frac{1}{3} - \frac{1}{x^2}$$

Derivation of the Product Rule   The rules you have seen in this section can all be obtained from the definition of the derivative. Proofs of the sum rule and the constant multiple rule are fairly straightforward (see Problems 46 and 35). The product rule and the quotient rule are not as "natural" as the other rules, and it may be helpful to see a proof of these results. We shall prove the product rule here and outline the proof of the quotient rule in Problem 47.

To see why the product rule is true, consider a typical secant line

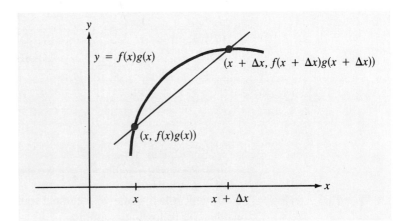

Figure 2.11 A secant through the curve $y = f(x)g(x)$.

through the graph of the product function fg, as shown in Figure 2.11.

To show that $\dfrac{d}{dx}(fg) = f\dfrac{dg}{dx} + g\dfrac{df}{dx}$, begin with the appropriate difference quotient and rewrite it using the algebraic "trick" of subtracting and then adding the quantity $f(x + \Delta x)g(x)$ in the numerator as follows:

$$\text{Slope of secant} = \frac{f(x + \Delta x)g(x + \Delta x) - f(x)g(x)}{\Delta x}$$

$$= \frac{f(x + \Delta x)g(x + \Delta x) - f(x + \Delta x)g(x)}{\Delta x}$$

$$+ \frac{f(x + \Delta x)g(x) - f(x)g(x)}{\Delta x}$$

$$= f(x + \Delta x)\frac{g(x + \Delta x) - g(x)}{\Delta x}$$

$$+ g(x)\frac{f(x + \Delta x) - f(x)}{\Delta x}$$

Now let Δx approach zero. Since

$$\lim_{\Delta x \to 0} \frac{f(x + \Delta x) - f(x)}{\Delta x} = \frac{df}{dx}$$

$$\lim_{\Delta x \to 0} \frac{g(x + \Delta x) - g(x)}{\Delta x} = \frac{dg}{dx}$$

and

$$\lim_{\Delta x \to 0} f(x + \Delta x) = f(x)$$

it follows that

$$\frac{d}{dx}(fg) = f\frac{dg}{dx} + g\frac{df}{dx}$$

and the product rule is proved.

Problems 2.2

In Problems 1 through 24, differentiate the given function. Do as much of the computation as possible in your head and simplify your answers.

1. $y = x^2 + 2x + 3$

2. $y = 3x^5 - 4x^3 + 9x - 6$

3. $f(x) = x^9 - 5x^8 + x + 12$

4. $f(x) = \frac{1}{4}x^8 - \frac{1}{2}x^6 - x + 2$

5. $y = \frac{1}{x} + \frac{1}{x^2} - \frac{1}{\sqrt{x}}$

6. $y = \frac{3}{x} - \frac{2}{x^2} + \frac{2}{3x^3}$

7. $f(x) = \sqrt{x^3} + \frac{1}{\sqrt{x^3}}$

8. $f(x) = 2\sqrt{x^3} + \frac{4}{\sqrt{x}} - \sqrt{2}$

9. $y = -\frac{x^2}{16} + \frac{2}{x} - x^{3/2} + \frac{1}{3x^2} + \frac{x}{3}$

10. $y = -\frac{2}{x^2} + x^{2/3} + \frac{1}{2\sqrt{x}} + \frac{x^2}{4} + \sqrt{5} + \frac{x+2}{3}$

11. $f(x) = (2x + 1)(3x - 2)$ (Use the product rule.)

12. $f(x) = (x^2 - 5)(1 - 2x)$ (Use the product rule.)

13. $y = 10(3x + 1)(1 - 5x)$

14. $y = 400(15 - x^2)(3x - 2)$

15. $f(x) = \frac{1}{3}(x^5 - 2x^3 + 1)$

16. $f(x) = -3(5x^3 - 2x + 5)$

17. $y = \frac{x+1}{x-2}$

18. $y = \frac{2x-3}{5x+4}$

19. $f(x) = \frac{x}{x^2-2}$

20. $f(x) = \frac{1}{x-2}$

21. $y = \frac{3}{x+5}$

22. $y = \frac{x^2+1}{1-x^2}$

23. $f(x) = \frac{x^2 - 3x + 2}{2x^2 + 5x - 1}$

24. $f(x) = \frac{x^2 + 2x + 1}{3}$

In Problems 25 through 28, find the equation of the line that is tangent to the graph of the given function at the specified point.

25. $y = x^5 - 3x^3 - 5x + 2$; $(1, -5)$

26. $y = (x^2 + 1)(1 - x^3)$; $(1, 0)$

27. $f(x) = \frac{x+1}{x-1}$; $(0, -1)$

28. $f(x) = 1 - \dfrac{1}{x} + \dfrac{2}{\sqrt{x}}; \left(4, \dfrac{7}{4}\right)$

In Problems 29 through 32, find the equation of the line that is tangent to the graph of the given function at the point $(x, f(x))$ for the specified value of x.

29. $f(x) = x^4 - 3x^3 + 2x^2 - 6; x = 2$

30. $f(x) = x - \dfrac{1}{x^2}; x = 1$

31. $f(x) = \dfrac{x^2 + 2}{x^2 - 2}; x = -1$

32. $f(x) = (x^3 - 2x^2 + 3x - 1)(x^5 - 4x^2 + 2); x = 0$

33. (a) Differentiate the function $y = 2x^2 - 5x - 3$.
    (b) Now factor the function in part (a) as $y = (2x + 1)(x - 3)$ and differentiate using the product rule. Show that the two answers are the same.

34. (a) Use the quotient rule to differentiate the function $y = \dfrac{2x - 3}{x^3}$.
    (b) Rewrite the function as $y = x^{-3}(2x - 3)$ and differentiate using the product rule.
    (c) Rewrite the function as $y = 2x^{-2} - 3x^{-3}$ and differentiate.
    (d) Show that your answers to parts (a), (b), and (c) are the same.

35. The product rule tells you how to differentiate the product of any two functions, while the constant multiple rule tells you how to differentiate products in which one of the factors is constant. Show that the two rules are consistent. In particular, use the product rule to show that $\dfrac{d}{dx}(cf) = c\dfrac{df}{dx}$ if c is a constant.

36. Sketch the graph of the function $f(x) = x^2 - 4x - 5$ and use calculus to determine its lowest point.

37. Sketch the graph of the function $f(x) = 3 - 2x - x^2$ and use calculus to determine its highest point.

38. Find numbers a and b such that the lowest point on the graph of the function $f(x) = ax^2 + bx$ is $(3, -8)$.

39. Find numbers a, b, and c such that the graph of the function $f(x) = ax^2 + bx + c$ will have x intercepts at $(0, 0)$ and $(5, 0)$, and a tangent with slope 1 when $x = 2$.

40. Find the equations of all the tangents to the graph of the function $f(x) = x^2 - 4x + 25$ that pass through the origin $(0, 0)$.

41. Find all the points (x, y) on the graph of the function $y = 4x^2$ with the property that the tangent to the graph at (x, y) passes through the point $(2, 0)$.

Consumer Expenditure   42. The consumer demand for a certain commodity is $D(p) = -200p + 12,000$ units per month when the market price is p dollars per unit.
  (a) Express consumers' total monthly expenditure for the commodity as a function of p and draw the graph.
  (b) Use calculus to determine the market price for which the consumer expenditure is greatest.

Landscaping   43. A landscaper wishes to enclose a rectangular flower garden with 20 feet of fencing. Use calculus to find the largest area that can be enclosed by such a garden.

44. (a) Show that

$$\frac{d}{dx}(fgh) = fg\frac{dh}{dx} + fh\frac{dg}{dx} + gh\frac{df}{dx}$$

(*Hint:* Apply the product rule twice.)

(b) Find $\frac{dy}{dx}$, where $y = (2x + 1)(x - 3)(1 - 4x)$.

45. (a) By combining the product rule and the quotient rule, find an expression for $\frac{d}{dx}\left(\frac{fg}{h}\right)$.

(b) Find $\frac{dy}{dx}$, where $y = \frac{(2x + 7)(x^2 + 3)}{3x + 5}$.

46. Prove the sum rule for derivatives. [*Hint:* Note that the difference quotient for $f + g$ can be written as

$$\frac{(f + g)(x + \Delta x) - (f + g)(x)}{\Delta x} = \frac{[f(x + \Delta x) + g(x + \Delta x)] - [f(x) + g(x)]}{\Delta x}$$

47. Derive the quotient rule. [*Hint:* Show that the difference quotient is

$$\frac{1}{\Delta x}\left[\frac{f(x + \Delta x)}{g(x + \Delta x)} - \frac{f(x)}{g(x)}\right] = \frac{g(x)f(x + \Delta x) - f(x)g(x + \Delta x)}{g(x + \Delta x)g(x)\Delta x}$$

Before letting Δx approach zero, rewrite this quotient using the trick of subtracting and adding $g(x)f(x)$ in the numerator.]

3   The Derivative as a Rate of Change

In this section, you will see how the derivative can be interpreted as a rate of change. Viewed in this way, derivatives may represent such quantities as the rate at which population grows, the speed of a moving object, a manufacturer's marginal cost, the rate of inflation, and the rate at which natural resources are being depleted.

At the beginning of this chapter, there was a brief discussion suggesting the relationship between rates of change and derivatives. The reasoning was that the rate of change of a function with respect to its independent variable is equal to the steepness of its graph, which is measured by the slope of its tangent line at the point in question. Since the slope of the tangent line is given by the derivative of the function, it follows that the rate of change is equal to the derivative.

The purpose of this section is to make the relationship between derivatives and rates of change more precise. The discussion begins with a familiar practical situation which will serve as a model for the general case.

Average and Instantaneous Speed

Imagine that a car is moving along a straight road and that $D(t)$ is its distance from its starting point after t hours. Suppose you want to determine the speed of the car at a particular time t but do not have access to the car's speedometer. Here's what you can do.

Record the position of the car first at time t and again at some later time $t + \Delta t$. That is, determine $D(t)$ and $D(t + \Delta t)$. Then compute the **average speed** of the car between the times t and $t + \Delta t$ as follows:

$$\text{Average speed} = \frac{\text{change in distance}}{\text{change in time}} = \frac{D(t + \Delta t) - D(t)}{\Delta t}$$

The situation is illustrated in Figure 2.12.

Figure 2.12
Average speed
$= \dfrac{\text{change in distance}}{\text{change in time}}$
$= \dfrac{D(t + \Delta t) - D(t)}{\Delta t}.$

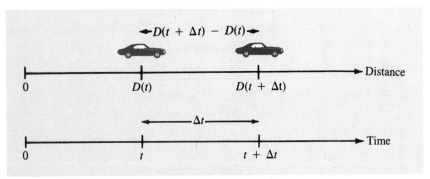

Since the speed of the car may fluctuate during the time interval from t to $t + \Delta t$, it is unlikely that this average speed is equal to the **instantaneous speed** (the speed shown on the speedometer) at time t. However, if Δt is small, the possibility of drastic changes in speed is small, and the average speed may be a fairly good approximation to the instantaneous speed. Indeed, you can find the instantaneous speed at time t by letting Δt approach zero in the expression for the average speed. Thus

$$\text{Instantaneous speed} = \lim_{\Delta t \to 0} \frac{D(t + \Delta t) - D(t)}{\Delta t}$$

This limit expression for instantaneous speed should look familiar. It is exactly the formula defining the derivative $D'(t)$ of the function $D(t)$. Thus, the instantaneous speed (or rate of change of distance) is equal to the derivative of distance.

Average and Instantaneous Rate of Change

These ideas can be extended to more general situations. Suppose that y is a function of x, say $y = f(x)$. Corresponding to a change from x to $x + \Delta x$, the variable y changes by an amount $\Delta y = f(x + \Delta x) - f(x)$. The resulting **average rate of change of y with respect to x** is the difference quotient

$$\text{Average rate of change} = \frac{\text{change in } y}{\text{change in } x} = \frac{f(x + \Delta x) - f(x)}{\Delta x}$$

As the interval over which you are averaging becomes shorter (that is, as Δx approaches zero), the average rate of change approaches what you would intuitively call the **instantaneous rate of change of y with respect to x** and the difference quotient approaches the derivative $f'(x)$ or $\frac{dy}{dx}$. That is,

$$\text{Instantaneous rate of change} = \lim_{\Delta x \to 0} \frac{f(x + \Delta x) - f(x)}{\Delta x}$$

$$= f'(x) = \frac{dy}{dx}$$

Instantaneous Rate of Change

If $y = f(x)$, the instantaneous rate of change of y with respect to x is given by the derivative of f. That is,

$$\text{Rate of change} = f'(x) = \frac{dy}{dx}$$

EXAMPLE 3.1

It is estimated that x months from now, the population of a certain community will be $P(x) = x^2 + 20x + 8,000$.

(a) At what rate will the population be changing with respect to time 15 months from now?

(b) By how much will the population actually change during the 16th month?

Solution

(a) The rate of change of the population with respect to time is the derivative of the population function. That is,

$$\text{Rate of change} = P'(x) = 2x + 20$$

The rate of change of the population 15 months from now will be

$$P'(15) = 2(15) + 20 = 50 \text{ people per month}$$

(b) The actual change in the population during the 16th month is the difference between the population at the end of 16 months and the population at the end of 15 months. That is,

$$\text{Change in population} = P(16) - P(15) = 8{,}576 - 8{,}525$$
$$= 51 \text{ people}$$

Comment: The reason for the difference between the actual change in population during the 16th month in part (b) and the monthly rate of change at the beginning of the month in part (a) in Example 3.1 is that the rate of change of the population varied during the month. The instantaneous rate of change in part (a) can be thought of as the change in population that would occur during the 16th month if the rate of change of the population remained constant.

Percentage Rate of Change

In many practical situations, the rate of change of a quantity is not as significant as its **percentage rate of change.** For example, a yearly rate of change in population of 500 people in a city of 5 million would be negligible, while the same rate of change would have an enormous impact in a town of 2,000. The percentage rate of change compares the rate of change of a quantity with the size of that quantity:

$$\text{Percentage rate of change} = 100 \frac{\text{rate of change of quantity}}{\text{size of quantity}}$$

Therefore, a rate of change of 500 people per year in the population of a city of 5 million amounts to a percentage rate of change of only $\frac{100(500)}{5{,}000{,}000} = 0.01$ percent of the population per year. On the other hand, the same rate of change in a town of 2,000 is equal to a percentage rate of change of $\frac{100(500)}{2{,}000} = 25$ percent of the population per year.

Here is the formula for percentage rate of change written in terms of the derivative.

> **Percentage Rate of Change**
>
> If $y = f(x)$, the percentage rate of change of y with respect to x is given by the formula
>
> $$\text{Percentage rate of change} = 100\frac{f'(x)}{f(x)} = 100\frac{dy/dx}{y}$$

EXAMPLE 3.2

The gross national product (GNP) of a certain country was $N(t) = t^2 + 5t + 106$ billion dollars t years after 1980.

(a) At what rate was the GNP changing with respect to time in 1988?
(b) At what percentage rate was the GNP changing with respect to time in 1988?

Solution

(a) The rate of change of the GNP is the derivative $N'(t) = 2t + 5$. The rate of change in 1988 was $N'(8) = 2(8) + 5 = 21$ billion dollars per year.
(b) The percentage rate of change of the GNP in 1988 was

$$100\frac{N'(8)}{N(8)} = 100\frac{21}{210} = 10 \text{ percent per year}$$

Rectilinear Motion   Rectilinear motion is motion along a straight line. For example, the motion of a rocket early in its flight can be regarded as rectilinear. When studying rectilinear motion, you may assume that an object is moving along a coordinate axis. The distance or **displacement** of the object from the origin on the axis is a function of time t and is often denoted by $s(t)$. The rate of change of the displacement with respect to time is the object's **velocity** $v(t)$, and the rate of change of the velocity with respect to t is its **acceleration** $a(t)$.

Expressed as derivatives, the velocity is $v(t) = \dfrac{ds}{dt}$ and the acceleration is $a(t) = \dfrac{dv}{dt}$. If $v(t) > 0$, the object is said to be **advancing,** and if $v(t) < 0$, it is **retreating.** If $v(t) = 0$, the object is neither advancing nor retreating and is said to be **stationary** (see Figure 2.13).

Finally, if $a(t) > 0$, the object is **speeding up,** and if $a(t) < 0$, it is **slowing down.**

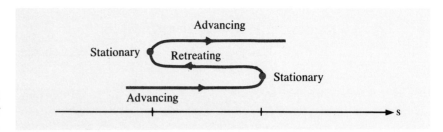

Figure 2.13 A diagram indicating rectilinear motion.

Rectilinear Motion

An object that moves along a straight line with **displacement** $s(t)$ has **velocity** $v(t) = \dfrac{ds}{dt}$ and **acceleration** $a(t) = \dfrac{dv}{dt}$ when these derivatives exist.

The object is **advancing** when $v(t) > 0$, **retreating** when $v(t) < 0$, and **stationary** when $v(t) = 0$. It is **speeding up** when $a(t) > 0$ and **slowing down** when $a(t) < 0$.

If distance is measured in meters and time in seconds, velocity is measured in meters per second (m/sec) and acceleration in meters per second per second (written as m/sec²). Similarly, if distance is measured in feet, velocity is measured in feet per second (ft/sec) and acceleration in feet per second per second (ft/sec²).

EXAMPLE 3.3

The displacement at time t of an object moving along a line is given by $s(t) = t^3 - 6t^2 + 9t + 5$.

(a) Find the velocity of the object and discuss its motion between times $t = 0$ and $t = 4$.
(b) Find the total distance traveled by the object between times $t = 0$ and $t = 4$.
(c) Find the acceleration of the object and determine when the object is speeding up and slowing down between times $t = 0$ and $t = 4$.

Solution

(a) The velocity is $v(t) = \dfrac{ds}{dt} = 3t^2 - 12t + 9$. The object will be stationary when

$$v(t) = 3t^2 - 12t + 9 = 3(t - 1)(t - 3) = 0$$

that is, at times $t = 1$ and $t = 3$. Otherwise, the object is either advancing or retreating, as described in the following table.

Interval	Sign of $v(t)$	Description of the motion
$0 < t < 1$	$+$	Advances from $s(0) = 5$ to $s(1) = 9$
$1 < t < 3$	$-$	Retreats from $s(1) = 9$ to $s(3) = 5$
$3 < t < 4$	$+$	Advances from $s(3) = 5$ to $s(4) = 9$

The motion of the object is summarized in the diagram in Figure 2.14.

Figure 2.14 The motion of an object with displacement $s(t) = t^3 - 6t^2 + 9t + 5$.

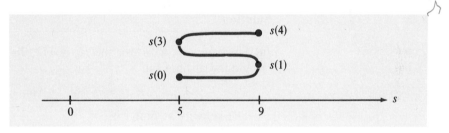

(b) The object travels from $s(0) = 5$ to $s(1) = 9$, then back to $s(3) = 5$, and finally to $s(4) = 9$. Thus, the total distance traveled by the object is

$$D = \underbrace{|9 - 5|}_{0 < t < 1} + \underbrace{|5 - 9|}_{1 < t < 3} + \underbrace{|9 - 5|}_{3 < t < 4} = 12$$

(c) The acceleration of the object is

$$a(t) = \frac{dv}{dt} = 6t - 12 = 6(t - 2)$$

The object will be speeding up ($a(t) > 0$) when $2 < t < 4$ and slowing down ($a(t) < 0$) when $0 < t < 2$.

The Motion of a Projectile
An important example of rectilinear motion is the motion of a projectile. Suppose an object is projected (e.g., thrown, fired, or dropped) vertically in such a way that the only acceleration acting on the object is the constant downward acceleration g due to gravity. Near sea level, g is approximately 32 ft/sec^2 (or 9.8 m/sec^2). It can be shown that at time t, the height of the object is given by the formula

$$H(t) = \frac{-1}{2} gt^2 + V_0 t + H_0$$

where H_0 and V_0 are the initial height and velocity of the object, respectively. Here is an example.

EXAMPLE 3.4

Suppose a person standing at the top of a building 112 feet high throws a ball vertically upward with an initial velocity of 96 ft/sec.

(a) Find the ball's height and velocity at time t.
(b) When does the ball hit the ground and what is its impact velocity?
(c) When is the velocity 0? What is the significance of this time?
(d) How far does the ball travel during its flight?

Solution

(a) Since $g = 32$, $V_0 = 96$, and $H_0 = 112$, the height of the ball above the ground at time t is

$$H(t) = -16t^2 + 96t + 112 \text{ feet}$$

The velocity at time t is

$$v(t) = \frac{dH}{dt} = -32t + 96 \text{ (ft/sec)}$$

(b) The ball hits the ground when $H = 0$. Solve the equation $-16t^2 + 96t + 112 = 0$ to find that this occurs when $t = 7$ and $t = -1$. (Verify). Disregarding the negative time $t = -1$, which is not meaningful in this context, conclude that impact occurs when $t = 7$ seconds and that the impact velocity is

$$v(7) = -32(7) + 96 = -128 \text{ ft/sec}$$

(The negative sign means the ball is coming down at the moment of impact.)

Figure 2.15 The motion of a ball thrown upward from the top of a building.

$H(3) = 258$
$v(3) = 0$

$H(0) = 112$
$v(0) = 96$

$H(t)$

$H(7) = 0$
$v(7) = -128$

Fig. 2.15

(c) The velocity is zero when $v(t) = -32t + 96 = 0$, which occurs when $t = 3$ seconds. For $t < 3$, the velocity is positive and the ball is rising, and for $t > 3$, the ball is falling (see Figure 2.15). Thus, the ball is at its highest point when $t = 3$ seconds.

(d) The ball starts at $H(0) = 112$ feet and rises to a maximum height of $H(3) = 256$ feet before falling to the ground. Thus,

$$\text{Total distance traveled} = \underbrace{(256 - 112)}_{\text{up}} + \underbrace{256}_{+ \text{ down}} = 400 \text{ feet}$$

Problems 2.3

In Problems 1 through 6, find the rate of change of the given function $f(x)$ with respect to x for the prescribed value of x.

1. $f(x) = x^3 - 3x + 5$; $x = 2$

2. $f(x) = \sqrt{x} + 5x$; $x = 4$

3. $f(x) = (x^2 + 2)(x + \sqrt{x})$; $x = 4$

4. $f(x) = (x^2 + 3)(5 - 2x^3)$; $x = 1$

5. $f(x) = \dfrac{2x - 1}{3x + 5}$; $x = 1$

6. $f(x) = x + \dfrac{3}{2 - 4x}$; $x = 0$

Newspaper Circulation

7. It is estimated that t years from now, the circulation of a local newspaper will be $C(t) = 100t^2 + 400t + 5{,}000$.

(a) Derive an expression for the rate at which the circulation will be changing with respect to time t years from now.

(b) At what rate will the circulation be changing with respect to time 5 years from now? Will the circulation be increasing or decreasing at that time?

(c) By how much will the circulation actually change during the 6th year?

Speed of a Moving Object

8. An object moves along a straight line so that after t minutes, its distance from its starting point is $D(t) = 10t + \dfrac{5}{t + 1}$ meters.

(a) At what speed is the object moving at the end of 4 minutes?

(b) How far does the object actually travel during the 5th minute?

Worker Efficiency

9. An efficiency study of the morning shift at a certain factory indicates that an average worker who arrives on the job at 8:00 A.M. will have assembled $f(x) = -x^3 + 6x^2 + 15x$ transistor radios x hours later.

(a) Derive a formula for the rate at which the worker will be assembling radios after x hours.

(b) At what rate will the worker be assembling radios at 9:00 A.M.?

(c) How many radios will the worker actually assemble between 9:00 and 10:00 A.M.?

Air Pollution   10. An environmental study of a certain suburban community suggests that t years from now, the average level of carbon monoxide in the air will be $Q(t) = 0.05t^2 + 0.1t + 3.4$ parts per million.
(a) At what rate will the carbon monoxide level be changing with respect to time 1 year from now?
(b) By how much will the carbon monoxide level change this year?
(c) By how much will the carbon monoxide level change over the next 2 years?

Population Growth   11. It is estimated that t years from now, the population of a certain suburban community will be $P(t) = 20 - \dfrac{6}{t + 1}$ thousand.
(a) Derive a formula for the rate at which the population will be changing with respect to time t years from now.
(b) At what rate will the population be growing 1 year from now?
(c) By how much will the population actually increase during the 2nd year?
(d) At what rate will the population be growing 9 years from now?
(e) What will happen to the rate of population growth in the long run?

SAT Scores   12. It is estimated that x years from now, the average SAT score of the incoming students at an eastern liberal arts college will be $f(x) = -6x + 582$.
(a) Derive an expression for the rate at which the average SAT score will be changing with respect to time x years from now.
(b) What is the significance of the fact that the expression in part (a) is a constant? What is the significance of the fact that the constant in part (a) is negative?

Distance   13. Two cars leave an intersection at the same time. One travels east at a constant speed of 60 kilometers per hour, while the other goes north at a constant speed of 80 kilometers per hour. Find an expression for the rate at which the distance between the cars is changing with respect to time.

Percentage Rate of Change   14. Find the percentage rate of change in the function $f(t) = 3t^2 - 7t + 5$ with respect to t, when $t = 2$.

Percentage Rate of Change   15. Find the percentage rate of change in the function $f(x) = x(x + 3)^2$ with respect to x, when $x = 3$.

Population Growth   16. It is projected that x months from now, the population of a certain town will be $P(x) = 2x + 4x^{3/2} + 5,000$.
(a) At what rate will the population be changing with respect to time 9 months from now?
(b) At what percentage rate will the population be changing with respect to time 9 months from now?

Annual Earnings   17. The gross annual earnings of a certain company were $A(t) = 0.1t^2 + 10t + 20$ thousand dollars t years after its formation in 1987.
(a) At what rate were the gross annual earnings of the company growing with respect to time in 1991?
(b) At what percentage rate were the gross annual earnings growing with respect to time in 1991?

Property Tax   18. Records indicate that x years after 1985, the average property tax on a three-bedroom home in a certain community was $T(x) = 20x^2 + 40x + 600$ dollars.
(a) At what rate was the property tax increasing with respect to time in 1991?
(b) At what percentage rate was the property tax increasing with respect to time in 1991?

Population Growth   19. It is estimated that t years from now, the population of a certain town will be $P(t) = t^2 + 200t + 10,000$.
(a) Express the percentage rate of change of the population as a function of t, simplify this function algebraically, and draw its graph.
(b) What will happen to the percentage rate of change of the population in the long run?

Salary Increases   20. Your starting salary will be $24,000, and you will get a raise of $2,000 each year.
(a) Express the percentage rate of change of your salary as a function of time and draw the graph.
(b) At what percentage rate will your salary be increasing after 1 year?
(c) What will happen to the percentage rate of change of your salary in the long run?

Gross National Product   21. The gross national product (GNP) of a certain country is growing at a constant rate. In 1986 the GNP was 125 billion dollars, and in 1988 the GNP was 155 billion dollars. At what percentage rate was the GNP growing in 1991?

Rectilinear Motion   In Problems 22 through 27, $s(t)$ is the displacement of an object moving along a line. In each case, for the specified time interval:
(a) Find the velocity of the object and discuss its motion.
(b) Find the total distance traveled by the object.
(c) Find the acceleration of the object and determine when the object is speeding up and slowing down.

22. $s(t) = t^2 - 2t + 6$ for $0 \leq t \leq 2$

23. $s(t) = 3t^2 + 2t - 5$ for $0 \leq t \leq 1$

24. $s(t) = t^3 - 9t^2 + 15t + 25$ for $0 \le t \le 6$

25. $s(t) = t^4 - 4t^3 + 8t$ for $0 \le t \le 4$

26. $s(t) = 2t^3 + 3t^2 - 36t + 40$ for $0 \le t \le 3$

27. $s(t) = t^3 - 9t^2 + 15t + 25$ for $0 \le t \le 6$

Motion of a Projectile   28. A stone is dropped from a height of 144 feet.
(a) When will the stone hit the ground?
(b) With what velocity does it hit the ground?

Motion of a Projectile   29. A stone is thrown downward with an initial velocity of -50 ft/sec from the top of a building (the minus sign indicates it is thrown downward), and it hits the ground 3 seconds later.
(a) How high is the building?
(b) With what velocity does the stone hit the ground?

Motion of a Projectile   30. A projectile is launched vertically upward from ground level with an initial velocity of 160 ft/sec.
(a) When will the projectile hit the ground?
(b) What is the impact velocity?
(c) When will the projectile reach its maximum height?
(d) What is the maximum height reached by the projectile?

Motion of a Projectile   31. You are standing on the top of a building and throw a ball vertically upward. After 2 seconds, the ball passes you on the way down, and 2 seconds after that, it hits the ground below.
(a) What is the initial velocity of the ball?
(b) How high is the building?
(c) What is the velocity of the ball when it passes you on the way down?
(d) What is the velocity of the ball as it hits the ground?

Spy Story   32. Our friend, the spy who escaped from the diamond smugglers in Chapter 1 (Problem 48 of Section 4), is on a secret mission in space. An encounter with an enemy agent leaves him with a mild concussion and temporary amnesia. Fortunately, he has a book that gives the formula for the motion of a projectile and the values of g for various heavenly bodies (32 ft/sec² on earth, 5.5 ft/sec² on the moon, 12 ft/sec² on Mars, and 28 ft/sec² on Venus). To deduce his whereabouts, he throws a rock vertically upward (from ground level) and notes that it reaches a maximum height of 37.5 ft and hits the ground 5 seconds after it leaves his hand. What conclusion does he reach?

33. Use calculus to prove that if y is a linear function of x, the rate of change of y with respect to x is constant.

34. If y is a linear function of x, what will happen to the percentage rate of change of y with respect to x as x increases without bound? Explain.

Motion of a Projectile

35. A ball is thrown vertically upward from ground level with an initial velocity V_0.
   (a) Derive a formula for the time at which the ball hits the ground.
   (b) Use the formula in part (a) to prove that the impact velocity of the ball is $-V_0$.

Manufacturing Cost

36. Suppose the total manufacturing cost C at a certain factory is a function of the number q of units produced, which in turn is a function of the number t of hours during which the factory has been operating.
   (a) What quantity is represented by the derivative $\dfrac{dC}{dq}$? In what units is this quantity measured?
   (b) What quantity is represented by the derivative $\dfrac{dq}{dt}$? In what units is this quantity measured?
   (c) What quantity is represented by the product $\dfrac{dC}{dq}\dfrac{dq}{dt}$? In what units is this quantity measured?

4 Approximation by Differentials; Marginal Analysis

In this section, you will learn a general procedure for using calculus to estimate how a function is affected by a small change in its variable. You will also be introduced to an important application of this approximation procedure to economics called **marginal analysis.**

The procedure is based on the fact that if y is a function of x, then

$$\begin{bmatrix} \text{Change} \\ \text{in } y \end{bmatrix} \approx \begin{pmatrix} \text{rate of change of } y \\ \text{with respect to } x \end{pmatrix}(\text{change in } x)$$

If the rate of change of y with respect to x happens to be constant, the approximation is exact. If the rate of change is not constant but the change in x is small, large variations in the rate of change are unlikely, and hence the approximation will be a good one.

Since the rate of change is given by the derivative, the approximation formula can be written more compactly as follows.

Approximation Formula

If $y = f(x)$ and Δx is a small change in x, then the corresponding change in y is

$$\Delta y \simeq \frac{dy}{dx}\Delta x$$

or, in functional notation, the corresponding change in f is

$$\Delta f = f(x + \Delta x) - f(x) \simeq f'(x)\Delta x$$

That is, the change in the function is approximately the derivative of the function times the change in its variable.

Notice that if you divide both sides of the approximation formula

$$f(x + \Delta x) - f(x) \simeq f'(x)\Delta x$$

by Δx, you get

$$\frac{f(x + \Delta x) - f(x)}{\Delta x} \simeq f'(x)$$

which shows that the approximation formula is simply a restatement of the fact that the difference quotient is close to the derivative when Δx is small.

The use of the approximation formula is illustrated in the following example.

EXAMPLE 4.1

Suppose the total cost in dollars of manufacturing q units of a certain commodity is $C(q) = 3q^2 + 5q + 10$. If the current level of production is 40 units, estimate how the total cost will change if 40.5 units are produced.

Solution

In this problem, the current value of the variable is $q = 40$ and the change in the variable is $\Delta q = 0.5$. By the approximation formula, the corresponding change in cost is

$$\Delta C = C(40.5) - C(40) \simeq C'(40)\Delta q = C'(40)(0.5)$$

Since

$$C'(q) = 6q + 5 \quad \text{and} \quad C'(40) = 6(40) + 5 = 245$$

it follows that

$$\Delta C \simeq C'(40)(0.5) = 245(0.5) = \$122.50$$

For practice, compute the actual change in cost caused by the increase in the level of production from 40 to 40.5 and compare your answer with the approximation. Is the approximation a good one?

In the next example, the approximation formula is used to study **propagation of error.** In particular, the derivative is used to estimate the maximum error in a calculation that is based on figures obtained through imperfect measurement.

EXAMPLE 4.2

You measure the side of a cube to be 12 centimeters long and conclude that the volume of the cube is $12^3 = 1,728$ cubic centimeters. If your measurement of the side is accurate to within 2 percent, approximately how accurate is your calculation of the volume?

Solution

The volume of the cube is $V(x) = x^3$, where x is the length of a side. The error you make in computing the volume if you take the length of the side to be 12 when it is really $12 + \Delta x$ is

$$\Delta V = V(12 + \Delta x) - V(12) \simeq V'(12)\Delta x$$

Your measurement of the side can be off by as much as 2 percent, that is, by as much as $0.02(12) = 0.24$ centimeter in either direction. Hence, the maximum error in your measurement of the side is $\Delta x = \pm 0.24$, and the corresponding maximum error in your calculation of the volume is

$$\text{Maximum error in volume} = \Delta V \simeq V'(12)(\pm 0.24)$$

Since

$$V'(x) = 3x^2 \quad \text{and} \quad V'(12) = 3(12)^2 = 432$$

it follows that

$$\text{Maximum error in volume} \simeq 432(\pm 0.24) = \pm 103.68$$

This says that, at worst, your calculation of the volume as 1,728 cubic centimeters is off by approximately 103.68 cubic centimeters.

In the next example, the desired change in the function is given, and the goal is to estimate the necessary corresponding change in the variable.

EXAMPLE 4.3

The daily output at a certain factory is $Q(L) = 900L^{1/3}$ units, where L denotes the size of the labor force measured in worker-hours. Currently, 1,000 worker-hours of labor are used each day. Use calculus to estimate the number of additional worker-hours of labor that will be needed to increase daily output by 15 units.

Solution

Solve for ΔL using the approximation formula

$$\Delta Q \simeq Q'(L)\Delta L$$

with $\Delta Q = 15$ $L = 1,000$ and $Q'(L) = 300L^{-2/3}$

to get $15 \simeq 300(1,000)^{-2/3}\,\Delta L$

or $\Delta L \simeq \dfrac{15}{300}(1,000)^{2/3} = \dfrac{15}{300}(10)^2 = 5$ worker-hours

Approximation of Percentage Change   The **percentage change** of a quantity expresses the change in that quantity as a percentage of its size prior to the change. In particular,

$$\text{Percentage change} = 100\,\frac{\text{change in quantity}}{\text{size of quantity}}$$

This formula can be combined with the approximation formula and written in functional notation as follows.

Approximation Formula for Percentage Change

If Δx is a (small) change in x, the corresponding percentage change in the function $f(x)$ is

$$\text{Percentage change in } f = 100\,\frac{\Delta f}{f(x)} \simeq 100\,\frac{f'(x)\Delta x}{f(x)}$$

EXAMPLE 4.4

The GNP of a certain country was $N(t) = t^2 + 5t + 200$ billion dollars t years after 1990. Use calculus to estimate the percentage change in the GNP during the first quarter of 1998.

Solution

Use the formula

$$\text{Percentage change in } N \simeq 100\,\frac{N'(t)\Delta t}{N(t)}$$

with $t = 8$ $\Delta t = 0.25$ and $N'(t) = 2t + 5$

to get   Percentage change in $N \simeq 100 \dfrac{N'(8)0.25}{N(8)}$

$$= 100 \frac{[2(8) + 5](0.25)}{(8)^2 + 5(8) + 200}$$

$$\simeq 1.73 \text{ percent}$$

The next example illustrates how the percentage change can sometimes be estimated even though the numerical value of the variable is not known.

EXAMPLE 4.5

At a certain factory, the daily output is $Q(K) = 4,000K^{1/2}$ units, where K denotes the firm's capital investment. Use calculus to estimate the percentage increase in output that will result from a 1 percent increase in capital investment.

Solution

The derivative is $Q'(K) = 2,000K^{-1/2}$. The fact that K increases by 1 percent means that $\Delta K = 0.01K$. Hence,

Percentage change in $Q \simeq 100 \dfrac{Q'(K)\Delta K}{Q(K)}$

$$= 100 \frac{2,000K^{-1/2}(0.01K)}{4,000K^{1/2}}$$

$$= \frac{2,000K^{1/2}}{4,000K^{1/2}} \qquad \text{since } K^{-1/2}K = K^{1/2}$$

$$= \frac{2,000}{4,000} = 0.5 \text{ percent}$$

Marginal Analysis in Economics   In economics, the use of the derivative to approximate the change in a function produced by a 1-unit change in its variable is called **marginal analysis**. For example, if $C(x)$ is the total production cost incurred by a manufacturer when x units are produced and $R(x)$ is the total revenue derived from the sale of x units, then $C'(x)$ is called the **marginal cost** and $R'(x)$ is called the **marginal revenue**. If production (or sales) is increased by 1 unit, then $\Delta x = 1$ and the approximation formula

$$\Delta C = C(x + \Delta x) - C(x) \approx C'(x)\,\Delta x$$

becomes
$$\Delta C = C(x + 1) - C(x) \approx C'(x)$$

while
$$\Delta R = R(x + \Delta x) - R(x) \approx R'(x)\,\Delta x$$

becomes
$$\Delta R = R(x + 1) - R(x) \approx R'(x)$$

That is, the marginal cost $C'(x)$ is an approximation to the cost $C(x + 1)$ $- C(x)$ of producing the $x + 1$st unit, and similarly, the marginal revenue $R'(x)$ is an approximation to the revenue derived from the sale of the $x + 1$st unit. To summarize:

> **Approximation by Marginal Cost and Marginal Revenue**
>
> If $C(x)$ is the total cost of producing x units and $R(x)$ is the total revenue derived from the sale of x units, then
>
> Marginal cost $= C'(x) \approx$ the cost of producing the $x + 1$st unit
>
> Marginal revenue $= R'(x) \approx$ the revenue derived from the sale of the $x + 1$st unit

EXAMPLE 4.6

A manufacturer estimates that when x units of a particular commodity are produced, the total cost will be $C(x) = \frac{1}{8}x^2 + 3x + 98$ dollars, and that $p(x) = \frac{1}{3}(75 - x)$ dollars per unit is the price at which all x units will be sold.

(a) Find the marginal cost and the marginal revenue.
(b) Use marginal cost to estimate the cost of producing the 9th unit.
(c) What is the actual cost of producing the 9th unit?
(d) Use marginal revenue to estimate the revenue derived from the sale of the 9th unit.
(e) What is the actual revenue derived from the sale of the 9th unit?

Solution

(a) The marginal cost is $C'(x) = \frac{1}{4}x + 3$. Since x units of the commodity are sold at a price of $p(x) = \frac{1}{3}(75 - x)$ dollars per unit, the total revenue is

$$R(x) = \text{(number of units sold)(price per unit)}$$

$$= xp(x) = x\left[\frac{1}{3}(75 - x)\right] = 25x - \frac{1}{3}x^2$$

The marginal revenue is

$$R'(x) = 25 - \frac{2}{3}x$$

(b) The cost of producing the 9th unit is the change in cost as x increases from 8 to 9 and can be estimated by the marginal cost

$$C'(8) = \frac{1}{4}(8) + 3 = \$5$$

(c) The actual cost of producing the 9th unit is

$$\Delta C = C(9) - C(8) = \$5.13$$

which is reasonably well approximated by the marginal cost $C'(8) = \$5$.

(d) The revenue obtained from the sale of the 9th unit is approximated by the marginal revenue

$$R'(8) = 25 - \frac{2}{3}(8) = \$19.67$$

(e) The actual revenue obtained from the sale of the 9th unit is

$$\Delta R = R(9) - R(8) = \$19.33$$

Differentials The expression $f'(x)\Delta x$ on the right-hand side of the approximation formula $\Delta f \simeq f'(x)\Delta x$ is sometimes called the **differential** of f and is denoted by df. Similarly, the expression $\frac{dy}{dx}\Delta x$ on the right-hand side of the other form of the approximation formula $\Delta y \simeq \frac{dy}{dx}\Delta x$ is known as the differential of y and is denoted by dy. Thus, if Δx is small,

$$\Delta y \simeq dy \qquad \text{where} \qquad dy = \frac{dy}{dx}\Delta x$$

Geometric Interpretation The approximation of Δy by its differential dy has a simple geometric interpretation that is illustrated in Figure 2.16.

Since the slope of the tangent is $\frac{dy}{dx}$, the differential $dy = \frac{dy}{dx}\Delta x$ is the change in the height of the tangent corresponding to a change from x to $x + \Delta x$. On the other hand, Δy is the change in the height of the curve corresponding to this change in x. Hence, approximating Δy by the differential dy is the same as approximating the change in height of a

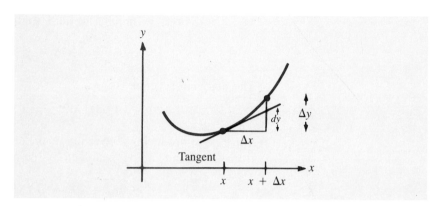

Figure 2.16 Approximation of Δy by the differential dy.

curve by the change in height of its tangent. If Δx is small, this approximation is a good one.

Problems 2.4

1. Estimate how much the function $f(x) = x^2 - 3x + 5$ will change as x increases from 5 to 5.3.

2. Estimate how much the function $f(x) = \dfrac{x}{x + 1} - 3$ will change as x decreases from 4 to 3.8.

3. Estimate the percentage change in the function $f(x) = x^2 + 2x - 9$ as x increases from 4 to 4.3.

4. Estimate the percentage change in the function $f(x) = 3x + \dfrac{2}{x}$ as x decreases from 5 to 4.6.

In each of the following problems, use calculus to obtain the required estimate.

Manufacturing

5. A manufacturer's total cost is $C(q) = 0.1q^3 - 0.5q^2 + 500q + 200$ dollars when the level of production is q units. The current level of production is 4 units, and the manufacturer is planning to increase this to 4.1 units. Estimate how the total cost will change as a result.

Manufacturing

6. A manufacturer's total monthly revenue is $R(q) = 240q + 0.05q^2$ dollars when q units are produced during the month. Currently, the manufacturer is producing 80 units a month and is planning to decrease the monthly output by 0.65 unit. Estimate how the total monthly revenue will change as a result.

Newspaper Circulation

7. It is projected that t years from now, the circulation of a local newspaper will be $C(t) = 100t^2 + 400t + 5,000$. Estimate the amount

by which the circulation will increase during the next 6 months. (*Hint:* The current value of the variable is $t = 0$.)

Population Growth   8. It is projected that t years from now, the population of a certain suburban community will be $P(t) = 20 - \dfrac{6}{t + 1}$ thousand. By approximately how much will the population increase during the next quarter year?

Air Pollution   9. An environmental study of a certain community suggests that t years from now, the average level of carbon monoxide in the air will be $Q(t) = 0.05t^2 + 0.1t + 3.4$ parts per million. By approximately how much will the carbon monoxide level change during the coming 6 months?

Efficiency   10. An efficiency study of the morning shift at a certain factory indicates that an average worker who arrives on the job at 8:00 A.M. will have assembled $f(x) = -x^3 + 6x^2 + 15x$ transistor radios x hours later. Approximately how many radios will the worker assemble between 9:00 and 9:15 A.M.?

Production   11. At a certain factory, the daily output is $Q(K) = 600K^{1/2}$ units, where K denotes the capital investment measured in units of $1,000. The current capital investment is $900,000. Estimate the effect that an additional capital investment of $800 will have on the daily output.

Production   12. At a certain factory, the daily output is $Q(L) = 60,000L^{1/3}$ units, where L denotes the size of the labor force measured in worker-hours. Currently 1,000 worker-hours of labor are used each day. Estimate the effect on output that will be produced if the labor force is cut to 940 worker-hours.

Marginal Analysis   13. At a certain factory, the daily output is $Q = 3,000K^{1/2}L^{1/3}$ units, where K denotes the firm's capital investment measured in units of $1,000 and L denotes the size of the labor force measured in worker-hours. Suppose that the current capital investment is $400,000 and that 1,331 worker-hours of labor are used each day. Use marginal analysis to estimate the effect that an additional capital investment of $1,000 will have on the daily output if the size of the labor force is not changed.

Circulation of Blood   14. The speed of blood flowing along the central axis of a certain artery is $S(R) = 1.8 \times 10^5 R^2$ centimeters per second, where R is the radius of the artery. A medical researcher measures the radius of the artery to be 1.2×10^{-2} centimeter and makes an error of 5×10^{-4} centimeter. Estimate the amount by which the calculated value of the speed of the blood will differ from the true speed if the incorrect value of the radius is used in the formula.

Area   15. You measure the radius of a circle to be 12 centimeters and use the formula $A = \pi r^2$ to calculate the area. If your measurement of the radius is accurate to within 3 percent, approximately how accurate is your calculation of the area?

Volume   16. You measure the radius of a sphere to be 6 inches and use the formula $V = \frac{4}{3}\pi r^3$ to calculate the volume. If your measurement of the radius is accurate to within 1 percent, approximately how accurate is your calculation of the volume?

Production   17. The daily output at a certain factory is $Q(L) = 300L^{2/3}$ units, where L denotes the size of the labor force measured in worker-hours. Currently 512 worker-hours of labor are used each day. Estimate the number of additional worker-hours of labor that will be needed to increase daily output by 12.5 units.

Manufacturing   18. A manufacturer's total cost is $C(q) = \frac{1}{6}q^3 + 642q + 400$ dollars when q units are produced. The current level of production is 4 units. Estimate the amount by which the manufacturer should decrease production to reduce the total cost by $130.

Property Tax   19. Records indicate that x years after 1988, the average property tax on a three-bedroom home in a certain community was $T(x) = 60x^{3/2} + 40x + 1,200$ dollars. Estimate the percentage by which the property tax increased during the first half of 1992.

Annual Earnings   20. The gross annual earnings of a company were $A(t) = 0.1t^2 + 10t + 20$ thousand dollars t years after its formation in 1988. Estimate the percentage change in the gross annual earnings during the third quarter of 1992.

Volume   21. Estimate what will happen to the volume of a cube if the length of each side is decreased by 2 percent. Express your answer as a percentage and verify that your answer is consistent with the calculation in Example 4.2.

Area   22. Estimate what will happen to the area of a circle if the radius is increased by 1 percent. Express your answer as a percentage.

Volume   23. Estimate what will happen to the volume of a sphere if the radius is increased by 1 percent. Express your answer as a percentage.

Circulation of Blood   24. According to Poiseuille's law, the speed of blood flowing along the central axis of an artery of radius R is $S(R) = cR^2$, where c is a constant. Estimate the percentage error you will make in the calculation of $S(R)$ from this formula if you make a 1 percent error in the measurement of R.

Production   25. The output at a certain factory is $Q(K) = 400K^{1/2}$ units, where K denotes the firm's capital investment. Estimate the percentage increase in output that will result from a 1 percent increase in capital investment.

Production   26. The output at a certain factory is $Q(L) = 9{,}000L^{1/3}$ units, where L denotes the size of the labor force. Estimate the percentage increase in output that will result from a 1.5 percent increase in the size of the labor force.

Production   27. The output at a certain factory is $Q = 600K^{1/2}L^{1/3}$ units, where K denotes the capital investment and L the size of the labor force. Estimate the percentage increase in output that will result from a 2 percent increase in the size of the labor force if capital investment is not changed.

Production   28. At a certain factory, the daily output is $Q(K) = 1{,}200K^{1/2}$ units, where K denotes the firm's capital investment. Estimate the percentage increase in capital investment that is needed to produce a 1.2 percent increase in output.

Volume   29. Estimate the largest percentage error you can allow in the measurement of the radius of a sphere if you want the error in the calculation of its volume using the formula $V = \frac{4}{3}\pi r^3$ to be no greater than 8 percent.

Volume   30. A soccer ball made of leather $\frac{1}{8}$ inch thick has an inner diameter of $8\frac{1}{2}$ inches. Estimate the volume of its leather shell. (*Hint:* Think of the volume of the shell as a certain change ΔV in volume.)

Volume   31. A melon in the form of a sphere has a rind $\frac{1}{5}$ inch thick and an inner diameter of 8 inches. Estimate what percentage of the total volume of the melon is rind.

Marginal Analysis   32. Suppose the total cost in dollars of manufacturing q units is $C(q) = 3q^2 + q + 500$.
   (a) Use marginal analysis to estimate the cost of manufacturing the 41st unit.
   (b) Compute the actual cost of manufacturing the 41st unit.

Marginal Analysis   33. A manufacturer's total cost is $C(q) = 0.1q^3 - 0.5q^2 + 500q + 200$ dollars, where q is the number of units produced.
   (a) Use marginal analysis to estimate the cost of manufacturing the 4th unit.
   (b) Compute the actual cost of manufacturing the 4th unit.

Marginal Analysis 34. A manufacturer's total monthly revenue is $R(q) = 240q + 0.05q^2$ dollars when q units are produced and sold during the month. Currently, the manufacturer is producing 80 units a month and is planning to increase the monthly output by 1 unit.
(a) Use marginal analysis to estimate the additional revenue that will be generated by the production and sale of the 81st unit.
(b) Use the revenue function to compute the actual additional revenue that will be generated by the production and sale of the 81st unit.

Marginal Analysis In Problems 35 and 36, $C(x)$ is the total cost of producing x units of a particular commodity and $p(x)$ is the price at which all x units will be sold.
(a) Find the marginal cost and the marginal revenue.
(b) Use marginal cost to estimate the cost of producing the 4th unit.
(c) Find the actual cost of producing the 4th unit.
(d) Use marginal revenue to estimate the revenue derived from the sale of the 4th unit.
(e) Find the actual revenue derived from the sale of the 4th unit.

35. $C(x) = \dfrac{1}{5}x^2 + 4x + 57$; $p(x) = \dfrac{1}{4}(36 - x)$

36. $C(x) = \dfrac{1}{4}x^2 + 3x + 67$; $p(x) = \dfrac{1}{5}(45 - x)$

5 The Chain Rule

In many practical situations, a quantity is given as a function of one variable which, in turn, can be thought of as a function of a second variable. In such cases, the rate of change of the quantity with respect to the second variable is equal to the rate of change of the quantity with respect to the first variable times the rate of change of the first variable with respect to the second.

For example, suppose the total manufacturing cost at a certain factory is a function of the number of units produced, which in turn is a function of the number of hours during which the factory has been operating. Let C, q, and t denote the cost (in dollars), the number of units, and the number of hours, respectively. Then,

$$\frac{dC}{dq} = \begin{array}{l}\text{rate of change of cost}\\ \text{with respect to output}\end{array} \qquad \text{(dollars per unit)}$$

and $$\frac{dq}{dt} = \begin{array}{l}\text{rate of change of output}\\ \text{with respect to time}\end{array} \qquad \text{(units per hour)}$$

The product of these two rates is the rate of change of cost with respect to time. That is,

$$\frac{dC}{dq}\frac{dq}{dt} = \begin{array}{c}\text{rate of change of cost}\\ \text{with respect to time}\end{array} \qquad \text{(dollars per hour)}$$

Since the rate of change of cost with respect to time is also given by the derivative $\frac{dC}{dt}$, it follows that

$$\frac{dC}{dt} = \frac{dC}{dq}\frac{dq}{dt}$$

This formula is a special case of an important rule called the **chain rule**.

The Chain Rule

Suppose y is a differentiable function of u and u is a differentiable function of x. Then y can be regarded as a function of x and

$$\frac{dy}{dx} = \frac{dy}{du}\frac{du}{dx}$$

That is, the derivative of y with respect to x is the derivative of y with respect to u times the derivative of u with respect to x.

Comment: Notice that one way to remember the chain rule is to pretend that the derivatives $\frac{dy}{du}$ and $\frac{du}{dx}$ are quotients and cancel du, reducing the expression $\frac{dy}{du}\frac{du}{dx}$ on the right-hand side of the equation to the expression $\frac{dy}{dx}$ on the left-hand side.

Here are two examples illustrating the use of the chain rule.

EXAMPLE 5.1

Find $\frac{dy}{dx}$ if $y = u^3 - 3u^2 + 1$ and $u = x^2 + 2$.

Solution

Since $\qquad \frac{dy}{du} = 3u^2 - 6u \qquad$ and $\qquad \frac{du}{dx} = 2x$

it follows that $\qquad \frac{dy}{dx} = \frac{dy}{du}\frac{du}{dx} = (3u^2 - 6u)(2x)$

Notice that this derivative is expressed in terms of the variables x and

u. Since you are thinking of y as a function of x, it is more natural to express $\dfrac{dy}{dx}$ in terms of x alone. To do this, substitute $x^2 + 2$ for u in the expression for $\dfrac{dy}{dx}$ and simplify the answer as follows.

$$\frac{dy}{dx} = [3(x^2 + 2)^2 - 6(x^2 + 2)](2x)$$
$$= 6x(x^2 + 2)[(x^2 + 2) - 2]$$
$$= 6x(x^2 + 2)(x^2) = 6x^3(x^2 + 2)$$

For practice, check this answer by first substituting $u = x^2 + 2$ into the original expression for y and then differentiating with respect to x.

In the next example, you will see how to use the chain rule to calculate a derivative for a particular value of the independent variable.

EXAMPLE 5.2

Find $\dfrac{dy}{dx}$ when $x = 1$ if $y = \dfrac{u}{u + 1}$ and $u = 3x^2 - 1$.

Solution

Since

$$\frac{dy}{du} = \frac{(u + 1)(1) - u(1)}{(u + 1)^2} = \frac{1}{(u + 1)^2} \qquad \text{by the quotient rule}$$

and

$$\frac{du}{dx} = 6x$$

it follows that

$$\frac{dy}{dx} = \frac{dy}{du}\frac{du}{dx} = \left[\frac{1}{(u + 1)^2}\right](6x) = \frac{6x}{(u + 1)^2}$$

The goal is to evaluate this derivative when $x = 1$. One way to do this is to replace u by its algebraic formula as in Example 5.1 and then evaluate the resulting expression when $x = 1$. However, it is easier to substitute numbers than algebraic expressions, and so it is preferable to compute the numerical value of u first and then substitute. In particular, when $x = 1$, the original formula $u = 3x^2 - 1$ gives $u = 3(1)^2 - 1 = 2$. Now substitute $x = 1$ and $u = 2$ in the formula for $\dfrac{dy}{dx}$ to conclude that when $x = 1$, $$\frac{dy}{dx} = \frac{6(1)}{(2 + 1)^2} = \frac{6}{9} = \frac{2}{3}$$

Composite Functions   Recall from Chapter 1, Section 1, that the composite function $g[h(x)]$ is the function formed from functions $g(u)$ and $h(x)$ by substituting $h(x)$ for u in the formula for $g(u)$. The chain rule is actually a rule for differentiating composite functions and can be rewritten using functional notation as follows.

> **The Chain Rule in Functional Notation**
>
> If $g(u)$ and $h(x)$ are differentiable functions,
>
> $$\frac{d}{dx}g[h(x)] = g'[h(x)]h'(x)$$

To see that this is nothing more than a restatement of the previous version of the chain rule, suppose that $y = g[h(x)]$. Then,

$$y = g(u) \qquad \text{where} \qquad u = h(x)$$

and, by the chain rule,

$$\frac{dy}{dx} = \frac{dy}{du}\frac{du}{dx} = g'(u)h'(x) = g'[h(x)]h'(x)$$

The use of this form of the chain rule is illustrated in the next example.

EXAMPLE 5.3

Differentiate the function $f(x) = \sqrt{x^2 + 3x + 2}$.

Solution

Think of $f(x)$ as the composite function $g[h(x)]$, where

$$g(u) = \sqrt{u} = u^{1/2} \qquad \text{and} \qquad h(x) = x^2 + 3x + 2$$

Then,

$$g'(u) = \frac{1}{2}u^{-1/2} \qquad \text{and} \qquad h'(x) = 2x + 3$$

and, by the chain rule,

$$f'(x) = g'[h(x)]h'(x)$$

$$= \frac{1}{2}(x^2 + 3x + 2)^{-1/2}(2x + 3)$$

$$= \frac{2x + 3}{2\sqrt{x^2 + 3x + 2}}$$

The Chain Rule for Powers In Section 2, you learned the rule

$$\frac{d}{dx}(x^n) = nx^{n-1}$$

for differentiating power functions. By combining this rule with the chain rule, you obtain the following rule for differentiating functions of the general form $[h(x)]^n$.

The Chain Rule for Powers

For any real number n and differentiable function h,

$$\frac{d}{dx}[h(x)]^n = n[h(x)]^{n-1}\frac{d}{dx}[h(x)]$$

To derive the chain rule for powers, think of $[h(x)]^n$ as the composite function

$$[h(x)]^n = g[h(x)] \qquad \text{where} \qquad g(u) = u^n$$

Then, $\qquad g'(u) = nu^{n-1} \qquad$ and $\qquad h'(x) = \frac{d}{dx}[h(x)]$

and, by the chain rule,

$$\frac{d}{dx}[h(x)]^n = \frac{d}{dx}g[h(x)] = g'[h(x)]h'(x) = n[h(x)]^{n-1}\frac{d}{dx}[h(x)]$$

The use of the chain rule for powers is illustrated in the following examples.

EXAMPLE 5.4

Differentiate the function $f(x) = (2x^4 - x)^3$.

Solution

One way to do this problem is to expand the function and rewrite it as

$$f(x) = 8x^{12} - 12x^9 + 6x^6 - x^3$$

and then differentiate this polynomial term by term to get

$$f'(x) = 96x^{11} - 108x^8 + 36x^5 - 3x^2$$

But see how much easier it is to use the chain rule for powers. According to this rule,

$$f'(x) = 3(2x^4 - x)^2 \frac{d}{dx}(2x^4 - x) = 3(2x^4 - x)^2(8x^3 - 1)$$

Not only is this method easier, but the answer even comes out in factored form!

In the next example, the solution to Example 5.3 is written more compactly with the aid of the chain rule for powers.

EXAMPLE 5.5

Differentiate the function $f(x) = \sqrt{x^2 + 3x + 2}$.

Solution

Rewrite the function as $f(x) = (x^2 + 3x + 2)^{1/2}$ and apply the chain rule for powers:

$$f'(x) = \frac{1}{2}(x^2 + 3x + 2)^{-1/2} \frac{d}{dx}(x^2 + 3x + 2)$$

$$= \frac{1}{2}(x^2 + 3x + 2)^{-1/2}(2x + 3)$$

$$= \frac{2x + 3}{2\sqrt{x^2 + 3x + 2}}$$

EXAMPLE 5.6

Differentiate the function $f(x) = \dfrac{1}{(2x + 3)^5}$.

Solution

Do *not* use the quotient rule! It's much easier to rewrite the function as

$$f(x) = (2x + 3)^{-5}$$

and apply the chain rule for powers to get

$$f'(x) = -5(2x + 3)^{-6}\frac{d}{dx}(2x + 3) = -5(2x + 3)^{-6}(2) = -\frac{10}{(2x + 3)^6}$$

The chain rule is often used in combination with the other rules you learned in Section 2. The next example involves the product rule.

EXAMPLE 5.7

Differentiate the function $f(x) = (3x + 1)^4(2x - 1)^5$ and simplify your answer.

Solution

First apply the product rule to get

$$f'(x) = (3x + 1)^4\frac{d}{dx}[(2x - 1)^5] + (2x - 1)^5\frac{d}{dx}[(3x + 1)^4]$$

Continue by applying the chain rule for powers twice:

$$f'(x) = (3x + 1)^4[5(2x - 1)^4(2)] + (2x - 1)^5[4(3x + 1)^3(3)]$$
$$= 10(3x + 1)^4(2x - 1)^4 + 12(2x - 1)^5(3x + 1)^3$$

Finally, simplify your answer by factoring:

$$f'(x) = 2(3x + 1)^3(2x - 1)^4[5(3x + 1) + 6(2x - 1)]$$
$$= 2(3x + 1)^3(2x - 1)^4[15x + 5 + 12x - 6]$$
$$= 2(3x + 1)^3(2x - 1)^4(27x - 1)$$

EXAMPLE 5.8

Differentiate the function $f(x) = \sqrt{\dfrac{x + 1}{x - 1}}$ and simplify your answer.

Solution

First rewrite the function as

$$f(x) = \left(\frac{x + 1}{x - 1}\right)^{1/2}$$

and then apply the chain rule for powers to get

$$f'(x) = \frac{1}{2}\left(\frac{x+1}{x-1}\right)^{-1/2} \frac{d}{dx}\left(\frac{x+1}{x-1}\right)$$

Now use the quotient rule to find that

$$\frac{d}{dx}\left(\frac{x+1}{x-1}\right) = \frac{(x-1)(1) - (x+1)(1)}{(x-1)^2} = -\frac{2}{(x-1)^2}$$

and substitute the result into the equation for $f'(x)$:

$$f'(x) = \frac{1}{2}\left(\frac{x+1}{x-1}\right)^{-1/2}\left[-\frac{2}{(x-1)^2}\right]$$

$$= -\frac{(x+1)^{-1/2}}{(x-1)^{-1/2}} \frac{1}{(x-1)^2}$$

$$= -\frac{(x+1)^{-1/2}}{(x-1)^{3/2}}$$

$$= -\frac{1}{(x+1)^{1/2}(x-1)^{3/2}}$$

Applications of the Chain Rule

In many practical problems, a quantity is given as a function of one variable which, in turn, can be written as a function of a second variable, and the goal is to find the rate of change of the original quantity with respect to the second variable. Such problems can be solved by means of the chain rule. Here is an example.

EXAMPLE 5.9

An environmental study of a certain suburban community suggests that the average daily level of carbon monoxide in the air will be $c(p) = \sqrt{0.5p^2 + 17}$ parts per million when the population is p thousand. It is estimated that t years from now, the population of the community will be $p(t) = 3.1 + 0.1t^2$ thousand. At what rate will the carbon monoxide level be changing with respect to time 3 years from now?

Solution

The goal is to find $\frac{dc}{dt}$ when $t = 3$. Since

$$\frac{dc}{dp} = \frac{1}{2}(0.5p^2 + 17)^{-1/2}[0.5(2p)] = \frac{1}{2}p(0.5p^2 + 17)^{-1/2}$$

and

$$\frac{dp}{dt} = 0.2t$$

it follows from the chain rule that

$$\frac{dc}{dt} = \frac{dc}{dp}\frac{dp}{dt} = \frac{1}{2}p(0.5p^2 + 17)^{-1/2}(0.2t) = \frac{0.1pt}{\sqrt{0.5p^2 + 17}}$$

When $t = 3$,

$$p = p(3) = 3.1 + 0.1(3)^2 = 4$$

and so

$$\frac{dc}{dt} = \frac{0.1(4)(3)}{\sqrt{0.5(4)^2 + 17}}$$

$$= \frac{1.2}{\sqrt{25}} = \frac{1.2}{5} = 0.24 \text{ part per million per year}$$

Problems 2.5

In Problems 1 through 10, use the chain rule to compute the derivative $\frac{dy}{dx}$ and simplify your answer.

1. $y = u^2 + 1, u = 3x - 2$ 2. $y = 2u^2 - u + 5, u = 1 - x^2$

3. $y = \sqrt{u}, u = x^2 + 2x - 3$ 4. $y = u^2 + 2u - 3, u = \sqrt{x}$

5. $y = \frac{1}{u^2}, u = x^2 + 1$ 6. $y = \frac{1}{u}, u = 3x^2 + 5$

7. $y = \frac{1}{\sqrt{u}}, u = x^2 - 9$ 8. $y = u^2 + u - 2, u = \frac{1}{x}$

9. $y = \frac{1}{u - 1}, u = x^2$ 10. $y = u^2, u = \frac{1}{x - 1}$

In Problems 11 through 16, use the chain rule to compute the derivative $\frac{dy}{dx}$ for the given value of x.

11. $y = 3u^4 - 4u + 5, u = x^3 - 2x - 5; x = 2$

12. $y = u^5 - 3u^2 + 6u - 5, u = x^2 - 1; x = 1$

13. $y = \sqrt{u}, u = x^2 - 2x + 6; x = 3$

14. $y = 3u^2 - 6u + 2, u = \frac{1}{x^2}; x = \frac{1}{3}$

15. $y = \frac{1}{u}, u = 3 - \frac{1}{x^2}; x = \frac{1}{2}$

16. $y = \frac{1}{u + 1}, u = x^3 - 2x + 5; x = 0$

In Problems 17 through 36, differentiate the given function and simplify your answer.

17. $f(x) = (2x + 1)^4$

18. $f(x) = \sqrt{5x^6 - 12}$

19. $f(x) = (x^5 - 4x^3 - 7)^8$

20. $f(x) = (3x^4 - 7x^2 + 9)^5$

21. $f(x) = \dfrac{1}{5x^2 - 6x + 2}$

22. $f(x) = \dfrac{2}{(6x^2 + 5x + 1)^2}$

23. $f(x) = \dfrac{1}{\sqrt{4x^2 + 1}}$

24. $f(x) = \dfrac{1}{\sqrt{5x^3 + 2}}$

25. $f(x) = \dfrac{3}{(1 - x^2)^4}$

26. $f(x) = \dfrac{2}{3(5x^4 + 1)^2}$

27. $f(x) = (1 + \sqrt{3x})^5$

28. $f(x) = \sqrt{1 + \dfrac{1}{3x}}$

29. $f(x) = (x + 2)^3(2x - 1)^5$

30. $f(x) = 2(3x + 1)^4(5x - 3)^2$

31. $f(x) = \sqrt{\dfrac{3x + 1}{2x - 1}}$

32. $f(x) = \left(\dfrac{x + 2}{2 - x}\right)^3$

33. $f(x) = \dfrac{(x + 1)^5}{(1 - x)^4}$

34. $f(x) = \dfrac{(1 - 2x)^2}{(3x + 1)^3}$

35. $f(x) = \dfrac{3x + 1}{\sqrt{1 - 4x}}$

36. $f(x) = \dfrac{1 - 5x^2}{\sqrt[3]{3 + 2x}}$

In Problems 37 through 40, find an equation of the line that is tangent to the graph of f for the given value of x.

37. $f(x) = (3x^2 + 1)^2; x = -1$

38. $f(x) = (x^2 - 3)^5(2x - 1)^3; x = 2$

39. $f(x) = \dfrac{1}{(2x - 1)^6}; x = 1$

40. $f(x) = \left(\dfrac{x + 1}{x - 1}\right)^3; x = 3$

In Problems 41 and 42, differentiate the given function $f(x)$ by two different methods, first by using the chain rule for powers and then by using the product rule. Show that the two answers are the same.

41. $f(x) = (3x + 5)^2$

42. $f(x) = (7 - 4x)^2$

Annual Earnings 43. The gross annual earnings of a certain company were $f(t) = \sqrt{10t^2 + t + 236}$ thousand dollars t years after its formation in January 1988.

(a) At what rate were the gross annual earnings of the company growing in January 1992?

(b) At what percentage rate were the gross annual earnings growing in January 1992?

Manufacturing Cost   44. At a certain factory, the total cost of manufacturing q units during the daily production run is $C(q) = 0.2q^2 + q + 900$ dollars. From experience it has been determined that approximately $q(t) = t^2 + 100t$ units are manufactured during the first t hours of a production run. Compute the rate at which the total manufacturing cost is changing with respect to time 1 hour after production commences.

Air Pollution   45. It is estimated that t years from now, the population of a certain suburban community will be $p(t) = 20 - \dfrac{6}{t+1}$ thousand. An environmental study indicates that the average daily level of carbon monoxide in the air will be $c(p) = 0.5\sqrt{p^2 + p + 58}$ parts per million when the population is p thousand. Find the rate at which the level of carbon monoxide will be changing with respect to time 2 years from now.

Consumer Demand   46. When electric blenders are sold for p dollars apiece, local consumers will buy $D(p) = \dfrac{8{,}000}{p}$ blenders a month. It is estimated that t months from now, the price of the blenders will be $p(t) = 0.04t^{3/2} + 15$ dollars. Compute the rate at which the monthly demand for the blenders will be changing with respect to time 25 months from now. Will the demand be increasing or decreasing?

Consumer Demand   47. An importer of Brazilian coffee estimates that local consumers will buy approximately $D(p) = \dfrac{4{,}374}{p^2}$ pounds of the coffee per week when the price is p dollars per pound. It is estimated that t weeks from now, the price of Brazilian coffee will be $p(t) = 0.02t^2 + 0.1t + 6$ dollars per pound. At what rate will the weekly demand for the coffee be changing with respect to time 10 weeks from now? Will the demand be increasing or decreasing?

Consumer Demand   48. When a certain commodity is sold for p dollars per unit, consumers will buy $D(p) = \dfrac{40{,}000}{p}$ units per month. It is estimated that t months from now, the price of the commodity will be $p(t) = 0.4t^{3/2} + 6.8$ dollars per unit. At what percentage rate will the monthly demand for the commodity be changing with respect to time 4 months from now?

Air Pollution   49. It is estimated that t years from now, the population of a certain community will be $p(t) = 12 - \dfrac{6}{t+1}$ thousand. An environmental study indicates that the average daily level of carbon monoxide in the air will be $c(p) = 0.6\sqrt{p^2 + 2p + 24}$ units when the population is p

thousand. At what percentage rate will the level of carbon monoxide be changing with respect to time 2 years from now?

50. Suppose $L(x)$ is a function with the property that $L'(x) = \dfrac{1}{x}$. Use the chain rule to find the derivatives of the following functions and simplify your answers:

(a) $f(x) = L(x^2)$

(c) $f(x) = L\left(\dfrac{2}{3\sqrt{x}}\right)$

(b) $f(x) = L\left(\dfrac{1}{x}\right)$

(d) $f(x) = L\left(\dfrac{2x+1}{1-x}\right)$

51. Prove the chain rule for powers for $n = 2$ by using the product rule to compute $\dfrac{dy}{dx}$ if $y = [h(x)]^2$.

52. Prove the chain rule for powers for $n = 3$ by using the product rule and the result of Problem 51 to compute $\dfrac{dy}{dx}$ if $y = [h(x)]^3$. {*Hint:* Begin by writing y as $h(x)[h(x)]^2$.}

6 Implicit Differentiation

Explicit and Implicit Functions

The functions you have worked with so far have all been given by equations of the form $y = f(x)$ in which the dependent variable y on the left is given explicitly by an expression on the right involving the independent variable. A function in this form is said to be in **explicit form.** For example, the functions

$$y = x^2 + 3x + 1 \qquad y = \frac{x^3 + 1}{2x - 3} \qquad \text{and} \qquad y = \sqrt{1 - x^2}$$

are all functions in explicit form.

Sometimes practical problems will lead to equations in which the function y is not written explicitly in terms of the independent variable x, equations such as

$$x^2 y^3 - 6 = 5y^3 + x \qquad \text{and} \qquad x^2 y + 2y^3 = 3x + 2y$$

for example. Since it has not been solved for y, such an equation is said to **define y implicitly as a function of x** and the function y is said to be in **implicit form.**

Differentiation of Functions in Implicit Form

Suppose you have an equation that defines y implicitly as a function of x and you want to find the derivative $\dfrac{dy}{dx}$. For instance, you may be interested in the slope of a line that is tangent to the graph of the equation at a particular point. One approach might be to solve the equation for y

explicitly and then differentiate using the techniques you already know. Unfortunately, it is not always possible to find y explicitly. For example, there is no obvious way to solve for y in the equation $x^2y + 2y^3 = 3x + 2y$. Moreover, even when you can solve for y explicitly, the resulting formula is often complicated and unpleasant to differentiate. For example, the equation

$$x^2y^3 - 6 = 5y^3 + x$$

can be solved for y to give

$$x^2y^3 - 5y^3 = x + 6$$

$$y^3(x^2 - 5) = x + 6$$

$$y = \left(\frac{x + 6}{x^2 - 5}\right)^{1/3}$$

The computation of $\dfrac{dy}{dx}$ for this function in explicit form would be tedious, involving both the chain rule and the quotient rule.

Fortunately, there is a simple technique based on the chain rule that you can use to find $\dfrac{dy}{dx}$ without first solving for y explicitly. This technique is known as **implicit differentiation.** It consists of differentiating both sides of the (unsolved) equation with respect to x and then solving algebraically for $\dfrac{dy}{dx}$. Here is an example illustrating the technique.

EXAMPLE 6.1

Find $\dfrac{dy}{dx}$ if $x^2y + 2y^3 = 3x + 2y$.

Solution

You are going to differentiate both sides of the given equation with respect to x. So that you won't forget that y is actually a function of x, temporarily replace y by the symbol $f(x)$ and begin by rewriting the equation as

$$x^2f(x) + 2[f(x)]^3 = 3x + 2f(x)$$

Now differentiate both sides of this equation term by term with respect to x. By the product rule,

$$\frac{d}{dx}[x^2f(x)] = x^2f'(x) + 2xf(x)$$

By the chain rule for powers,

$$\frac{d}{dx}\{2[f(x)]^3\} = 6[f(x)]^2 f'(x)$$

and by the constant multiple rule,

$$\frac{d}{dx}(3x) = 3 \quad \text{and} \quad \frac{d}{dx}[2f(x)] = 2f'(x)$$

Putting it all together, you get

$$x^2 f'(x) + 2xf(x) + 6[f(x)]^2 f'(x) = 3 + 2f'(x)$$

Since $f(x) = y$ and $f'(x) = \dfrac{dy}{dx}$, you can now rewrite this equation as

$$x^2 \frac{dy}{dx} + 2xy + 6y^2 \frac{dy}{dx} = 3 + 2\frac{dy}{dx}$$

Finally, solve this equation for $\dfrac{dy}{dx}$ to get

$$x^2 \frac{dy}{dx} + 6y^2 \frac{dy}{dx} - 2\frac{dy}{dx} = 3 - 2xy$$

$$(x^2 + 6y^2 - 2)\frac{dy}{dx} = 3 - 2xy$$

$$\frac{dy}{dx} = \frac{3 - 2xy}{x^2 + 6y^2 - 2}$$

Comment: Notice that the formula for $\dfrac{dy}{dx}$ contains both the independent variable x and the dependent variable y. This is usual when derivatives are computed implicitly.

So that you would not forget to use the chain rule for powers when first learning implicit differentiation, it was suggested in Example 6.1 that you temporarily replace y by $f(x)$. As soon as you feel comfortable with the technique, try to leave out this unnecessary step and differentiate the given equation directly. Just keep in mind that y is really a function of x and remember to use the chain rule when it is appropriate. Here's what the solution to Example 6.1 looks like without the unnecessary substitution of $f(x)$.

EXAMPLE 6.2

Find $\dfrac{dy}{dx}$ if $x^2 y + 2y^3 = 3x + 2y$.

Solution

Differentiate both sides of the equation as it stands with respect to x. Remember that y is really a function of x and that you will have to use the chain rule to differentiate powers of y. In particular,

$$x^2 \frac{dy}{dx} + 2xy + 6y^2 \frac{dy}{dx} = 3 + 2\frac{dy}{dx}$$

Now solve for $\frac{dy}{dx}$ as before to get

$$x^2 \frac{dy}{dx} + 6y^2 \frac{dy}{dx} - 2\frac{dy}{dx} = 3 - 2xy$$

$$(x^2 + 6y^2 - 2)\frac{dy}{dx} = 3 - 2xy$$

$$\frac{dy}{dx} = \frac{3 - 2xy}{x^2 + 6y^2 - 2}$$

Here is an outline of the procedure.

Implicit Differentiation

Suppose an equation defines y implicitly as a function of x. To find $\frac{dy}{dx}$:

1. Differentiate both sides of the equation with respect to x. Remember that y is really a function of x and use the chain rule when differentiating terms containing y.
2. Solve the differentiated equation algebraically for $\frac{dy}{dx}$.

The technique is further illustrated in the next example.

EXAMPLE 6.3

Find $\frac{dy}{dx}$ if $(x^2 - 3y^2)^4 = x^2 y^3$.

Solution

Differentiate both sides of the equation with respect to x, using the chain rule for powers on the left and the product rule on the right to get

$$4(x^2 - 3y^2)^3 \frac{d}{dx}(x^2 - 3y^2) = x^2 \frac{d}{dx}(y^3) + y^3 \frac{d}{dx}(x^2)$$

$$4(x^2 - 3y^2)^3 \left(2x - 6y\frac{dy}{dx}\right) = x^2 \left(3y^2 \frac{dy}{dx}\right) + y^3(2x)$$

$$4(x^2 - 3y^2)^3 \left(2x - 6y\frac{dy}{dx}\right) = 3x^2y^2 \frac{dy}{dx} + 2xy^3$$

Now solve for $\frac{dy}{dx}$. Whatever you do, *don't* expand $(x^2 - 3y^2)^3$! Instead, use the distributive law on the left-hand side to find that

$$4(x^2 - 3y^2)^3(2x) - 4(x^2 - 3y^2)^3 \left(6y\frac{dy}{dx}\right) = 3x^2y^2 \frac{dy}{dx} + 2xy^3$$

or $\qquad 8x(x^2 - 3y^2)^3 - 24y(x^2 - 3y^2)^3 \dfrac{dy}{dx} = 3x^2y^2 \dfrac{dy}{dx} + 2xy^3$

Now bring all the terms containing $\frac{dy}{dx}$ to the left and all the others to the right:

$$-24y(x^2 - 3y^2)^3 \frac{dy}{dx} - 3x^2y^2 \frac{dy}{dx} = 2xy^3 - 8x(x^2 - 3y^2)^3$$

Finally, factor out $\frac{dy}{dx}$ on the left and divide to get

$$[-24y(x^2 - 3y^2)^3 - 3x^2y^2]\frac{dy}{dx} = 2xy^3 - 8x(x^2 - 3y^2)^3$$

or $\qquad \dfrac{dy}{dx} = \dfrac{2xy^3 - 8x(x^2 - 3y^2)^3}{-24y(x^2 - 3y^2)^3 - 3x^2y^2}$

Computing the Slope of a Tangent Line by Implicit Differentiation   In the next example, you will see how to use implicit differentiation to find the slope of a tangent line.

EXAMPLE 6.4

Find the slope of the line that is tangent to the curve $x^2y^3 - 6 = 5y^3 + x$ when $x = 2$.

Solution

Differentiate both sides of the equation with respect to x (using the product rule on the left-hand side) to get

$$3x^2y^2 \frac{dy}{dx} + 2xy^3 = 15y^2 \frac{dy}{dx} + 1$$

and solve for $\frac{dy}{dx}$:

$$(3x^2y^2 - 15y^2)\frac{dy}{dx} = 1 - 2xy^3$$

or

$$\frac{dy}{dx} = \frac{1 - 2xy^3}{3x^2y^2 - 15y^2}$$

The desired slope is the value of this derivative when $x = 2$. Before you can compute this value, you have to find the value of y that corresponds to $x = 2$. To do this, substitute $x = 2$ into the original equation and solve, getting

$$4y^3 - 6 = 5y^3 + 2$$
$$y^3 = -8$$
$$y = -2$$

Now substitute $x = 2$ and $y = -2$ into the formula for $\frac{dy}{dx}$ to conclude that

$$\text{Slope of tangent} = \frac{dy}{dx} = \frac{1 - 2(2)(-2)^3}{3(2)^2(-2)^2 - 15(-2)^2} = -\frac{11}{4}$$

Application to Economics   Implicit differentiation is used in economics in both practical and theoretical work. In Chapter 3, Section 5, you will see it used to derive certain theoretical relationships. A more practical application of implicit differentiation is given in the next example, which is a preview of the discussion of level curves of a function of two variables given in Section 1 of Chapter 9.

EXAMPLE 6.5

Suppose the output at a certain factory is $Q = 2x^3 + x^2y + y^3$ units, where x is the number of hours of skilled labor used and y the number of hours of unskilled labor. The current labor force consists of 30 hours of skilled labor and 20 hours of unskilled labor. Use calculus to estimate the change in unskilled labor y that should be made to offset a 1-hour increase in skilled labor x so that output will be maintained at its current level.

Solution

The current level of output is the value of Q when $x = 30$ and $y = 20$. That is, $\qquad Q = 2(30)^3 + (30)^2(20) + (20)^3 = 80,000$ units

If output is to be maintained at this level, the relationship between skilled labor x and unskilled labor y is given by the equation

$$80,000 = 2x^3 + x^2y + y^3$$

which defines y implicitly as a function of x.

The goal is to estimate the change in y that corresponds to a 1-unit increase in x when x and y are related by this equation. As you saw in Section 4, the change in y caused by a 1-unit increase in x can be approximated by the derivative $\dfrac{dy}{dx}$. To find this derivative, use implicit differentiation. (Remember that the derivative of the constant 80,000 on the left-hand side is zero.)

$$0 = 6x^2 + x^2\frac{dy}{dx} + y\frac{d}{dx}(x^2) + 3y^2\frac{dy}{dx}$$

$$0 = 6x^2 + x^2\frac{dy}{dx} + 2xy + 3y^2\frac{dy}{dx}$$

$$-(x^2 + 3y^2)\frac{dy}{dx} = 6x^2 + 2xy$$

$$\frac{dy}{dx} = -\frac{6x^2 + 2xy}{x^2 + 3y^2}$$

Now evaluate this derivative when $x = 30$ and $y = 20$ to conclude that

$$\text{Change in } y \approx \frac{dy}{dx} = -\frac{6(30)^2 + 2(30)(20)}{(30)^2 + 3(20)^2} \approx -3.14 \text{ hours}$$

That is, to maintain the current level of output, unskilled labor should be decreased by approximately 3.14 hours to offset a 1-hour increase in skilled labor.

Comment: Did you notice that the solution to Example 6.5 contains an unnecessary step? The calculation of the current output to be 80,000 units could be omitted and the equation relating x and y written as

$$C = 2x^3 + x^2y + y^3$$

where C is a constant that stands for the (unspecified) current level of output. Since the derivative of any constant is zero, the differentiated equations will be the same.

Problems 2.6

In Problems 1 through 10, find $\dfrac{dy}{dx}$ by implicit differentiation.

1. $x^2 + y^2 = 25$

2. $x^2 + y = x^3 + y^2$

3. $x^3 + y^3 = xy$

4. $5x - x^2y^3 = 2y$

5. $y^2 + 2xy^2 - 3x + 1 = 0$

6. $\dfrac{1}{x} + \dfrac{1}{y} = 1$

7. $(2x + y)^3 = x$

8. $(x - 2y)^2 = y$

9. $(x^2 + 3y^2)^5 = 2xy$

10. $(3xy^2 + 1)^4 = 2x - 3y$

In Problems 11 through 17, find the slope of the tangent line to the given curve at the specified value of x.

11. $x^2 = y^3;\ x = 8$

12. $\dfrac{1}{x} - \dfrac{1}{y} = 2;\ x = \dfrac{1}{4}$

13. $xy = 2;\ x = 2$

14. $x^2y^3 - 2xy = 6x + y + 1;\ x = 0$

15. $(1 - x + y)^3 = x + 7;\ x = 1$

16. $(x^2 - 2y)^3 = 2xy^2 + 64;\ x = 0$

17. $(2xy^3 + 1)^3 = 2x - y^3;\ x = 0$

In Problems 18 through 21, find $\dfrac{dy}{dx}$ in two ways: by implicit differentiation of the given equation and by differentiation of an explicit formula for y. In each case, show that the two answers are really the same.

18. $x^2 + y^3 = 12$

19. $xy + 2y = x^2$

20. $x + \dfrac{1}{y} = 5$

21. $xy - x = y + 2$

Manufacturing 22. The output at a certain plant is $Q = 0.08x^2 + 0.12xy + 0.03y^2$ units per day, where x is the number of hours of skilled labor used and y the number of hours of unskilled labor used. Currently, 80 hours of skilled labor and 200 hours of unskilled labor are used each day. Use calculus to estimate the change in unskilled labor that should be made to offset a 1-hour increase in skilled labor so that output will be maintained at its current level.

Manufacturing 23. At a certain factory, output Q is related to inputs x and y by the equation $Q = 2x^3 + 3x^2y^2 + (1 + y)^3$. If the current levels of input are $x = 30$ and $y = 20$, use calculus to estimate the change in input y

that should be made to offset a decrease of 0.8 unit in input x so that output will be maintained at its current level.

Manufacturing   24. At a certain factory, output Q is related to inputs u and v by the equation $Q = 3u^2 + \dfrac{2u + 3v}{(u + v)^2}$. If the current levels of input are $u = 10$ and $v = 25$, use calculus to estimate the change in input v that should be made to offset a decrease of 0.7 unit in input u so that output will be maintained at its current level.

Manufacturing   25. The output of a certain plant is $Q = 0.06x^2 + 0.14xy + 0.05y^2$ units per day, where x is the number of hours of skilled labor used and y is the number of hours of unskilled labor used. Currently, 60 hours of skilled labor and 300 hours of unskilled labor are used each day. Use calculus to estimate the change in unskilled labor that should be made to offset a 1-hour decrease in skilled labor so that output will be maintained at its current level.

26. Show that the tangent line to the curve

$$\frac{x^2}{a^2} + \frac{y^2}{b^2} = 1$$

at the point (x_0, y_0) is

$$\frac{x_0 x}{a^2} + \frac{y_0 y}{b^2} = 1$$

27. Find an equation for the tangent line to the curve

$$\frac{x^2}{a^2} - \frac{y^2}{b^2} = 1$$

at the point (x_0, y_0).

28. An object projected from a point P moves along a straight line. It is known that the velocity of the object is directly proportional to the product of the time the object has been moving and the distance it has moved from P. It is also known that at the end of 5 seconds, the object is 20 feet from P and is moving at the rate of 4 ft/sec. Find the acceleration of the object at this time (when $t = 5$).

29. (a) Show that there are no points (x, y) that satisfy the equation $x^2 + y^2 = 6y - 10$. (*Hint:* Complete the square.)
   (b) Show that if implicit differentiation is applied blindly to the equation $x^2 + y^2 = 6y - 10$, you obtain the derivative $\dfrac{dy}{dx} = \dfrac{x}{3 - y}$.

   The point of this exercise is to show that one must be careful when applying implicit differentiation. Just because it is possible to find a derivative by implicit differentiation does not mean that the derivative has any meaning.

In Problems 30 and 31, find all points on the given curve where the tangent line is either horizontal or vertical.

30. $(x^2 + y^2)^2 = 4(x^2 - y^2)$ 31. $x^2 + y^2 = \sqrt{x^2 + y^2} + x$

7 Higher-Order Derivatives

The Rate of Change of a Rate of Change

This section is about the rate of change of the rate of change of a quantity. Such rates arise in a variety of situations. For example, the acceleration of a car is the rate of change with respect to time of its speed, which is itself the rate of change with respect to time of its distance. If distance is measured in miles and time in hours, the speed (or rate of change of distance) is measured in miles per hour, and the acceleration (or rate of change of speed) is measured in miles per hour per hour.

Statements about the rate of change of a rate of change are used frequently in economics. In inflationary times, for example, you may hear a government economist assure the nation that although the inflation rate is increasing, the rate at which it is doing so is decreasing. That is, prices are still going up, but not as quickly as they were before.

The Second Derivative

The rate of change with respect to x of a function $f(x)$ is given by its derivative $f'(x)$. Similarly, the rate of change of $f'(x)$ is given by its derivative, that is, by the derivative of the derivative of the original function. The derivative of the derivative of a function $f(x)$ is known as the **second derivative** of the function and is denoted by the symbol $f''(x)$, which is read "f double prime of x." The derivative $f'(x)$ is sometimes called the **first derivative** to distinguish it from the second derivative $f''(x)$.

If the function is denoted by y instead of $f(x)$, the symbol $\dfrac{d^2y}{dx^2}$ (which is read "the second derivative of y with respect to x") is often used instead of $f''(x)$ to denote the second derivative.

The Second Derivative

The second derivative of a function is the derivative of its derivative. If $y = f(x)$, the second derivative is denoted by

$$\frac{d^2y}{dx^2} \quad \text{or} \quad f''(x)$$

The second derivative gives the rate of change of the rate of change of the original function.

Computation of the Second Derivative

You don't have to use any new rules to find the second derivative of a function. Just find the first derivative and then differentiate again.

EXAMPLE 7.1

Find the second derivative of the function $f(x) = 5x^4 - 3x^2 - 3x + 7$.

Solution

Compute the first derivative

$$f'(x) = 20x^3 - 6x - 3$$

and then differentiate again to get

$$f''(x) = 60x^2 - 6$$

EXAMPLE 7.2

Find the second derivative of the function $y = (x^2 + 1)^5$.

Solution

Compute the first derivative, using the chain rule, to get

$$\frac{dy}{dx} = 5(x^2 + 1)^4(2x) = 10x(x^2 + 1)^4$$

Then differentiate again, using the product rule, to get

$$\frac{d^2y}{dx^2} = 10x[4(x^2 + 1)^3(2x)] + 10(x^2 + 1)^4$$
$$= 80x^2(x^2 + 1)^3 + 10(x^2 + 1)^4$$
$$= 10(x^2 + 1)^3[8x^2 + (x^2 + 1)]$$
$$= 10(x^2 + 1)^3(9x^2 + 1)$$

EXAMPLE 7.3

Find the second derivative of the function $f(x) = \dfrac{3x - 2}{(x - 1)^2}$.

Solution

By the quotient rule,

$$f'(x) = \frac{(x - 1)^2(3) - (3x - 2)[2(x - 1)(1)]}{(x - 1)^4}$$

$$= \frac{(x - 1)[3(x - 1) - 2(3x - 2)]}{(x - 1)^4}$$

$$= \frac{3x - 3 - 6x + 4}{(x - 1)^3}$$

$$= \frac{1 - 3x}{(x - 1)^3}$$

By the quotient rule again,

$$f''(x) = \frac{(x - 1)^3(-3) - (1 - 3x)[3(x - 1)^2(1)]}{(x - 1)^6}$$

$$= \frac{-3(x - 1)^2[(x - 1) + (1 - 3x)]}{(x - 1)^6}$$

$$= \frac{-3(-2x)}{(x - 1)^4} = \frac{6x}{(x - 1)^4}$$

A Word of Advice   Before computing the second derivative of a function, always take the time to simplify the first derivative as much as possible. The more complicated the form of the first derivative is, the more tedious the computation of the second derivative will be.

The Second Derivative by Implicit Differentiation   The next example illustrates how to find second derivatives by implicit differentiation.

EXAMPLE 7.4

Use implicit differentiation to find $\frac{d^2y}{dx^2}$ if $4x^2 - 2y^2 = 9$. Express your answer in terms of x and y and simplify.

Solution

First differentiate both sides of the equation implicitly with respect to x to get

$$8x - 4y\frac{dy}{dx} = 0 \quad \text{or} \quad \frac{dy}{dx} = \frac{8x}{4y} = \frac{2x}{y}$$

Now differentiate implicitly with respect to x again, this time using the quotient rule.

$$\frac{d^2y}{dx^2} = \frac{d}{dx}\left(\frac{2x}{y}\right)$$

$$= \frac{y\frac{d}{dx}(2x) - 2x\left(\frac{dy}{dx}\right)}{y^2}$$

$$= \frac{2y - 2x\frac{dy}{dx}}{y^2}$$

Use the fact that $\frac{dy}{dx} = \frac{2x}{y}$ to replace the symbol $\frac{dy}{dx}$ so that the second derivative will be expressed in terms of just x and y.

$$\frac{d^2y}{dx^2} = \frac{2y - 2x\left(\frac{2x}{y}\right)}{y^2}$$

Simplify the resulting quotient by multiplying the numerator and denominator by y to get

$$\frac{d^2y}{dx^2} = \frac{2y^2 - 4x^2}{y^3}$$

Finally, notice from the original equation that

$$2y^2 - 4x^2 = -(4x^2 - 2y^2) = -9$$

and make this substitution in the numerator to get

$$\frac{d^2y}{dx^2} = -\frac{9}{y^3}$$

Applications of the Second Derivative

The second derivative will be used in Chapter 3, Section 2, to obtain information about the shapes of graphs. In Sections 3 and 4 of that chapter, the second derivative will appear again, this time in the solution of optimization problems. Here is a more elementary application illustrating the interpretation of the second derivative as the rate of change of a rate of change.

EXAMPLE 7.5

An efficiency study of the morning shift at a certain factory indicates that an average worker who arrives on the job at 8:00 A.M. will have produced $Q(t) = -t^3 + 6t^2 + 24t$ units t hours later.

(a) Compute the worker's rate of production at 11:00 A.M.
(b) At what rate is the worker's rate of production changing with respect to time at 11:00 A.M.?

$x'(x)\Delta x$ (c) Use calculus to estimate the change in the worker's rate of production between 11:00 and 11:10 A.M.

(d) Compute the actual change in the worker's rate of production between 11:00 and 11:10 A.M.

Solution

(a) The worker's rate of production is the first derivative

$$Q'(t) = -3t^2 + 12t + 24$$

of the output function $Q(t)$. At 11:00 A.M., $t = 3$ and the rate of production is

$$Q'(3) = -3(3)^2 + 12(3) + 24 = 33 \text{ units per hour}$$

(b) The rate of change of the rate of production is the second derivative

$$Q''(t) = -6t + 12$$

of the output function. At 11:00 A.M., this rate is

$$Q''(3) = -6(3) + 12 = -6 \text{ units per hour per hour}$$

The minus sign indicates that the worker's rate of production is decreasing; that is, the worker is slowing down. The rate of this decrease in efficiency at 11:00 A.M. is 6 units per hour per hour.

(c) Note that 10 minutes is $\frac{1}{6}$ hour. To estimate the change in the production rate $Q'(t)$ due to a change in t of $\Delta t = \frac{1}{6}$ hour, apply the approximation formula from Section 4 to the function $Q'(t)$ to get

$$\text{Change in rate of production} = \Delta Q' \simeq Q''(t)\,\Delta t$$

Evaluate this expression when $t = 3$ and $\Delta t = \frac{1}{6}$ to conclude that

$$\begin{matrix}\text{Change in rate} \\ \text{of production}\end{matrix} \simeq Q''(3)\Delta t = -6\left(\frac{1}{6}\right) = -1 \text{ unit per hour}$$

That is, the worker's rate of production (which was 33 units per hour at 11:00 A.M.) will decrease by approximately 1 unit per hour (to approximately 32 units per hour) during the subsequent 10 minutes.

(d) The actual change in the worker's rate of production between 11:00 and 11:10 A.M. is the difference between the values of the rate $Q'(t)$ when $t = 3$ and when $t = 3\frac{1}{6} = \frac{19}{6}$. That is,

$$\begin{matrix}\text{Actual change in} \\ \text{rate of production}\end{matrix} = Q'\left(\frac{19}{6}\right) - Q'(3)$$

$$= \left[-3\left(\frac{19}{6}\right)^2 + 12\left(\frac{19}{6}\right) + 24 \right]$$

$$- [-3(3)^2 + 12(3) + 24]$$

$$\approx 31.92 - 33 = -1.08 \text{ units per hour}$$

That is, by 11:10 A.M., the worker's rate of production, which was 33 units per hour at 11:00 A.M., will actually have decreased by 1.08 units per hour to 31.92 units per hour.

Recall from Section 3 that the **acceleration** $a(t)$ of an object moving along a straight line is the derivative of the velocity $v(t)$, which in turn is the derivative of the displacement function $s(t)$. Thus, the acceleration may be thought of as the **second** derivative of the displacement; that is,

$$a(t) = \frac{d^2s}{dt^2}$$

This notation is used in the following example.

EXAMPLE 7.6

An object moves along a straight line with displacement $s(t) = t^3 - 3t^2 + 4t$. Find the acceleration of the object at time t.

Solution

The velocity of the object is

$$v(t) = \frac{ds}{dt} = 3t^2 - 6t + 4$$

and its acceleration is

$$a(t) = \frac{dv}{dt} = \frac{d^2s}{dt^2} = 6t - 6$$

Higher-Order Derivatives   If you differentiate the second derivative $f''(x)$ of a function $f(x)$ one more time, you get the third derivative $f'''(x)$. Differentiate again and you get the fourth derivative, which is denoted by $f^{(4)}(x)$ since the prime notation $f''''(x)$ begins to get cumbersome. In general, the derivative obtained from $f(x)$ after n successive differentiations is called the **nth derivative** or **derivative of order n** and is denoted by $f^{(n)}(x)$.

> **The *n*th Derivative**
>
> For any positive integer n, the nth derivative of a function is obtained from the function by differentiating successively n times. If the original function is $y = f(x)$, the nth derivative is denoted by
>
> $$\frac{d^n y}{dx^n} \quad \text{or} \quad f^{(n)}(x)$$

EXAMPLE 7.7

Find the 5th derivative of each of the following functions:

(a) $f(x) = 4x^3 + 5x^2 + 6x - 1$

(b) $y = \dfrac{1}{x}$

Solution

(a) $f'(x) = 12x^2 + 10x + 6$

$\quad f''(x) = 24x + 10$

$\quad f'''(x) = 24$

$\quad f^{(4)}(x) = 0$

$\quad f^{(5)}(x) = 0$

(b) $\dfrac{dy}{dx} = \dfrac{d}{dx}(x^{-1}) = -x^{-2} = -\dfrac{1}{x^2}$

$\dfrac{d^2 y}{dx^2} = \dfrac{d}{dx}(-x^{-2}) = 2x^{-3} = \dfrac{2}{x^3}$

$\dfrac{d^3 y}{dx^3} = \dfrac{d}{dx}(2x^{-3}) = -6x^{-4} = -\dfrac{6}{x^4}$

$\dfrac{d^4 y}{dx^4} = \dfrac{d}{dx}(-6x^{-4}) = 24x^{-5} = \dfrac{24}{x^5}$

$\dfrac{d^5 y}{dx^5} = \dfrac{d}{dx}(24x^{-5}) = -120x^{-6} = -\dfrac{120}{x^6}$

Derivatives of higher order will play a central role in Chapter 11, Section 3, in which a powerful technique for approximating functions will be developed.

Problems 2.7

In Problems 1 through 16, find the second derivative of the given function. In each case, use the appropriate notation for the second derivative and simplify your answer. (Don't forget to simplify the first derivative as much as possible before computing the second derivative.)

1. $f(x) = 5x^{10} - 6x^5 - 27x + 4$

2. $f(x) = \frac{2}{5}x^5 - 4x^3 + 9x^2 - 6x - 2$

3. $y = 5\sqrt{x} + \frac{3}{x^2} + \frac{1}{3\sqrt{x}} + \frac{1}{2}$

4. $y = \frac{2}{3x} - \sqrt{2x} + \sqrt{2}\,x - \frac{1}{6\sqrt{x}}$

5. $f(x) = (3x + 1)^5$

6. $f(x) = \frac{2}{5x + 1}$

7. $y = (x^2 + 5)^8$

8. $y = (1 - 2x^3)^4$

9. $f(x) = \sqrt{1 + x^2}$

10. $f(x) = \frac{1}{(3x^2 - 1)^2}$

11. $y = \frac{2}{1 + x^2}$

12. $y = \frac{x}{(x + 1)^2}$

13. $f(x) = x(2x + 1)^4$ (Use the product rule.)

14. $f(x) = 2x(x + 4)^3$ (Use the product rule.)

15. $y = \left(\frac{x}{x + 1}\right)^2$

16. $y = \frac{(x - 2)^3}{x^2}$

In Problems 17 through 20, use implicit differentiation to find $\frac{d^2y}{dx^2}$. Express your answer in terms of x and y and simplify.

17. $2y^2 - 5x^2 = 3$ 18. $2x^2 - 3y^2 = 7$

19. $ax^2 + by^2 = 1$, where a and b are constants.

20. $\dfrac{x^2}{a^2} - \dfrac{y^2}{b^2} = 1$, where a and b are constants.

Worker Efficiency 21. An efficiency study of the morning shift at a certain factory indicates that an average worker who arrives on the job at 8:00 A.M. will have produced $Q(t) = -t^3 + 8t^2 + 15t$ units t hours later.
   (a) Compute the worker's rate of production at 9:00 A.M.
   (b) At what rate is the worker's rate of production changing with respect to time at 9:00 A.M.?
   (c) Use calculus to estimate the change in the worker's rate of production between 9:00 and 9:15 A.M.
   (d) Compute the actual change in the worker's rate of production between 9:00 and 9:15 A.M.

Inflation 22. It is projected that t months from now, the average price per unit for goods in a certain sector of the economy will be $P(t) = -t^3 + 7t^2 + 200t + 300$ dollars.
   (a) At what rate will the price per unit be increasing with respect to time 5 months from now?
   (b) At what rate will the rate of price increase be changing with respect to time 5 months from now?
   (c) Use calculus to estimate the change in the rate of price increase during the first half of the 6th month.
   (d) Compute the actual change in the rate of price increase during the first half of the 6th month.

Population Growth 23. Suppose that a 5-year projection of population trends suggests that t years from now, the population of a certain community will be $P(t) = -t^3 + 9t^2 + 48t + 200$ thousand.
   (a) At what rate will the population be growing 3 years from now?
   (b) At what rate will the rate of population growth be changing with respect to time 3 years from now?
   (c) Use calculus to estimate the change in the rate of population growth during the first month of the 4th year.
   (d) Compute the actual change in the rate of population growth during the first month of the 4th year.

Acceleration 24. An object moves along a straight line so that after t seconds, its distance from its starting point is $D(t) = t^3 - 12t^2 + 100t + 12$ meters. Find the acceleration of the object after 3 seconds. Is the object slowing down or speeding up at this time?

Acceleration 25. If an object is dropped or thrown vertically, its height (in feet) after t seconds is $H(t) = -16t^2 + S_0 t + H_0$, where S_0 is the initial speed of the object and H_0 its initial height.
   (a) Derive an expression for the acceleration of the object.

(b) How does the acceleration vary with time?

(c) What is the significance of the fact that the answer to part (a) is negative?

Acceleration   26. After t hours of an 8-hour trip, a car has gone $D(t) = 64t + \dfrac{10}{3}t^2 - \dfrac{2}{9}t^3$ kilometers.

(a) Derive a formula expressing the acceleration of the car as a function of time.

(b) At what rate is the speed of the car changing with respect to time at the end of 6 hours? Is the speed increasing or decreasing at this time?

(c) By how much does the speed of the car actually change during the 7th hour?

27. Find $f^{(4)}(x)$ if $f(x) = x^5 - 2x^4 + x^3 - 3x^2 + 5x - 6$.

28. Find $\dfrac{d^3y}{dx^3}$ if $y = \sqrt{x} - \dfrac{1}{2x} + \dfrac{x}{\sqrt{2}}$.

29. Find $f'''(x)$ if $f(x) = \dfrac{1}{\sqrt{3x}} - \dfrac{2}{x^2} + \sqrt{2}$.

30. Find $xf''(x) - 2f'(x) - \dfrac{4}{x}f(x)$ if $f(x) = x^3 - \sqrt{x} + \dfrac{1}{x}$.

Chapter Summary and Review Problems

Important Terms, Symbols, and Formulas

Secant line; tangent line

Derivative: $f'(x) = \lim\limits_{\Delta x \to 0} \dfrac{f(x + \Delta x) - f(x)}{\Delta x}$

Geometric interpretation: derivative = slope of the tangent

Limits, continuity, differentiable functions

Power rule: $\dfrac{d}{dx}(x^n) = nx^{n-1}$

Derivative of a constant: $\dfrac{d}{dx}(c) = 0$

Constant multiple rule: $\dfrac{d}{dx}(cf) = c\dfrac{df}{dx}$

Sum rule: $\dfrac{d}{dx}(f + g) = \dfrac{df}{dx} + \dfrac{dg}{dx}$

Product rule: $\dfrac{d}{dx}(fg) = f\dfrac{dg}{dx} + g\dfrac{df}{dx}$

Quotient rule: $\dfrac{d}{dx}\left(\dfrac{f}{g}\right) = \dfrac{g\dfrac{df}{dx} - f\dfrac{dg}{dx}}{g^2}$

Average rate of change $= \dfrac{f(x + \Delta x) - f(x)}{\Delta x}$

Instantaneous rate of change $= f'(x) = \dfrac{dy}{dx}$

Rectilinear motion: velocity and acceleration

Percentage rate of change of y with respect to $x = 100\dfrac{dy/dx}{y}$

Approximation formula: $\Delta y \simeq \dfrac{dy}{dx}\Delta x$; $\Delta f = f(x + \Delta x) - f(x) \simeq f'(x)\Delta x$

Percentage change in $f = 100\dfrac{\Delta f}{f} \simeq 100\dfrac{f'(x)\Delta x}{f(x)}$

Marginal cost $=$ derivative of cost $\approx$ cost of $x + $ 1st unit

Marginal revenue $=$ derivative of revenue
$\approx$ revenue derived from sale of $x + $ 1st unit

Differential: $dy = \dfrac{dy}{dx}\Delta x =$ change in y along the tangent line

Chain rule: $\dfrac{dy}{dx} = \dfrac{dy}{du}\dfrac{du}{dx}$; $\dfrac{d}{dx}g[h(x)] = g'[h(x)]h'(x)$

Chain rule for powers: $\dfrac{d}{dx}[h(x)] = n[h(x)]^{n-1}h'(x)$

Implicit differentiation

Second derivative: $\dfrac{d^2y}{dx^2} = f''(x) =$ rate of change of rate of change

nth derivative: $\dfrac{d^n y}{dx^n} = f^{(n)}(x)$

Review Problems   1. Use the definition to find the derivative of the given function.

  (a) $f(x) = x^2 - 3x + 1$

  (b) $f(x) = \dfrac{1}{x - 2}$

2. Differentiate the following functions using the rules. Simplify your answers.

(a) $f(x) = 6x^4 - 7x^3 + 2x + \sqrt{2}$

(b) $f(x) = x^3 - \dfrac{1}{3x^5} + 2\sqrt{x} - \dfrac{3}{x} + \dfrac{1 - 2x}{x^3}$

(c) $y = \dfrac{2 - x^2}{3x^2 + 1}$

(d) $y = (2x + 5)^3(x + 1)^2$

(e) $f(x) = (5x^4 - 3x^2 + 2x + 1)^{10}$

(f) $f(x) = \sqrt{x^2 + 1}$

(g) $y = \left(x + \dfrac{1}{x}\right)^2 - \dfrac{5}{\sqrt{3x}}$

(h) $y = \left(\dfrac{x + 1}{1 - x}\right)^2$

(i) $f(x) = (3x + 1)\sqrt{6x + 5}$

(j) $f(x) = \dfrac{(3x + 1)^3}{(1 - 3x)^4}$

(k) $y = \sqrt{\dfrac{1 - 2x}{3x + 2}}$

3. Find the equation of the line that is tangent to the graph of f at the point $(x, f(x))$ for the given value of x.

find slope (derivative)
the y coordinate
$$y - y_1 = m(x - x_1)$$

(a) $f(x) = x^2 - 3x + 2;\ x = 1$

(b) $f(x) = \dfrac{4}{x - 3};\ x = 1$

(c) $f(x) = \dfrac{x}{x^2 + 1};\ x = 0$

(d) $f(x) = \sqrt{x^2 + 5};\ x = -2$

derivative
then plug
in t

4. In each of the following cases, find the rate of change of $f(t)$ with respect to t at the given value of t.

(a) $f(t) = t^3 - 4t^2 + 5t\sqrt{t} - 5$ at $t = 4$

(b) $f(t) = t^3(t^2 - 1)$ at $t = 0$

5. In each of the following cases, find the percentage rate of change of the function $f(t)$ with respect to t.

(a) $f(t) = t^2 - 3t + \sqrt{t}$

(b) $f(t) = t^2(3 - 2t)^3$

(c) $f(t) = \dfrac{1}{t + 1}$

6. After x weeks, the number of people using a new rapid transit system was approximately $N(x) = 6x^3 + 500x + 8,000$.

   (a) At what rate was the use of the system changing with respect to time after 8 weeks?

   (b) By how much did the use of the system change during the 8th week?

7. It is estimated that the weekly output at a certain plant is $Q(x) = 50x^2 + 9,000x$ units, where x is the number of workers employed at the plant. Currently there are 30 workers employed at the plant.

   (a) Use calculus to estimate the change in the weekly output that will result from the addition of 1 worker to the force.

   (b) Compute the actual change in output that will result from the addition of 1 worker.

8. It is projected that t months from now, the population of a certain town will be $P(t) = 3t + 5t^{3/2} + 6,000$. At what percentage rate will the population be changing with respect to time 4 months from now?

9. At a certain factory, the daily output is $Q(L) = 20,000L^{1/2}$ units, where L denotes the size of the labor force measured in worker-hours. Currently 900 worker-hours of labor are used each day. Use calculus to estimate the effect on output that will be produced if the labor force is cut to 885 worker-hours.

10. The gross national product of a certain country was $N(t) = t^2 + 6t + 300$ billion dollars t years after 1990. Use calculus to estimate the percentage change in the GNP during the second quarter of 1998.

11. The level of air pollution in a certain city is proportional to the square of the population. Use calculus to estimate the percentage by which the air pollution level will increase if the population increases by 5 percent.

12. The output at a certain factory is $Q(L) = 600L^{2/3}$ units, where L is the size of the labor force. The manufacturer wishes to increase output by 1 percent. Use calculus to estimate the percentage increase in labor that will be required.

13. Use the chain rule to find $\dfrac{dy}{dx}$.

   (a) $y = 5u^2 + u - 1; u = 3x + 1$

   (b) $y = \dfrac{1}{u^2}; u = 2x + 3$

14. Use the chain rule to find $\dfrac{dy}{dx}$ for the given value of x.

    (a) $y = u^3 - 4u^2 + 5u + 2$, $u = x^2 + 1$; $x = 1$

    (b) $y = \sqrt{u}$, $u = x^2 + 2x - 4$; $x = 2$

15. At a certain factory, approximately $q(t) = t^2 + 50t$ units are manufactured during the first t hours of a production run, and the total cost of manufacturing q units is $C(q) = 0.1q^2 + 10q + 400$ dollars. Find the rate at which the manufacturing cost is changing with respect to time 2 hours after production commences.

In Problems 16 and 17, $s(t)$ denotes the displacement of an object moving along a line.
(a) Find the velocity and acceleration of the object and describe its motion during the indicated time interval.
(b) Compute the total distance traveled by the object during the indicated time interval.

16. $s(t) = 2t^3 - 21t^2 + 60t - 25$; $1 \le t \le 6$.

17. $s(t) = \dfrac{2t + 1}{t^2 + 12}$; $0 \le t \le 3$

18. A car is traveling at 88 ft/sec when the driver applies the brakes to avoid hitting a child. After t seconds, the car is $s = 88t - 8t^2$ feet from the point where the brakes were applied. How long does it take for the car to come to a stop, and how far does it travel before stopping?

19. It is estimated that t years from now, the population of a certain suburban community will be $p(t) = 10 - \dfrac{20}{(t + 1)^2}$ thousand. An environmental study indicates that the average daily level of carbon monoxide in the air will be $c(p) = 0.8\sqrt{p^2 + p + 139}$ units when the population is p thousand. At what percentage rate will the level of carbon monoxide be changing with respect to time 1 year from now?

20. Find $\dfrac{dy}{dx}$ by implicit differentiation.

    (a) $5x + 3y = 12$ (b) $x^2y = 1$

    (c) $(2x + 3y)^5 = x + 1$ (d) $(1 - 2xy^3)^5 = x + 4y$

21. Use implicit differentiation to find the slope of the line that is tangent to the given curve for the specified value of x.

    (a) $xy^3 = 8$; $x = 1$

    (b) $x^2y - 2xy^3 + 6 = 2x + 2y$; $x = 0$

22. The output Q at a certain factory is related to inputs x and y by the equation $Q = x^3 + 2xy^2 + 2y^3$. If the current levels of input are $x = 10$ and $y = 20$, use calculus to estimate the change in input y that should be made to offset an increase of 0.5 in input x so that output will be maintained at its current level.

23. Find the second derivative of each of the following functions:

    (a) $f(x) = 6x^5 - 4x^3 + 5x^2 - 2x + \dfrac{1}{x}$

    (b) $y = (3x^2 + 2)^4$

    (c) $f(x) = \dfrac{x - 1}{(x + 1)^2}$

24. Use implicit differentiation to find $\dfrac{d^2y}{dx^2}$ if $3x^2 - 2y^2 = 6$.

25. An efficiency study of the morning shift at a certain factory indicates that an average worker who arrives on the job at 8:00 A.M. will have produced $Q(t) = -t^3 + 9t^2 + 12t$ units t hours later.
    (a) Compute the worker's rate of production at 9:00 A.M.
    (b) At what rate is the worker's rate of production changing with respect to time at 9:00 A.M.?
    (c) Use calculus to estimate the change in the worker's rate of production between 9:00 and 9:06 A.M.
    (d) Compute the actual change in the worker's rate of production between 9:00 and 9:06 A.M.

26. Find the 4th derivative of each of the following functions:

    (a) $y = 2x^5 + 5x^4 - 2x + \dfrac{1}{x}$

    (b) $f(x) = \sqrt{3x} + \dfrac{3}{2x^2}$

CHAPTER 3
ADDITIONAL APPLICATIONS OF THE DERIVATIVE

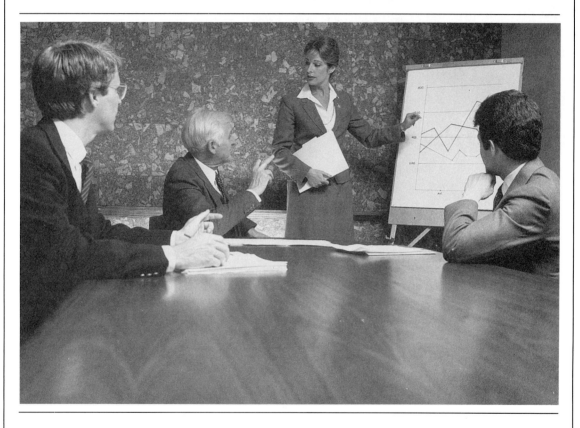

1   Relative Extrema; Curve Sketching with the First Derivative

In Example 1.4 of Chapter 2, Section 1, we used calculus to maximize a profit function like the one shown in Figure 3.1. In particular we observed that the maximum value corresponds to the unique point on the graph at which the slope of the tangent is zero, and we therefore set the derivative equal to zero and solved for x.

Figure 3.1 A profit function.

The simplicity of this example is misleading. In general, not every point at which the derivative of a function is zero is a peak of its graph. Two functions whose derivatives are zero when $x = 0$ are sketched in Figure 3.2. Both have horizontal tangents at $(0, 0)$, but the function $y = x^2$ in Figure 3.2a reaches its *lowest* point at $(0, 0)$, while the function $y = x^3$ in Figure 3.2b has neither a maximum nor a minimum at this point.

The situation is further complicated by the existence of functions with

Figure 3.2 Two functions with horizontal tangents when $x = 0$.

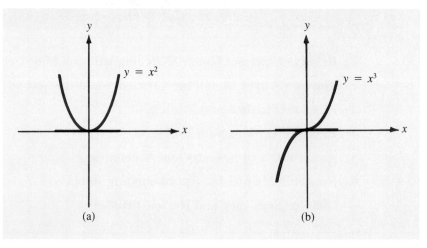

(a)   (b)

maxima or minima at points where the derivative is not even defined. Two such functions are sketched in Figure 3.3.

Figure 3.3 Two functions with extrema where the derivative is undefined.

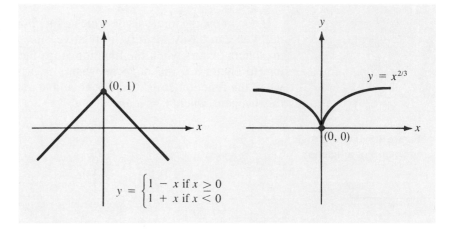

$$y = \begin{cases} 1 - x \text{ if } x \geq 0 \\ 1 + x \text{ if } x < 0 \end{cases}$$

In this section, you will learn a systematic procedure you can use to locate and identify maxima and minima of differentiable functions. In the process, you will also see how to use derivatives to help you sketch the graphs of functions.

Relative Maxima and Minima

A **relative maximum** of a function is a peak, a point on the graph of the function that is higher than any neighboring point on the graph. A **relative minimum** is the bottom of a valley, a point on the graph that is lower than any neighboring point. (The relative maxima and minima of a function are collectively called **relative extrema**.) The function sketched in Figure 3.4 has a relative maximum at $x = b$ and relative minima at $x = a$ and $x = c$. Notice that a relative maximum need not be the highest point on a graph. It is maximal only *relative to* neighboring points. Similarly, a relative minimum need not be the lowest point on the graph.

Figure 3.4 Relative maxima and minima.

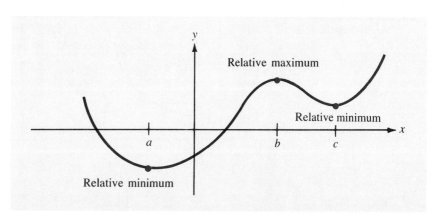

Increase and Decrease of Functions

A function is said to be **increasing** if its graph is rising as x increases and **decreasing** if its graph is falling as x increases. The function in Figure 3.5 is increasing for $a < x < b$ and for $x > c$. It is decreasing for $x < a$ and for $b < x < c$.

If you know the intervals on which a function is increasing and decreasing, you can easily identify its relative maxima and minima. A relative maximum occurs when the function stops increasing and starts decreasing. In Figure 3.5, this happens when $x = b$. A relative minimum occurs when the function stops decreasing and starts increasing. In Figure 3.5, this happens when $x = a$ and $x = c$.

Figure 3.5 Increase and decrease of a function.

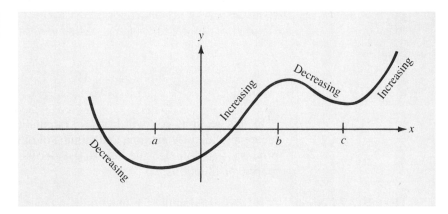

The Sign of the Derivative

You can find out where a differentiable function is increasing or decreasing by checking the sign of its (first) derivative. This is because the derivative is the slope of the tangent. When the derivative is positive, the slope of the tangent is positive and the function is increasing. When the derivative is negative, the slope of the tangent is negative and the function is decreasing. This situation is illustrated in Figure 3.6.

Figure 3.6 The geometric significance of the sign of the first derivative.

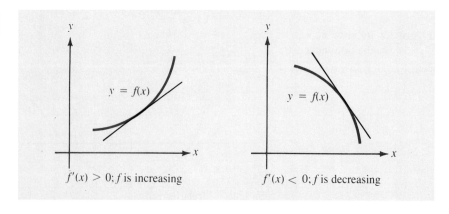

$f'(x) > 0; f$ is increasing $f'(x) < 0; f$ is decreasing

Here is a more precise statement of the situation.

> **Derivative Test for Intervals of Increase and Decrease**
>
> If $f'(x) > 0$ on the interval $a < x < b$, then f is increasing on this interval.
>
> If $f'(x) < 0$ on the interval $a < x < b$, then f is decreasing on this interval.

Critical Points   Since a function f is increasing when its derivative is positive and decreasing when its derivative is negative, the only points where the function can have a relative maximum or minimum are those where its derivative either is zero or is undefined. A number c in the domain of f for which $f'(c) = 0$ or $f'(c)$ is undefined is said to be a **critical value** of f. The corresponding point on the graph is called a **critical point** of f.

Three functions with critical points at which the derivative is zero are shown in Figure 3.7. In each case, the line that is tangent to the graph at the critical point is horizontal because the derivative (the slope of the tangent line) is zero.

Three functions with critical points at which the derivative is undefined are shown in Figure 3.8. In Figures 3.8b and 3.8c, the tangent line is

Figure 3.7 Three critical points where $f'(x) = 0$.

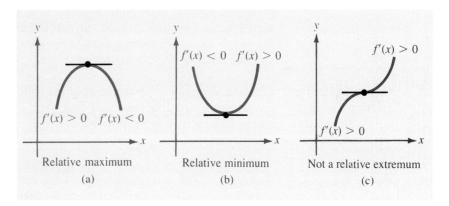

Figure 3.8 Three critical points where $f'(x)$ is undefined.

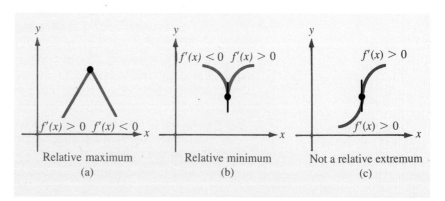

vertical and hence the derivative (or slope of the tangent) is undefined. In Figure 3.8a, the derivative is undefined at the critical point because the function does not even have a (uniquely defined) tangent line at this point.

Critical Points

A **critical point** of a function is a point on its graph where either

(1) the derivative is zero, or
(2) the derivative is undefined.

The relative maxima and minima of the function can occur only at critical points.

Classification of Critical Points

Every relative extremum is a critical point. However, as you saw in Figures 3.7c and 3.8c, not every critical point is necessarily a relative extremum. If the derivative is positive to the left of a critical point and negative to the right of it, the graph changes from increasing to decreasing and the critical point is a relative maximum (Figures 3.7a and 3.8a). If the derivative is negative to the left of a critical point and positive to the right of it, the graph changes from decreasing to increasing and the critical point is a relative minimum (Figures 3.7b and 3.8b). If the sign of the derivative is the same on both sides of the critical point, the direction of the graph does not change and the critical point is neither a relative maximum nor a relative minimum (Figures 3.7c and 3.8c).

Curve Sketching Using the First Derivative

The preceding observations suggest the following general procedure you can use to graph functions and find their intervals of increase and decrease and their relative extrema.

How to Use the First Derivative to Graph a Function

Step 1. Compute the derivative $f'(x)$ and put it in factored form if possible.

Step 2. Find the critical values of $f(x)$ (values for which $f'(x) = 0$ or $f'(x)$ is undefined). For each such value x in the domain of f, substitute x into the function $f(x)$ to get the y coordinate of the corresponding critical point.

Step 3. Plot the critical points on the graph. These are the only points where relative extrema can possibly occur.

Step 4. The critical values divide the x axis into a number of intervals. Determine whether $f(x)$ is increasing or decreasing on each such interval by checking the sign of the derivative $f'(x)$ in the interval.

> Step 5. Sketch the graph so that it increases on intervals where $f'(x) > 0$, decreases where $f'(x) < 0$, and has a horizontal tangent where $f'(x) = 0$.

Here are some examples that illustrate this graphing procedure.

Graphs of Polynomials

EXAMPLE 1.1

Find the intervals of increase and decrease and the relative extrema of the function $f(x) = 2x^3 + 3x^2 - 12x - 7$ and sketch the graph.

Solution

Begin by computing and factoring the derivative

$$f'(x) = 6x^2 + 6x - 12 = 6(x^2 + x - 2) = 6(x + 2)(x - 1)$$

From the factored form of the derivative you can see that $f'(x) = 0$ when $x = -2$ and when $x = 1$. Since $f(-2) = 13$ and $f(1) = -14$, the corresponding critical points are $(-2, 13)$ and $(1, -14)$. Begin the sketch (Figure 3.9a) by plotting these critical points. (To help you remember that the graph should have horizontal tangents at these points, you may wish to draw a short horizontal line segment through each critical point.)

To find the intervals of increase and decrease of the function, check the sign of the derivative for $x < -2$, for $-2 < x < 1$, and for $x > 1$.

On the interval $x < -2$ (at $x = -3$, for example), both factors $x + 2$ and $x - 1$ are negative. Hence, the derivative $f'(x) = 6(x + 2)(x - 1)$ is positive and the function is increasing on this interval.

On the interval $-2 < x < 1$ (at $x = 0$, for example), the factor $x + 2$ is positive while the factor $x - 1$ is still negative. Hence, the derivative is negative and the function is decreasing on this interval.

Finally, on the interval $x > 1$ (at $x = 2$, for example), both factors $x + 2$ and $x - 1$ are positive. Hence, the derivative is positive and the function is increasing on this interval.

These observations are summarized in the following table.

Interval	$(x + 2)$	$(x - 1)$	Sign of f'	Increase or decrease of f
$x < -2$	$-$	$-$	$+$	Increasing
$-2 < x < 1$	$+$	$-$	$-$	Decreasing
$x > 1$	$+$	$+$	$+$	Increasing

Complete the sketch in Figure 3.9b by drawing the curve so that it

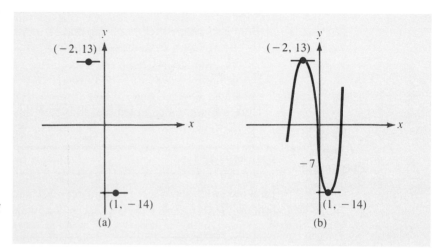

Figure 3.9 The graph of $y = 2x^3 + 3x^2 - 12x - 7$.

increases for $x < -2$, decreases for $-2 < x < 1$, increases again for $x > 1$, and levels off at the critical points. Notice that the function has a relative maximum at the critical point $(-2, 13)$ and a relative minimum at the critical point $(1, -14)$. Notice also that $f(0) = -7$, so the y intercept is $(0, -7)$.

Comment: It is important to realize that a relative extremum is a *local* property. That is, a relative maximum (minimum) is only the highest (lowest) of all *nearby* points, and points that are farther away may be higher or lower.

EXAMPLE 1.2

Find the intervals of increase and decrease and the relative extrema of the function $f(x) = x^4 + 8x^3 + 18x^2 - 8$ and sketch the graph.

Solution

The derivative is

$$f'(x) = 4x^3 + 24x^2 + 36x = 4x(x^2 + 6x + 9) = 4x(x + 3)^2$$

which is zero when $x = 0$ and $x = -3$. Since $f(0) = -8$ and $f(-3) = 19$, the corresponding critical points are $(0, -8)$ and $(-3, 19)$.

To find the intervals of increase and decrease of the function, check the sign of the derivative from the table at the top of p. 175.

Sketch the graph using this information, as shown in Figure 3.10. Notice that since the graph is decreasing on both sides of the critical point $(-3, 19)$, this point is neither a relative maximum nor a relative minimum. On the other hand, the critical point $(0, -8)$ is a relative minimum.

Interval	x	$(x + 3)^2$	Sign of f'	Increase or decrease of f
$x < -3$	–	+	–	Decreasing
$-3 < x < 0$	–	+	–	Decreasing
$x > 0$	+	+	+	Increasing

Figure 3.10 The graph of $y = x^4 + 8x^3 + 18x^2 - 8$.

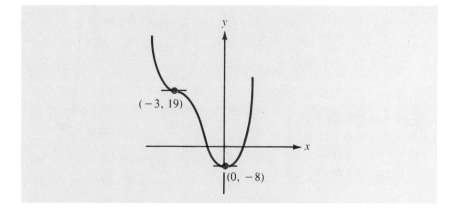

EXAMPLE 1.3

Find the intervals of increase and decrease and the relative extrema of the function $f(x) = 3x^4 - 4x^3 - 12x^2 + 17$ and sketch the graph.

Solution

The derivative is

$$f'(x) = 12x^3 - 12x^2 - 24x = 12x(x + 1)(x - 2)$$

which is zero when $x = -1$, $x = 0$, and $x = 2$. Since $f(-1) = 12, f(0) = 17$, and $f(2) = -15$, the corresponding critical points are $(-1, 12)$, $(0, 17)$, and $(2, -15)$. To find the intervals of increase and decrease of $f(x)$, check the sign of the derivative for the intervals $x < -1$, $-1 < x < 0$, $0 < x < 2$, and $x > 2$, as indicated in the table below.

Interval	x	$x + 1$	$x - 2$	Sign of f'	Increase or decrease of f
$x < -1$	–	–	–	–	Decreasing
$-1 < x < 0$	–	+	–	+	Increasing
$0 < x < 2$	+	+	–	–	Decreasing
$x > 2$	+	+	+	+	Increasing

Sketch the graph using this information, as shown in Figure 3.11.

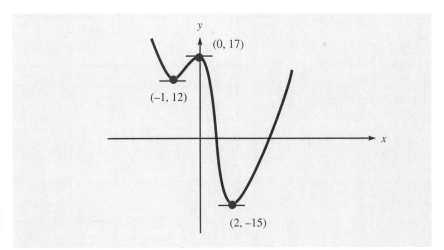

Figure 3.11 The graph of $f(x) = 3x^4 - 4x^3 - 12x^2 + 17.$

Notice that $(-1, 12)$ and $(2, -15)$ are relative minima and that $(0, 17)$ is a relative maximum.

In the next example, you will need the quadratic formula to find the critical values of the given function. You may want to use a calculator to help with the computations.

EXAMPLE 1.4

Find the intervals of increase and decrease and the relative extrema of the function $f(x) = x^3 + 2x^2 - x + 1$ and sketch the graph.

Solution

The derivative is $\qquad f'(x) = 3x^2 + 4x - 1$

which has no obvious factors. According to the quadratic formula, $3x^2 + 4x - 1 = 0$ when

$$x = \frac{-4 \pm \sqrt{4^2 - 4(3)(-1)}}{2(3)} = \frac{-4 \pm \sqrt{28}}{6}$$

Hence, the critical values are

$$x = \frac{-4 + \sqrt{28}}{6} \simeq 0.22 \qquad \text{and} \qquad x = \frac{-4 - \sqrt{28}}{6} \simeq -1.55$$

and the corresponding y coordinates are

$$y \simeq f(0.22) \simeq 0.89 \quad \text{and} \quad y \simeq f(-1.55) \simeq 3.63$$

For each of the intervals $x < -1.55$, $-1.55 < x < 0.22$, and $x > 0.22$, use any convenient value of x in the interval to check the sign of the derivative. The values $x = -2$, $x = 0$, and $x = 1$ would be reasonable choices.

Interval	Sign of f'	Increase or decrease of f
$x < -1.55$	$+$	Increasing
$-1.55 < x < 0.22$	$-$	Decreasing
$x > 0.22$	$+$	Increasing

The graph is sketched in Figure 3.12. Notice that the critical point $(-1.55, 3.63)$ is a relative maximum and that the critical point $(0.22, 0.89)$ is a relative minimum. Notice also that since $f(0) = 1$, the y intercept is $(0, 1)$.

Figure 3.12 The graph of $y = x^3 + 2x^2 - x + 1$.

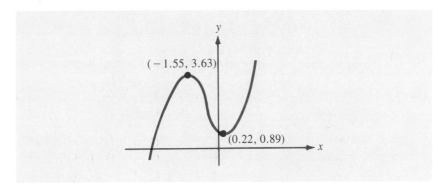

$(-1.55, 3.63)$

$(0.22, 0.89)$

Graphs of Rational Functions

EXAMPLE 1.5

Find the intervals of increase and decrease and the relative extrema of the rational function $f(x) = \dfrac{x^2}{x - 2}$ and sketch the graph.

Solution

Since the denominator of $f(x)$ is zero when $x = 2$, the function is undefined and has a discontinuity for this value of x. To remind yourself that there is no point on the graph whose x coordinate is 2, begin your sketch by drawing a broken vertical line at $x = 2$. Your graph should not cross this line.

By the quotient rule, the derivative is

$$f'(x) = \frac{(x - 2)(2x) - x^2(1)}{(x - 2)^2} = \frac{2x^2 - 4x - x^2}{(x - 2)^2} = \frac{x(x - 4)}{(x - 2)^2}$$

which is zero when its numerator is zero, that is, when $x = 0$ and $x = 4$.

Since $f(0) = 0$ and $f(4) = 8$, the corresponding critical points are $(0, 0)$ and $(4, 8)$. The derivative is also undefined at $x = 2$, but this is not a critical value since $x = 2$ is not in the domain of the function.

To identify the intervals of increase and decrease, check the sign of the derivative on the intervals determined by the critical values. Since $f'(x)$ is undefined at $x = 2$, its sign for $0 < x < 2$ could be different from its sign for $2 < x < 4$, and so you will have to check these two intervals separately.

Interval	x	$(x - 4)$	$(x - 2)^2$	Sign of f'	Increase or decrease of f
$x < 0$	−	−	+	+	Increasing
$0 < x < 2$	+	−	+	−	Decreasing
$2 < x < 4$	+	−	+	−	Decreasing
$x > 4$	+	+	+	+	Increasing

Now sketch the graph using this information as shown in Figure 3.13.

Figure 3.13 The graph of $y = \dfrac{x^2}{x - 2}$.

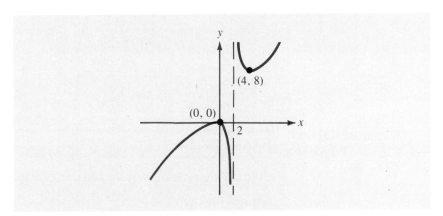

Don't forget that the graph has a discontinuity at $x = 2$. Notice that f has a relative maximum at the critical point $(0, 0)$ and a relative minimum at the critical point $(4, 8)$.

Comment: Notice that the relative minimum $(4, 8)$ in Figure 3.13 is actually higher than the relative maximum at $(0, 0)$. This further illustrates the fact that relative maxima and minima are *local* features.

Critical Values at Which $f'(x)$ Is Undefined

In the next example, the relative extremum occurs at a critical value at which the derivative is not defined.

EXAMPLE 1.6

Find the intervals of increase and decrease and the relative extrema of the function $f(x) = x^{2/3}$ and draw the graph.

Solution

The derivative is

$$f'(x) = \frac{2}{3}x^{-1/3} = \frac{2}{3\sqrt[3]{x}}$$

is never zero, but it is undefined at $x = 0$, which is in the domain of f. Thus, since $f(0) = 0$, the corresponding point $(0, 0)$ is the only critical point.

To find the intervals of increase and decrease, check the sign of the derivative for $x < 0$ and $x > 0$. Notice that $(0, 0)$ is a relative minimum.

Interval	Sign of f'	Increase or decrease of f
$x < 0$	$-$	Decreasing
$x > 0$	$+$	Increasing

The graph of f is sketched in Figure 3.14. Since the slope of the tangent line at $(0, 0)$ is undefined (infinite), the tangent line is vertical.

Figure 3.14 The graph of $y = x^{2/3}$.

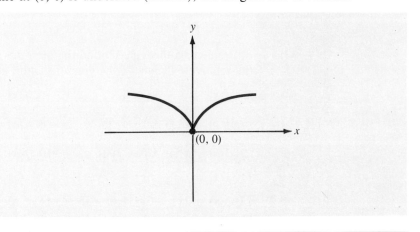

Problems 3.1

In Problems 1 through 4, specify the intervals on which the derivative of the given function is positive and those on which it is negative.

1.

2.

3.

4.

In Problems 5 through 32, find the intervals of increase and decrease and the relative extrema of the given function and sketch the graph.

5. $f(x) = x^2 - 4x + 5$ 6. $f(x) = x^3 + 3x^2 + 1$

7. $f(x) = x^3 - 3x - 4$ 8. $f(x) = \frac{1}{3}x^3 - 9x + 2$

9. $f(x) = x^5 - 5x^4 + 100$ 10. $f(x) = 3x^5 - 5x^3$

11. $f(x) = 3x^4 - 8x^3 + 6x^2 + 2$ 12. $f(x) = 324x - 72x^2 + 4x^3$

13. $f(x) = 2x^3 + 6x^2 + 6x + 5$ 14. $f(x) = 10x^6 + 24x^5 + 15x^4 + 3$

15. $f(x) = (x - 1)^5$ 16. $f(x) = 3 - (x + 1)^3$

17. $f(x) = (x^2 - 1)^5$ 18. $f(x) = (x^2 - 1)^4$

19. $f(x) = (x^3 - 1)^4$

20. $f(x) = \dfrac{x^2}{x - 1}$

21. $f(x) = \dfrac{x^2}{x + 2}$

22. $f(x) = \dfrac{x^2 - 3x}{x + 1}$

23. $f(x) = \dfrac{1}{x^2 - 9}$

24. $f(x) = x + \dfrac{1}{x}$

25. $f(x) = 2x + \dfrac{18}{x} + 1$

26. $f(x) = 6x^2 + \dfrac{12{,}000}{x}$

27. $f(x) = 1 + x^{1/3}$

28. $f(x) = x^{3/5}$

29. $f(x) = 2 + (x - 1)^{2/3}$

30. $f(x) = x^3 - 2x^2 - 3x + 2$

31. $f(x) = 3x^5 - 5x^3 + 4$

32. $f(x) = \dfrac{1}{50}(3x^4 - 8x^3 - 90x^2 + 70)$

33. Sketch a graph of a function that has all of the following properties:
    (a) $f'(x) > 0$ when $x < -5$ and when $x > 1$
    (b) $f'(x) < 0$ when $-5 < x < 1$
    (c) $f(-5) = 4$ and $f(1) = -1$

34. Sketch a graph of a function that has all of the following properties:
    (a) $f'(x) < 0$ when $x < -1$
    (b) $f'(x) > 0$ when $-1 < x < 3$ and when $x > 3$
    (c) $f'(-1) = 0$ and $f'(3) = 0$

35. Sketch a graph of a function that has all of the following properties:
    (a) $f'(x) > 0$ when $x > 2$
    (b) $f'(x) < 0$ when $x < 0$ and when $0 < x < 2$
    (c) $x = 0$ is not in the domain of f

36. Sketch a graph of a function that has all of the following properties:
    (a) $f'(x) > 0$ when $-1 < x < 3$ and when $x > 6$
    (b) $f'(x) < 0$ when $x < -1$ and when $3 < x < 6$
    (c) $f'(-1) = 0$ and $f'(6) = 0$
    (d) $f'(x)$ is undefined when $x = 3$

37. Find constants a, b, and c so that the graph of the function $f(x) = ax^2 + bx + c$ has a relative maximum at $(5, 12)$ and crosses the y axis at $(0, 3)$.

38. Find constants a, b, c, and d so that the graph of the function $f(x) = ax^3 + bx^2 + cx + d$ has a relative maximum at $(-2, 8)$ and a relative minimum at $(1, -19)$.

39. Use calculus to prove that the relative extremum of the quadratic function $y = ax^2 + bx + c$ occurs when $x = -\dfrac{b}{2a}$.

40. Use calculus to prove that the relative extremum of the quadratic function $y = (x - p)(x - q)$ occurs midway between its x intercepts.

2   More on Curve Sketching; Concavity and the Second Derivative Test

In the preceding section, you saw that the sign of the first derivative of a function tells you where its graph is increasing and where it is decreasing. In this section, you will see that the sign of the second derivative also gives useful information about the shape of the graph. By way of introduction, here is a brief description of a situation from industry that can be analyzed with the aid of the second derivative.

The Efficiency of a Worker   The number of units that a factory worker can produce in t hours is often given by a function $Q(t)$ like the one whose graph is shown in Figure 3.15.

Figure 3.15 The output of a factory worker.

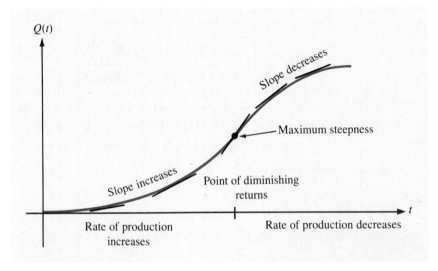

Notice that at first the graph is not very steep. The steepness increases, however, until the graph reaches a point of maximum steepness, after which the steepness begins to decrease. This reflects the fact that at first the worker's rate of production is low. The rate of production increases, however, as the worker settles into a routine and continues to increase until the worker is performing at maximum efficiency, after which fatigue sets in and the rate of production begins to decrease. The moment of maximum efficiency is known in economics as the **point of diminishing returns.**

The behavior of the graph of this production function on either side of the point of diminishing returns can be described in terms of its tangent lines. To the left of this point, the slope of the tangent increases as t increases. To the right of this point, the slope of the tangent decreases as t increases. It is this increase and decrease of slopes that we shall examine in this section with the aid of the second derivative. (We shall return to the questions of worker efficiency and diminishing returns in Section 4 in the discussion of optimization.)

Concavity   The following notions of **concavity** are used to describe the increase and decrease of the slope of the tangent to a curve.

Concavity

A curve is said to be **concave upward** if the slope of its tangent increases as it moves along the curve from left to right.

A curve is said to be **concave downward** if the slope of its tangent decreases as it moves along the curve from left to right.

In Figure 3.15, for example, the production curve was concave upward to the left of the point of diminishing returns and concave downward to the right of this point. Concavity is illustrated further in Figure 3.16, in which the curve is concave upward to the left of $x = a$ and concave downward to the right. Note that the graph can be described as "spilling water" where it is concave downward and as "holding water" where it is concave upward.

Figure 3.16 Concavity and the slope of the tangent.

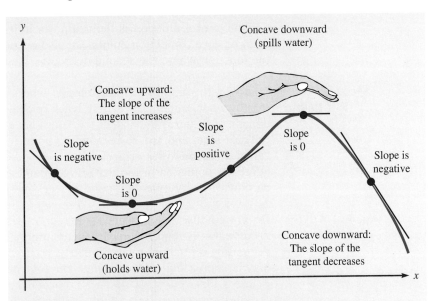

The Sign of the Second Derivative

There is a simple characterization of concavity in terms of the sign of the second derivative. It is based on the fact (established in the preceding section) that a quantity increases when its derivative is positive and decreases when its derivative is negative. The second derivative comes into the picture when this fact is applied to the first derivative (or slope of the tangent). Here is the argument.

Suppose the second derivative f'' is positive on an interval. This implies that the first derivative f' must be increasing on the interval. But f' is the slope of the tangent. Hence, the slope of the tangent is increasing, and so the graph of f is concave upward on the interval. On the other hand, if f'' is negative on an interval, then f' is decreasing. This implies that the slope of the tangent is decreasing, and so the graph of f is concave downward on the interval. To summarize:

> **Test for Concavity**
>
> If $f''(x) > 0$ on the interval $a < x < b$, then f is concave upward on this interval.
>
> If $f''(x) < 0$ on the interval $a < x < b$, then f is concave downward on this interval.

Inflection Points

A point on the graph of a function at which the concavity of the function changes is called an **inflection point.** The production function in Figure 3.15 has an inflection point at the point of diminishing returns. The function in Figure 3.16 has an inflection point when $x = a$. If the second derivative of a function is defined at an inflection point, its value there must be zero. Inflection points can also occur at points in the domain of the function where the second derivative is undefined.

Second-Order Critical Points

Points on the graph of a function $f(x)$ where $f'(x) = 0$ or $f'(x)$ does not exist are sometimes called **first-order critical points.** A number c in the domain of f where $f''(c) = 0$ or $f''(c)$ does not exist is called a **second-order critical value**, and the corresponding point $(c, f(c))$ on the graph of f is called a **second-order critical point.** Second-order critical points are to inflection points as first-order critical points are to relative extrema. In particular, every inflection point is a second-order critical point.

A Word of Advice

Do not confuse the concavity of a curve with its increase or decrease. A curve that is concave upward on an interval may be either increasing or decreasing on that interval. Similarly, a curve that is concave downward may be increasing or decreasing. The four possibilities are illustrated in Figure 3.17.

Figure 3.17 Possible combinations of increase and decrease, and concavity.

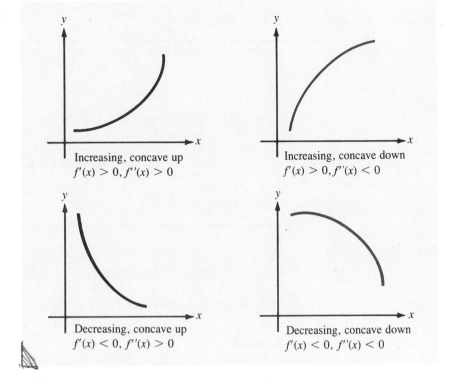

Increasing, concave up
$f'(x) > 0, f''(x) > 0$

Increasing, concave down
$f'(x) > 0, f''(x) < 0$

Decreasing, concave up
$f'(x) < 0, f''(x) > 0$

Decreasing, concave down
$f'(x) < 0, f''(x) < 0$

Curve Sketching Using the First and Second Derivatives

The preceding observations can be combined with the techniques developed in Section 1 of this chapter to get the following procedure for obtaining detailed graphs of functions using calculus.

How to Use Calculus to Graph a Function

Step 1. Compute the first derivative $f'(x)$, factor it if possible, find the first-order critical points, and plot them on the graph.

Step 2. Compute the second derivative $f''(x)$, factor it if possible, find the second-order critical points, and plot them on the graph.

Step 3. Use the x coordinates of the first- and second-order critical points (and of any discontinuities) to divide the x axis into a collection of intervals. Check the signs of the first and second derivatives on each of these intervals.

Step 4. Draw the graph on each interval according to the following table:

Sign of f'	Sign of f''	Increase or decrease of f	Concavity of f	Shape of f
+	+	Increasing	C-up	
+	−	Increasing	C-down	
−	+	Decreasing	C-up	
−	−	Decreasing	C-down	

Here are two examples.

EXAMPLE 2.1

Determine where the function $f(x) = x^4 + 8x^3 + 18x^2 - 8$ is increasing, decreasing, concave upward, and concave downward. Find the relative extrema and inflection points and draw the graph.

Solution

The first derivative is

$$f'(x) = 4x^3 + 24x^2 + 36x = 4x(x^2 + 6x + 9) = 4x(x + 3)^2$$

which is zero when $x = 0$ and $x = -3$. Since $f(0) = -8$ and $f(-3) = 19$, the first-order critical points are $(0, -8)$ and $(-3, 19)$.

The second derivative (computed from the unfactored form of the first derivative) is

$$f''(x) = 12x^2 + 48x + 36 = 12(x^2 + 4x + 3) = 12(x + 3)(x + 1)$$

which is 0 when $x = -3$ and $x = -1$. Since $f(-3) = 19$ and $f(-1) = 3$, the corresponding second-order critical points are $(-3, 19)$ and $(-1, 3)$.

Plot these critical points and check the signs of $f'(x)$ and $f''(x)$ on each of the intervals defined by their x coordinates.

Interval	$f'(x)$	$f''(x)$	Increase or decrease of f	Concavity of f	Shape of graph
$x < -3$	−	+	Decreasing	C-up	
$-3 < x < -1$	−	−	Decreasing	C-down	
$-1 < x < 0$	−	+	Decreasing	C-up	
$x > 0$	+	+	Increasing	C-up	

Figure 3.18 The graph of $y = x^4 + 8x^3 + 18x^2 - 8$.

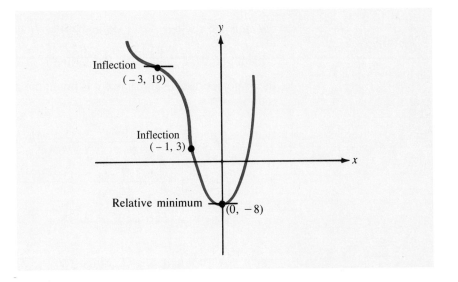

Draw the graph as shown in Figure 3.18 by connecting the critical points with a curve of appropriate shape on each interval.

Notice that the first-order critical point $(0, -8)$ is a relative minimum, while the first-order critical point $(-3, 19)$ is neither a relative minimum nor a relative maximum, and that both of the second-order critical points $(-3, 19)$ and $(-1, 3)$ are inflection points.

EXAMPLE 2.2

Determine where the function $f(x) = \dfrac{x}{(x + 1)^2}$ is increasing, decreasing, concave upward, and concave downward. Find the relative extrema and inflection points and draw the graph.

Solution

Since the denominator of $f(x)$ is zero when $x = -1$, the function is undefined and has a discontinuity for this value of x. To remind yourself of this, begin your sketch by drawing a broken vertical line at $x = -1$. Your graph should not cross this line.

By the quotient rule, the first derivative is

$$f'(x) = \frac{(x + 1)^2(1) - x[2(x + 1)(1)]}{(x + 1)^4}$$

$$= \frac{(x + 1)[(x + 1) - 2x]}{(x + 1)^4} = \frac{1 - x}{(x + 1)^3}$$

which is zero when $x = 1$. Since $f(1) = \frac{1}{4}$, the corresponding first-order critical point is $\left(1, \frac{1}{4}\right)$. (The derivative is also undefined at $x = -1$, but this is not a critical value since it is not in the domain of f.) By the quotient rule again.

$$f''(x) = \frac{(x + 1)^3(-1) - (1 - x)[3(x + 1)^2(1)]}{(x + 1)^6}$$

$$= \frac{(x + 1)^2[-(x + 1) - 3(1 - x)]}{(x + 1)^6}$$

$$= \frac{-x - 1 - 3 + 3x}{(x + 1)^4} = \frac{2x - 4}{(x + 1)^4} = \frac{2(x - 2)}{(x + 1)^4}$$

which is zero when $x = 2$. Since $f(2) = \frac{2}{9}$, the corresponding second-order critical point is $\left(2, \frac{2}{9}\right)$.

Plot these critical points and check the signs of $f'(x)$ and $f''(x)$ on each of the intervals defined by the x coordinates of the critical points and the discontinuity.

Interval	$f'(x)$	$f''(x)$	Increase or decrease of f	Concavity of f	Shape of graph
$x < -1$	$-$	$-$	Decreasing	C-down	⌐
$-1 < x < 1$	$+$	$-$	Increasing	C-down	⌐
$1 < x < 2$	$-$	$-$	Decreasing	C-down	⌐
$x > 2$	$-$	$+$	Decreasing	C-up	⌐

Next, pass a smooth curve through the points on the left of the vertical line $x = -1$ and another smooth curve through the points on the right of this line. Notice that $\frac{x}{(x + 1)^2}$ approaches zero as x increases without bound. The graph is shown in Figure 3.19.

Notice that the first-order critical point $\left(1, \frac{1}{4}\right)$ is a relative maximum and that the second-order critical point $\left(2, \frac{2}{9}\right)$ is an inflection point.

Notice also that the y intercept is $(0, f(0)) = (0, 0)$. The x intercepts of a

Figure 3.19 The graph of
$$y = \frac{x}{(x + 1)^2}.$$

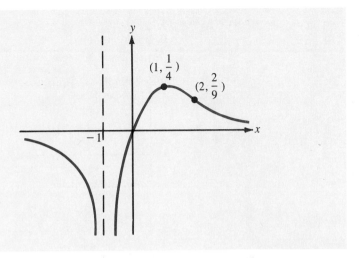

rational function occur when the numerator is zero, which happens in this case only when $x = 0$. Hence, $(0, 0)$ is the only x intercept and so the graph crosses the x axis only at this point.

The Second Derivative Test

The second derivative can be used to classify the first-order critical points of a function as relative maxima or relative minima. Here is a statement of the procedure, which is known as the **second derivative test.**

The Second Derivative Test

Suppose $f'(a) = 0$.

If $f''(a) > 0$, then f has a relative minimum at $x = a$.
If $f''(a) < 0$, then f has a relative maximum at $x = a$.

However, if $f''(a) = 0$, the test is inconclusive and f may have a relative maximum, a relative minimum, or no relative extremum at all at $x = a$.

To see why the second derivative test works, look at Figure 3.20, which shows the four possibilities that can occur when $f'(a) = 0$.

Figure 3.20a suggests that at a relative maximum, f must be concave downward and so $f''(a) \leq 0$. Figure 3.20b suggests that at a relative minimum, f must be concave upward and so $f''(a) \geq 0$. On the other hand, Figure 3.20c and 3.20d suggests that if a point at which $f'(a) = 0$ is not a relative extremum, it must be an inflection point and so $f''(a)$, if it is defined, must be zero. It follows that if $f'(a) = 0$ and $f''(a) < 0$, the

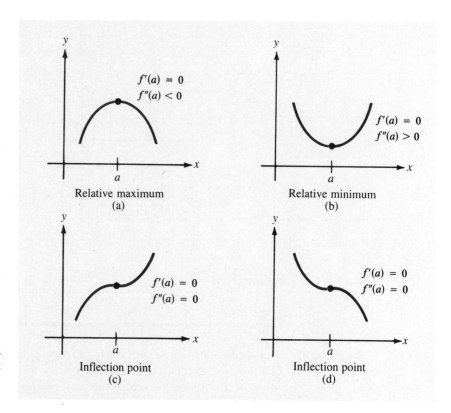

Figure 3.20 Behavior of a
graph when the first
derivative is zero.

corresponding critical point must be a relative maximum, while if $f'(a) = 0$ and $f''(a) > 0$, the corresponding critical point must be a relative minimum.

The use of the second derivative test is illustrated in the following example.

EXAMPLE 2.3

Use the second derivative test to find the relative maxima and minima of the function $f(x) = 2x^3 + 3x^2 - 12x - 7$.

Solution

Since the first derivative

$$f'(x) = 6x^2 + 6x - 12 = 6(x + 2)(x - 1)$$

is zero when $x = -2$ and $x = 1$, the corresponding points $(-2, 13)$ and $(1, -14)$ are the first-order critical points of f. To test these points, compute the second derivative

$$f''(x) = 12x + 6$$

and evaluate it at $x = -2$ and $x = 1$. Since

$$f''(-2) = -18 < 0$$

it follows that the critical point $(-2, 13)$ is a relative maximum, and since

$$f''(1) = 18 > 0$$

it follows that the critical point $(1, -14)$ is a relative minimum. For reference, the graph of f is sketched in Figure 3.21.

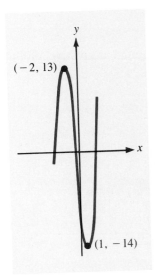

Figure 3.21 The graph of $y = 2x^3 + 3x^2 - 12x - 7$.

Comment: The function in Example 2.3 is the same one you analyzed in Example 1.1 of the preceding section using only the first derivative. Notice the relative ease with which you can now identify the extrema. Using the second derivative test, you check the sign of $f''(x)$ only at the critical point itself. Using the first derivative, you had to check the sign of $f'(x)$ over entire intervals on either side of the critical point.

There are, however, some limitations of the second derivative test. For some functions, the work involved in computing the second derivative is time-consuming and may diminish the appeal of the test. Moreover, the test applies only to critical points at which the derivative is zero and not to those at which the derivative is undefined. Finally, if both $f'(a)$ and $f''(a)$ are zero, the second derivative test tells you nothing whatsoever about the nature of the critical point. This is illustrated in Figure 3.22, which

Figure 3.22 Three functions whose first and second derivatives are zero at $x = 0$.

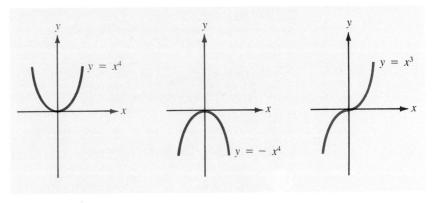

shows the graphs of three functions whose first and second derivatives are both zero when $x = 0$. When it is impractical or impossible to apply the second derivative test, you can always return to the methods of the preceding section and use the first derivative to check the increase or decrease of the function on either side of the critical point.

Problems 3.2

In Problems 1 and 2, determine where the second derivative of the function is positive and where it is negative.

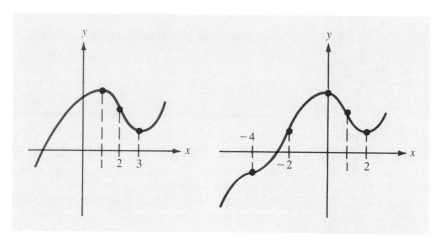

In Problems 3 through 26, determine where the given function is increasing, decreasing, concave upward, and concave downward. Find the relative extrema and inflection points and draw the graph.

3. $f(x) = \frac{1}{3}x^3 - 9x + 2$

4. $f(x) = x^3 + 3x^2 + 1$

5. $f(x) = x^4 - 4x^3 + 10$

6. $f(x) = x^3 - 3x^2 + 3x + 1$

7. $f(x) = (x - 2)^3$

8. $f(x) = x^5 - 5x$

9. $f(x) = (x^2 - 5)^3$

10. $f(x) = (x - 2)^4$

11. $f(x) = x + \frac{1}{x}$

12. $f(x) = (x^2 - 3)^2$

13. $f(x) = \frac{x^2}{x - 3}$

14. $f(x) = 1 + 2x + \frac{18}{x}$

15. $f(x) = (x + 1)^{1/3}$

16. $f(x) = \frac{x^2 - 3x}{x + 1}$

17. $f(x) = (x + 1)^{4/3}$

18. $f(x) = (x + 1)^{2/3}$

19. $f(x) = \sqrt{x^2 + 1}$

20. $f(x) = (x + 1)^{5/3}$

21. $f(x) = 2x(x + 4)^3$

22. $f(x) = \left(\frac{x}{x + 1}\right)^2$

23. $f(x) = \frac{2}{1 + x^2}$

24. $f(x) = \frac{x + 1}{(x - 1)^2}$

25. $f(x) = \dfrac{(x - 2)^3}{x^2}$

26. $f(x) = \dfrac{(x + 3)^3}{(x - 1)^2}$

27. Sketch the graph of a function that has all of the following properties:
  (a) $f'(x) > 0$ when $x < -1$ and when $x > 3$
  (b) $f'(x) < 0$ when $-1 < x < 3$
  (c) $f''(x) < 0$ when $x < 2$
  (d) $f''(x) > 0$ when $x > 2$

28. Sketch the graph of a function that has all of the following properties:
  (a) $f'(x) > 0$ when $x < 2$ and when $2 < x < 5$
  (b) $f'(x) < 0$ when $x > 5$
  (c) $f'(2) = 0$
  (d) $f''(x) < 0$ when $x < 2$ and when $4 < x < 7$
  (e) $f''(x) > 0$ when $2 < x < 4$ and when $x > 7$

29. Sketch the graph of a function that has all of the following properties:
  (a) $f'(x) > 0$ when $x < 1$
  (b) $f'(x) < 0$ when $x > 1$
  (c) $f''(x) > 0$ when $x < 1$ and when $x > 1$
  What can you say about the first derivative of f when $x = 1$?

30. Sketch the graph of a function f that has all of the following properties:
  (a) $f'(x) > 0$ when $-1 < x < 0$ and when $2 < x < 4$
  (b) $f'(x) < 0$ when $x < -1$, when $0 < x < 2$, and when $x > 4$
  (c) $f''(x) < 0$ when $0.5 < x < 1.5$ and when $x > 3$
  (d) $f''(x) > 0$ when $x < 0.5$ and when $1.5 < x < 3$

31. Sketch the graph of a function f that has all of the following properties:
  (a) The graph has a discontinuity at $x = 2$.
  (b) $f'(x) > 0$ for $x > 0$, $x \neq 2$.
  (c) $f'(x) < 0$ for $x < 0$.
  (d) $f''(x) > 0$ for $x < 2$ and $f''(x) < 0$ for $x > 2$.

32. Sketch the graph of a function f that has all of the following properties:
  (a) The graph has discontinuities at $x = -1$ and at $x = 3$.
  (b) $f'(x) > 0$ for $x < 1$, $x \neq -1$.
  (c) $f'(x) < 0$ for $x > 1$, $x \neq 3$.
  (d) $f''(x) > 0$ for $x < -1$ and $x > 3$ and $f''(x) < 0$ for $-1 < x < 3$.
  (e) $f(0) = 0 = f(2)$, $f(1) = 3$.

33. The first derivative of a certain function is $f'(x) = x^2 - 4x$.
  (a) On what intervals is f increasing? Decreasing?
  (b) On what intervals is f concave upward? Concave downward?
  (c) Find the x coordinates of the relative extrema and inflection points of f.

34. The first derivative of a certain function is $f'(x) = x^2 - 2x - 8$.
    (a) On what intervals is f increasing? Decreasing?
    (b) On what intervals is f concave upward? Concave downward?
    (c) Find the x coordinates of the relative extrema and inflection points of f.

In Problems 35 through 38, the *first derivative* of a certain function is sketched. In each case, determine where the *function* itself is increasing, decreasing, concave upward, and concave downward and find the x coordinates of its relative extrema and inflection points.

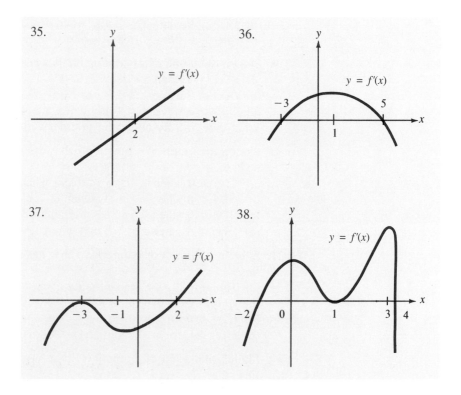

39. Use calculus to show that the graph of the quadratic function $y = ax^2 + bx + c$ is concave upward if a is positive and concave downward if a is negative.

In Problems 40 through 45, use the second derivative test to find the relative maxima and minima of the given function.

40. $f(x) = x^3 + 3x^2 + 1$ 41. $f(x) = x^4 - 2x^2 + 3$

42. $f(x) = (x^2 - 9)^2$

43. $f(x) = x + \dfrac{1}{x}$

44. $f(x) = 2x + 1 + \dfrac{18}{x}$

45. $f(x) = \dfrac{x^2}{x - 2}$

3 Absolute Maxima and Minima

In most practical optimization problems, the goal is to find the **absolute maximum** or **absolute minimum** of a particular function on some relevant interval. The absolute maximum of a function on an interval is the largest value of the function on the interval. The absolute minimum is the smallest value. Absolute extrema often coincide with relative extrema, but not always. For example, in Figure 3.23, the absolute maximum and relative maximum on the interval $a \leq x \leq b$ are the same. The absolute minimum, on the other hand, occurs at the endpoint $x = a$, which is not a relative minimum.

Figure 3.23 Absolute extrema.

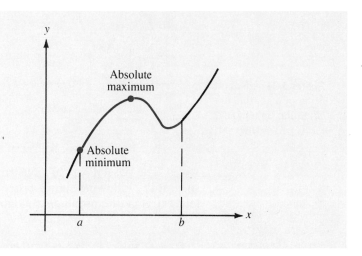

In this section, you will learn how to find absolute extrema of functions on intervals. Particular attention will be given to two special cases, which include virtually all the practical optimization problems you are likely to encounter. In Sections 4 and 5, these optimization techniques will be applied to a variety of practical situations.

Absolute Extrema on Closed Intervals

A **closed interval** is an interval of the form $a \leq x \leq b$, that is, an interval that contains both of its endpoints. It can be shown that a function that is continuous on a closed interval attains an absolute maximum and an absolute minimum on that interval. An absolute extremum can occur either at a relative extremum in the interval or at an endpoint $x = a$ or

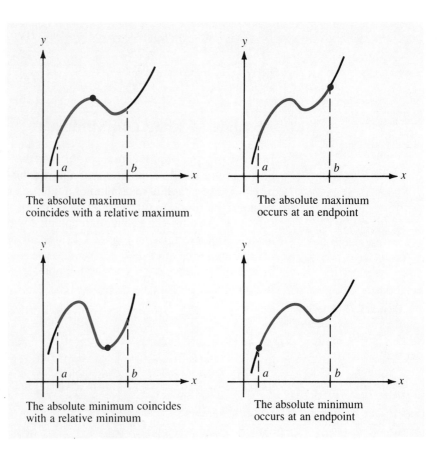

The absolute maximum
coincides with a relative maximum

The absolute maximum
occurs at an endpoint

The absolute minimum coincides
with a relative minimum

The absolute minimum
occurs at an endpoint

Figure 3.24 Absolute extrema of a continuous function on a closed interval.

$x = b$. The possibilities are illustrated in Figure 3.24. These observations suggest the following simple procedure for locating and identifying absolute extrema of continuous functions on closed intervals.

How to Find the Absolute Extrema of a Continuous Function f on a Closed Interval $a \leq x \leq b$

Step 1. Find the x coordinates of all the first-order critical points of f in the interval $a \leq x \leq b$.

Step 2. Compute $f(x)$ at these critical values and at the endpoints $x = a$ and $x = b$.

Step 3. Select the largest and smallest values of $f(x)$ obtained in Step 2. These are the absolute maximum and absolute minimum, respectively.

The procedure is illustrated in the following example.

Figure 3.25 The absolute extrema of $y = 2x^3 + 3x^2 - 12x - 7$ on $-3 \le x \le 0$.

EXAMPLE 3.1

Find the absolute maximum and absolute minimum of the function $f(x) = 2x^3 + 3x^2 - 12x - 7$ on the interval $-3 \le x \le 0$.

Solution

From the derivative

$$f'(x) = 6x^2 + 6x - 12 = 6(x + 2)(x - 1)$$

you see that the first-order critical points occur when $x = -2$ and $x = 1$. Of these, only $x = -2$ lies in the interval $-3 \le x \le 0$. Compute $f(x)$ at $x = -2$ and at the endpoints $x = -3$ and $x = 0$.

$$f(-2) = 13 \quad f(-3) = 2 \quad f(0) = -7$$

Compare these values to conclude that the absolute maximum of f on the interval $-3 \le x \le 0$ is $f(-2) = 13$ and the absolute minimum is $f(0) = -7$.

Notice that you did not have to classify the critical points or draw the graph to locate the absolute extrema. The sketch in Figure 3.25 is presented only for the sake of illustration.

Applications Here are two applications of this technique.

EXAMPLE 3.2

For several weeks, the highway department has been recording the speed of freeway traffic flowing past a certain downtown exit. The data suggest that between 1:00 and 6:00 P.M. on a normal weekday, the speed of the traffic at the exit is approximately $S(t) = t^3 - 10.5t^2 + 30t + 20$ miles per hour, where t is the number of hours past noon. At what time between 1:00 and 6:00 P.M. is the traffic moving the fastest, and at what time is it moving the slowest?

Solution

The goal is to find the absolute maximum and absolute minimum of the function $S(t)$ on the interval $1 \le t \le 6$. From the derivative

$$S'(t) = 3t^2 - 21t + 30 = 3(t^2 - 7t + 10) = 3(t - 2)(t - 5)$$

you get the t coordinates $t = 2$ and $t = 5$ of the first-order critical points, both of which lie in the interval $1 \leq t \leq 6$.

Now compute $S(t)$ for these values of t and at the endpoints $t = 1$ and $t = 6$ to get

$$S(1) = 40.5 \qquad S(2) = 46 \qquad S(5) = 32.5 \qquad S(6) = 38$$

Since the largest of these values is $S(2) = 46$ and the smallest is $S(5) = 32.5$, you can conclude that the traffic is moving fastest at 2:00 P.M., when its speed is 46 miles per hour, and slowest at 5:00 P.M., when its speed is 32.5 miles per hour.

Figure 3.26 Traffic speed: $S(t) = t^3 - 10.5t^2 + 30t + 20$.

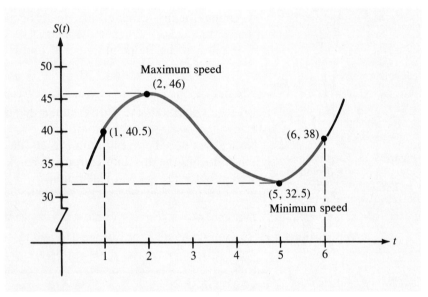

For reference, the graph of S is sketched in Figure 3.26.

The next example comes from biology.

EXAMPLE 3.3

When you cough, the radius of your trachea (windpipe) decreases, affecting the speed of the air in the trachea. If r_0 is the normal radius of the trachea, the relationship between the speed S of the air and the radius r of the trachea during a cough is given by a function of the form $S(r) = ar^2(r_0 - r)$, where a is a positive constant. Find the radius r for which the speed of the air is greatest.

Solution

The radius r of the contracted trachea cannot be greater than the normal radius r_0 or less than zero. Hence, the goal is to find the absolute maximum of $S(r)$ on the interval $0 \leq r \leq r_0$.

First differentiate $S(r)$ with respect to r using the product rule and factor the derivative as follows. (Note that a and r_0 are constants.)

$$S'(r) = -ar^2 + (r_0 - r)(2ar) = ar[-r + 2(r_0 - r)] = ar(2r_0 - 3r)$$

Then set the factored derivative equal to zero and solve to get the r coordinates of the first-order critical points.

$$ar(2r_0 - 3r) = 0$$

$$r = 0 \quad \text{or} \quad r = \frac{2}{3}r_0$$

Both of these values of r lie in the interval $0 \leq r \leq r_0$, and one is actually an endpoint of the interval. Compute $S(r)$ for these two values of r and at the other endpoint $r = r_0$ to get

$$S(0) = 0 \qquad S\left(\frac{2}{3}r_0\right) = \frac{4a}{27}r_0^3 \qquad S(r_0) = 0$$

Compare these values and conclude that the speed of the air is greatest when the radius of the contracted trachea is $\frac{2}{3}r_0$, that is, when it is two-thirds the radius of the uncontracted trachea.

A graph of the function $S(r)$ is sketched in Figure 3.27. Notice that the r

Figure 3.27 The speed of air during a cough: $S(r) = ar^2(r_0 - r)$.

intercepts of the graph are obvious from the factored function $S(r) =$

$ar^2(r_0 - r)$. Notice also that the graph has a horizontal tangent when $r = 0$, reflecting the fact that $S'(0) = 0$.

Absolute Extrema on Intervals That Are Not Closed

When the interval on which you wish to maximize or minimize a continuous function is not of the form $a \leq x \leq b$, the procedure illustrated in the preceding examples no longer applies. This is because there is no longer any guarantee that the function actually has an absolute maximum or minimum on the interval in question. On the other hand, if an absolute extremum does exist and the function is continuous on the interval, the absolute extremum will still occur at a relative extremum or endpoint contained in the interval. Two of the possibilities for functions on unbounded intervals are illustrated in Figure 3.28.

Figure 3.28 Extrema of functions on unbounded intervals.

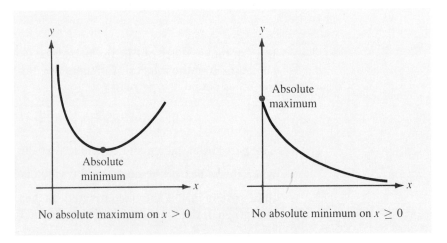

No absolute maximum on $x > 0$ No absolute minimum on $x \geq 0$

To find the absolute extrema of a continuous function on an interval that is not of the form $a \leq x \leq b$, you still evaluate the function at all the critical points and endpoints that are contained in the interval. However, before you can draw any final conclusions, you must find out if the function actually has relative extrema on the interval. One way to do this is to use the first derivative to determine where the function is increasing and where it is decreasing and sketch the graph. The technique is illustrated in the next example.

EXAMPLE 3.4

Find the absolute maximum and absolute minimum (if any) of the function $f(x) = x^2 + \dfrac{16}{x}$ on the interval $x > 0$.

Solution

The function is continuous on the interval $x > 0$ since its only discontinuity occurs at $x = 0$. The derivative is

$$f'(x) = 2x - \frac{16}{x^2} = \frac{2x^3 - 16}{x^2} = \frac{2(x^3 - 8)}{x^2}$$

which is zero when

$$x^3 - 8 = 0 \qquad x^3 = 8 \qquad \text{or} \qquad x = 2$$

Since the interval has no right-hand endpoint and does not contain its left-hand endpoint at $x = 0$, the only possible absolute extremum is

$$f(2) = (2)^2 + \frac{16}{2} = 12$$

at the first-order critical value.

Since $f'(x) < 0$ for $0 < x < 2$ and $f'(x) > 0$ for $x > 2$, the graph of f is decreasing for $0 < x < 2$ and increasing for $x > 2$, as shown in Figure 3.29.

Figure 3.29 The function $y = x^2 + \dfrac{16}{x}$ on the interval $x > 0$.

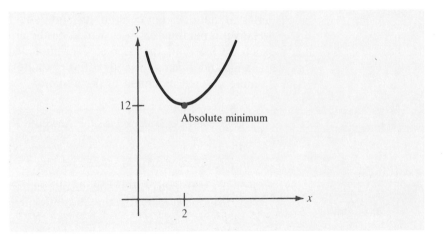

Absolute minimum

It follows that $f(2) = 12$ is the absolute minimum of f on the interval $x > 0$ and that there is no absolute maximum.

Absolute Extrema on Intervals Containing Only One Critical Value

In many practical optimization problems, the interval with which you are working will contain only one first-order critical value of the function. When this happens, you can use the second derivative test from Section 2 to identify its absolute extremum, even though this test is really a test for only relative extrema. The reason is that, in this special case, every *relative* maximum or minimum is necessarily also an *absolute* maximum or minimum.

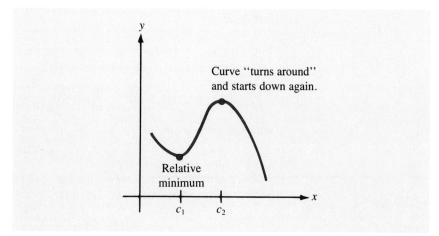

Figure 3.30 The relative minimum is not the absolute minimum because of the effect of another critical point.

To see why, look at Figure 3.30, in which the curve is eventually lower than its relative minimum at $x = c_1$. What allows this to happen is the existence of the relative maximum (i.e., another first-order critical value) at $x = c_2$, where the curve "turns around" and starts down again. On an interval that does not contain this additional first-order critical value, the relative minimum at $x = c_1$ would be the absolute minimum. Hence, on such an interval, the second derivative test, which identifies relative maxima and minima, also identifies absolute maxima and minima. Here is a more precise statement of the situation.

The Second Derivative Test for Absolute Extrema

Suppose that f is continuous on an interval on which $x = a$ is the only first-order critical value and that $f'(a) = 0$.

If $f''(a) > 0$, then $f(a)$ is the absolute minimum of f on the interval.
If $f''(a) < 0$, then $f(a)$ is the absolute maximum of f on the interval.

The second derivative test for absolute extrema can be applied to any interval, whether closed or not. The only requirement is that the function be continuous and have only one critical value on the interval.

Here are two examples.

EXAMPLE 3.5

Suppose that when q units of a certain commodity are produced, the total manufacturing cost is $C(q) = 3q^2 + 5q + 75$ dollars. At what level of production will the average cost per unit be smallest?

Solution

The average cost per unit is the total cost divided by the number of units produced. Thus, if q units are produced, the average cost is

$$A(q) = \frac{C(q)}{q} = \frac{3q^2 + 5q + 75}{q} = 3q + 5 + \frac{75}{q}$$

dollars per unit. Since only positive values of q are meaningful in this context, the goal is to find the absolute minimum of the function $A(q)$ on the unbounded interval $q > 0$.

The derivative of A is

$$A'(q) = 3 - \frac{75}{q^2}$$

which is zero when

$$3 - \frac{75}{q^2} = 0 \qquad q^2 = 25 \qquad \text{or} \qquad q = \pm 5$$

Of these two values of q, only $q = 5$ is in the relevant interval $q > 0$. Since $q = 5$ is the only first-order critical value in this interval, and since $A(q) = 3q + 5 + \dfrac{75}{q}$ is continuous on the interval (its only discontinuity occurring at $q = 0$, which is not in the interval), you can apply the second derivative test for absolute extrema.

The second derivative is $A''(q) = \dfrac{150}{q^3}$. Since $A''(5) = \dfrac{150}{(5)^3} > 0$, it follows that $A(q)$ has its absolute minimum at $q = 5$. That is, the average cost per unit is minimal when 5 units are produced and is equal to $A(5) = 35$ dollars per unit at this level of production.

For reference, the relevant portion of the graph of the average cost function is sketched in Figure 3.31. Notice that it is decreasing for

Figure 3.31 The average cost:

$$A(q) = 3q + 5 + \frac{75}{q}.$$

$0 < q < 5$ and increasing for $q > 5$ and has a relative and absolute minimum when $q = 5$.

Applied Optimality Conditions

In the next example, the goal is not to find the absolute maximum or minimum of a particular function, but rather to prove that when the absolute extremum occurs, a certain condition must be satisfied.

EXAMPLE 3.6

Suppose that at a certain factory, setup cost is directly proportional to the number of machines used and operating cost is inversely proportional to the number of machines used. Show that when the total cost is minimal, the setup cost is equal to the operating cost.

Solution

Let x be the number of machines used and $C(x)$ the corresponding total cost. Since

$$\text{Total cost} = \text{setup cost} + \text{operating cost}$$

and

$$\text{setup cost} = k_1 x \quad \text{and} \quad \text{operating cost} = \frac{k_2}{x}$$

where k_1 and k_2 are positive constants of proportionality, it follows that

$$C(x) = k_1 x + \frac{k_2}{x}$$

Since only positive values of x are meaningful in this context, the goal is to minimize $C(x)$ on the interval $x > 0$.

The derivative of C is

$$C'(x) = k_1 - \frac{k_2}{x^2}$$

which is zero when

$$k_1 - \frac{k_2}{x^2} = 0 \quad \text{or} \quad k_1 = \frac{k_2}{x^2}$$

At this point in the calculation, notice that if you multiply both sides of the equation by x, you will get

$$k_1 x = \frac{k_2}{x}$$

Since $k_1 x = $ setup cost   and $\dfrac{k_2}{x} = $ operating cost

it follows from this equation that $C'(x) = 0$ when

$$\text{setup cost} = \text{operating cost}$$

which is exactly what you were trying to show!

All that remains is to verify that the function $C(x)$ actually has its absolute minimum when its derivative is zero. To do this, solve the equation $C'(x) = 0$ for x to get

$$k_1 - \frac{k_2}{x^2} = 0 \qquad x^2 = \frac{k_2}{k_1} \qquad \text{or} \qquad x = \pm\sqrt{\frac{k_2}{k_1}}$$

Only the positive square root is in the relevant interval $x > 0$. Since this is the only first-order critical value in the interval and since the function $C(x) = k_1 x + \dfrac{k_2}{x}$ is continuous on this interval (its only discontinuity occurring at $x = 0$), you can apply the second derivative test for absolute extrema. In particular, the second derivative is $C''(x) = \dfrac{2k_2}{x^3}$, which is clearly positive when $x > 0$ (since k_2 is a positive constant). It follows that the point at which the derivative is zero is indeed the absolute minimum you were seeking.

For reference, the graph of the cost function is sketched in Figure 3.32.

Figure 3.32 Total cost: $C(x) = k_1 x + \dfrac{k_2}{x}$.

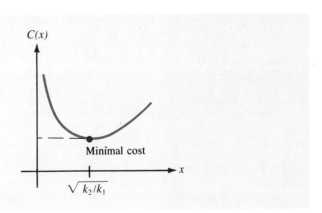

Problems 3.3

In Problems 1 through 16, find the absolute maximum and absolute minimum (if any) of the given function on the specified interval.

1. $f(x) = x^2 + 4x + 5$; $-3 \le x \le 1$

2. $f(x) = x^3 + 3x^2 + 1$; $-3 \le x \le 2$

3. $f(x) = \frac{1}{3}x^3 - 9x + 2$; $0 \le x \le 2$

4. $f(x) = x^5 - 5x^4 + 1$; $0 \le x \le 5$

5. $f(x) = 3x^5 - 5x^3$; $-2 \le x \le 0$

6. $f(x) = 10x^6 + 24x^5 + 15x^4 + 3$; $-1 \le x \le 1$

7. $f(x) = (x^2 - 4)^5$; $-3 \le x \le 2$

8. $f(x) = \frac{x^2}{x - 1}$; $-2 \le x \le -\frac{1}{2}$

9. $f(x) = x + \frac{1}{x}$; $\frac{1}{2} \le x \le 3$

10. $f(x) = \frac{1}{x^2 - 9}$; $0 \le x \le 2$ 11. $f(x) = x + \frac{1}{x}$; $x > 0$

12. $f(x) = 2x + \frac{32}{x}$; $x > 0$ 13. $f(x) = \frac{1}{x}$; $x > 0$

14. $f(x) = \frac{1}{x^2}$; $x > 0$ 15. $f(x) = \frac{1}{x + 1}$; $x \ge 0$

16. $f(x) = \frac{1}{(x + 1)^2}$; $x \ge 0$

In Problems 17 through 30, solve the practical optimization problem and use one of the techniques from this section to verify that you have actually found the desired absolute extremum.

Group Membership   17. Suppose that x years after its founding in 1978, a certain national consumers' association had a membership of $f(x) = 100(2x^3 - 45x^2 + 264x)$.
(a) At what time between 1978 and 1992 was the membership of the association largest? What was the membership at that time?
(b) At what time between 1978 and 1992 was the membership of the association smallest? What was the membership at that time?

Broadcasting   18. An all-news radio station has made a survey of the listening habits of local residents between the hours of 5:00 P.M. and midnight. The survey indicates that the percentage of the local adult population that is tuned in to the station x hours after 5:00 P.M. is $f(x) = \frac{1}{8}(-2x^3 + 27x^2 - 108x + 240)$.
(a) At what time between 5:00 P.M. and midnight are the most people listening to the station? What percentage of the population is listening at this time?

(b) At what time between 5:00 P.M. and midnight are the fewest people listening? What percentage of the population is listening at this time?

Profit   19. A manufacturer can produce radios at a cost of $5 apiece and estimates that if they are sold for x dollars apiece, consumers will buy $20 - x$ radios a day. At what price should the manufacturer sell the radios to maximize profit?

Consumer Expenditure   20. The demand function for a certain commodity is $D(p) = 160 - 2p$, where p is the price at which the commodity is sold. At what price is the total consumer expenditure for the commodity greatest?

Circulation   21. Poiseuille's law asserts that the speed of blood that is r centimeters from the central axis of an artery of radius R is $S(r) = c(R^2 - r^2)$, where c is a positive constant. Where is the speed of the blood greatest?

Respiration   22. Biologists define the flow F of air in the trachea by the formula $F = SA$, where S is the speed of the air and A the area of a cross section of the trachea.

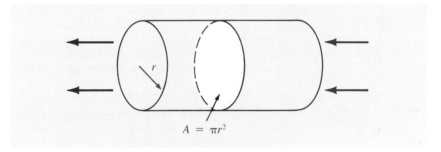

$A = \pi r^2$

(a) Assume that the cross section of the trachea is circular and use the formula from Example 3.3 for the speed of air in the trachea during a cough to express the flow F as a function of the radius r.
(b) Find the radius r for which the flow is greatest.

Population Growth   23. When environmental factors impose an upper bound on its possible size, population grows at a rate that is jointly proportional to its current size and the difference between its current size and the upper bound. (Recall from Chapter 1, Section 4, that joint proportionality means direct proportionality to the product of the two quantities.) Show that the rate of population growth will be greatest when the population has climbed to 50 percent of its upper bound.

The Spread of an Epidemic   24. The rate at which an epidemic spreads through a community is jointly proportional to the number of people who have caught the disease and the number who have not. Show that the epidemic is spreading most rapidly when half the people have caught the disease.

Transportation Cost   25. A truck is hired to transport goods from a factory to a warehouse. The driver's wages are figured by the hour and so are inversely proportional to the speed at which the truck is driven. The amount of gasoline used is directly proportional to the speed at which the truck is driven, and the price of gasoline remains constant during the trip. Show that the total cost is smallest at the speed for which the driver's wages are equal to the cost of the gasoline used.

Microeconomics   26. An economic law states that under normal circumstances, profit is maximized when marginal revenue equals marginal cost. (Recall that marginal revenue and marginal cost are the derivatives of total revenue and total cost, respectively.) Use the theory of extrema to show why this law is true. (You may assume without proof that the critical point is actually the desired absolute maximum.)

Microeconomics   27. An economic law states that under normal circumstances, average cost is smallest when it equals marginal cost. Derive this law using calculus. [*Hint:* If $C(q)$ represents the total cost of manufacturing q units, then average cost is $A(q) = \dfrac{C(q)}{q}$. Assume that average cost will be smallest when $A'(q) = 0$. Use the quotient rule to compute $A'(q)$; then set $A'(q)$ equal to zero, and the economic law will emerge.]

Average Cost   28. Suppose the total cost in dollars of manufacturing q units is given by the function $C(q) = 3q^2 + q + 48$.
   (a) Express the average manufacturing cost per unit as a function of q.
   (b) For what value of q is the average cost the smallest?
   (c) For what value of q is the average cost equal to the marginal cost? Compare this value with your answer in part (b) and look at the general economic law stated in Problem 27.
   (d) On the same set of axes, graph the total cost, marginal cost, and average cost functions.

Speed of Flight   29. In a paper published in 1969 ("The Mechanics of Bird Migration," *Ibis* 111, pp. 525–556), C. J. Pennycuick provided experimental evidence to show that the power P required by a bird to maintain flight is given by the formula

$$P = \frac{w^2}{2\rho S v} + \frac{1}{2}\rho A v^3$$

where v is the relative speed of the bird, w is its weight, ρ is the density of air, and S and A are positive constants associated with the bird's size and shape. What relative speed v will minimize the power required by the bird?

Learning   30. In a certain learning model, two responses (A and B) are possible for each of a series of observations. If there is a probability p of getting response A in any individual observation, the probability of getting response A exactly n times in a series of m observations is $F(p) = p^n(1 - p)^{m-n}$. The **maximum likelihood estimate** is the value of p that maximizes $F(p)$ for $0 \le p \le 1$. For what value of p does this occur?

4 Practical Optimization Problems

In this section, you will learn how to combine the techniques of model building from Chapter 1, Section 4, with the optimization techniques of Chapter 3, Section 3, to solve practical optimization problems.

The first step in solving such a problem is to decide precisely what you are to optimize. Once you have identified this quantity, choose a letter to represent it. You may find it helpful to choose a letter that is closely related to the quantity, such as R for revenue or A for area.

Your goal is to represent the quantity to be optimized as a function of some other variable so that you can apply calculus. It is usually a good idea to express the desired function in words before trying to represent it mathematically.

Once the function has been expressed in words, the next step is to choose an appropriate variable. Often the choice is obvious. Sometimes you will be faced with a choice among several natural variables. When this happens, think ahead and try to choose the variable that leads to the simplest functional representation. In some problems, the quantity to be optimized is expressed most naturally in terms of two variables. If so, you will have to find a way to write one of these variables in terms of the other.

The next step is to express the quantity to be optimized as a function of the variable you have chosen. In most problems, the function has a practical interpretation only when the variable lies in a certain interval. Once you have written the function and identified the appropriate interval, the difficult part is done and the rest is routine. To complete the solution, simply apply the techniques from the preceding section to optimize your function on the specified interval.

Here are several examples to illustrate the procedure.

EXAMPLE 4.1

A manufacturer can produce radios at a cost of $2 apiece. The radios have been selling for $5 apiece, and at this price, consumers have been buying 4,000 radios a month. The manufacturer is planning to raise the price of the radios and estimates that for each $1 increase in the price, 400 fewer radios will be sold each month. At what price should the manufacturer sell the radios to maximize profit?

Solution

Let x denote the new price at which the radios will be sold and $P(x)$ the corresponding profit. The goal is to maximize the profit. As in Example 4.1 of Chapter 1, Section 4, begin by stating the formula for profit in words.

$$\text{Profit} = (\text{number of radios sold})(\text{profit per radio})$$

Since 4,000 radios are sold each month when the price is $5 and 400 fewer will be sold each month for each $1 increase in the price, it follows that

$$\text{Number of radios sold} = 4{,}000 - 400(\text{number of \$1 increases})$$

The number of $1 increases in the price is the difference $x - 5$ between the new and old selling prices. Hence,

$$\begin{aligned} \text{Number of radios sold} &= 4{,}000 - 400(x - 5) \\ &= 400[10 - (x - 5)] \\ &= 400(15 - x) \end{aligned}$$

The profit per radio is simply the difference between the selling price x and the cost $2. That is,

$$\text{Profit per radio} = x - 2$$

Putting it all together,

$$P(x) = 400(15 - x)(x - 2)$$

The goal is to find the absolute maximum of the profit function $P(x)$. You have some leeway in the selection of the relevant interval for this problem. Since the new price x is to be at least as high as the old price of $5, a reasonable choice is the interval $x \geq 5$. On the other hand, the number of radios sold is $400(15 - x)$, which will drop to zero when $x = 15$. If you assume that the manufacturer will not price the radios so high that no one will buy them, you can restrict the optimization problem to the closed interval $5 \leq x \leq 15$. Of the two choices, this closed interval is probably preferable since you have a special technique for maximizing functions on closed intervals.

To find the critical points, compute the derivative (using the product and constant multiple rules) to get

$$\begin{aligned} P'(x) &= 400[(15 - x)(1) + (x - 2)(-1)] \\ &= 400(15 - x - x + 2) = 400(17 - 2x) \end{aligned}$$

which is zero when

$$17 - 2x = 0 \quad \text{or} \quad x = 8.5$$

Comparing the values of the profit function

$$P(5) = 12,000 \qquad P(8.5) = 16,900 \qquad \text{and} \qquad P(15) = 0$$

at this critical value and at the endpoints of the interval, you can conclude that the maximum possible profit is $16,900, which will be generated if the radios are sold for $8.50 apiece.

Figure 3.33 The profit function $P(x) = 400(15 - x)(x - 2)$.

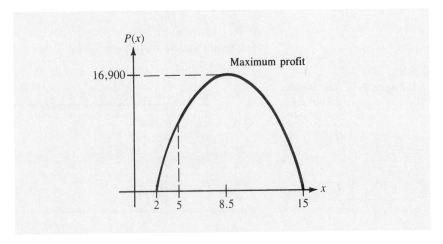

For reference, the graph of the profit function is sketched in Figure 3.33.

An Alternative Solution to the Profit Maximization Problem

Since the number of radios sold in Example 4.1 is described in terms of the number of $1 increases in price, you may prefer to let the variable x denote the number of $1 increases in price rather than the new price itself. If you make this choice, you should find that the corresponding profit function is

$$P(x) = 400(10 - x)(3 + x)$$

and that the relevant interval is $0 \le x \le 10$. The absolute maximum in this case will occur when $x = 3.5$, that is, when the old price is increased from $5 by $3.50 to a new price of $8.50. For practice, work out the details of the solution in this case.

Elimination of Variables

In the next example, the quantity to be optimized is expressed most naturally in terms of two variables. Before you can apply calculus, you will have to write one of these variables in terms of the other to get a function of just one variable.

EXAMPLE 4.2

The highway department is planning to build a picnic area for motorists along a major highway. It is to be rectangular with an area of 5,000 square yards and is to be fenced off on the three sides not adjacent to the highway. What is the least amount of fencing that will be needed to complete the job?

Solution

As in Example 4.2 of Chapter 1, Section 4, label the sides of the picnic area as indicated in Figure 3.34 and let F denote the amount of fencing

Figure 3.34 Rectangular picnic area.

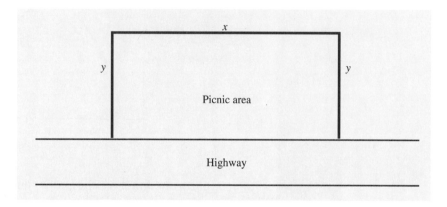

required. Then, $F = x + 2y$

The fact that the area is to be 5,000 tells you that

$$xy = 5,000 \quad \text{or} \quad y = \frac{5,000}{x}$$

To rewrite F in terms of the single variable x, substitute this expression for y into the formula for F:

$$F(x) = x + 2\left(\frac{5,000}{x}\right) = x + \frac{10,000}{x}$$

Since $F(x)$ has a practical interpretation for any positive value of x, your goal is to find the absolute minimum of $F(x)$ on the interval $x > 0$.

To find the critical value, set the derivative

$$F'(x) = 1 - \frac{10,000}{x^2}$$

equal to zero and solve for x, getting

$$1 - \frac{10,000}{x^2} = 0 \quad x^2 = 10,000 \quad \text{or} \quad x = \pm 100$$

Only the positive value $x = 100$ lies in the interval $x > 0$. Since this is the only critical value in the interval, you can apply the second derivative test for absolute extrema. In particular, the second derivative is $F''(x) = \dfrac{20,000}{x^3}$, which is positive when $x > 0$. Hence, the critical point at $x = 100$ is the absolute minimum of F on the interval. That is, the least amount of fencing needed to complete the job is $F(100) = 200$ yards.

Figure 3.35 The graph of
$$F(x) = x + \frac{10,000}{x}$$
for $x > 0$.

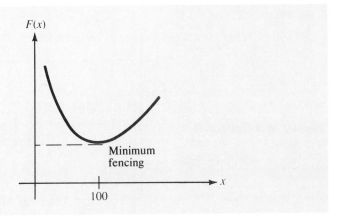

For reference, the graph of $F(x)$ is sketched in Figure 3.35.

Optimization Problems Involving the Pythagorean Theorem

In the next example, the function to be optimized is obtained geometrically with the aid of the pythagorean theorem.

EXAMPLE 4.3

A cable is to be run from a power plant on one side of a river 900 meters wide to a factory on the other side, 3,000 meters downstream. The cost of running the cable under the water is $5 per meter, while the cost over land is $4 per meter. What is the most economical route over which to run the cable?

Solution

To help you visualize the situation, begin by drawing a diagram as shown in Figure 3.36. (Notice that in drawing the diagram in Figure 3.36, we have already assumed that the cable should be run in a *straight line* from the power plant to some point P on the opposite bank. Do you see why this assumption is justified?)

The goal is to minimize the cost of installing the cable. Let C denote this cost and represent C as follows:

Figure 3.36 Relative positions of factory, river, and power plant.

$$C = 5(\text{number of meters of cable under water})$$
$$+ \ 4(\text{number of meters of cable over land})$$

Since you wish to describe the optimal route over which to run the cable, it will be convenient to choose a variable in terms of which you can easily locate the point P. Two reasonable choices for the variable x are illustrated in Figure 3.37.

Before plunging into the calculations, take a minute to decide which choice of variables is more advantageous. In Figure 3.37a, the distance

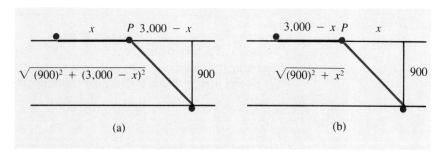

Figure 3.37 Two choices for the variable x.

across the water from the power plant to the point P is (by the pythagorean theorem) $\sqrt{(900)^2 + (3{,}000 - x)^2}$, and the corresponding total cost function is

$$C(x) = 5\sqrt{(900)^2 + (3{,}000 - x)^2} + 4x$$

In Figure 3.37b, the distance across the water is $\sqrt{(900)^2 + x^2}$, and the total cost function is

$$C(x) = 5\sqrt{(900)^2 + x^2} + 4(3{,}000 - x)$$

The second of these functions is the more attractive since the term $3,000 - x$ is merely multiplied by 4, while in the first function it is squared and appears under the radical. Hence, you should choose x as in Figure 3.37b and work with the total cost function

$$C(x) = 5\sqrt{(900)^2 + x^2} + 4(3,000 - x)$$

Since the distances x and $3,000 - x$ cannot be negative, the relevant interval is $0 \leq x \leq 3,000$, and your goal is to find the absolute minimum of the function $C(x)$ on this closed interval.

To find the critical values, compute the derivative

$$C'(x) = \frac{5}{2}[(900)^2 + x^2]^{-1/2}(2x) - 4 = \frac{5x}{\sqrt{(900)^2 + x^2}} - 4$$

and set it equal to zero to get

$$\frac{5x}{\sqrt{(900)^2 + x^2}} - 4 = 0 \quad \text{or} \quad \sqrt{(900)^2 + x^2} = \frac{5}{4}x$$

Square both sides of this equation and solve for x to get

$$(900)^2 + x^2 = \frac{25}{16}x^2$$

$$x^2 = \frac{16}{9}(900)^2$$

$$x = \pm\frac{4}{3}(900) = \pm 1,200$$

Since only the positive value $x = 1,200$ is in the interval $0 \leq x \leq 3,000$, compute $C(x)$ at this critical value and at the endpoints $x = 0$ and $x = 3,000$. Since

$$C(0) = 5\sqrt{(900)^2 + 0} + 4(3,000 - 0) = 16,500$$

$$C(1,200) = 5\sqrt{(900)^2 + (1,200)^2} + 4(3,000 - 1,200) = 14,700$$

$$C(3,000) = 5\sqrt{(900)^2 + (3,000)^2} + 4(3,000 - 3,000) \approx 15,660$$

it follows that the minimal installation cost is \$14,700, which will occur if the cable reaches the opposite bank 1,200 meters downstream from the power plant.

Cylinder Problems The next two problems involve the volume and surface area of a cylindrical can. Such problems generally fall into one of two basic categories:

A. Given a fixed capacity (volume) of the can, minimize the amount of material (surface area), or the cost of construction.

B. Given a fixed amount of material (or fixed construction cost), maximize the capacity.

To deal with such problems, you need to know a few geometric formulas regarding a cylinder with radius r and height h:

$$\text{Area of top or bottom (circular disk)} = \pi r^2$$

$$\text{Lateral area (area of the curved side)} = 2\pi rh$$

$$\text{Volume} = \text{(area of the circular base)(height)} = \pi r^2 h$$

The formulas for the area of the top or bottom and the volume of the cylinder are easy to understand and remember. To see why the lateral area is $2\pi rh$, imagine the top and bottom of the cylinder removed and the side cut and spread out to form a rectangle, as shown in Figure 3.38. The

Figure 3.38 Cylindrical can and its curved side spread out.

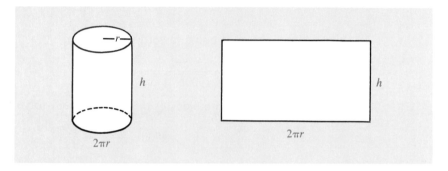

height of the rectangle is the height h of the cylinder, and the length of the rectangle is the circumference $2\pi r$ of the circular top (or bottom). Hence, the area of the rectangle (and the lateral area of the cylinder) is $2\pi rh$.

EXAMPLE 4.4

A cylindrical can is to be constructed to hold a fixed volume of liquid. The cost of the material used for the top and bottom of the can is 3 cents per square inch, and the cost of the material used for the curved side is 2 cents per square inch. Use calculus to derive a simple relationship between the radius and height of the can that is the least costly to construct.

Solution

Let r denote the radius, h the height, C the cost (in cents), and V the (fixed) volume. The goal is to minimize cost:

$$\text{Cost} = \text{cost of top} + \text{cost of bottom} + \text{cost of side}$$

where, for each component of the cost,

$$\text{Cost} = \text{(cost per square inch)(area)}$$

Hence,   Cost of top = cost of bottom = $3\pi r^2$

and   Cost of side = $2(2\pi rh) = 4\pi rh$

and the total cost is

$$C = 3\pi r^2 + 3\pi r^2 + 4\pi rh = 6\pi r^2 + 4\pi rh$$

Before you can apply calculus, you must write the cost in terms of just one variable. To do this, use the fact that the can is to have a fixed volume V and solve the equation

$$V = \pi r^2 h$$

for h to get $h = \dfrac{V}{\pi r^2}$ (where V is a constant)

Substitute this expression for h into the formula for C to get

$$C(r) = 6\pi r^2 + 4\pi r\left(\frac{V}{\pi r^2}\right) = 6\pi r^2 + \frac{4V}{r}$$

Since the radius can be any positive number, the goal is to find the absolute minimum of $C(r)$ on the interval $r > 0$.

The derivative is

$$C'(r) = 12\pi r - \frac{4V}{r^2}   \text{(remember that V is a constant)}$$

which is zero when

$$12\pi r = \frac{4V}{r^2}   \text{or}   3r = \frac{V}{\pi r^2}$$

But $$\frac{V}{\pi r^2} = h$$

and so this equation says

$$3r = h$$

That is, assuming that the cost function is minimized when its derivative is zero, the height of the least expensive can must be 3 times its radius.

Finally, to verify that this actually corresponds to the absolute minimum, convince yourself that $C(r)$ has only one first-order critical point, at $r = \sqrt[3]{\dfrac{V}{3\pi}}$, on the interval $r > 0$ and apply the second derivative test for absolute extrema. Since the second derivative

$$C''(r) = 12\pi + \frac{8V}{r^3}$$

is clearly positive when $r > 0$, it follows that the critical point that led to the relationship $3r = h$ corresponds to the absolute minimum of the cost

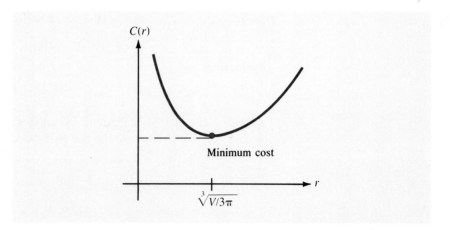

Figure 3.39 The cost function
$$C(r) = 6\pi r^2 + \frac{4V}{r}$$
for $r > 0$.

function on the interval $r > 0$.

For reference, the relevant portion of the graph of $C(r)$ is sketched in Figure 3.39.

The Maximum Efficiency of a Worker

In the next example, we return to the questions of worker efficiency and diminishing returns, which were introduced in Section 2 of this chapter. Our goal will be to maximize a worker's rate of production. The moment at which this maximum efficiency occurs is known in economics as the **point of diminishing returns.** As usual, the strategy will be to set the first derivative of the function that is to be maximized equal to zero. In this case, the function to be maximized is the rate of change or first derivative of the worker's output function. Hence, it will be the second derivative of the output function that will be set equal to zero to find the point of diminishing returns.

EXAMPLE 4.5

An efficiency study of the morning shift at a factory indicates that an average worker who arrives on the job at 8:00 A.M. will have produced $Q(t) = -t^3 + 9t^2 + 12t$ units t hours later. At what time during the morning is the worker performing most efficiently?

Solution

The worker's rate of production is the derivative
$$R(t) = Q'(t) = -3t^2 + 18t + 12$$

of the output function $Q(t)$. Assuming that the morning shift runs from

8:00 A.M. until noon, the goal is to find the absolute maximum of the function $R(t)$ on the closed interval $0 \le t \le 4$. The derivative of R is

$$R'(t) = Q''(t) = -6t + 18$$

which is zero when $t = 3$. Comparing the values

$$R(0) = 12 \qquad R(3) = 39 \qquad R(4) = 36$$

of the production-rate function at the critical value and endpoints, you can conclude that the rate of production will be greatest and the worker performing most efficiently when $t = 3$, that is, at 11:00 A.M., when the worker is performing at the rate of 39 units per hour.

Figure 3.40 A production curve and corresponding rate of production.

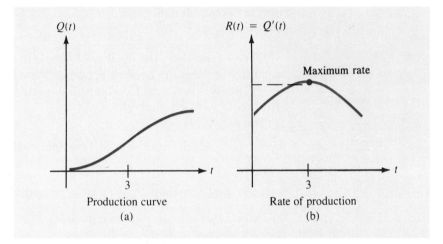

Production curve
(a)

Rate of production
(b)

The graphs of the output function $Q(t)$ and its derivative, the rate of production $R(t)$, are sketched in Figure 3.40. Notice that the production curve is steepest and the rate of production greatest when $t = 3$.

Optimization of Discrete Functions   In the next example, the function to be maximized has practical meaning only when its independent variable is a whole number. However, the optimization procedure leads to a fractional value of this variable, and additional analysis is needed to obtain a meaningful solution.

EXAMPLE 4.6

A bus company will charter a bus that holds 50 people to groups of 35 or more. If a group contains exactly 35 people, each person pays $60. In larger groups, everybody's fare is reduced by $1 for each person in excess of 35. Determine the size of the group for which the bus company's revenue will be greatest.

Solution

Let R denote the bus company's revenue. Then,

$$R = \text{(number of people in the group)(fare per person)}$$

You could let x denote the total number of people in the group, but it is slightly more convenient to let x denote the number of people in excess of 35. Then,

$$\text{Number of people in the group} = 35 + x$$

and $\qquad\qquad\qquad$ Fare per person $= 60 - x$

so the revenue function is

$$R(x) = (35 + x)(60 - x)$$

Since x represents the number of people in excess of 35, and since the total size of the group must be between 35 and 50, it follows that x must be in the interval $0 \leq x \leq 15$. The goal is to find the absolute maximum of $R(x)$ on this closed interval.

The derivative is

$$R'(x) = (35 + x)(-1) + (60 - x)(1) = 25 - 2x$$

which is zero when $x = 12.5$. Since

$$R(0) = 2,100 \qquad R(12.5) = 2,256.25 \qquad R(15) = 2,250$$

it follows that the absolute maximum of $R(x)$ on the interval $0 \leq x \leq 15$ occurs when $x = 12.5$.

But x represents a certain number of people and must be a whole number. Hence, $x = 12.5$ cannot be the solution to this practical optimization problem. To find the optimal *integer* value of x, observe that R is increasing for $0 < x < 12.5$ and decreasing for $x > 12.5$, as shown in Figure 3.41.

Figure 3.41 The revenue function $R(x) = (35 + x)(60 - x)$.

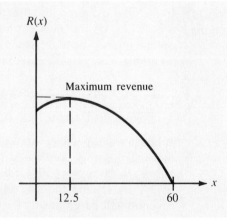

It follows that the optimal integer value of x is either $x = 12$ or $x = 13$. Since

$$R(12) = 2{,}256 \qquad \text{and} \qquad R(13) = 2{,}256$$

you can conclude that the bus company's revenue will be greatest when the group contains either 12 or 13 people in excess of 35, that is, for groups of 47 or 48. The revenue in either case will be $2,256.

Perspective: In the preceding example, the graph of revenue as a function of x was actually a collection of discrete points corresponding to the integer values of x, as indicated in Figure 3.42a. Since calculus cannot be

Figure 3.42 A discrete function and its continuous model.

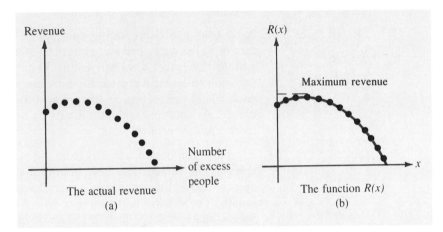

used to study such a function, we worked with the differentiable function $R(x) = (35 + x)(60 - x)$, which is defined for all values of x and whose graph (Figure 3.42b) "connects" the points in Figure 3.42a. After applying calculus to this continuous model, we obtained a mathematical solution that was not the solution of the discrete practical problem but that did suggest where to look for the practical solution.

Problems 3.4

1. What number exceeds its square by the largest amount? [*Hint:* Find the number x that maximizes $f(x) = x - x^2$.]

2. What number is exceeded by its square root by the largest amount?

3. Find two positive numbers whose sum is 50 and whose product is as large as possible.

4. Find two positive numbers x and y whose sum is 30 and are such that xy^2 is as large as possible.

Retail Sales 5. A store has been selling skateboards at the price of $40 per board, and at this price, skaters have been buying 50 boards a month. The owner of the store wishes to raise the price and estimates that for each $1 increase in price, 3 fewer boards will be sold each month. If each board costs the store $25, at what price should the store sell the boards in order to maximize profit?

Retail Sales 6. A bookstore can obtain a certain gift book from the publisher at a cost of $3 per book. The bookstore has been offering the book at a price of $15 per copy and, at this price, has been selling 200 copies a month. The bookstore is planning to lower its price to stimulate sales and estimates that for each $1 reduction in the price, 20 more books will be sold each month. At what price should the bookstore sell the book to generate the greatest possible profit?

Retail Sales 7. A baseball card store can obtain Mel Schlabotnik rookie cards at a cost of $5 per card. The store has been offering the cards at $10 apiece and, at this price, has been selling 50 cards per month. The store is planning to lower the price to stimulate sales and estimates that for each 50-cent reduction in the price, 5 more cards will be sold each month. At what price should the cards be sold in order to maximize total monthly profit?

Retail Sales 8. A retailer can obtain cameras from the manufacturer at a cost of $50 apiece. The retailer has been selling the cameras at a price of $80 apiece, and at this price, consumers have been buying 40 cameras a month. The retailer is planning to lower the price to stimulate sales and estimates that for each $5 reduction in the price, 10 more cameras will be sold each month. At what price should the retailer sell the cameras to maximize profit?

Agricultural Yield 9. A Florida citrus grower estimates that if 60 orange trees are planted, the average yield per tree will be 400 oranges. The average yield will decrease by 4 oranges per tree for each additional tree planted on the same acreage. How many trees should the grower plant to maximize the total yield?

Harvesting 10. Farmers can get $2 per bushel for their potatoes on July first, and after that, the price drops by 2 cents per bushel per day. On July first, a farmer has 80 bushels of potatoes in the field and estimates that the crop is increasing at the rate of 1 bushel per day. When should the farmer harvest the potatoes to maximize revenue?

Recycling 11. To raise money, a service club has been collecting used bottles that it plans to deliver to a local glass company for recycling. Since the project began 80 days ago, the club has collected 24,000 pounds of glass for which the glass company currently offers 1 cent per pound. However, because bottles are accumulating faster than they can be recycled, the company plans to reduce by 1 cent each day the price it will pay for 100 pounds of used glass. Assume that the club can continue to collect bottles at the same rate and that transportation costs make more than one trip to the glass company unfeasible. What is the most advantageous time for the club to conclude its project and deliver the bottles?

Fencing 12. A city recreation department plans to build a rectangular playground having an area of 3,600 square meters and surround it by a fence. How can this be done using the least amount of fencing?

Fencing 13. There are 320 yards of fencing available to enclose a rectangular field. How should this fencing be used so that the enclosed area is as large as possible?

Geometry 14. Prove that of all rectangles with a given perimeter, the square has the largest area.

Geometry 15. Prove that of all rectangles with a given area, the square has the smallest perimeter.

Geometry 16. A rectangle is inscribed in a right triangle, as shown in the accompanying figure. If the triangle has sides of length 5, 12, and 13, what are the dimensions of the inscribed rectangle of greatest area?

Geometry 17. A triangle is positioned with its hypotenuse on a diameter of a circle, as shown in the accompanying figure. If the circle has radius 4, what are the dimensions of the triangle of greatest area?

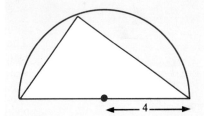

Problem 16   Problem 17

Construction Cost 18. A carpenter has been asked to build an open box with a square base. The sides of the box will cost $3 per square meter, and the base will cost $4 per square meter. What are the dimensions of the box of greatest volume that can be constructed for $48?

Construction Cost 19. A closed box with a square base is to have a volume of 250 cubic meters. The material for the top and bottom of the box costs $2 per square meter, and the material for the sides costs $1 per square meter. Can the box be constructed for less than $300?

Postal Regulations   20. According to postal regulations, the girth plus length of parcels sent by fourth-class mail may not exceed 108 inches. What is the largest possible volume of a rectangular parcel with two square sides that can be sent by fourth-class mail?

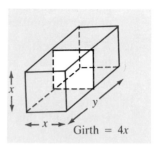

Girth = 4x

Construction   21. An open box is to be made from a square piece of cardboard, 18 inches by 18 inches, by removing a small square from each corner and folding up the flaps to form the sides. What are the dimensions of the box of greatest volume that can be constructed in this way?

Distance between Moving Objects   22. A truck is 300 miles due east of a car and is traveling west at the constant speed of 30 miles per hour. Meanwhile, the car is going north at the constant speed of 60 miles per hour. At what time will the car and truck be closest to each other? (*Hint:* You will simplify the calculation if you minimize the *square* of the distance between the car and truck rather than the distance itself. Can you explain why this simplification is justified?)

Installation Cost   23. A cable is to be run from a power plant on one side of a river 1,200 meters wide to a factory on the other side, 1,500 meters downstream. The cost of running the cable under the water is $25 per meter, while the cost over land is $20 per meter. What is the most economical route over which to run the cable?

Installation Cost   24. Find the most economical route in Problem 23 if the power plant is 2,000 meters downstream from the factory.

Installation Cost   25. For the summer, the company that is installing the cable in Example 4.3 has hired a temporary employee with a Ph.D. in mathematics. The mathematician, recalling a problem from first-year calculus, asserts that no matter how far downstream the factory is located (beyond 1,200 meters), it would be most economical to have the cable reach the opposite bank 1,200 meters downstream from the power plant. The supervisor, amused by the naïveté of the overeducated employee, replies, "Any fool can see that if the factory is further away, the cable should reach the opposite bank further downstream. It's just common sense." Who is right? And why?

Spy Story   26. It is noon, and the spy is back from space (see Problem 32 in Section 3 of Chapter 2) and driving a jeep through the sandy desert in the tiny principality of Alta Loma. He is 32 kilometers from the nearest point on a straight paved road. Down the road 16 kilometers is a power plant in which a band of international terrorists has placed a time bomb set to explode at 12:50 P.M. The jeep can travel 48 km/hr in the sand and 80 km/hr on the paved road. If he arrives at the power plant in the shortest possible time, how long will our hero have to defuse the bomb? (*Hint:* The goal is to minimize time, which is distance divided by speed.)

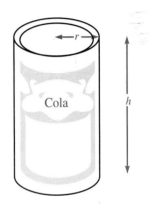

Problem 26   Problem 27

Packaging   27. Use the fact that 12 fluid ounces is approximately 6.89π cubic inches to find the dimensions of the 12-ounce soda can that can be constructed using the least amount of metal. Compare these dimensions with those of one of the soda cans in your refrigerator. What do you think accounts for the difference?

Packaging   28. A cylindrical can is to hold 4π cubic inches of frozen orange juice. The cost per square inch of constructing the metal top and bottom is twice the cost per square inch of constructing the cardboard side. What are the dimensions of the least expensive can?

Volume   29. What is the maximum possible volume of a cylindrical can with no top that can be made from 27π square inches of metal?

Volume   30. A cylindrical can (with top) is to be constructed using a fixed amount of metal. Use calculus to derive a simple relationship between the radius and height of the can having the greatest volume.

Construction Cost   31. A cylindrical container with no top is to be constructed to hold a fixed volume of liquid. The cost of the material used for the bottom is 3 cents per square inch, and that for the curved side is 2 cents per square inch. Use calculus to derive a simple relationship between the radius and height of the least expensive container.

Poster Design   **32.** A printer receives an order to produce a rectangular poster containing 25 square centimeters of print surrounded by margins of 2 centimeters on each side and 4 centimeters on the top and bottom. What are the dimensions of the smallest piece of paper that can be used to make the poster? (*Hint:* An unwise choice of variables will make the calculations unnecessarily complicated.)

Production Cost   **33.** A plastics firm has received an order from the city recreation department to manufacture 8,000 special Styrofoam kickboards for its summer swimming program. The firm owns 10 machines, each of which can produce 30 kickboards an hour. The cost of setting up the machines to produce the kickboards is $20 per machine. Once the machines have been set up, the operation is fully automated and can be overseen by a single production supervisor earning $4.80 per hour.
(a) How many of the machines should be used to minimize the cost of production?
(b) How much will the supervisor earn during the production run if the optimal number of machines is used?
(c) How much will it cost to set up the optimal number of machines?

Production Cost   **34.** A manufacturing firm receives an order for q units of a certain commodity. Each of the firm's machines can produce n units per hour. The setup cost is s dollars per machine, and the operating cost is p dollars per hour.
(a) Derive a formula for the number of machines that should be used to keep total cost as low as possible.
(b) Prove that when the total cost is minimal, the cost of setting up the machines is equal to the cost of operating the machines.

Efficiency   **35.** An efficiency study of the morning shift (from 8:00 A.M. to 12:00 noon) at a certain factory indicates that an average worker who arrives on the job at 8:00 A.M. will have assembled $Q(t) = -t^3 + 6t^2 + 15t$ transistor radios t hours later.
(a) At what time during the morning is the worker performing most efficiently?
(b) At what time during the morning is the worker performing least efficiently?

Efficiency   **36.** An efficiency study of the morning shift (from 8:00 A.M. to 12:00 noon) at a factory indicates that an average worker who arrives on the job at 8:00 A.M. will have produced $Q(t) = -t^3 + \frac{9}{2}t^2 + 15t$ units t hours later.

(a) At what time during the morning is the worker performing most efficiently?

(b) At what time during the morning is the worker performing least efficiently?

The Steepness of a Curve

37. At what point does the tangent to the curve $y = 2x^3 - 3x^2 + 6x$ have the smallest slope? What is the slope of the tangent at this point?

The Steepness of a Curve

38. For what value of x in the interval $-1 \leq x \leq 4$ is the graph of the function $f(x) = 2x^2 - \frac{1}{3}x^3$ steepest? What is the slope of the tangent at this point?

Population Growth

39. A 5-year projection of population trends suggests that t years from now, the population of a certain community will be $P(t) = -t^3 + 9t^2 + 48t + 50$ thousand.

(a) At what time during the 5-year period will the population be growing most rapidly?

(b) At what time during the 5-year period will the population be growing least rapidly?

Production Cost

40. Each machine at a certain factory can produce 50 units per hour. The setup cost is $80 per machine, and the operating cost is $5 per hour. How many machines should be used to produce 8,000 units at the least possible cost? (Remember that the answer should be a whole number.)

Construction Cost

41. It is estimated that the cost of constructing an office building that is n floors high is $C(n) = 2n^2 + 500n + 600$ thousand dollars. How many floors should the building have in order to minimize the average cost per floor? (Remember that your answer should be a whole number.)

Retail Sales

42. A retailer has bought several cases of a certain imported wine. As the wine ages, its value initially increases, but eventually the wine will pass its prime and its value will decrease. Suppose that x years from now, the value of a case will be changing at the rate of $53 - 10x$ dollars per year. Suppose, in addition, that storage rates will remain fixed at $3 per case per year. When should the retailer sell the wine to obtain the greatest possible profit?

Worker Efficiency

43. An efficiency study of the morning shift at a certain factory indicates that an average worker who arrives on the job at 8:00 A.M. will have assembled $f(x) = -x^3 + 6x^2 + 15x$ transistor radios x hours later. The study indicates further that after a 15-minute coffee break, the worker can assemble $g(x) = -\frac{1}{3}x^3 + x^2 + 23x$ radios in x hours.

Determine the time between 8:00 A.M. and noon at which a 15-minute coffee break should be scheduled so that the worker will assemble the maximum number of radios by lunchtime at 12:15 P.M. (*Hint:* If the coffee break begins x hours after 8:00 A.M., $4 - x$ hours will remain after the break.)

Transportation Cost   44. For speeds between 40 and 65 miles per hour, a truck gets $\dfrac{480}{x}$ miles per gallon when driven at a constant speed of x miles per hour. Gasoline costs \$1.12 per gallon, and the driver is paid \$8.40 per hour. What is the most economical constant speed between 40 and 65 miles per hour at which to drive the truck?

5 Applications to Business and Economics

In this section, you will see three applications of calculus to business and economics. The first is an optimization problem from business in which inventory size is selected to minimize cost. The second involves an important relationship between marginal and average quantities in economics. And the third deals with the economic concept of elasticity of demand, which is closely related to the analysis of total revenue.

An Inventory Problem   For each shipment of raw materials, a manufacturer must pay an ordering fee to cover handling and transportation. When the raw materials arrive, they must be stored until needed, and storage costs result. If each shipment of raw materials is large, few shipments will be needed and ordering costs will be low. Storage costs, however, will be high. On the other hand, if each shipment is small, ordering costs will be high because many shipments will be needed, but storage costs will be low. A manufacturer would like to determine the shipment size that will minimize total cost. The problem can be solved using calculus. Here is an example.

EXAMPLE 5.1

A bicycle manufacturer buys 6,000 tires a year from a distributor. The ordering fee is \$20 per shipment, the storage cost is 96 cents per tire per year, and each tire costs 25 cents. Suppose that the tires are used at a constant rate throughout the year and that each shipment arrives just as the preceding shipment has been used up. How many tires should the manufacturer order each time to minimize cost?

Solution

The goal is to minimize the total cost, which can be written as

Total cost = storage cost + ordering cost + purchase cost

Let x denote the number of tires in each shipment and $C(x)$ the corresponding total cost in dollars. Then,

Ordering cost = (ordering cost per shipment)(number of shipments)

Since 6,000 tires are ordered during the year and each shipment contains x tires, the number of shipments is $\dfrac{6,000}{x}$ and so

$$\text{Ordering cost} = 20 \left(\frac{6,000}{x}\right) = \frac{120,000}{x}$$

Moreover,

Purchase cost = (total number of tires ordered)(cost per tire)

$$= 6,000(0.25) = 1,500$$

The storage cost is slightly more complicated. When a shipment arrives, all x tires are placed in storage and then withdrawn for use at a constant rate. The inventory decreases linearly until there are no tires left, at which time the next shipment arrives. The situation is illustrated in Figure 3.43a.

Figure 3.43 Inventory graphs.

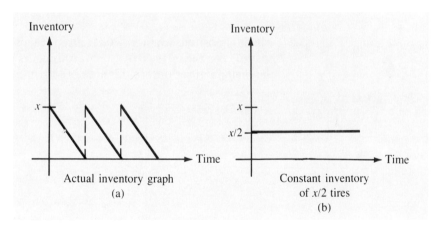

Actual inventory graph
(a)

Constant inventory of $x/2$ tires
(b)

The average number of tires in storage during the year is $\dfrac{x}{2}$, and the total yearly storage cost is the same as if $\dfrac{x}{2}$ tires were kept in storage for the entire year (Figure 3.43b). (This assertion, although reasonable, is not really obvious, and you have every right to be unconvinced. In Chapter 5, you will learn how to prove this fact mathematically using integral calculus.) It follows that

$$\text{Storage cost} = \frac{x}{2}(\text{cost of storing 1 tire for 1 year})$$

$$= \frac{x}{2}(0.96) = 0.48x$$

Putting it all together, the total cost is

$$C(x) = 0.48x + \frac{120,000}{x} + 1,500$$

The goal is to find the absolute minimum of $C(x)$ on the interval $0 < x \leq 6,000$. The derivative is

$$C'(x) = 0.48 - \frac{120,000}{x^2}$$

which is zero when

$$x^2 = \frac{120,000}{0.48} = 250,000 \quad \text{or} \quad x = \pm 500$$

Only the positive value $x = 500$ is in the relevant interval $0 < x \leq 6,000$. Since this is the only critical value in the interval, you can apply the second derivative test for absolute extrema. In particular, the second derivative of the cost function is

$$C''(x) = \frac{240,000}{x^3}$$

which is positive when $x > 0$. Hence, the absolute minimum of the total cost $C(x)$ on the interval $0 < x \leq 6,000$ occurs when $x = 500$, that is, when the manufacturer orders the tires in lots of 500.

Figure 3.44 Total cost: $C(x) = 0.48x + \frac{120,000}{x} + 1,500.$

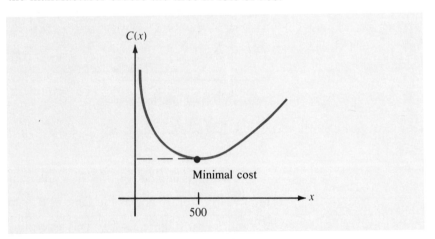

For reference, the graph of the total cost function is sketched in Figure 3.44.

Comment: Since the derivative of the (constant) purchase price of $1,500 in Example 5.1 was zero, this component of the total cost had no bearing on the optimization problem. In general, economists distinguish between **fixed costs** (such as the total purchase price) and **variable costs** (such as the storage and ordering costs). To minimize total cost, it is sufficient to minimize the sum of all the variable components of cost.

Average Cost and Marginal Cost   The purpose of this discussion is to derive an important relationship between average cost and marginal cost. This relationship generalizes to any pair of average and marginal quantities in economics. To begin, consider the following computational example.

EXAMPLE 5.2

Suppose the total cost in dollars of manufacturing q units of a certain commodity is $C(q) = 3q^2 + q + 48$.

(a) At what level of production is the average cost per unit the smallest?
(b) At what level of production is the average cost per unit equal to the marginal cost?
(c) On the same set of axes, graph the average cost and marginal cost functions.

Solution

(a) The average cost per unit $A(q)$ is the total cost divided by the number of units produced. That is,

$$A(q) = \frac{C(q)}{q} = \frac{3q^2 + q + 48}{q} = 3q + 1 + \frac{48}{q}$$

The goal is to find the absolute minimum of $A(q)$ on the interval $q > 0$. The first derivative is

$$A'(q) = 3 - \frac{48}{q^2} = \frac{3q^2 - 48}{q^2} = \frac{3(q^2 - 16)}{q^2} = \frac{3(q + 4)(q - 4)}{q^2}$$

which is zero on the interval $q > 0$ only when $q = 4$. Since the second derivative $A''(q) = \frac{96}{q^3}$ is positive when $q > 0$, it follows from the second derivative test that the average cost $A(q)$ is minimal on $q > 0$ when $q = 4$, that is, when 4 units are produced.

(b) The marginal cost is the derivative $C'(q) = 6q + 1$ of the total cost function and equals average cost when

$$C'(q) = A(q)$$

$$6q + 1 = 3q + 1 + \frac{48}{q}$$

$$3q = \frac{48}{q}$$

$$q^2 = 16 \quad \text{or} \quad q = 4$$

which is the same level of production as that in part (a) for which average cost is minimal.

(c) The marginal cost $C'(q) = 6q + 1$ is a linear function, and its graph is a straight line with slope 6 and vertical intercept 1.

To graph the average cost function $A(q) = 3q + 1 + \dfrac{48}{q}$, observe from part (a) that its derivative

$$A'(q) = \frac{3(q + 4)(q - 4)}{q^2}$$

is negative for $0 < q < 4$ and positive for $q > 4$. Hence, $A(q)$ is decreasing for $0 < q < 4$, is increasing for $q > 4$, and has a relative minimum at $q = 4$.

The graphs of the marginal and average cost functions for $q > 0$ are sketched in Figure 3.45.

Notice that the average cost in Figure 3.45 is decreasing when the marginal cost is less than the average cost. The average cost is increasing when the marginal cost is greater than the average cost. And the average cost has a first-order critical point (a relative minimum in this case) when the marginal cost is equal to the average cost.

Figure 3.45 Average cost and marginal cost curves for Example 5.2.

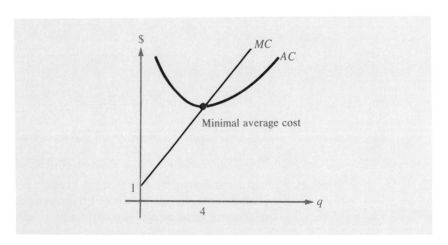

This relationship between average cost and marginal cost is not accidental but holds in general. Here is a summary of the situation.

The Relationship between Average and Marginal Cost

Suppose AC and MC denote the average cost and marginal cost, respectively. Then

AC is decreasing when $MC < AC$.
AC is increasing when $MC > AC$.
AC has a first-order critical point (usually a relative minimum) when $MC = AC$.

Economic Explanation

Here is the informal explanation of this relationship that is often given in economics texts.

The marginal cost is (approximately) the cost of producing one additional unit. If the additional unit costs less to produce than the average cost of the existing units (i.e., if $MC < AC$), then this less expensive unit will cause the average cost per unit to decrease. On the other hand, if the additional unit costs more than the average cost of the existing units (i.e., if $MC > AC$), then this more expensive unit will cause the average cost per unit to increase. If the cost of the additional unit is equal to the average cost of the existing units (i.e., if $MC = AC$), then the average cost will neither increase nor decrease; that is, it will have a first-order critical point.

Mathematical Derivation

The relationship between average cost and marginal cost can be derived mathematically using calculus. If $C(q)$ is the total cost of manufacturing q units, then the average cost per unit is

$$A(q) = \frac{C(q)}{q}$$

By the quotient rule, the first derivative is

$$A'(q) = \frac{qC'(q) - C(q)(1)}{q^2}$$

which is zero when its numerator is zero, that is, when

$$qC'(q) - C(q) = 0$$
$$qC'(q) = C(q)$$
$$C'(q) = \frac{C(q)}{q}$$

or
$$MC = AC$$

That is, when $A(q)$ has a first-order critical value, marginal cost is equal to average cost.

If $MC < AC$, then

$$C'(q) < \frac{C(q)}{q} \quad \text{or} \quad qC'(q) < C(q)$$

and $\quad A'(q) = \dfrac{qC'(q) - C(q)}{q^2} < \dfrac{C(q) - C(q)}{q^2} = 0$

which implies that average cost $A(q)$ is decreasing.

Similarly, if $MC > AC$, then

$$C'(q) > \frac{C(q)}{q} \quad \text{or} \quad qC'(q) > C(q)$$

and $\quad A'(q) = \dfrac{qC'(q) - C(q)}{q^2} > \dfrac{C(q) - C(q)}{q^2} = 0$

which implies that average cost $A(q)$ is increasing.

Generalization   The relationship between average cost and marginal cost is actually a special case of a general relationship between any pair of average and marginal quantities in economics (average revenue and marginal revenue, for example). The only possible change is the nature of the first-order critical point that occurs when the average quantity equals the marginal quantity. (Average revenue, for example, usually has a relative maximum when average revenue equals marginal revenue.)

Here is a summary of the general situation. Its informal economic explanation and mathematical derivation are virtually identical to those for the special case of average and marginal cost.

General Relationship between Average and Marginal Quantities

Suppose AQ and MQ are the average and marginal values, respectively, of some quantity Q. Then

AQ is decreasing when $MQ < AQ$.
AQ is increasing when $MQ > AQ$.
AQ has a first-order critical point when $MQ = AQ$.

Elasticity of Demand   Consumer demand for a product is usually related to its price. In most cases, the demand decreases as the price increases. The sensitivity of demand to changes in price varies from one product to another. For some products, such as soap, flashlight batteries, and salt, small percentage changes in price have little effect on demand. For other products, such as airline tickets, designer furniture, and home loans, small percentage changes in price can have a considerable effect on demand.

A convenient measure of the sensitivity of demand to changes in price is the percentage change in demand that is generated by a 1 percent increase in price. If p denotes the price, q the corresponding number of units demanded, and Δp a (small) change in price, the approximation formula for percentage change (from Chapter 2, Section 4) gives

$$\text{Percentage change in } q \simeq 100\,\frac{(dq/dp)\,\Delta p}{q}$$

If, in particular, the change in p is a 1 percent increase, then $\Delta p = 0.01p$ and

$$\text{Percentage change in } q \simeq 100\,\frac{(dq/dp)(0.01p)}{q} = \frac{p}{q}\frac{dq}{dp}$$

The expression on the right-hand side of this approximation is known in economics as the **elasticity of demand.**

Elasticity of Demand

If q denotes the demand for a commodity and p its price, the elasticity of demand, denoted by the Greek letter η (eta), is given by the formula

$$\eta = \frac{p}{q}\frac{dq}{dp}$$

and has the following interpretation:

$$\eta \simeq \left(\begin{array}{c}\text{percentage change in demand due to}\\ \text{a 1 percent increase in price}\end{array}\right)$$

Here is an example.

EXAMPLE 5.3

Suppose the demand q and price p for a certain commodity are related by the linear equation $q = 240 - 2p$ (for $0 \le p \le 120$).

(a) Express the elasticity of demand as a function of p.
(b) Calculate the elasticity of demand when the price is $p = 100$. Interpret your answer.
(c) Calculate the elasticity of demand when the price is $p = 50$. Interpret your answer.
(d) At what price is the elasticity of demand equal to -1? What is the economic significance of this price?

Solution

(a) The elasticity of demand is

$$\eta = \frac{p}{q}\frac{dq}{dp} = \frac{p}{q}(-2) = -\frac{2p}{240 - 2p} = -\frac{p}{120 - p}$$

(b) When $p = 100$, the elasticity of demand is

$$\eta = -\frac{100}{120 - 100} = -5$$

That is, when the price is $p = 100$, a 1 percent increase in price will produce a decrease in demand of approximately 5 percent.

(c) When $p = 50$, the elasticity of demand is

$$\eta = -\frac{50}{120 - 50} \simeq -0.71$$

That is, when the price is $p = 50$, a 1 percent increase in price will produce a decrease in demand of approximately 0.71 percent.

(d) The elasticity of demand will be equal to -1 when

$$-1 = -\frac{p}{120 - p} \qquad 120 - p = p \qquad 2p = 120 \qquad \text{or} \qquad p = 60$$

At this price, a 1 percent increase in price will result in a decrease in demand of approximately the same percent.

Levels of Elasticity of Demand

In general, the elasticity of demand η is negative, since demand decreases as price increases. If $|\eta| > 1$, the percentage decrease in demand is greater than the percentage increase in price that caused it. This was the case in part (b) of the preceding example, in which $\eta = -5$, implying that an increase in price of 1 percent produces a decrease in demand of approximately 5 percent. At prices for which $|\eta| > 1$, the demand is relatively sensitive to changes in price, and economists say that demand is **elastic** with respect to price.

If $|\eta| < 1$, the percentage decrease in demand is less than the percentage increase in price that caused it. This was the case in part (c) of the preceding example, in which $\eta \simeq -0.71$, implying that an increase in price of 1 percent produces a decrease in demand of approximately only 0.71 percent. At prices for which $|\eta| < 1$, the demand is relatively insensitive to changes in price, and economists say that demand is **inelastic** with respect to price.

If $|\eta| = 1$, as in part (d) of the preceding example, the percentage changes in price and demand are (approximately) equal, and the demand is said to be of **unit elasticity.**

> **Levels of Elasticity**
>
> If $|\eta| > 1$, demand is said to be elastic with respect to price.
>
> If $|\eta| < 1$, demand is said to be inelastic with respect to price.
>
> If $|\eta| = 1$, demand is said to be of unit elasticity with respect to price.

Elasticity and the Total Revenue

The total revenue generated by the sale of a commodity is the price per unit times the number of units sold. That is,

$$R = pq$$

where R is the revenue, p the price per unit, and q the number of units sold (i.e., the demand).

The level of the elasticity of demand with respect to price gives useful information about the total revenue obtained from the sale of the product. In particular, if the demand is inelastic ($|\eta| < 1$), the total revenue increases as the price increases (even though demand drops). The idea is that, in this case, the relatively small percentage decrease in demand is offset by the larger percentage increase in price, and hence the revenue, which is price times demand, increases. If the demand is elastic ($|\eta| > 1$), the total revenue decreases as the price increases. In this case, the relatively large percentage decrease in demand is not offset by the smaller percentage increase in price.

Here is a summary of the situation.

> **Elasticity and Revenue**
>
> If demand is inelastic ($|\eta| < 1$), total revenue increases as price increases.
>
> If demand is elastic ($|\eta| > 1$), total revenue decreases as price increases.

A graph showing the usual relationship between revenue and price is drawn in Figure 3.46. The graph is increasing when the demand is inelastic ($|\eta| < 1$), is decreasing when the demand is elastic ($|\eta| > 1$), and has a maximum at the price for which the demand is of unit elasticity ($|\eta| = 1$).

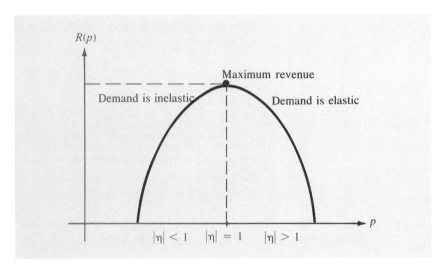

Figure 3.46 Revenue as a function of price.

Derivation of the Relationship between Elasticity and Revenue

To see why the level of elasticity of demand determines the increase or decrease of total revenue, start with the revenue equation

$$R = pq$$

and differentiate both sides implicitly with respect to p to get

$$\frac{dR}{dp} = p\frac{dq}{dp} + q \qquad \text{(by the product rule)}$$

To get the elasticity $\eta = \dfrac{p}{q}\dfrac{dq}{dp}$ into the picture, simply multiply the expression on the right-hand side by $\dfrac{q}{q}$ as follows:

$$\frac{dR}{dp} = \frac{q}{q}\left(p\frac{dq}{dp} + q\right) = q\left(\frac{p}{q}\frac{dq}{dp} + 1\right) = q(\eta + 1)$$

If $|\eta| < 1$, then $\eta > -1$ (since η is negative) and

$$\frac{dR}{dp} > q(-1 + 1) = 0$$

That is, the derivative of R with respect to p is positive, and so revenue is an increasing function of price.

If $|\eta| > 1$, then $\eta < -1$ (again, since η is negative) and

$$\frac{dR}{dp} < q(-1 + 1) = 0$$

which implies that revenue is a decreasing function of price.

The relationship between elasticity of demand and total revenue is illustrated in the next example.

EXAMPLE 5.4

Suppose the demand q and price p for a certain commodity are related by the equation $q = 300 - p^2$ (for $0 \leq p \leq \sqrt{300}$).

(a) Determine where the demand is elastic, inelastic, and of unit elasticity with respect to price.
(b) Use the results of part (a) to describe the behavior of the total revenue as a function of price.
(c) Find the total revenue function explicitly and use its first derivative to determine its intervals of increase and decrease and the price at which revenue is maximized.

Solution

(a) The elasticity of demand is

$$\eta = \frac{p}{q}\frac{dq}{dp} = \frac{p}{300 - p^2}(-2p) = -\frac{2p^2}{300 - p^2}$$

The demand is of unit elasticity when $|\eta| = 1$, that is, when

$$\frac{2p^2}{300 - p^2} = 1 \qquad 2p^2 = 300 - p^2 \qquad 3p^2 = 300$$

$$p^2 = 100 \qquad \text{or} \qquad p = \pm 10$$

of which only $p = 10$ is in the relevant interval $0 \leq p \leq \sqrt{300}$. If $0 \leq p < 10$,

$$|\eta| = \frac{2p^2}{300 - p^2} < \frac{2(10)^2}{300 - (10)^2} = 1$$

and hence the demand is inelastic. If $10 < p \leq \sqrt{300}$,

$$|\eta| = \frac{2p^2}{300 - p^2} > \frac{2(10)^2}{300 - (10)^2} = 1$$

and hence the demand is elastic.

(b) The total revenue is an increasing function of p when demand is inelastic, that is, on the interval $0 \leq p < 10$, and a decreasing function of p when demand is elastic, that is, on the interval $10 < p \leq \sqrt{300}$. At the price $p = 10$ of unit elasticity, the revenue function has a relative maximum.

(c) The revenue function is

$$R = pq$$

or $\qquad\qquad R(p) = p(300 - p^2) = 300p - p^3$

Its derivative is

$$R'(p) = 300 - 3p^2 = 3(10 - p)(10 + p)$$

which is zero when $p = \pm 10$, of which only $p = 10$ is in the relevant interval $0 \leq p \leq \sqrt{300}$.

On the interval $0 \leq p < 10$, $R'(p)$ is positive and so $R(p)$ is increasing. On the interval $10 < p \leq \sqrt{300}$, $R'(p)$ is negative and so $R(p)$ is decreasing. At the critical value $p = 10$, $R(p)$ stops increasing and starts decreasing and hence has a relative maximum.

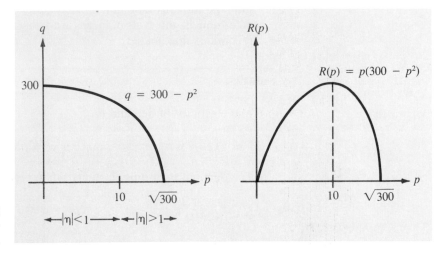

Figure 3.47 Demand and revenue curves for Example 5.4.

The relevant portions of the graphs of the demand and revenue functions are sketched in Figure 3.47.

Perspective: The analysis of elasticity and revenue presented here differs slightly from that usually found in economics texts. For technical reasons, economists prefer to treat demand rather than price as the independent variable in the discussion of these concepts. The approach adopted in this section (in which price was the independent variable) was chosen because it is the more straightforward and natural for those not already accustomed to mathematical economics. However, so that you will have some familiarity and experience with the other approach, the development of this material from that point of view is outlined in problems 20 through 23 of the following exercise set.

Problems 3.5

Inventory   1. An electronics firm uses 600 cases of transistors each year. The cost of storing one case for a year is 90 cents, and the ordering fee is $30 per shipment. How many cases should the firm order each time to keep total cost at a minimum? (Assume that the transistors are used at a constant rate throughout the year and that each shipment arrives just as the preceding shipment has been used up.)

Inventory   2. A store expects to sell 800 bottles of perfume this year. The perfume costs $20 per bottle, the ordering fee is $10 per shipment, and the cost of storing the perfume is 40 cents per bottle per year. The perfume is consumed at a constant rate throughout the year, and each shipment arrives just as the preceding shipment has been used up.
   (a) How many bottles should the store order in each shipment to minimize cost?
   (b) How often should the store order the perfume?

Inventory   3. Through its franchised stations, an oil company gives out 16,000 road maps per year. The cost of setting up a press to print the maps is $100 for each production run. In addition, production costs are 6 cents per map and storage costs are 20 cents per map per year. The maps are distributed at a uniform rate throughout the year and are printed in equal batches timed so that each arrives just as the preceding batch has been used up. How many maps should the oil company print in each batch to minimize cost?

Inventory   4. A manufacturing firm receives raw materials in equal shipments arriving at regular intervals throughout the year. The cost of storing the raw materials is directly proportional to the size of each shipment, while the total yearly ordering cost is inversely proportional to the shipment size. Show that the total cost is lowest when the total storage cost and total ordering cost are equal.

Average Cost   5. Suppose the total cost in dollars of manufacturing q units of a certain commodity is $C(q) = 3q^2 + 5q + 75$.
   (a) At what level of production is the average cost per unit the smallest?
   (b) At what level of production is the average cost per unit equal to the marginal cost?
   (c) On the same set of axes, graph the average cost and marginal cost functions for $q > 0$.

Average Cost   6. Suppose the total cost in dollars of manufacturing q units of a certain commodity is $C(q) = q^3 + 5q + 162$.
   (a) At what level of production is the average cost per unit smallest?
   (b) At what level of production is the average cost per unit equal to the marginal cost?
   (c) On the same set of axes, graph the average cost and marginal cost functions for $q > 0$.

Average Revenue   7. Suppose the total revenue in dollars from the sale of q units of a certain commodity is $R(q) = -2q^2 + 68q - 128$.

(a) At what level of sales is the average revenue per unit equal to the marginal revenue?

(b) Verify that the average revenue is increasing if the level of sales is less than the level in part (a) and decreasing if the level of sales is greater than the level in part (a).

(c) On the same set of axes, graph the relevant portions of the average and marginal revenue functions.

Average Productivity 8. The output Q at a certain factory is a function of the number L of worker-hours of labor that are used. Use calculus to prove that when the average output per worker-hour is greatest, the average output is equal to the marginal output per worker-hour. (*Hint:* The marginal output per worker-hour is the derivative of output Q with respect to labor L.) You may assume without proof that the first-order critical point of the average output function is actually the desired absolute maximum.

Minimal Cost 9. A manufacturer finds that in producing x units per day (for $0 < x < 100,000$), three different kinds of cost are involved:

(a) A fixed cost of $1,200 per day in wages

(b) A production cost of $1.20 per day for each unit produced

(c) An ordering cost of $\dfrac{100}{x^2}$ dollars per day.

Express the total cost as a function of x and determine the level of production that results in minimal total cost.

Demand 10. A manufacturer finds that the demand function for a certain product is $x(p) = \dfrac{73}{\sqrt{p}}$ when the price is p dollars per unit. Should the price be raised or lowered in order to increase consumer expenditure? Why?

National Consumption 11. Assume that total national consumption is given by a function $C(x)$, where x is the total national income. The derivative $C'(x)$ is called the **marginal propensity to consume,** and if $S = x - C$ represents total national savings, then $S'(x)$ is called the **marginal propensity to save.** Suppose the consumption function is $C(x) = 8 + 0.8x + 0.8\sqrt{x}$. Find the marginal propensity to consume and determine the value of x that results in the smallest total savings.

Inventory 12. The inventory model analyzed in this section is not the only such model possible. Suppose a company must supply N units per time period at a uniform rate. Assume that the storage cost per unit is D_1 dollars per time period and that the setup cost is D_2 dollars. If production is at a uniform rate of m units per time period (with no items in inventory at the end of each period), it can be shown that the total storage cost is $C_1 = \dfrac{D_1 x}{2}\left(1 - \dfrac{N}{m}\right)$, where x is the number of items produced in each run.

(a) Show that the total average cost per period is

$$C = \frac{D_1 x}{2}\left(1 - \frac{N}{m}\right) + \frac{D_2 N}{x}$$

(b) Find an expression for the number of items that should be produced in each run to minimize the total average cost per time period.

Elasticity   13. Suppose that the demand equation for a certain commodity is $q = 60 - 0.1p$ (for $0 \le p \le 600$).

(a) Express the elasticity of demand as a function of p.
(b) Calculate the elasticity of demand when the price is $p = 200$. Interpret your answer.
(c) At what price is the elasticity of demand equal to -1?

Elasticity   14. Suppose that the demand equation for a certain commodity is $q = 200 - 2p^2$ (for $0 \le p \le 10$).

(a) Express the elasticity of demand as a function of p.
(b) Calculate the elasticity of demand when the price is $p = 6$. Interpret your answer.
(c) At what price is the elasticity of demand equal to -1?

Elasticity and Revenue   15. Suppose that the demand equation for a certain commodity is $q = 500 - 2p$ (for $0 \le p \le 250$).

(a) Determine where the demand is elastic, inelastic, and of unit elasticity with respect to price.
(b) Use the results of part (a) to determine the intervals of increase and decrease of the revenue function and the price at which revenue is maximized.
(c) Find the total revenue function explicitly and use its first derivative to determine its intervals of increase and decrease and the price at which revenue is maximized.
(d) Graph the relevant portions of the demand and revenue functions.

Elasticity and Revenue   16. Suppose that the demand equation for a certain commodity is $q = 120 - 0.1p^2$ (for $0 \le p \le \sqrt{1{,}200}$).

(a) Determine where the demand is elastic, inelastic, and of unit elasticity with respect to price.
(b) Use the results of part (a) to determine the intervals of increase and decrease of the revenue function and the price at which revenue is maximized.
(c) Find the total revenue function explicitly and use its first deriva-

tive to determine its intervals of increase and decrease and the price at which revenue is maximized.

(d) Graph the relevant portions of the demand and revenue functions.

Elasticity
17. Suppose the demand equation for a certain commodity is $q = \dfrac{a}{p^m}$, where a and m are positive constants. Show that the elasticity of demand is equal to $-m$ for all values of p. Interpret this result.

Elasticity and Revenue
18. Suppose the demand equation for a certain commodity is linear, that is, $q = b - ap$ $\left(\text{for } 0 \le p \le \dfrac{b}{a}\right)$, where a and b are positive constants.

(a) Express the elasticity of demand as a function of p.

(b) Show that the demand is of unit elasticity at the midpoint $p = \dfrac{b}{2a}$
of the relevant interval $0 \le p \le \dfrac{b}{a}$.

(c) Find the intervals on which the demand is elastic and on which it is inelastic.

(d) Find the total revenue function explicitly and use its first derivative to determine its intervals of increase and decrease.

(e) Graph the relevant portions of the demand and revenue functions.

Effect of Taxation on a Monopoly
19. **A monopolist** is a manufacturer who can manipulate the price of a commodity and usually does so with an eye toward maximizing profit. When the government taxes output, the tax effectively becomes an additional cost item, and the monopolist is forced to decide how much of the tax to absorb and how much to pass on to the consumer.

Suppose a particular monopolist estimates that when x units are produced, the total cost will be $C(x) = \dfrac{7}{8}x^2 + 5x + 100$ dollars and the market price of the commodity will be $p(x) = 15 - \dfrac{3}{8}x$ dollars per unit. Further assume that the government imposes a tax of t dollars on each unit produced.

(a) Show that profit is maximized when $x = \dfrac{2}{5}(10 - t)$.

(b) Suppose that government assumes that the monopolist will always act so as to maximize total profit. What value of t should be chosen so as to guarantee maximum total tax revenue?

(c) If the government chooses the optimum rate of taxation found in part (b), how much of this tax will be absorbed by the monopolist and how much will be passed on to the consumer?

Elasticity and Revenue as Functions of Demand

For technical reasons, economists often treat elasticity and revenue as functions of demand q rather than price p. In this context, the elasticity of demand is defined as

$$\eta = \frac{p/q}{dp/dq}$$

which can be shown to be equivalent to the definition with which you are already familiar. (The proof of this equivalence involves properties of inverse functions, which are discussed in more advanced texts.)

In Problems 20 through 23, use this alternative definition of elasticity of demand to carry out the required calculations and derivations.

Elasticity

20. Suppose the demand q and price p for a certain commodity are related by the equation $p = 60 - 2q$ (for $0 \le q \le 30$).
    (a) Express the elasticity of demand as a function of q.
    (b) Calculate the elasticity of demand when $q = 10$. Interpret your answer.
    (c) Substitute for q in the formula in part (a) to express the elasticity of demand as a function of p.
    (d) Use our original definition of η to express the elasticity of demand as a function of p. [The answer should be the same as in part (c).]

Elasticity and Revenue

21. Prove that the derivative of total revenue R with respect to demand q is

$$\frac{dR}{dq} = p\left(1 + \frac{1}{\eta}\right)$$

(*Hint:* As in the derivation given in the text, start with the equation $R = pq$, but this time differentiate with respect to q.)

Elasticity and Revenue

22. Use the formula derived in Problem 21 to show that revenue R is an increasing function of demand q when $|\eta| > 1$ and a decreasing function of q when $|\eta| < 1$.

Elasticity and Revenue

23. Suppose that the demand equation for a certain commodity is $p = 600 - 2q^2$ (for $0 \le q \le \sqrt{300}$).
    (a) Express the elasticity of demand as a function of q and determine where $|\eta| = 1$, $|\eta| < 1$, and $|\eta| > 1$.
    (b) Use the results of part (a) to determine the intervals of increase and decrease of revenue as a function of demand q. At what level of demand is the revenue greatest?
    (c) Express total revenue explicitly as a function of q and use its first derivative to determine its intervals of increase and decrease and the demand at which revenue is maximized.
    (d) Graph the relevant portions of the functions expressing price and revenue in terms of q.

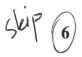

6 Newton's Method for Approximating Roots

The Roots of an Equation

A value of a variable that satisfies an equation is said to be a **solution** or **root** of that equation. In geometric terms, a root of an equation $f(x) = 0$ is an x intercept of the graph of f. In this section, you will see a technique known as **Newton's method** which uses a formula involving the derivative to approximate roots of equations of the form $f(x) = 0$.

Like most techniques for approximating a root of an equation, Newton's method requires that you start with a rough estimate of the location of the root. This initial estimate is often obtained by applying the following result, which is an immediate consequence of the intermediate value property of continuous functions introduced in Problems 62 through 65 in Section 5 of Chapter 1.

> **Estimating Roots with the Intermediate Value Property**
>
> If $f(x)$ is continuous on the interval $a \leq x \leq b$, and if $f(a)$ and $f(b)$ have opposite signs, then the equation $f(x) = 0$ has at least one root between $x = a$ and $x = b$.

In geometric terms, this says that if f is continuous and if the points $(a, f(a))$ and $(b, f(b))$ lie on opposite sides of the x axis, then the graph of f (which is an unbroken curve) must cross the x axis somewhere between $x = a$ and $x = b$. The situation is illustrated in Figure 3.48.

The use of the intermediate value property to estimate the location of roots is illustrated in the following example.

EXAMPLE 6.1

Make a rough sketch of the function $f(x) = x^3 - x^2 - 1$ and use the intermediate value property to estimate the location of the roots of the equation $f(x) = 0$.

Solution

The graph (obtained with the aid of the first derivative) is sketched in Figure 3.49. It indicates that the equation $f(x) = 0$ has only one root. Moreover, since

$$f(1) = -1 < 0 \quad \text{and} \quad f(2) = 3 > 0$$

it follows from the intermediate value property that the root lies between $x = 1$ and $x = 2$.

Figure 3.48 Estimating a root with the intermediate value property.

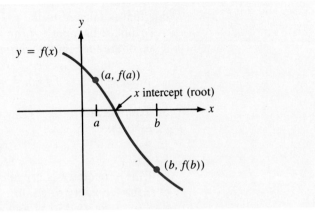

Figure 3.49 The graph of $y = x^3 - x^2 - 1$.

Newton's Method   The basic idea behind Newton's method is illustrated in Figure 3.50a, in which r is a root of the equation $f(x) = 0$, x_0 is an initial approximation to

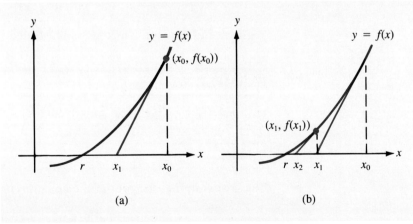

(a)   (b)

Figure 3.50 Newton's method.

r, and x_1 is a better approximation obtained by taking the x intercept of the line that is tangent to the graph of f at $(x_0, f(x_0))$.

To find a formula for the improved approximation x_1, recall that the slope of the tangent line through $(x_0, f(x_0))$ is the derivative $f'(x_0)$. Hence,

$$f'(x_0) = \text{slope} = \frac{\Delta y}{\Delta x} = \frac{f(x_0) - 0}{x_0 - x_1}$$

or, equivalently,

$$x_1 = x_0 - \frac{f(x_0)}{f'(x_0)}$$

If the procedure is repeated using x_1 as the initial approximation, an even better approximation is obtained (Figure 3.50b). This approximation, x_2, is related to x_1 as x_1 was to x_0. That is,

$$x_2 = x_1 - \frac{f(x_1)}{f'(x_1)}$$

The process can be continued until the desired degree of accuracy is obtained. In general, the nth approximation x_n is related to the $(n - 1)$st by the formula

$$x_n = x_{n-1} - \frac{f(x_{n-1})}{f'(x_{n-1})}$$

Newton's Method

To approximate a root of the equation $f(x) = 0$, start with an initial estimate x_0 and generate a sequence of increasingly accurate approximations $x_1, x_2, x_3, \ldots$ using the formula

$$x_n = x_{n-1} - \frac{f(x_{n-1})}{f'(x_{n-1})}$$

The use of Newton's method is illustrated in the following example.

EXAMPLE 6.2

Use three repetitions of Newton's method to approximate $\sqrt{5}$.

Solution

The goal is to find the root of the equation $f(x) = 0$, where

$$f(x) = x^2 - 5$$

The derivative of f is $f'(x) = 2x$, and so

$$x - \frac{f(x)}{f'(x)} = x - \frac{x^2 - 5}{2x} = \frac{x^2 + 5}{2x}$$

Thus, for $n = 1, 2, 3, \ldots$

$$x_n = \frac{x_{n-1}^2 + 5}{2x_{n-1}}$$

A convenient choice for the initial estimate is $x_0 = 2$ (since $2^2 = 4$, which is close to 5). Then,

$$x_1 = \frac{x_0^2 + 5}{2x_0} = 2.250 \qquad \text{using } x_0 = 2$$

$$x_2 = \frac{x_1^2 + 5}{2x_1} \approx 2.2361111 \qquad \text{using } x_1 = 2.250$$

$$x_3 = \frac{x_2^2 + 5}{2x_2} \approx 2.2360680 \qquad \text{using } x_2 = 2.2361111$$

Thus, $\sqrt{5}$ is approximately 2.236. (Actually, rounded off to seven decimal places, $\sqrt{5} = 2.2360680$, which is exactly the value obtained by Newton's method after only three repetitions!)

Estimating the Accuracy of Newton's Method

A systematic analysis of the error involved in approximation by Newton's method is beyond the scope of this text. However, there is a simple rule of thumb which states that, to use Newton's method to approximate a root to a given number of decimal places, round off all calculations to one more decimal place and stop when there is no change from one approximation to the next. This is illustrated in the following examples.

EXAMPLE 6.3

Use Newton's method to approximate the roots of the equation $x^3 - x^2 - 1 = 0$ to three decimal places.

Solution

Let $f(x) = x^3 - x^2 - 1$. Then, $f'(x) = 3x^2 - 2x$ and

$$x - \frac{f(x)}{f'(x)} = x - \frac{x^3 - x^2 - 1}{3x^2 - 2x} = \frac{2x^3 - x^2 + 1}{3x^2 - 2x}$$

and so, for $n = 1, 2, 3, \ldots$

$$x_n = \frac{2x_{n-1}^3 - x_{n-1}^2 + 1}{3x_{n-1}^2 - 2x_{n-1}}$$

The goal is to find the roots of the equation $f(x) = 0$. As you saw in Example 6.1, there is only one such root, and it must lie between 1 and 2. Either of these two values is a reasonable choice for the initial estimate x_0. If you take $x_0 = 1$, you get

$$x_1 = \frac{2x_0^3 - x_0^2 + 1}{3x_0^2 - 2x_0} = 2 \qquad \text{using } x_0 = 1$$

$$x_2 = \frac{2x_1^3 - x_1^2 + 1}{3x_1^2 - 2x_1} = 1.6250 \qquad \text{using } x_1 = 2$$

$$x_3 = \frac{2x_2^3 - x_2^2 + 1}{3x_2^2 - 2x_2} \simeq 1.4858 \qquad \text{using } x_2 = 1.6250$$

$$x_4 = \frac{2x_3^3 - x_3^2 + 1}{3x_3^2 - 2x_3} \simeq 1.4660 \qquad \text{using } x_3 = 1.4858$$

$$x_5 = \frac{2x_4^3 - x_4^2 + 1}{3x_4^2 - 2x_4} \simeq 1.4656 \qquad \text{using } x_4 = 1.4660$$

$$x_6 = \frac{2x_5^3 - x_5^2 + 1}{3x_5^2 - 2x_5} \simeq 1.4656 \qquad \text{using } x_5 = 1.4656$$

Since x_5 and x_6 are identical (to four decimal places), no further computation is necessary. You now simply round off the number 1.4656 to conclude that, to three-decimal-place accuracy, the desired root is 1.466.

When Newton's Method Fails   It can be shown that if the approximations $x_1, x_2, x_3, \ldots$ generated by Newton's method converge to some (finite) number, this number must be a root of the equation $f(x) = 0$. However, there are cases in which the approximations do not converge, even though the equation has a root. For instance, this occurs if the initial estimate x_0 is not "close enough" to the root. This phenomenon is illustrated in Figure 3.51, which shows successive "approximations" that fail to converge to the root but instead increase without bound. In such cases, it is usually possible to

Figure 3.51 Newton's method fails because x_0 is not close enough to r.

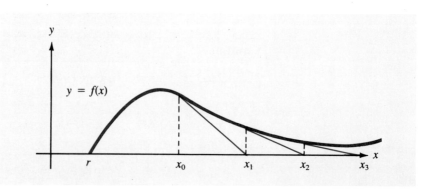

achieve convergence by reworking the problem using a better initial estimate.

Problems 3.6

In Problems 1 through 8, use Newton's method to find all the roots of the given equation to three-decimal-place accuracy.

1. $x^2 - 12 = 0$ 2. $x^3 + 49 = 0$

3. $x^3 + x^2 - 1 = 0$ 4. $x^3 + 2x^2 - x + 1 = 0$

5. $x^3 - 2x - 5 = 0$ 6. $3x^6 + x^2 = 2$

7. $x^2 - 5x + 1 = 0$ (*Hint:* There are two roots.)

8. $x^4 - 4x^3 + 10 = 0$ (*Hint:* There are two roots.)

In Problems 9 through 14, use Newton's method to compute the given number to five decimal places.

9. $\sqrt{2}$ 10. $\sqrt{7}$ 11. $\sqrt[3]{9}$

12. $\sqrt[3]{-20}$ 13. $\sqrt{\dfrac{1}{26}}$ 14. $\sqrt[4]{15}$

In Problems 15 and 16, sketch the graph of the function $f(x)$ and explain why Newton's method fails for the equation $f(x) = 0$ and the given initial value x_0.

15. $f(x) = 9x^2 + 2x + 3; x_0 = -\dfrac{1}{9}$ 16. $f(x) = 1 - \dfrac{1}{x}; x_0 = 2$

17. Can the equation $\sqrt{x} = 0$ be solved using Newton's method with the initial estimate $x_0 = 0.05$? Does it make any difference if another initial estimate is chosen? Explain.

18. The ancient Babylonians (circa 1700 B.C.) approximated $\sqrt{N}$ by applying the formula

$$x_{n+1} = \frac{1}{2}\left(x_n + \frac{N}{x_n}\right) \qquad \text{for } n = 1, 2, 3, \ldots$$

Show that this formula can be derived from Newton's method and then use it to estimate $\sqrt{1{,}265}$ correct to 5 decimal places.

19. Sometimes Newton's method fails no matter what initial value x_0 is chosen. Let $f(x) = \sqrt[3]{x}$ and choose x_0 arbitrarily.
    (a) Show that $x_{n+1} = -2x_n$ for $n = 0, 1, 2, \ldots$ so that the successive values generated by Newton's method are $x_0, -2x_0, 4x_0, \ldots$.
    (b) Sketch the graph of $f(x) = \sqrt[3]{x}$ and show the tangent lines that result in the successive approximations x_1, x_2, x_3, x_4. Why do these numbers fail to estimate a root of $f(x) = 0$?

Chapter Summary and Review Problems

Important Terms, Symbols, and Formulas

Relative maxima and minima

f is increasing: $f'(x) > 0$

f is decreasing: $f'(x) < 0$

First-order critical point: $f'(x) = 0$ or $f'(x)$ is undefined

Concave upward: Slope of tangent is increasing; $f''(x) > 0$

Concave downward: Slope of tangent is decreasing; $f''(x) < 0$

Inflection point: Concavity changes

Second-order critical point: $f''(x) = 0$ or $f''(x)$ is undefined

Second derivative test for relative extrema: Suppose $f'(a) = 0$.

If $f''(a) > 0$, then f has a relative minimum at $x = a$

If $f''(a) < 0$, then f has a relative maximum at $x = a$

Absolute maxima and minima

Absolute extrema on closed intervals: At critical values or endpoints

Second derivative test for absolute extrema: Suppose $f'(a) = 0$ and $x = a$ is the only critical value on the interval.

If $f''(a) > 0$, then $f(a)$ is the absolute minimum on the interval

If $f''(a) < 0$, then $f(a)$ is the absolute maximum on the interval

Point of diminishing returns

Inventory models

Average and marginal quantities

Elasticity of demand:

$$\eta = \frac{p}{q}\frac{dq}{dp} \simeq \left(\begin{array}{c}\text{percentage change in demand due to}\\ \text{a 1 percent increase in price}\end{array}\right)$$

Elastic demand, inelastic demand, unit elasticity

Elasticity and the revenue function

Root of an equation

Newton's method: $x_n = x_{n-1} - \dfrac{f(x_{n-1})}{f'(x_{n-1})}$

Review Problems

1. Determine where the given function is increasing and where it is decreasing, find its relative extrema, and sketch the graph.

(a) $f(x) = -2x^3 + 3x^2 + 12x - 5$

(b) $f(x) = 3x^5 - 20x^3$

(c) $f(x) = \dfrac{x^2}{x + 1}$

(d) $f(x) = 2x + \dfrac{8}{x} + 2$

2. Sketch the graph of a function that has all of the following properties:

(a) $f'(x) > 0$ when $x < 0$ and when $x > 5$

(b) $f'(x) < 0$ when $0 < x < 5$

(c) $f''(x) > 0$ when $-6 < x < -3$ and when $x > 2$

(d) $f''(x) < 0$ when $x < -6$ and when $-3 < x < 2$

3. Determine where the given function is increasing, decreasing, concave upward, and concave downward. Find the relative extrema and inflection points and draw the graph.

(a) $f(x) = x^2 - 6x + 1$ (b) $f(x) = x^3 - 3x^2 + 2$

(c) $f(x) = \dfrac{x^2 + 3}{x - 1}$ (d) $f(x) = \dfrac{x - 1}{(x + 1)^2}$

4. Use the second derivative test to find the relative maxima and minima of the given function.

(a) $f(x) = -2x^3 + 3x^2 + 12x - 5$

(b) $f(x) = \dfrac{x^2}{x + 1}$

(c) $f(x) = 2x + \dfrac{8}{x} + 2$

5. Find the absolute maximum and absolute minimum (if any) of the given function on the specified interval.

(a) $f(x) = -2x^3 + 3x^2 + 12x - 5$; $-3 \le x \le 3$

(b) $f(x) = -3x^4 + 8x^3 - 10$; $0 \le x \le 3$

(c) $f(x) = \dfrac{x^2}{x + 1}$; $-\dfrac{1}{2} \le x \le 1$

(d) $f(x) = 2x + \dfrac{8}{x} + 2$; $x > 0$

6. It is estimated that between the hours of noon and 7:00 P.M., the speed of highway traffic flowing past a certain downtown exit is

approximately $S(t) = t^3 - 9t^2 + 15t + 45$ miles per hour, where t is the number of hours past noon. At what time between noon and 7:00 P.M. is the traffic moving the fastest, and at what time between noon and 7:00 P.M. is it moving the slowest?

7. The rate at which a rumor spreads through a community is jointly proportional to the number of people who have heard the rumor and the number who have not. Show that the rumor is spreading most rapidly when half the people have heard it.

8. A manufacturer has been selling lamps at $6 apiece, and at this price, consumers have been buying 3,000 lamps per month. The manufacturer wishes to raise the price and estimates that for each $1 increase in the price, 1,000 fewer lamps will be sold each month. The manufacturer can produce the lamps at a cost of $4 per lamp. At what price should the manufacturer sell the lamps to generate the greatest possible profit?

9. You wish to use 300 meters of fencing to surround two identical adjacent rectangular plots, as shown in the accompanying figure. How should you do this to make the combined area of the plots as large as possible?

10. A farmer wishes to enclose a rectangular pasture with 320 feet of fence. What dimensions give the maximum area if
    (a) The fence is on all four sides of the pasture.
    (b) The fence is on three sides of the pasture and the fourth side is bounded by a wall.

11. A cylindrical container with no top is to be constructed for a fixed amount of money. The cost of the material used for the bottom is 3 cents per square inch, and the cost of the material used for the curved side is 2 cents per square inch. Use calculus to derive a simple relationship between the radius and height of the container having the greatest volume.

12. A postal clerk spends 4 hours each morning sorting mail. During that time, the clerk can sort approximately $f(t) = -t^3 + 7t^2 + 200t$ letters

in t hours. At what time during this period is the clerk performing at peak efficiency?

13. A citrus grower estimates that if 60 lemon trees are planted in a grove, the average yield per tree will be 475 lemons. The average yield will decrease by 5 lemons per tree for each additional tree planted in the grove. How many trees should the grower plant to maximize the total yield? (Remember that the answer should be a whole number.)

14. A manufacturing firm has received an order to make 400,000 souvenir medals commemorating the 25th anniversary of the landing of Apollo 11 on the moon. The firm owns 20 machines, each of which can produce 200 medals per hour. The cost of setting up the machines to produce the medals is $80 per machine, and the total operating cost is $5.76 per hour. How many machines should be used to minimize the cost of producing the 400,000 medals?

15. You are standing on the bank of a river that is 1 mile wide and want to get to a town on the opposite bank, 1 mile upstream. You plan to row on a straight line to some point P on the opposite bank and then walk the remaining distance along the bank. To what point P should you row in order to reach the town in the shortest possible time if you can row 4 miles per hour and walk 5 miles per hour?

16. Find constants A, B, and C so that the function $f(x) = Ax^3 + Bx^2 + C$ will have a relative extremum at $(2, 11)$ and an inflection point at $(1, 5)$. Sketch the graph of f.

17. A manufacturer estimates that if x units of a particular commodity are produced, the total cost will be $C(x) = x^3 - 24x^2 + 350x + 400$ dollars.
    (a) At what level of production will the marginal cost $C'(x)$ be minimized?
    (b) At what level of production will the average cost $A(x) = \dfrac{C(x)}{x}$ be minimized?

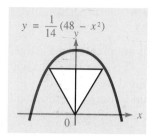

$y = \dfrac{1}{14}(48 - x^2)$

18. An isosceles triangle is drawn with its vertex at the origin, its base parallel to and above the x axis, and the vertices of its base on the curve $14y = 48 - x^2$. Determine the area of the largest such triangle.

19. A toy manufacturer produces an inexpensive doll (Floppsy) and an expensive doll (Moppsy) in units of x hundreds and y hundreds, respectively. Suppose that it is possible to produce the dolls in such a way that $y = \dfrac{82 - 10x}{10 - x}$ for $0 \le x \le 8$ and that the company receives *twice* as much for selling a Moppsy doll as for selling a Floppsy doll. Find the level of production (both x and y) for which the total revenue derived from selling these dolls is maximized. You may assume that the company sells every doll it produces.

20. Oil from an offshore rig located 3 miles from the shore is to be pumped to a location on the edge of the shore that is 8 miles east of the rig. The cost of constructing a pipe in the ocean from the rig to the shore is 1.5 times more expensive than the cost of construction on land. How should the pipe be laid to minimize cost?

21. A window is in the form of an equilateral triangle surmounted on a rectangle. The rectangle is of clear glass and transmits twice as much light as does the triangle, which is made of stained glass. If the entire window has a perimeter of 20 feet, find the dimensions of the window that will admit the most light.

22. Use Newton's method to find $\sqrt{55}$ to five decimal places.

In Problems 23 and 24, use Newton's method to find all the roots of the given equation to three decimal places.

23. $x^3 + 3x^2 + 1 = 0$ 24. $2x^4 - 8x^3 + 8x^2 - 1 = 0$

CHAPTER 4
EXPONENTIAL AND
LOGARITHMIC FUNCTIONS

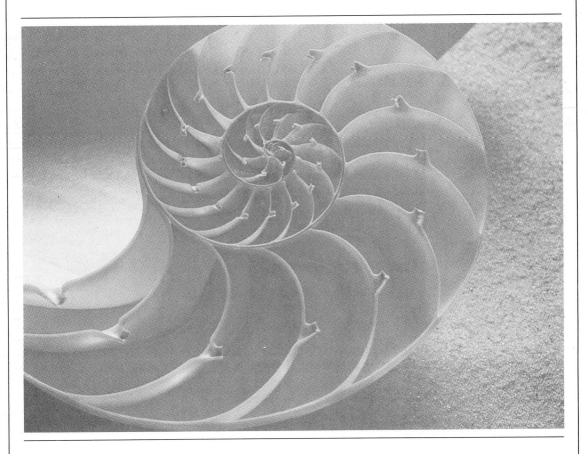

1 Exponential Functions

An **exponential function** is a function of the form $f(x) = a^x$, where a is a positive constant. In an exponential function, the independent variable x is the **exponent** of a positive constant known as the **base** of the function. Thus, an exponential function is fundamentally different from a power function $f(x) = x^n$ in which the base is the variable and the exponent is the constant.

Exponential functions play a central role in applied mathematics. They are used in demography to forecast population size, in finance to calculate the value of investments, in archaeology to date ancient artifacts, in psychology to study learning phenomena, in public health to analyze the spread of epidemics, and in industry to estimate the reliability of products. You will see some of these applications in Section 2.

The Algebra of Exponents Working with exponential functions requires the use of exponential notation and the algebraic laws of exponents. For reference, the major facts are summarized here. A more complete review, including worked examples and practice problems, is in the Algebra Review at the back of the book. Be sure that you are comfortable with this material before you proceed with this chapter.

Definition of a^n for Rational Values of n (and $a > 0$)

Integer powers: If n is a positive integer,

$$a^n = \underbrace{a \cdot a \cdots a}_{n \text{ factors}}$$

Fractional powers: If n and m are positive integers,

$$a^{n/m} = (\sqrt[m]{a})^n = \sqrt[m]{a^n}$$

where $\sqrt[m]{}$ denotes the positive mth root.

Negative powers: $a^{-n} = \dfrac{1}{a^n}$

Zero power: $a^0 = 1$

For example,

$$3^4 = 3 \cdot 3 \cdot 3 \cdot 3 = 81 \qquad 3^{-4} = \frac{1}{3^4} = \frac{1}{81}$$

$$4^{1/2} = \sqrt{4} = 2 \qquad 4^{3/2} = (\sqrt{4})^3 = 2^3 = 8$$

$$4^{-3/2} = \frac{1}{4^{3/2}} = \frac{1}{8} \qquad 27^{-2/3} = \frac{1}{(\sqrt[3]{27})^2} = \frac{1}{3^2} = \frac{1}{9}$$

The Laws of Exponents

The product law: $a^r a^s = a^{r+s}$ The power law: $(a^r)^s = a^{rs}$

The quotient law: $\dfrac{a^r}{a^s} = a^{r-s}$

For example,

$$(2^{-2})^3 = 2^{-6} = \frac{1}{2^6} = \frac{1}{64}$$

$$\frac{3^3}{3^{1/3}(3^{2/3})} = \frac{3^3}{3^{(1/3+2/3)}} = \frac{3^3}{3^1} = 3^2 = 9$$

$$2^{7/4}(8^{-1/4}) = 2^{7/4}(2^3)^{-1/4} = 2^{7/4}(2^{-3/4}) = 2^{(7/4-3/4)} = 2^1 = 2$$

The Graphs of Exponential Functions

The graphs of four exponential functions are shown in Figure 4.1.

Figure 4.1 The graphs of four exponential functions.

x	2^x	3^x
-2	$\dfrac{1}{4}$	$\dfrac{1}{9}$
-1	$\dfrac{1}{2}$	$\dfrac{1}{3}$
0	1	1
1	2	3
2	4	9

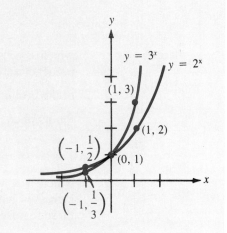

x	$\left(\dfrac{1}{2}\right)^x$	$\left(\dfrac{1}{3}\right)^x$
-2	4	9
-1	2	3
0	1	1
1	$\dfrac{1}{2}$	$\dfrac{1}{3}$
2	$\dfrac{1}{4}$	$\dfrac{1}{9}$

One way to obtain a rough sketch of an exponential function very quickly is to find its y intercept and determine its behavior as x increases without bound and as x decreases without bound. The technique is illustrated in the following example.

EXAMPLE 1.1

Sketch the function $f(x) = a^x$ if $0 < a < 1$ and if $a > 1$.

Solution

In both cases, the y intercept is $(0, 1)$ since $f(0) = a^0 = 1$. There are no x intercepts since, for positive a, a^x is always positive. To determine the behavior of the graph as x increases or decreases without bound, consider the two cases $0 < a < 1$ and $a > 1$ separately.

If $0 < a < 1$, the value of the product $a^n = a \cdot a \cdots a$ approaches zero as the number n of factors increases. For example,

$$\left(\frac{1}{10}\right)^n = \left(\frac{1}{10}\right)\left(\frac{1}{10}\right) \cdots \left(\frac{1}{10}\right) = \frac{1}{10^n}$$

which is very close to zero when n is large. This suggests that a^x approaches zero as x increases without bound. On the other hand, the value of the product $a^{-n} = \frac{1}{a} \cdot \frac{1}{a} \cdots \frac{1}{a}$ increases without bound (since $\frac{1}{a} > 1$) as the number n of factors increases. For example,

$$\left(\frac{1}{10}\right)^{-n} = 10 \cdot 10 \cdots 10 = 10^n$$

which is a very large number when n is large. This suggests that a^x increases without bound as x decreases without bound. A graph with these features is sketched in Figure 4.2a.

Figure 4.2 The graph of the exponential function $y = a^x$.

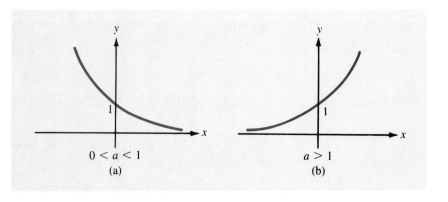

$0 < a < 1$
(a)

$a > 1$
(b)

If $a > 1$, a^x increases without bound as x increases without bound and a^x approaches zero as x decreases without bound. The corresponding graph is sketched in Figure 4.2b.

Irrational Exponents   The graphs in Figure 4.2 should not really have been drawn as unbroken curves since a^x has been defined only for *rational* values of x. However, it can be shown (using techniques beyond the scope of this book) that because there are "so many" rational numbers, there can be only *one* unbroken curve that passes through all the points (x, a^x) for which x is rational. That is, there exists a unique continuous function $f(x)$ that is defined for all real numbers x and that is equal to a^x when x is rational. When x is irrational, one *defines* a^x to be the value $f(x)$ of this function. Fortunately, in practical work you will rarely, if ever, have to deal explicitly with a^x for an irrational value of x.

The Number e   First impressions can be deceiving. At first glance, the expression $\left(1 + \dfrac{1}{n}\right)^n$ may look no more interesting than many other algebraic expressions. Yet the number it approaches as n increases without bound turns out to be one of the most important and useful numbers in mathematics. This special number will appear as the base in most of the exponential functions you will encounter.

To get a feel for what happens to the expression $\left(1 + \dfrac{1}{n}\right)^n$ as n increases without bound, use your calculator to complete the following table. (Round off your answers to three decimal places.)

n	1	2	5	10	100	1,000	10,000	100,000
$\left(1 + \dfrac{1}{n}\right)^n$	2.000	2.250	2.488	2.594	2.705			

It turns out that as n increases without bound, the expression $\left(1 + \dfrac{1}{n}\right)^n$ approaches an irrational number, traditionally denoted by the letter e, whose value is approximately 2.718. (A rigorous proof of this fact requires techniques beyond the scope of this book.)

Here is the definition of the number e, written compactly with the aid of limit notation. The symbol "$\lim\limits_{n \to \infty}$" is read "the limit as n approaches infinity" and denotes the behavior of the subsequent expression as n increases without bound.

The Number e $e = \lim\limits_{n \to \infty} \left(1 + \dfrac{1}{n}\right)^n \simeq 2.718$

To illustrate how the number e might arise in practice, here is a brief discussion of compound interest.

Compound Interest
Suppose a sum of money is invested and the interest is compounded only once. If P is the initial investment (the principal) and r is the interest rate (expressed as a decimal), the balance B after the interest is added will be

$$B = P + Pr = P(1 + r) \qquad \text{dollars}$$

That is, to compute the balance at the end of an interest period, you multiply the balance at the beginning of the period by the expression $1 + r$, where r is the interest rate per period.

At most banks, interest is compounded more than once a year. The interest that is added to the account during one period will itself earn interest during the subsequent periods. If the annual interest rate is r and interest is compounded k times per year, then the year is divided into k equal compounding periods and the interest rate in each period is $\frac{r}{k}$. Hence, the balance at the end of the first period is

$$P_1 = P + P\left(\frac{r}{k}\right) = P\left(1 + \frac{r}{k}\right)$$

$$\underset{\text{principal} \qquad \text{interest}}{}$$

At the end of the second period, the balance is

$$P_2 = P_1 + P_1\left(\frac{r}{k}\right) = P_1\left(1 + \frac{r}{k}\right)$$

$$= \left[P\left(1 + \frac{r}{k}\right)\right]\left(1 + \frac{r}{k}\right) = P\left(1 + \frac{r}{k}\right)^2$$

and in general, the balance at the end of the mth period is

$$P_m = P\left(1 + \frac{r}{k}\right)^m$$

Since there are k periods in a year, the balance after 1 year is

$$P\left(1 + \frac{r}{k}\right)^k$$

At the end of t years, interest has been compounded kt times and the balance is given by the function

$$B(t) = P\left(1 + \frac{r}{k}\right)^{kt}$$

To summarize:

Compound Interest

If P dollars are invested at an annual interest rate r (expressed as a decimal) and interest is compounded k times per year, the balance $B(t)$ after t years will be

$$B(t) = P\left(1 + \frac{r}{k}\right)^{kt} \quad \text{dollars}$$

Continuously Compounded Interest

As the frequency with which the interest is compounded increases, the corresponding balance $B(t)$ also increases. Hence, a bank that compounds interest frequently may attract more customers than does one that offers the same interest rate but compounds interest less often. The question arises: What happens to the balance at the end of t years as the frequency with which the interest is compounded increases without bound? That is, what will the balance be at the end of t years if interest is compounded not quarterly, not monthly, not daily, but continuously? In mathematical terms: What happens to the expression $P\left(1 + \frac{r}{k}\right)^{kt}$ as k increases without bound? The answer turns out to involve the number e. Here is the argument.

To simplify the calculation, let $n = \frac{k}{r}$. Then, $k = nr$ and so

$$P\left(1 + \frac{r}{k}\right)^{kt} = P\left(1 + \frac{1}{n}\right)^{nrt} = P\left[\left(1 + \frac{1}{n}\right)^{n}\right]^{rt}$$

Since n increases without bound as k does, and since $\left(1 + \frac{1}{n}\right)^{n}$ approaches e as n increases without bound, it follows that the balance after t years is

$$B(t) = \lim_{k \to \infty} P\left(1 + \frac{r}{k}\right)^{kt} = \lim_{n \to \infty} P\left[\left(1 + \frac{1}{n}\right)^{n}\right]^{rt} = Pe^{rt}$$

Here is a summary of the situation.

Continuously Compounded Interest

If P dollars are invested at an annual interest rate r (expressed as a decimal) and interest is compounded continuously, the balance $B(t)$ after t years will be

$$B(t) = Pe^{rt} \quad \text{dollars}$$

The following example illustrates what happens to a bank balance as the interest is compounded with increasing frequency. The numerical computations were done on a calculator. We will have much more to say about compound interest in Section 5, after we have developed the general properties of exponential and logarithmic functions.

EXAMPLE 1.2

Suppose $1,000 is invested at an annual interest rate of 6 percent. Compute the balance after 10 years if the interest is compounded

(a) Quarterly   (b) Monthly   (c) Daily   (d) Continuously

Solution

(a) To compute the balance after 10 years if the interest is compounded quarterly, use the formula $B(t) = P\left(1 + \dfrac{r}{k}\right)^{kt}$, with $t = 10$, $P = 1,000$, $r = 0.06$, and $k = 4$:

$$B(10) = 1,000\left(1 + \frac{0.06}{4}\right)^{40} \simeq \$1,814.02$$

(b) This time, take $t = 10$, $P = 1,000$, $r = 0.06$, and $k = 12$ to get

$$B(10) = 1,000\left(1 + \frac{0.06}{12}\right)^{120} \simeq \$1,819.40$$

(c) Take $t = 10$, $P = 1,000$, $r = 0.06$, and $k = 365$ to obtain

$$B(10) = 1,000\left(1 + \frac{0.06}{365}\right)^{3,650} \simeq \$1,822.03$$

(d) For continuously compounded interest use the formula $B(t) = Pe^{rt}$, with $t = 10$, $P = 1,000$, and $r = 0.06$:

$$B(10) = 1,000e^{0.6} \simeq \$1,822.12$$

This value, $1,822.12, is an upper bound for the possible balance. No matter how often interest is compounded, $1,000 invested at an annual interest rate of 6 percent cannot grow to more than $1,822.12 in 10 years.

Graphs of Functions Involving Powers of e   Since $e > 1$, the graph of the function $y = e^x$ (Figure 4.3a) resembles the graph of the general exponential function $y = a^x$ for $a > 1$, which you saw in Figure 4.2b. Since $\dfrac{1}{e} < 1$, the graph of the function $y = e^{-x} = \left(\dfrac{1}{e}\right)^x$

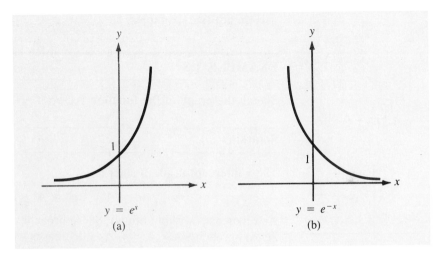

Figure 4.3 The graphs of the functions $y = e^x$ and $y = e^{-x}$.

(Figure 4.3b) resembles the graph of the general exponential function $y = a^x$ for $0 < x < 1$, which you saw in Figure 4.2a.

Notice that as x increases without bound, e^x increases without bound while e^{-x} approaches zero. Also, as x decreases without bound, e^x approaches zero while e^{-x} increases without bound. These statements may be written compactly in limit notation as follows:

Limits at Infinity for e^x and e^{-x}

$$\lim_{x \to \infty} e^x = \infty \quad \text{and} \quad \lim_{x \to -\infty} e^x = 0$$

$$\lim_{x \to \infty} e^{-x} = 0 \quad \text{and} \quad \lim_{x \to -\infty} e^{-x} = \infty$$

For example, the expression $\lim_{x \to -\infty} e^{-x} = \infty$ is read "the limit of e^{-x} as x approaches minus infinity is infinity" and expresses the fact that as x decreases without bound, e^{-x} increases without bound. Limits involving infinity will be discussed in more detail in Chapter 8.

More complicated functions involving powers of e arise frequently in applications. You can obtain a rough sketch of such a function using the following procedure.

How to Sketch Functions Involving Powers of e

Step 1. Find the y intercept (by setting x equal to zero).

Step 2. Find $\lim_{x \to \infty} f(x)$ and $\lim_{x \to -\infty} f(x)$.

Step 3. Sketch the graph using these clues.

Here are two examples.

EXAMPLE 1.3

Sketch the graph of the function $f(x) = 4 - 2e^{-x}$.

Solution

The y intercept in $(0, 2)$ since

$$f(0) = 4 - 2e^0 = 4 - 2 = 2$$

As x increases without bound, e^{-x} approaches zero and hence $f(x) = 4 - 2e^{-x}$ approaches $4 - 0 = 4$. As x decreases without bound, e^{-x} increases without bound, which implies that $f(x) = 4 - 2e^{-x}$ decreases without bound (because of the minus sign). In limit notation

$$\lim_{x \to \infty} f(x) = 4 \quad \text{and} \quad \lim_{x \to -\infty} f(x) = -\infty$$

A graph of $f(x)$ is sketched in Figure 4.4.

Fig. 4.4

Figure 4.4 The graph of $y = 4 - 2e^{-x}$.

EXAMPLE 1.4

Sketch the graph of the function $f(x) = \dfrac{3}{1 + 2e^{-5x}}$.

Solution

The y intercept is $(0, 1)$ since

$$f(0) = \frac{3}{1 + 2e^0} = \frac{3}{1 + 2} = 1$$

As x increases without bound, e^{-5x} approaches zero. Hence, the denominator $1 + 2e^{-5x}$ approaches 1 and $f(x) = \dfrac{3}{1 + 2e^{-5x}}$ approaches $\dfrac{3}{1} = 3$. Also, as x decreases without bound, e^{-5x} increases without bound, so that $1 + 2e^{-5x}$ increases without bound and $f(x) = \dfrac{3}{1 + 2e^{-5x}}$ approaches zero. In limit notation,

$$\lim_{x \to \infty} f(x) = 3 \quad \text{and} \quad \lim_{x \to -\infty} f(x) = 0$$

A graph of $f(x)$ is sketched in Figure 4.5.

Figure 4.5 The graph of $y = \dfrac{3}{1 + 2e^{-5x}}$.

**An Application
of the Power Law
of Exponents** To conclude this introductory section, here is an example illustrating a simple but useful computational technique based on the power law for exponents. You will use this technique several times in Section 2.

EXAMPLE 1.5

Find $f(6)$ if $f(x) = e^{kx}$ and $f(2) = 5$.

Solution

You do not have to know the value of k or of e to solve this problem! The fact that $f(2) = 5$ tells you that

$$e^{2k} = 5$$

and using the power law for exponents, you can rewrite the expression for $f(6)$ in terms of this quantity to get

$$f(6) = e^{6k} = (e^{2k})^3 = 5^3 = 125$$

Problems 4.1

In Problems 1 and 2, use your calculator to find the indicated powers of e. (Round off your answers to three decimal places.)

1. e^2, e^{-2}, $e^{0.05}$, $e^{-0.05}$, e^0, e, $\sqrt{e}$, and $\dfrac{1}{\sqrt{e}}$.

2. e^3, e^{-1}, $e^{0.01}$, $e^{-0.1}$, e^2, $e^{-1/2}$, $e^{1/3}$, and $\dfrac{1}{\sqrt[3]{e}}$.

3. Sketch the curves $y = 3^x$ and $y = 4^x$ on the same set of axes.

4. Sketch the curves $y = \left(\dfrac{1}{3}\right)^x$ and $y = \left(\dfrac{1}{4}\right)^x$ on the same set of axes.

In Problems 5 through 16, sketch the given function.

5. $f(x) = e^x$ 6. $f(x) = e^{-x}$

7. $f(x) = 2 + e^x$ 8. $f(x) = 3 + e^{-x}$

9. $f(x) = 2 - 3e^x$ 10. $f(x) = 3 - 2e^x$

11. $f(x) = 5 - 3e^{-x}$ 12. $f(x) = 3 - 5e^{-x}$

13. $f(x) = \dfrac{2}{1 + 3e^{-2x}}$ 14. $f(x) = \dfrac{2}{1 + 3e^{2x}}$

15. $f(x) = 1 - \dfrac{3}{1 + 2e^{-3x}}$

16. $f(x) = 1 - \dfrac{6}{2 + e^{3x}}$

17. Find $f(2)$ if $f(x) = e^{kx}$ and $f(1) = 20$.

18. Find $f(9)$ if $f(x) = e^{kx}$ and $f(3) = 2$.

19. Find $f(8)$ if $f(x) = Ae^{kx}$, $f(0) = 20$, and $f(2) = 40$.

20. Find $f(4)$ if $f(x) = 50 - Ae^{-kx}$, $f(0) = 20$, and $f(2) = 30$.

21. Find $f(2)$ if $f(x) = 50 - Ae^{kx}$, $f(0) = 30$, and $f(4) = 5$.

22. Find $f(3)$ if $f(x) = \dfrac{4}{1 + Ae^{-kx}}$, $f(0) = 1$, and $f(1) = \dfrac{1}{4}$.

Compound Interest 23. Suppose $1,000 is invested at an annual interest rate of 7 percent. Compute the balance after 10 years if the interest is compounded
    (a) Annually   (b) Quarterly
    (c) Monthly   (d) Continuously

Compound Interest 24. Suppose $5,000 is invested at an annual interest rate of 10 percent. Compute the balance after 10 years if the interest is compounded
    (a) Annually   (c) Daily (using 365 days per year)
    (b) Semiannually   (d) Continuously

Compound Interest 25. A sum of money is invested at a certain fixed interest rate, and the interest is compounded continuously. After 10 years, the money has doubled. How will the balance at the end of 20 years compare with the initial investment?

Compound Interest 26. A sum of money is invested at a certain fixed interest rate, and the interest is compounded quarterly. After 15 years, the money has doubled. How will the balance at the end of 30 years compare with the initial investment?

Compound Interest 27. (a) Solve the equation $B = Pe^{rt}$ for P.
    (b) Use the result from part (a) to determine how much money should be invested today at an annual interest rate of 6 percent compounded continuously so that 10 years from now it will be worth $10,000.

Compound Interest 28. How much money should be invested today at 7 percent compounded quarterly so that it will be worth $5,000 in 5 years?

Effective Interest Rate When a bank offers an annual interest rate of $100r$ percent (e.g., 7 percent interest corresponds to $r = 0.07$) and compounds interest more than once a year, the total interest earned during a year is greater than $100r$ percent of the balance at the beginning of that year. The actual percentage by which the balance grows during a year is sometimes called the **effective interest rate**, while the advertised rate of $100r$ percent is called the **nominal interest rate**. In Exercises 29 and 30, find the effective interest rate.

29. The nominal rate is 6 percent and interest is compounded

(a) Quarterly   (b) Continuously

30. The nominal rate is 10 percent and interest is compounded
    (a) Monthly   (b) Continuously

31. Program a computer or use a calculator to evaluate $\left(1 + \dfrac{1}{n}\right)^n$ for

    $n = 1{,}000,\ 2{,}000,\ \ldots,\ 50{,}000$.

32. Program a computer or use a calculator to evaluate $\left(1 + \dfrac{1}{n}\right)^n$ for

    $n = -1{,}000,\ -2{,}000,\ \ldots,\ -50{,}000$. On the basis of these calcula-

    tions, what can you conjecture about the behavior of $\left(1 + \dfrac{1}{n}\right)^n$ as

    n decreases without bound?

33. Program a computer or use a calculator to estimate $\lim\limits_{n \to \infty}\left(1 + \dfrac{3}{n}\right)^{2n}$.

34. Program a computer or use a calculator to estimate $\lim\limits_{n \to \infty}\left(2 - \dfrac{5}{2n}\right)^{n/3}$.

2   Exponential Models

Functions involving powers of e play a central role in applied mathematics. Here is a sampling of practical situations from the social, managerial,
and natural sciences that can be described mathematically in terms of
such functions.

Exponential Growth   A quantity $Q(t)$ that increases according to a law of the form $Q(t) = Q_0 e^{kt}$,
where Q_0 and k are positive constants, is said to experience **exponential
growth.** For example, if interest is compounded continuously, the resulting bank balance $B(t) = Pe^{rt}$ grows exponentially. Also, in the absence of
environmental constraints, population increases exponentially. As you
will see later in this book, quantities that grow exponentially are characterized by the fact that their rate of growth is proportional to their size and
that their percentage rate of change is constant.

Exponential Growth

$Q(t)$ grows exponentially if

$$Q(t) = Q_0 e^{kt}$$

where k is a positive constant and Q_0 is the initial value $Q(0)$.

To sketch the function $Q(t) = Q_0 e^{kt}$, observe that $Q(t)$ is always
positive, that $Q(0) = Q_0$, that $Q(t)$ increases without bound as t increases

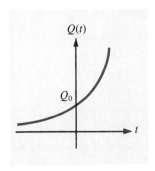

Figure 4.6 Exponential growth: $Q(t) = Q_0 e^{kt}$.

without bound, and that $Q(t)$ approaches zero as t decreases without bound. The graph is sketched in Figure 4.6.

Here is an example from biology.

EXAMPLE 2.1

Biologists have determined that under ideal conditions, the number of bacteria in a culture grows exponentially. Suppose that 2,000 bacteria are initially present in a certain culture and that 6,000 are present 20 minutes later. How many bacteria will be present at the end of 1 hour?

Solution

Let $Q(t)$ denote the number of bacteria present after t minutes. Since the number of bacteria grows exponentially and since 2,000 bacteria were initially present, you know that Q is a function of the form

$$Q(t) = 2,000 e^{kt}$$

Since 6,000 bacteria are present after 20 minutes, it follows that

$$6,000 = 2,000 e^{20k} \qquad \text{or} \qquad e^{20k} = 3$$

To find the number of bacteria present at the end of 1 hour (60 minutes), compute $Q(60)$ using the power law for exponents as follows:

$$Q(60) = 2,000 e^{60k} = 2,000 (e^{20k})^3 = 2,000 (3)^3 = 54,000$$

That is, 54,000 bacteria will be present at the end of 1 hour.

Exponential Decay   A quantity $Q(t)$ that decreases according to a law of the form $Q(t) = Q_0 e^{-kt}$, where Q_0 and k are positive constants, is said to experience **exponential decay** or, equivalently, to **decrease exponentially.** Radioactive substances decay exponentially. Sales of many products decrease exponentially when advertising is discontinued. Quantities that decrease exponentially are characterized by the fact that their rate of decrease is proportional to their size. A graph of the function $Q(t) = Q_0 e^{-kt}$ is sketched in Figure 4.7.

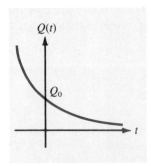

Figure 4.7 Exponential decay: $Q(t) = Q_0 e^{-kt}$.

Exponential Decay

$Q(t)$ decreases (or decays) exponentially if

$$Q(t) = Q_0 e^{-kt}$$

where k is a positive constant and Q_0 is the initial value $Q(0)$.

EXAMPLE 2.2

A certain industrial machine depreciates so that its value after t years is given by a function of the form $Q(t) = Q_0 e^{-0.04t}$. After 20 years, the machine is worth \$8,986.58. What was its original value?

Solution

The goal is to find Q_0. Since $Q(20) = 8{,}986.58$,

$$Q_0 e^{-0.8} = 8{,}986.58$$

Multiply both sides of this equation by $e^{0.8}$ to get

$$Q_0 = 8{,}986.58 e^{0.8} \simeq \$20{,}000$$

Learning Curves   The graph of a function of the form $Q(t) = B - Ae^{-kt}$, where B, A, and k are positive constants, is sometimes called a **learning curve.** The name arose when psychologists discovered that functions of this form often describe the relationship between the efficiency with which an individual performs a task and the amount of training or experience the individual has had.

To sketch the function $Q(t) = B - Ae^{-kt}$, observe that the vertical intercept is $Q(0) = B - A$. As t increases without bound, e^{-kt} approaches zero and so

$$\lim_{t \to \infty} Q(t) = B - 0 = B$$

As t decreases without bound, e^{-kt} increases without bound, and so $Q(t) = B - Ae^{-kt}$ *decreases* without bound (because of the minus sign preceding the A). That is

$$\lim_{t \to -\infty} Q(t) = -\infty$$

A graph with these features is sketched in Figure 4.8. The behavior of the graph as t increases without bound reflects the fact that eventually an individual will approach peak efficiency, and additional training will have little effect on performance.

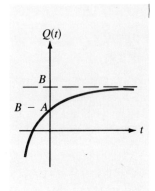

Figure 4.8 A learning curve: $Q(t) = B - Ae^{-kt}$.

EXAMPLE 2.3

The rate at which a postal clerk can sort mail is a function of the clerk's experience. Suppose the postmaster of a large city estimates that after t months on the job, the average clerk can sort $Q(t) = 700 - 400e^{-0.5t}$ letters per hour.

(a) How many letters can a new employee sort per hour?

(b) How many letters can a clerk with 6 months' experience sort per hour?

(c) Approximately how many letters will the average clerk ultimately be able to sort per hour?

Solution

(a) The number of letters a new employee can sort per hour is

$$Q(0) = 700 - 400 = 300$$

(b) After 6 months, the average clerk can sort

$$Q(6) = 700 - 400e^{-0.5(6)} = 700 - 400e^{-3} \simeq 680 \qquad \text{letters per hour}$$

(c) As t increases without bound, $Q(t)$ approaches 700. Hence, the average clerk will ultimately be able to sort approximately 700 letters per hour.

Figure 4.9 Worker efficiency: $Q(t) = 700 - 400e^{-0.5t}$.

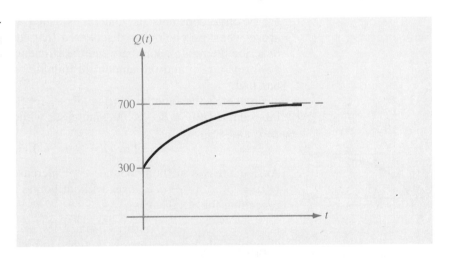

The graph of the function $Q(t)$ is sketched in Figure 4.9.

Logistic Curves   The graph of a function of the form $Q(t) = \dfrac{B}{1 + Ae^{-Bkt}}$, where B, A, and k are positive constants, is an S-shaped or **sigmoidal curve**. The term **logistic curve** is also used to refer to such a graph.

To sketch the logistic function $Q(t) = \dfrac{B}{1 + Ae^{-Bkt}}$, notice that the vertical intercept is

$$Q(0) = \frac{B}{1 + Ae^0} = \frac{B}{1 + A}$$

As t increases without bound, e^{-Bkt} approaches zero (since B and k are positive). Hence,

$$\lim_{t \to \infty} Q(t) = \lim_{t \to \infty} \frac{B}{1 + Ae^{-Bkt}} = \frac{B}{1 + 0} = B$$

As t decreases without bound, e^{-Bkt} increases without bound (because of the minus sign in the exponent). Hence, the denominator $1 + Ae^{-Bkt}$ increases without bound and

$$\lim_{t \to -\infty} Q(t) = \lim_{t \to -\infty} \frac{B}{1 + Ae^{-Bkt}} = 0$$

A graph with these properties is sketched in Figure 4.10.

Figure 4.10 A logistic curve:
$$Q(t) = \frac{B}{1 + Ae^{-Bkt}}.$$

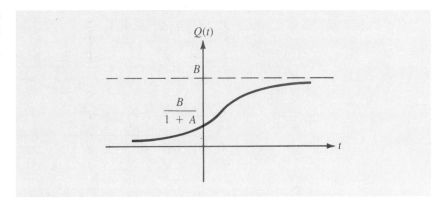

Logistic curves are rather accurate models of population growth when environmental factors impose an upper bound on the possible size of the population. They also describe the spread of epidemics and rumors in a community. Here is a typical example.

EXAMPLE 2.4

Public health records indicate that t weeks after the outbreak of a certain form of influenza, approximately $Q(t) = \frac{20}{1 + 19e^{-1.2t}}$ thousand people had caught the disease.

(a) How many people had the disease when it first broke out?
(b) How many had caught the disease by the end of the 2nd week?
(c) If the trend continues, approximately how many people in all will contract the disease?

Solution

(a) Since
$$Q(0) = \frac{20}{1 + 19} = 1$$

it follows that 1,000 people had the disease initially.

(b) Since
$$Q(2) = \frac{20}{1 + 19e^{-1.2(2)}} \simeq 7.343$$

it follows that approximately 7,343 people had caught the disease by the end of the 2nd week.

(c) Since $Q(t)$ approaches 20 as t increases without bound, it follows that approximately 20,000 people will eventually contract the disease.

Figure 4.11 The spread of an epidemic: $Q(t) = \dfrac{20}{1 + 19e^{-1.2t}}$.

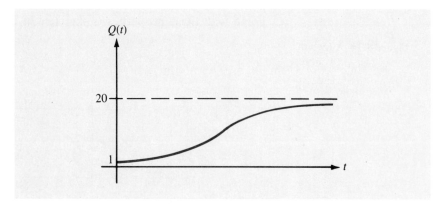

For reference, the graph of $Q(t)$ is sketched in Figure 4.11.

Problems 4.2

Population Growth

1. It is projected that t years from now, the population of a certain country will be $P(t) = 50e^{0.02t}$ million.
   (a) What is the current population?
   (b) What will the population be 30 years from now?

Compound Interest

2. How much money should be invested today at an annual interest rate of 7 percent compounded continuously so that 20 years from now it will be worth $20,000? [Recall that the balance after t years is $B(t) = Pe^{rt}$, where r is the interest rate expressed as a decimal and P is the initial deposit.]

Population Growth

3. It is estimated that the population of a certain country grows exponentially. If the population was 60 million in 1986 and 90 million in 1991, what will the population be in 2001?

Growth of Bacteria

4. The following data were compiled by a researcher during the first 10 minutes of an experiment designed to study the growth of bacteria:

Number of minutes	0	10
Number of bacteria	5,000	8,000

Assuming that the number of bacteria grows exponentially, how many bacteria will be present after 30 minutes?

Gross National Product   5. The gross national product (GNP) of a certain country was 100 billion dollars in 1980 and 180 billion dollars in 1990. Assuming that the GNP is growing exponentially, what will the GNP be in 2000?

Retail Sales   6. The total number of hamburgers sold by a national fast-food chain is growing exponentially. If 4 billion had been sold by 1986 and 12 billion had been sold by 1991, how many will have been sold by 1996?

Population Density   7. The population density x miles from the center of a certain city is $D(x) = 12e^{-0.07x}$ thousand people per square mile.
   (a) What is the population density at the center of the city?
   (b) What is the population density 10 miles from the center of the city?

Radioactive Decay   8. The amount of a sample of a radioactive substance remaining after t years is given by a function of the form $Q(t) = Q_0 e^{-0.0001t}$. At the end of 5,000 years, 2,000 grams of the substance remain. How many grams were present initially?

Radioactive Decay   9. A radioactive substance decays exponentially. If 500 grams of the substance were present initially and 400 grams are present 50 years later, how many grams will be present after 200 years?

Product Reliability   10. A statistical study indicates that the fraction of the electric toasters manufactured by a certain company that are still in working condition after t years of use is approximately $f(t) = e^{-0.2t}$.
   (a) What fraction of the toasters can be expected to work for at least 3 years?
   (b) What fraction of the toasters can be expected to fail during the 3rd year of use?
   (c) What fraction of the toasters can be expected to fail before 1 year of use?

Product Reliability   11. A manufacturer of toys has found that the fraction of its plastic battery-operated toy oil tankers that sink in fewer than t days is approximately $f(t) = 1 - e^{-0.03t}$.
   (a) Sketch this reliability function. What happens to the graph as t increases without bound?
   (b) What fraction of the tankers can be expected to float for at least 10 days?
   (c) What fraction of the tankers can be expected to sink between the 15th and 20th days?

Sales   12. Once the initial publicity surrounding the release of a new book is over, sales of the hardcover edition tend to decrease exponentially. At the time publicity was discontinued, a certain book was experiencing sales of 25,000 copies per month. One month later, sales of the book had dropped to 10,000 copies per month. What will the sales be after 1 more month?

Recall from Memory   13. Psychologists believe that when a person is asked to recall a set of facts, the number of facts recalled after t minutes is given by a function of the form $Q(t) = A(1 - e^{-kt})$, where k is a positive constant and A is the total number of relevant facts in the person's memory.
(a) Sketch the graph of $Q(t)$.
(b) What happens to the graph as t increases without bound? Explain this behavior in practical terms.

Advertising   14. When professors select texts for their courses, they usually choose from among the books already on their shelves. For this reason, most publishers send complimentary copies of new texts to professors teaching related courses. The mathematics editor at a major publishing house estimates that if x thousand complimentary copies are distributed, the first-year sales of a certain new mathematics text will be approximately $f(x) = 20 - 15e^{-0.2x}$ thousand copies.
(a) Sketch this sales function.
(b) How many copies can the editor expect to sell in the first year if no complimentary copies are sent out?
(c) How many copies can the editor expect to sell in the first year if 10,000 complimentary copies are sent out?
(d) If the editor's estimate is correct, what is the most optimistic projection for the first-year sales of the text?

Depreciation   15. When a certain industrial machine is t years old, its resale value will be $V(t) = 4,800e^{-t/5} + 400$ dollars.
(a) Sketch the graph of $V(t)$. What happens to the value of the machine as t increases without bound?
(b) How much was the machine worth when it was new?
(c) How much will the machine be worth after 10 years?

Efficiency   16. The daily output of a worker who has been on the job for t weeks is given by a function of the form $Q(t) = 40 - Ae^{-kt}$. Initially the worker could produce 20 units a day, and after 1 week the worker can produce 30 units a day. How many units will the worker produce per day after 3 weeks?

Newton's Law of Cooling   17. On a hot summer day, a cool drink is removed from a refrigerator whose temperature is 10° Celsius and is placed in a room where the temperature is 30° Celsius. According to Newton's law of cooling (in this case, "heating" might be the more appropriate term), the temperature of the drink after t minutes is given by a function of the form $f(t) = 30 - Ae^{-kt}$. If the temperature of the drink is 15° Celsius after 20 minutes, what will it be after 40 minutes?

Newton's Law of
Cooling

18. Instant coffee is made by adding boiling water (212° F) to coffee mix. If the air temperature is 70° F, Newton's law of cooling says that after t minutes, the temperature of the coffee will be given by a function of the form $f(t) = 70 - Ae^{-kt}$. Assuming that the coffee is too hot to drink when the temperature exceeds 150° F, how long will it take for the coffee to become drinkable?

The Spread of an
Epidemic

19. Public health records indicate that t weeks after the outbreak of a certain form of influenza, approximately $f(t) = \dfrac{2}{1 + 3e^{-0.8t}}$ thousand people had caught the disease.
(a) Sketch the graph of $f(t)$.
(b) How many people had the disease initially?
(c) How many had caught the disease by the end of 3 weeks?
(d) If the trend continues, approximately how many people in all will contract the disease?

The Spread of an
Epidemic

20. An epidemic spreads through a community so that t weeks after its outbreak, the number of people who have been infected is given by a function of the form $f(t) = \dfrac{B}{1 + Ce^{-kt}}$, where B is the number of residents in the community who are susceptible to the disease. If $\dfrac{1}{5}$ of the susceptible residents were infected initially and $\dfrac{1}{2}$ had been infected by the end of the 4th week, what fraction of the susceptible residents will have been infected by the end of the 8th week?

Population Growth

21. It is estimated that t years from now, the population of a certain country will be $P(t) = \dfrac{20}{2 + 3e^{-0.06t}}$ million.
(a) Sketch the graph of $P(t)$.
(b) What is the current population?
(c) What will the population be 50 years from now?
(d) What will happen to the population in the long run?

The Spread of a Rumor

22. A traffic accident was witnessed by $\dfrac{1}{10}$ of the residents of a small town. The number of residents who had heard about the accident t hours later is given by a function of the form $f(t) = \dfrac{B}{1 + Ce^{-kt}}$, where B is the population of the town. If $\dfrac{1}{4}$ of the residents had heard about the accident after 2 hours, how long did it take for $\dfrac{1}{2}$ of the residents to hear the news?

Gompertz Curve   23. Based on various projections, a company predicts that the number of employees it will have in t years will be

$$N = 500(0.03)^{(0.4)^t}$$

(a) How many employees are employed now ($t = 0$)?
(b) How many will be employed in 5 years?
(Note: A **Gompertz curve** is the graph of a function of the form $f(x) = Ca^{r^t}$ where r $(0 < r < 1)$ is a rate of growth and a and C are positive constants. Such curves are used by psychologists and others to describe things such as growth and learning.)

3 The Natural Logarithm

In many practical problems, a number a is known and the goal is to find the corresponding number b such that $a = e^b$. This number b is called the **natural logarithm** of a and is denoted by the symbol ln a. The letter l in the symbol ln stands for "logarithm," and the letter n stands for "natural." (The word "ln" is virtually unpronounceable. It can be read as "el en," "log," or "Lynn.")

Figure 4.12 $b = \ln a$ if and only if $a = e^b$.

The Natural Logarithm

Corresponding to each positive number a there is a unique power b such that $a = e^b$. This power b is called the natural logarithm of a and is denoted by ln a. Thus,

$$b = \ln a \quad \text{if and only if} \quad a = e^b$$

The situation is illustrated in Figure 4.12.

EXAMPLE 3.1

Find   (a) ln e (b) ln 1   (c) ln $\sqrt{e}$ (d) ln 2

Solution

(a) According to the definition, ln e is the unique number b such that $e = e^b$. Clearly this number is $b = 1$. Hence, ln $e = 1$.
(b) ln 1 is the unique number b such that $1 = e^b$. Since $e^0 = 1$, it follows that ln 1 $= 0$.

(c) $\ln \sqrt{e} = \ln e^{1/2}$ is the unique number b such that $e^{1/2} = e^b$; that is, $b = \dfrac{1}{2}$. Hence, $\ln \sqrt{e} = \dfrac{1}{2}$.

(d) $\ln 2$ is the unique number b such that $2 = e^b$. The value of this number is not obvious, and you will have to use your calculator (or Table II at the back of the book) to find that $\ln 2 \simeq 0.69315$.

The Relationship between e^x and $\ln x$

The next example establishes two important identities which show that logarithmic and exponential functions have a certain "neutralizing" effect on each other.

EXAMPLE 3.2

Simplify the following expressions:

(a) $e^{\ln x}$ (for $x > 0$)   (b) $\ln e^x$

Solution

(a) According to the definition, $\ln x$ is the unique number b for which $x = e^b$. Hence, $e^{\ln x} = e^b = x$.

(b) Similarly, $\ln e^x$ is the unique number b for which $e^x = e^b$. Clearly, this number b is x itself. Hence, $\ln e^x = x$.

$b = \ln a$

$e^b = a$

The two identities derived in Example 3.2 show that the composite functions $\ln e^x$ and $e^{\ln x}$ leave the variable x unchanged. In general, two functions f and g for which $f[g(x)] = x$ and $g[f(x)] = x$ are said to be **inverses** of one another. Thus the exponential function $y = e^x$ and the logarithmic function $y = \ln x$ are inverses.

> **The Inverse Relationship between e^x and $\ln x$**
>
> $$e^{\ln x} = x \quad \text{(if } x > 0) \quad \text{and} \quad \ln e^x = x$$

The next example illustrates how you can use the inverse relationship between e^x and $\ln x$ to solve equations.

EXAMPLE 3.3

Solve each of the following equations for x:

(a) $3 = e^{20x}$ (b) $2 \ln x = 1$

Solution

(a) Take the natural logarithm of each side of the equation to get

$$\ln 3 = \ln e^{20x} \quad \text{or} \quad \ln 3 = 20x$$

Solve for x (using a calculator or the natural logarithm table at the back of the book to find $\ln 3$).

$$x = \frac{\ln 3}{20} \simeq \frac{1.0986}{20} \simeq 0.0549$$

(b) First isolate $\ln x$ on the left-hand side of the equation by dividing both sides by 2.

$$\ln x = \frac{1}{2}$$

Then apply the exponential function to both sides of the equation to get

$$e^{\ln x} = e^{1/2} \quad \text{or} \quad x = e^{1/2} = \sqrt{e} \simeq 1.6487$$

The Graph of ln x   There is an easy way to obtain the graph of the logarithmic function $y = \ln x$ from the graph of the exponential function $y = e^x$. The method is based on the geometric fact that the point (b, a) is the reflection of the point (a, b) across the line $y = x$. This fact is illustrated in Figure 4.13, and a proof is outlined in Problem 45.

Figure 4.13 Reflection of points across the line $y = x$.

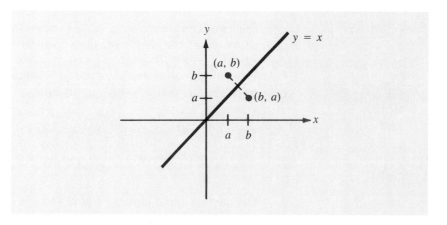

The idea is that since $y = \ln x$ means that $x = e^y$, the graph of $y = \ln x$ is the same as the graph of $y = e^x$ with the roles of x and y reversed. More precisely, suppose (a, b) is a point on the curve $y = \ln x$. Then $b = \ln a$, or equivalently, $a = e^b$. Hence, the reflected point (b, a) can be written as (b, e^b), which is a point on the curve $y = e^x$. Conversely, if $(a,$

b) is on the curve $y = e^x$, it follows that $b = e^a$ and $a = \ln b$. Hence, the reflected point is $(b, a) = (b, \ln b)$, which lies on the curve $y = \ln x$. To summarize:

> **Relationship between the Graphs of $y = \ln x$ and $y = e^x$**
>
> The graph of $y = \ln x$ is obtained by reflecting the graph of $y = e^x$ across the line $y = x$.

The graph of the exponential function $y = e^x$, the line $y = x$, and the graph of the logarithmic function $y = \ln x$ are sketched in Figure 4.14.

Figure 4.14 The graph of $y = \ln x$ is the reflection of the graph of $y = e^x$ across the line $y = x$.

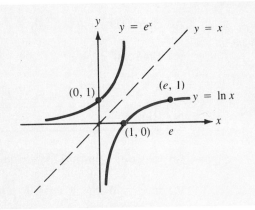

Notice that the natural logarithm function has the following properties:

(a) $\ln x$ is defined only for $x > 0$.

(b) $\ln 1 = 0$ and $\ln e = 1$.

(c) $\ln x$ increases without bound as x increases without bound and decreases without bound as x approaches zero from the right.

Properties of the Natural Logarithm   The laws of exponents can be used to derive the following important properties of logarithms.

> **Properties of Logarithms**
>
> The logarithm of a product: $\ln uv = \ln u + \ln v$
>
> The logarithm of a quotient: $\ln \dfrac{u}{v} = \ln u - \ln v$
>
> The logarithm of a power: $\ln u^v = v \ln u$

The first of these properties states that logarithms transform multiplication into the simpler operation of addition. The derivation of this property will be given in Example 3.5. The second property states that logarithms transform division into subtraction, and the third states that logarithms reduce exponentiation to multiplication. The derivations of these two properties are left as problems for you to do (see Problems 42 and 43).

The next example illustrates the use of these algebraic properties of logarithms.

EXAMPLE 3.4

(a) Find $\ln \sqrt{ab}$ if $\ln a = 3$ and $\ln b = 7$.

(b) Show that $\ln \dfrac{1}{x} = -\ln x$.

(c) Find x if $2^x = e^3$.

Solution

(a) $\ln \sqrt{ab} = \ln (ab)^{1/2} = \dfrac{1}{2} \ln ab = \dfrac{1}{2}(\ln a + \ln b) = \dfrac{1}{2}(3 + 7) = 5$

(b) $\ln \dfrac{1}{x} = \ln 1 - \ln x = 0 - \ln x = -\ln x$

(c) Take the natural logarithm of each side of the equation $2^x = e^3$ and solve for x to get

$$x \ln 2 = 3 \quad \text{or} \quad x = \frac{3}{\ln 2} \approx 4.33$$

The formula for the logarithm of a product is derived in the next example.

EXAMPLE 3.5

Derive the formula $\ln uv = \ln u + \ln v$.

Solution

The fact that e^x and $\ln x$ are inverses implies that

$$e^{\ln uv} = uv \qquad e^{\ln u} = u \qquad e^{\ln v} = v$$

Use these identities together with the product law for exponents to get

$$e^{\ln uv} = uv = e^{\ln u}e^{\ln v} = e^{\ln u + \ln v}$$

Compare the powers of e at each end of this string of equalities to conclude that

$$\ln uv = \ln u + \ln v$$

Logarithms with Other Bases

You may be wondering how natural logarithms are related to the logarithms you studied in high school. The following discussion will show you the connection.

The graph of the exponential function $y = a^x$ (shown in Figure 4.15 for

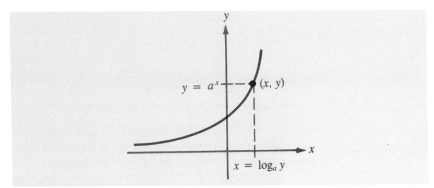

Figure 4.15 The graph of the function $y = a^x$ for $a > 1$.

$a > 1$) suggests that for each positive number y there corresponds a unique number x such that $y = a^x$. This power x is called the **logarithm of y to the base a** and is denoted by $\log_a y$. Thus,

$$x = \log_a y \quad \text{if and only if} \quad y = a^x$$

The logarithms you used in high school algebra to simplify numerical calculations were logarithms to the base 10. For numerical work, 10 is a particularly convenient base for logarithms because the standard decimal representation of numbers is based on powers of 10. Natural logarithms are logarithms to the base e. Because of the importance of the special exponential function e^x, natural logarithms arise frequently in applied work.

Applications Involving Logarithms

Next, we shall examine several practical applications in which the natural logarithm plays a natural role. In the next two examples, you will see how to use logarithms to determine how much time is required for a quantity that changes exponentially to either double or halve in size.

Doubling Time

EXAMPLE 3.6

The world's population grows at the rate of approximately 2 percent per year. If it is assumed that the population growth is exponential, it can be shown that the population t years from now will be given by a function of the form $P(t) = P_0 e^{0.02t}$, where P_0 is the current population. (This formula is derived in Chapter 7.) Assuming that this model of population growth is correct, how long does it take for the world's population to double?

Solution

The goal is to find the value of t for which $P(t) = 2P_0$, that is, for which

$$2P_0 = P_0 e^{0.02t}$$

To do this, divide each side of the equation by P_0 to get

$$2 = e^{0.02t}$$

and then take the logarithm of each side and solve for t:

$$\ln 2 = 0.02t \qquad \text{or} \qquad t = \frac{\ln 2}{0.02} \simeq 34.66$$

That is, the world's population will double in about $34\frac{2}{3}$ years.

Notice that the constant P_0 representing the current population was eliminated from the equation at an early stage and that the final answer is therefore independent of this quantity.

Half-Life   It has been experimentally determined that most radioactive substances decay exponentially, so that the amount of a sample of initial size Q_0 that is present after t years is given by a function of the form $Q(t) = Q_0 e^{-kt}$. The positive constant k measures the rate of decay, but this rate is usually given by specifying the amount of time t required for half a given sample to decay. This time is called the **half-life** of the radioactive substance. Example 3.7 shows how half-life is related to k.

EXAMPLE 3.7

Show that a radioactive substance that decays according to the formula $Q(t) = Q_0 e^{-kt}$ has half-life $t = \frac{\ln 2}{k}$.

Solution

The goal is to find the value of t for which $Q(t) = \frac{1}{2}Q_0$; that is,

$$\frac{1}{2}Q_0 = Q_0 e^{-kt}$$

Divide by Q_0 and take the natural logarithm of each side to get

$$\ln \frac{1}{2} = -kt$$

so that the half-life is $t = \dfrac{-\ln \frac{1}{2}}{k} = \dfrac{\ln 2}{k}$

as required.

Remark: Note that doubling time and half-life (Examples 3.6 and 3.7) both equal $\dfrac{\ln 2}{k}$. The only difference is that $k > 0$ for doubling time and $k < 0$ for half-life.

Exponential Curve Fitting   In the next example, you will see how to use logarithms to fit an exponential curve to a set of data.

EXAMPLE 3.8

The population density x miles from the center of a city is given by a function of the form $Q(x) = Ae^{-kx}$. Find this function if it is known that the population density at the center of the city is 15,000 people per square mile and the density 10 miles from the center is 9,000 people per square mile.

Solution

For simplicity, express the density in units of 1,000 people per square mile. The fact that $Q(0) = 15$ tells you that $A = 15$. The fact that $Q(10) = 9$ tells you that

$$9 = 15e^{-10k} \qquad \text{or} \qquad \frac{3}{5} = e^{-10k}$$

Taking the logarithm of each side of this equation, you get

$$\ln \frac{3}{5} = -10k \qquad \text{or} \qquad k = -\frac{\ln 3/5}{10} \approx 0.051$$

Hence the exponential function for the population density is (approximately) $Q(x) = 15e^{-0.051x}$.

Carbon Dating   In 1960, W. F. Libby won a Nobel prize for his discovery of **carbon dating**, a technique for determining the age of certain fossils and artifacts.

Here is an outline of the technique.

The carbon dioxide in the air contains the radioactive isotope ^{14}C ("carbon 14") as well as the stable isotope ^{12}C ("carbon 12"). Living plants absorb carbon dioxide from the air, which means that the ratio of ^{14}C to ^{12}C in a living plant (or in an animal that eats plants) is the same as that in the air itself. When a plant or an animal dies, the absorption of carbon dioxide ceases. The ^{12}C already in the plant or animal remains the same as at the time of death, while the ^{14}C decays, and the ratio of ^{14}C to ^{12}C decreases exponentially. It is reasonable to assume that the ratio R_0 of ^{14}C to ^{12}C in the atmosphere is the same today as it was in the past, so that the ratio of ^{14}C to ^{12}C in a sample (e.g., a fossil, or an artifact) is given by a function of the form $R(t) = R_0 e^{-kt}$. The half-life of ^{14}C is 5,730 years. By comparing $R(t)$ to R_0, archaeologists can estimate the age of the sample. Here is an example.

EXAMPLE 3.9

An archaeologist has found a fossil in which the ratio of ^{14}C to ^{12}C is $\dfrac{1}{5}$ the ratio found in the atmosphere. Approximately how old is the fossil?

Solution

The age of the fossil is the value of t for which $R(t) = \dfrac{1}{5}R_0$, that is, for which

$$\frac{1}{5}R_0 = R_0 e^{-kt}$$

Dividing by R_0 and taking logarithms, you find that

$$\frac{1}{5} = e^{-kt}$$

$$\ln \frac{1}{5} = -kt$$

and

$$t = \frac{-\ln \dfrac{1}{5}}{k} = \frac{\ln 5}{k}$$

To find k, note that since the half-life of ^{14}C is 5,730 years,

$$R(5{,}730) = \frac{1}{2}R_0 = R_0 e^{-k(5{,}730)}$$

so that

$$\frac{1}{2} = e^{-5{,}730k}$$

and
$$\ln \frac{1}{2} = -k(5,730)$$

or
$$k = \frac{-\ln \frac{1}{2}}{5,730} = \frac{\ln 2}{5,730}$$

Therefore, the age of the fossil is

$$t = \frac{\ln 5}{k} = \frac{\ln 5}{(\ln 2)5,730} = \frac{5,730 \ln 5}{\ln 2} \approx 13,304.65$$

That is, the fossil is approximately 13,305 years old.

Problems 4.3

In Problems 1 and 2, use your calculator to find the indicated natural logarithms.

1. Find $\ln 1$, $\ln 2$, $\ln e$, $\ln 5$, $\ln \frac{1}{5}$, and $\ln e^2$. What happens if you try to find $\ln 0$ or $\ln (-2)$? Why?

2. Find $\ln 7$, $\ln \frac{1}{3}$, $\ln e^{-3}$, $\ln \frac{1}{e^{2.1}}$, and $\ln \sqrt[5]{e}$.

In Problems 3 through 8, evaluate the given expression without using tables or a calculator.

3. $\ln e^3$

4. $\ln \sqrt{e}$

5. $e^{\ln 5}$

6. $e^{2\ln 3}$

7. $e^{3\ln 2 - 2\ln 5}$

8. $\ln \dfrac{e^3 \sqrt{e}}{e^{1/3}}$

In Problems 9 through 20, solve the given equation for x.

9. $2 = e^{0.06x}$

10. $\frac{1}{2}Q_0 = Q_0 e^{-1.2x}$

11. $3 = 2 + 5e^{-4x}$

12. $-2 \ln x = b$

13. $-\ln x = \dfrac{t}{50} + C$

14. $5 = 3 \ln x - \dfrac{1}{2} \ln x$

15. $\ln x = \dfrac{1}{3}(\ln 16 + 2 \ln 2)$

16. $\ln x = 2(\ln 3 - \ln 5)$

17. $3^x = e^2$

18. $a^k = e^{kx}$

19. $a^{x+1} = b$

20. $x^{\ln x} = e$

21. Find $\ln \dfrac{1}{\sqrt{ab^3}}$ if $\ln a = 2$ and $\ln b = 3$.

22. Find $\dfrac{1}{a} \ln \left(\dfrac{\sqrt{b}}{c}\right)^a$ if $\ln b = 6$ and $\ln c = -2$.

Compound Interest 23. How quickly will money double if it is invested at an annual interest rate of 6 percent compounded continuously?

Compound Interest 24. How quickly will money double if it is invested at an annual interest rate of 7 percent and interest is compounded
(a) Continuously (b) Annually (c) Quarterly

Compound Interest 25. Money deposited in a certain bank doubles every 13 years. The bank compounds interest continuously. What annual interest rate does the bank offer?

Population Growth 26. Based on the estimate that there are 10 billion acres of arable land on the earth and that each acre can produce enough food to feed 4 people, some demographers believe that the earth can support a population of no more than 40 billion people. The population of the earth was approximately 3 billion in 1960 and 4 billion in 1975. If the population of the earth were growing exponentially, when would it reach the theoretical limit of 40 billion?

Radioactive Decay 27. The amount of a certain radioactive substance remaining after t years is given by a function of the form $Q(t) = Q_0 e^{-0.003t}$. Find the half-life of the substance.

Radioactive Decay 28. Radium decays exponentially. Its half-life is 1,690 years. How long will it take for a 50-gram sample of radium to be reduced to 5 grams?

Radioactive Decay 29. A radioactive substance decays exponentially. Show that the amount $Q(t)$ of the substance remaining after t years is $Q(t) = Q_0 e^{-(\ln 2/\lambda)t}$, where Q_0 is the amount present initially and λ is the half-life of the substance.

Radioactive Decay 30. A radioactive substance decays exponentially. Show that the amount $Q(t)$ of the substance remaining after t years is $Q(t) = Q_0\left(\dfrac{1}{2}\right)^{t/\lambda}$, where Q_0 is the amount present initially and λ is the half-life of the substance. (*Hint:* Use the result of Problem 29.)

Advertising 31. The mathematics editor at a major publishing house estimates that if x thousand complimentary copies are distributed to instructors, the first-year sales of a new mathematics text will be approximately $f(x) = 20 - 15e^{-0.2x}$ thousand copies. According to this estimate, approximately how many complimentary copies should the editor send out to generate first-year sales of 12,000 copies?

Growth of Bacteria   32. A medical student studying the growth of bacteria in a certain culture has compiled the following data:

Number of minutes	0	20
Number of bacteria	6,000	9,000

Use these data to find an exponential function of the form $Q(t) = Q_0 e^{kt}$ expressing the number of bacteria in the culture as a function of time.

Gross National Product   33. An economist has compiled the following data on the gross national product (GNP) of a certain country:

Year	1980	1990
GNP (in billions)	100	180

Use these data to predict the GNP in 2000 if the GNP is growing
(a) Linearly   (b) Exponentially

Worker Efficiency   34. An efficiency expert hired by a manufacturing firm has compiled the following data relating workers' output to their experience:

Experience (months)	0	6
Output (units per hour)	300	410

The expert believes that the output Q is related to experience t by a function of the form $Q(t) = 500 - Ae^{-kt}$. Find the function of this form that fits the data.

Population Growth   35. According to a logistic model based on the assumption that the earth can support no more than 40 billion people, the world's population (in billions) t years after 1960 is given by a function of the form $P(t) = \dfrac{40}{1 + Ce^{-kt}}$, where C and k are positive constants. Find the function of this form that is consistent with the fact that the world's population was approximately 3 billion in 1960 and 4 billion in 1975. What does your model predict the population will be in the year 2000?

Carbon Dating   36. An archaeologist has found a fossil in which the ratio of ^{14}C to ^{12}C is $\dfrac{1}{3}$ the ratio found in the atmosphere. Approximately how old is the fossil?

Carbon Dating   37. How old is a fossil in which the ratio of ^{14}C to ^{12}C is $\dfrac{1}{2}$ the ratio found in the atmosphere?

Carbon Dating   38. A Cro-Magnon cave painting at Lascaux, France, is approximately 15,000 years old. Approximately what ratio of ^{14}C to ^{12}C would you expect to find in a fossil found in the same cave as the painting?

Carbon Dating   39. "Lucy," the famous prehuman whose skeleton was discovered in Africa by Donald Johanson, has been found to be approximately 3.8 million years old. If you try to apply carbon dating to Lucy, approximately what ratio of ^{14}C to ^{12}C do you obtain? Why would this be a problem if you were trying to find the age of Lucy? (In practice, carbon dating works well only for relatively "recent" samples—those which are less than approximately 50,000 years old. For older samples, such as Lucy, variations on carbon dating have been developed, such as potassium-argon dating and rubidium-strontium dating.)

Magnitude of   40. On the Richter scale, the magnitude of an earthquake of intensity I is
Earthquakes   given by $R = \dfrac{\ln I}{\ln 10}$

(a) Find the intensity of the 1906 San Francisco earthquake, which measured 8.3 on the Richter scale.
(b) How much more intense was the San Francisco earthquake of 1906 than the Bay Area World Series earthquake of 1989, which measured 6.9 on the Richter scale?

Air Pressure   41. The air pressure $f(s)$ at a height of s meters above sea level is given by

$$f(s) = e^{-1.25 \times 10^{-4}s} \text{ atmospheres}$$

(a) The atmospheric pressure outside an airplane is 0.25 atmosphere. How high is the plane?
(b) A mountain climber decides she will wear an oxygen mask once she has reached an altitude of 7,000 meters. What is the atmospheric pressure at this altitude?

42. Use one of the laws of exponents to prove that $\ln \dfrac{u}{v} = \ln u - \ln v$.

43. Use one of the laws of exponents to prove that $\ln u^v = v \ln u$.

44. Show that $\log x = \dfrac{\ln x}{\ln a}$ and $a^x = e^{x \ln a}$.

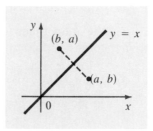

45. Show that the reflection of the point (a, b) in the line $y = x$ is (b, a). [*Hint:* Show that the line joining (a, b) and (b, a) is perpendicular to $y = x$ and that the distance from (a, b) to $y = x$ is the same as the distance from $y = x$ to (b, a).]

4   Differentiation of Logarithmic and Exponential Functions

Both the logarithmic function $y = \ln x$ and the exponential function $y = e^x$ have simple derivatives.

Logarithmic Functions

The Derivative of ln x

$$\frac{d}{dx}(\ln x) = \frac{1}{x} \qquad \text{for } x > 0$$

You will see an outline of the proof of this formula a little later in this section. First, here are some examples illustrating its use.

EXAMPLE 4.1

Differentiate the function $f(x) = x \ln x$.

Solution

Combine the product rule with the formula for the derivative of $\ln x$ to get

$$f'(x) = x\left(\frac{1}{x}\right) + \ln x = 1 + \ln x$$

EXAMPLE 4.2

Differentiate the function $f(x) = \ln (2x^3 + 1)$.

Solution

Think of the function $f(x) = \ln (2x^3 + 1)$ as the composite function $f(x) = g[h(x)]$, where

$$g(u) = \ln u \qquad \text{and} \qquad h(x) = 2x^3 + 1$$

Then, $\qquad g'(u) = \dfrac{1}{u} \qquad$ and $\qquad h'(x) = 6x^2$

and, by the chain rule,

$$f'(x) = g'[h(x)]h'(x) = \frac{1}{2x^3 + 1}(6x^2) = \frac{6x^2}{2x^3 + 1}$$

The preceding example illustrates the following general rule for differentiating the logarithm of a differentiable function.

The Chain Rule for Logarithmic Functions

$$\frac{d}{dx}[\ln h(x)] = \frac{h'(x)}{h(x)}$$

That is, to differentiate $\ln h$, simply divide the derivative of h by h itself.

Here is another example.

EXAMPLE 4.3

Differentiate the function $f(x) = \ln (x^2 + 1)^3$.

Solution

As a preliminary step, simplify $f(x)$ using a property of logarithms to get

$$f(x) = 3 \ln (x^2 + 1)$$

Now apply the chain rule for logarithms and conclude that

$$f'(x) = 3\frac{2x}{x^2 + 1} = \frac{6x}{x^2 + 1}$$

For practice, convince yourself that the final answer would have been the same if you had not made the initial simplification of $f(x)$.

The Proof that
$$\frac{d}{dx}(\ln x) = \frac{1}{x}$$

The proof that $\dfrac{d}{dx}(\ln x) = \dfrac{1}{x}$ is based on the fact that $\left(1 + \dfrac{1}{n}\right)^n$ approaches e as n increases (or decreases) without bound. To derive the formula for the derivative of $f(x) = \ln x$, form the difference quotient and rewrite it using the properties of logarithms as follows:

$$\frac{f(x + \Delta x) - f(x)}{\Delta x} = \frac{\ln (x + \Delta x) - \ln x}{\Delta x}$$

$$= \frac{1}{\Delta x} \ln \left(\frac{x + \Delta x}{x}\right)$$

$$= \ln \left(\frac{x + \Delta x}{x}\right)^{1/\Delta x}$$

$$= \ln \left(1 + \frac{\Delta x}{x}\right)^{1/\Delta x}$$

To find the derivative of $\ln x$, let Δx approach zero in the simplified difference quotient. This will be easier to do if you first let $n = \frac{x}{\Delta x}$. Then,

$$\frac{\Delta x}{x} = \frac{1}{n} \quad \text{and} \quad \frac{1}{\Delta x} = \frac{n}{x}$$

and so

$$\frac{f(x + \Delta x) - f(x)}{\Delta x} = \ln \left(1 + \frac{1}{n}\right)^{n/x} = \ln \left[\left(1 + \frac{1}{n}\right)^{n}\right]^{1/x}$$

As Δx approaches zero, $n = \frac{x}{\Delta x}$ increases or decreases without bound, depending on the sign of Δx. Since

$$\left(1 + \frac{1}{n}\right)^{n} \to e$$

as n increases or decreases without bound, it follows that

$$\ln \left[\left(1 + \frac{1}{n}\right)^{n}\right]^{1/x} \to \ln e^{1/x} = \frac{1}{x}$$

as n increases or decreases without bound. Hence,

$$\frac{d}{dx}(\ln x) = \frac{1}{x}$$

and the proof is complete.

Exponential Functions To find the formula for the derivative of e^x, differentiate both sides of the equation

$$\ln e^x = x$$

with respect to x, using the chain rule for logarithms to differentiate $\ln e^x$. You get

$$\frac{\frac{d}{dx}(e^x)}{e^x} = 1 \quad \text{or} \quad \frac{d}{dx}(e^x) = e^x$$

That is, e^x is its own derivative!

The Derivative of e^x $\dfrac{d}{dx}(e^x) = e^x$

Combining the formula for the derivative of e^x with the chain rule gives the following formula for the derivative of $e^{h(x)}$, where h is a differentiable function of x.

The Chain Rule for Exponential Functions

$$\frac{d}{dx}[e^{h(x)}] = h'(x)e^{h(x)}$$

That is, to compute the derivative of $e^{h(x)}$, simply multiply $e^{h(x)}$ by the derivative of the exponent $h(x)$.

Here are two examples that illustrate the use of these rules.

EXAMPLE 4.4

Differentiate the function $f(x) = e^{x^2+1}$.

Solution

By the chain rule,
$$f'(x) = \left[\frac{d}{dx}(x^2 + 1)\right]e^{x^2+1} = 2xe^{x^2+1}$$

EXAMPLE 4.5

Differentiate the function $f(x) = xe^{2x}$.

Solution

By the product rule,

$$f'(x) = x\frac{d}{dx}(e^{2x}) + e^{2x}\frac{d}{dx}(x) = x(2e^{2x}) + e^{2x}(1) = (2x + 1)e^{2x}$$

Implicit
Differentiation

In the next example, the chain rule for exponential functions is combined with the technique of implicit differentiation from Chapter 2, Section 6.

EXAMPLE 4.6

Find $\dfrac{dy}{dx}$ if $e^{xy} = 3xy^2$.

Solution

Differentiate both sides of the equation with respect to x (using the product rule and the chain rule). Thus,

$$\left[\frac{d}{dx}(xy)\right]e^{xy} = 3x\left(2y\frac{dy}{dx}\right) + 3y^2$$

so that

$$\left(x\frac{dy}{dx} + y\right)e^{xy} = 6xy\frac{dy}{dx} + 3y^2$$

Now solve for $\dfrac{dy}{dx}$ to get

$$xe^{xy}\frac{dy}{dx} + ye^{xy} = 6xy\frac{dy}{dx} + 3y^2$$

$$xe^{xy}\frac{dy}{dx} - 6xy\frac{dy}{dx} = 3y^2 - ye^{xy}$$

$$(xe^{xy} - 6xy)\frac{dy}{dx} = 3y^2 - ye^{xy}$$

or

$$\frac{dy}{dx} = \frac{3y^2 - ye^{xy}}{xe^{xy} - 6xy} = \frac{y(3y - e^{xy})}{x(e^{xy} - 6y)}$$

To simplify the answer further, you may substitute $e^{xy} = 3xy^2$ (from the original equation) to get

$$\frac{dy}{dx} = \frac{y(3y - 3xy^2)}{x(3xy^2 - 6y)} = \frac{3y^2(1 - xy)}{3xy(xy - 2)} = \frac{y(1 - xy)}{x(xy - 2)}$$

Exponential Growth

In Section 2 you saw that a quantity $Q(t)$ that increases according to a law of the form $Q(t) = Q_0e^{kt}$, where Q_0 and k are positive constants, is said to experience exponential growth. The next example shows that if a quantity grows exponentially, its rate of growth is proportional to its size and its percentage rate of growth is constant. In Problems 34 and 35 you will derive the analogous results for exponential decay. The converse of these facts will be established in Chapter 7.

EXAMPLE 4.7

Suppose $Q(t)$ grows exponentially.

(a) Show that the rate of change of Q with respect to t is proportional to its size.
(b) Show that the percentage rate of change of Q with respect to t is constant.

Solution

If $Q(t)$ grows exponentially, it is given by a function of the form $Q(t) = Q_0 e^{kt}$, where Q_0 and k are positive constants.

(a) The rate of change of Q with respect to t is the derivative

$$Q'(t) = kQ_0 e^{kt} = kQ(t)$$

This says that the rate of change, $Q'(t)$, is proportional to $Q(t)$ itself and that the constant k that appears in the exponent of $Q(t)$ is the constant of proportionality.

(b) The percentage rate of change of Q with respect to t is

$$100\frac{Q'(t)}{Q(t)} = 100\frac{kQ(t)}{Q(t)} = 100k$$

This says that the constant k which appears in the exponent of the function $Q(t) = Q_0 e^{kt}$ is the percentage rate of change of Q with respect to t (expressed as a decimal).

Perspective: Note that this is consistent with what you already know about compound interest, namely, that if interest is compounded continuously, the balance after t years is $B(t) = Pe^{rt}$, where r is the interest rate expressed as a decimal.

Exponential Growth

If $Q(t)$ grows exponentially according to the law $Q(t) = Q_0 e^{kt}$, then

1. The rate of change of Q with respect to t is proportional to its size. In particular,

$$\text{Rate of change} = Q'(t) = kQ(t)$$

2. The percentage rate of change of Q with respect to t is constant. In particular,

$$\text{Percentage rate of change} = 100\frac{Q'(t)}{Q(t)} = 100k$$

Optimization   Now that you know how to differentiate exponential functions, you can use calculus to find maximum and minimum values of functions involving powers of e. Here is an example.

EXAMPLE 4.8

The consumer demand for a certain commodity is $D(p) = 5,000e^{-0.02p}$ units per month when the market price is p dollars per unit. Determine the market price that will result in the greatest consumer expenditure.

Solution

The consumer expenditure $E(p)$ for the commodity is the price per unit times the number of units sold. That is,

$$E(p) = pD(p) = 5,000pe^{-0.02p}$$

Since only nonnegative values of p are meaningful in this context, the goal is to find the absolute maximum of $E(p)$ on the interval $p \geq 0$.

The derivative of E is

$$E'(p) = 5,000(-0.02pe^{-0.02p} + e^{-0.02p}) = 5,000(1 - 0.02p)e^{-0.02p}$$

Since $e^{-0.02p}$ is never zero, $E'(p) = 0$ if and only if

$$1 - 0.02p = 0 \quad \text{or} \quad p = \frac{1}{0.02} = 50$$

The easiest way to verify that $p = 50$ actually gives the absolute maximum of $E(p)$ for $p \geq 0$ is to check the sign of $E'(p)$ for $0 < p < 50$ and for $p > 50$. Since $5,000e^{-0.02p}$ is always positive, it follows that $E'(p) > 0$ if $0 < p < 50$ and $E'(x) < 0$ if $p > 50$. This implies that $E(p)$ is increasing for $0 < p < 50$, is decreasing for $p > 50$, and has an absolute maximum

Figure 4.16 Consumer expenditure: $E(p) = 5,000pe^{-0.02p}$.

when $p = 50$, as shown in Figure 4.16. Thus, consumer expenditure will be greatest when the market price is $50 per unit.

Curve Sketching   In the next example, the methods of calculus are used to obtain a detailed sketch of the function $f(x) = \dfrac{1}{\sqrt{2\pi}} e^{-x^2/2}$. This function is known as the **standard normal probability density function** and plays a vital role in probability and statistics. The graph of this function is the famous "bell-shaped" curve that is used by physical and social scientists to describe the distributions of IQ scores, measurements on large populations of living organisms, the velocity of a molecule in a gas, and numerous other important phenomena.

EXAMPLE 4.9

Determine where the function

$$f(x) = \frac{1}{\sqrt{2\pi}} e^{-x^2/2}$$

is increasing, decreasing, concave upward, and concave downward. Find the relative extrema and inflection points and draw the graph.

Solution

The first derivative is

$$f'(x) = \frac{-x}{\sqrt{2\pi}} e^{-x^2/2}$$

Since $e^{-x^2/2}$ is always positive, $f'(x)$ is zero if and only if $x = 0$. Since $f(0) = \dfrac{1}{\sqrt{2\pi}} \approx 0.4$, the only first-order critical point is (approximately) $(0, 0.4)$. By the product rule, the second derivative is

$$f''(x) = \frac{x^2}{\sqrt{2\pi}} e^{-x^2/2} - \frac{1}{\sqrt{2\pi}} e^{-x^2/2} = \frac{1}{\sqrt{2\pi}} (x^2 - 1) e^{-x^2/2}$$

which is zero if $x = \pm 1$. Since

$$f(1) = \frac{e^{-1/2}}{\sqrt{2\pi}} \approx 0.24 \qquad \text{and} \qquad f(-1) = \frac{e^{-1/2}}{\sqrt{2\pi}} \approx 0.24$$

the corresponding second-order critical points are (approximately) $(1, 0.24)$ and $(-1, 0.24)$.

Plot the critical points and then check the signs of the first and second derivatives on each of the intervals defined by the x coordinates of these points.

Complete the graph as shown in Figure 4.17 by connecting the critical

Interval	$f'(x)$	$f''(x)$	Increase or Decrease of f	Concavity of f	Shape of graph
$x < -1$	$+$	$+$	Increasing	C-up	
$-1 < x < 0$	$+$	$-$	Increasing	C-down	
$0 < x < 1$	$-$	$-$	Decreasing	C-down	
$x > 1$	$-$	$+$	Decreasing	C-up	

Figure 4.17 The standard normal density function:
$$f(x) = \frac{1}{\sqrt{2\pi}} e^{-x^2/2}.$$

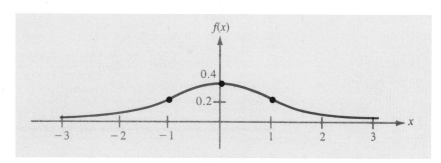

points with a curve of appropriate shape on each interval. Notice that the graph has no x intercepts since $e^{-x^2/2}$ is always positive and that the graph approaches the x axis as $|x|$ increases without bound since $e^{-x^2/2}$ approaches zero.

Logarithmic Differentiation   Sometimes you can simplify the work involved in differentiating a function if you first take its logarithm. This technique, called **logarithmic differentiation,** is illustrated in the following example.

EXAMPLE 4.10

Differentiate the function $f(x) = \dfrac{\sqrt[3]{x + 1}}{(1 - 3x)^4}$.

Solution

You could do this problem using the quotient rule and the chain rule, but the resulting computation would be somewhat tedious. (Try it!)

A more efficient approach is to take logarithms of both sides of the expression for f:

$$\ln f(x) = \ln\left[\frac{\sqrt[3]{x+1}}{(1-3x)^4}\right] = \ln\sqrt[3]{(x+1)} - \ln(1-3x)^4$$

$$= \frac{1}{3}\ln(x+1) - 4\ln(1-3x)$$

(Notice that by introducing the logarithm, you eliminate the quotient, the cube root, and the fourth power.)

Now use the chain rule for logarithms to differentiate both sides of this equation to get

$$\frac{f'(x)}{f(x)} = \frac{1}{3}\frac{1}{(x+1)} - 4\left(\frac{-3}{1-3x}\right) = \frac{1}{3}\frac{1}{x+1} + \frac{12}{1-3x}$$

so that

$$f'(x) = f(x)\left[\frac{1}{3}\frac{1}{x+1} + \frac{12}{1-3x}\right]$$

$$= \left[\frac{\sqrt[3]{x+1}}{(1-3x)^4}\right]\left[\frac{1}{3}\frac{1}{x+1} + \frac{12}{1-3x}\right]$$

The Proof of the Power Rule   In the next example, logarithmic differentiation is used to prove the power rule from Chapter 2, Section 2.

EXAMPLE 4.11

Use logarithmic differentiation to prove the power rule:

$$\frac{d}{dx}(x^n) = nx^{n-1}$$

Solution

Let $f(x) = x^n$. Then,

$$\ln f(x) = n \ln x$$

and differentiation gives

$$\frac{f'(x)}{f(x)} = \frac{n}{x}$$

or

$$f'(x) = \frac{n}{x}f(x) = \frac{n}{x}(x^n) = nx^{n-1}$$

and the power rule is proved.

Problems 4.4

In Problems 1 through 26, differentiate the given function.

1. $f(x) = e^{5x}$

2. $f(x) = 3e^{4x+1}$

3. $f(x) = e^{x^2+2x-1}$

4. $f(x) = e^{1/x}$

5. $f(x) = 30 + 10e^{-0.05x}$

6. $f(x) = x^2 e^x$

7. $f(x) = (x^2 + 3x + 5)e^{6x}$

8. $f(x) = xe^{-x^2}$

9. $f(x) = \dfrac{x}{e^x}$

10. $f(x) = \dfrac{x^2}{e^x}$

11. $f(x) = (1 - 3e^x)^2$

12. $f(x) = \sqrt{1 + e^x}$

13. $f(x) = e^{\sqrt{3x}}$

14. $f(x) = e^{-1/(2x)}$

15. $f(x) = \ln x^3$

16. $f(x) = \ln 2x$

17. $f(x) = \ln (x^2 + 5x - 2)$

18. $f(x) = \ln \sqrt{x^2 + 1}$

19. $f(x) = x^2 \ln x$

20. $f(x) = x \ln \sqrt{x}$

21. $f(x) = \dfrac{\ln x}{x}$

22. $f(x) = \dfrac{x}{\ln x}$

23. $f(x) = \ln\left(\dfrac{x + 1}{x - 1}\right)$

24. $f(x) = e^x \ln x$

25. $f(x) = \ln e^{2x}$

26. $f(x) = e^{\ln x + 2\ln 3x}$

In Problems 27 through 30, find $\dfrac{dy}{dx}$ by implicit differentiation. Simplify your answers.

27. $e^{xy} = xy^3$

28. $e^{x/y} = 3y^5$

29. $\ln \dfrac{y}{x} = x^2 y^3$

30. $\ln xy^2 = \dfrac{x}{3y}$

Population Growth   31. It is projected that t years from now, the population of a certain country will be $P(t) = 50e^{0.02t}$ million.

   (a) At what rate will the population be changing with respect to time 10 years from now?

   (b) At what percentage rate will the population be changing with respect to time t years from now? Does this percentage rate depend on t or is it constant?

Compound Interest   32. Money is deposited in a bank offering interest at an annual rate of 6 percent compounded continuously. Find the percentage rate of change of the balance with respect to time.

Depreciation   33. A certain industrial machine depreciates so that its value after t years is $Q(t) = 20,000e^{-0.4t}$ dollars.
   (a) At what rate is the value of the machine changing with respect to time after 5 years?
   (b) At what percentage rate is the value of the machine changing with respect to time after t years? Does this percentage rate depend on t or is it constant?

Exponential Decay   34. Show that if a quantity decreases exponentially, its rate of decrease is proportional to its size.

Exponential Decay   35. Show that if a quantity decreases exponentially, its percentage rate of change is constant.

The Spread of an Epidemic   36. Public health records indicate that t weeks after the outbreak of a certain form of influenza, approximately $Q(t) = \dfrac{80}{4 + 76e^{-1.2t}}$ thousand people had caught the disease. At what rate was the disease spreading at the end of the 2nd week?

Population Growth   37. According to a logistic model based on the assumption that the earth can support no more than 40 billion people, the world's population (in billions) t years after 1960 will be approximately $P(t) = \dfrac{40}{1 + 12e^{-0.08t}}$.
   (a) If this model is correct, at what rate will the world's population be increasing with respect to time in 1995?
   (b) If this model is correct, at what percentage rate will the world's population be increasing with respect to time in 1995?

Newton's Law of Cooling   38. A cool drink is removed from a refrigerator on a hot summer day and placed in a room whose temperature is 30° Celsius. According to a law of physics, the temperature of the drink t minutes later is given by a function of the form $f(t) = 30 - Ae^{-kt}$. Show that the rate of change of the temperature of the drink with respect to time is proportional to the difference between the temperature of the room and that of the drink.

Marginal Analysis   39. The mathematics editor at a major publishing house estimates that if x thousand complimentary copies are distributed to professors, the first-year sales of a certain new text will be $f(x) = 20 - 15e^{-0.2x}$ thousand copies. Currently, the editor is planning to distribute 10,000 complimentary copies.
   (a) Use marginal analysis to estimate the increase in first-year sales that will result if 1,000 additional complimentary copies are distributed.
   (b) Calculate the actual increase in first-year sales that will result from the distribution of the additional 1,000 complimentary copies. Is the estimate in part (a) a good one?

Consumer Expenditure 40. The consumer demand for a certain commodity is $D(p) = 3,000e^{-0.01p}$ units per month when the market price is p dollars per unit. Express consumers' total monthly expenditure for the commodity as a function of p and determine the market price that will result in the greatest consumer expenditure.

Profit Maximization 41. A manufacturer can produce radios at a cost of $5 apiece and estimates that if they are sold for x dollars apiece, consumers will buy approximately $1,000e^{-0.1x}$ radios per week. At what price should the manufacturer sell the radios to maximize profit?

In Problems 42 through 51, determine where the given function is increasing, decreasing, concave upward, and concave downward. Find the relative extrema and inflection points and draw the graph.

42. $f(x) = xe^x$

43. $f(x) = xe^{-x}$

44. $f(x) = xe^{2-x}$

45. $f(x) = e^{-x^2}$

46. $f(x) = x^2e^{-x}$

47. $f(x) = e^x + e^{-x}$

48. $f(x) = \dfrac{6}{1 + e^{-x}}$

49. $f(x) = x - \ln x$ (for $x > 0$)

50. $f(x) = (\ln x)^2$ (for $x > 0$)

51. $f(x) = \dfrac{\ln x}{x}$ (for $x > 0$)

In Problems 52 through 61, use logarithmic differentiation to find $f'(x)$.

52. $f(x) = \dfrac{(x + 2)^5}{\sqrt[9]{3x - 5}}$

53. $f(x) = \sqrt[4]{\dfrac{2x + 1}{1 - 3x}}$

54. $f(x) = (x + 1)^3(6 - x)^2\sqrt[3]{2x + 1}$

55. $f(x) = \dfrac{e^{-3x}\sqrt{2x - 5}}{(6 - 5x)^4}$

56. $f(x) = 2^x$

57. $f(x) = a^x$ (for $a > 0$)

58. $f(x) = (2^x)^{\ln x}$ (for $x > 0$)

59. $f(x) = \dfrac{e^{2x}}{x^x}$ (for $x > 0$)

60. $f(x) = xe^{3x}$ (for $x > 0$)

61. $f(x) = x^x(e^x)$ (for $x > 0$)

Population Growth 62. It is estimated that t years from now the population of a certain country will be $P(t) = \dfrac{160}{1 + 8e^{-0.01t}}$ million. When will the population be growing most rapidly?

The Spread of an Epidemic 63. An epidemic spreads through a community so that t weeks after its outbreak, the number of residents who have been infected is given by a function of the form $f(t) = \dfrac{A}{1 + Ce^{-kt}}$, where A is the total number of susceptible residents. Show that the epidemic is spreading most rapidly when half of the susceptible residents have been infected.

Percentage Rate of Change

64. Show that the percentage rate of change of f with respect to x is

$$100\frac{d}{dx}[\ln f(x)].$$

In Problems 65 and 66, use the formula from Problem 64 to find the specified percentage rate of change.

Percentage Rate of Change

65. A quantity grows according to the law $Q(t) = Q_0\dfrac{e^{kt}}{t}$. Find the percentage rate of change of Q with respect to t.

Population Growth

66. It is projected that x years from now the population of a certain town will be approximately $P(x) = 5,000\sqrt{x^2 + 4x + 19}$. At what percentage rate will the population be changing with respect to time 3 years from now?

5 Applications Involving Compound Interest

In this section, you will learn more about the concept of compound interest, which was introduced in Section 1. To begin the discussion, here is a summary of what you should already know.

When money is invested, it (usually) earns **interest.** The amount of money invested is called the **principal.** Interest that is computed on the principal alone is called **simple interest,** and the balance after t years is given by the following formula.

Simple Interest

If P dollars are invested at an annual simple interest rate r (expressed as a decimal), the balance $B(t)$ after t years will be

$$B(t) = P(1 + rt) \qquad \text{dollars}$$

Interest that is computed on the principal plus the previous interest is called **compound interest.** If the annual interest rate is r and interest is compounded k times per year, the year is divided into k equal **interest periods** and the interest rate during each period is $\dfrac{r}{k}$. The balance at the end of t years is given by the following formula.

Compound Interest

If P dollars are invested at an annual interest rate r and interest is compounded k times per year, the balance $B(t)$ after t years will be

$$B(t) = P\left(1 + \frac{r}{k}\right)^{kt} \quad \text{dollars}$$

If the number of times interest is compounded per year is allowed to increase without bound, the interest is said to be **compounded continuously.** In this case, the balance at the end of t years is given by the following formula.

Continuously Compounded Interest

If P dollars are invested at an annual interest rate r and interest is compounded continuously, the balance $B(t)$ after t years will be

$$B(t) = Pe^{rt} \quad \text{dollars}$$

The use of these formulas is illustrated in the next example.

EXAMPLE 5.1

Suppose \$1,000 is invested at an annual interest rate of 8 percent. Compute the balance after 10 years in each of the following cases:

(a) The interest is simple interest.
(b) The interest is compounded quarterly.
(c) The interest is compounded continuously.

Solution

(a) Use the formula $B(t) = P(1 + rt)$ with $P = 1,000$, $r = 0.08$, and $t = 10$:

$$B(10) = 1,000(1.8) = \$1,800$$

(b) Use the formula $B(t) = P\left(1 + \frac{r}{k}\right)^{kt}$ with $P = 1,000$, $r = 0.08$, $k = 4$, and $t = 10$:

$$B(10) = 1,000(1.02)^{40} \approx \$2,208.04$$

(c) Use the formula $B(t) = Pe^{rt}$ with $P = 1,000$, $r = 0.08$, and $t = 10$:

$$B(10) = 1,000e^{0.8} \approx \$2,225.54$$

Doubling Time   Investors often want to know how long it will take for a given investment to grow to a particular size. In the next example, you will see how to use the compound interest formulas to calculate the time it takes for money to double.

EXAMPLE 5.2

How quickly will money double if it is invested at an annual interest rate of 8 percent and interest is compounded

(a) Quarterly   (b) Continuously

Solution

(a) The balance after t years is

$$B(t) = P\left(1 + \frac{r}{k}\right)^{kt} = P\left(1 + \frac{0.08}{4}\right)^{4t} = P(1.02)^{4t}$$

The goal is to find the value of t for which $B(t) = 2P$, that is, for which

$$2P = P(1.02)^{4t} \quad \text{or} \quad 2 = (1.02)^{4t}$$

Take the natural logarithm of each side and solve for t to get

$$\ln 2 = 4t \ln 1.02$$

or

$$t = \frac{\ln 2}{4 \ln 1.02} \approx 8.75 \text{ years}$$

That is, at 8 percent compounded quarterly, money will double in approximately 8.75 years.

(b) This time, use the formula

$$B(t) = Pe^{rt} = Pe^{0.08t}$$

and solve the equation

$$2P = Pe^{0.08t}$$

to get

$$2 = e^{0.08t} \quad \ln 2 = 0.08t$$

or

$$t = \frac{\ln 2}{0.08} \approx 8.66 \text{ years}$$

That is, at 8 percent compounded continuously, money will double in approximately 8.66 years.

Effective Interest Rate   Money deposited in a bank offering interest at an annual rate of 6 percent will increase in value by more than 6 percent in a year if the interest is compounded more than once. This is because interest compounded during one period will itself earn interest during subsequent periods. The actual percentage by which an investment grows during a year is called the **effective interest rate,** while the corresponding annual compound interest rate is called the **nominal interest rate.** In other words, the effective interest rate is the *simple* interest rate that is equivalent to the nominal compound interest rate. In the next example, you will see how to derive a formula for the effective interest rate.

EXAMPLE 5.3

Find a formula for the effective interest rate if the nominal rate is r and interest is compounded k times per year.

Solution

If P dollars are invested at a nominal rate r and interest is compounded k times per year, the balance at the end of 1 year will be

$$B = P\left(1 + \frac{r}{k}\right)^k$$

On the other hand, if x is the simple interest rate, the corresponding balance at the end of 1 year will be

$$B = P(1 + x)$$

The goal is to find the value of x for which these two expressions for the balance are equal, that is,

$$P\left(1 + \frac{r}{k}\right)^k = P(1 + x)$$

$$\left(1 + \frac{r}{k}\right)^k = 1 + x$$

or

$$x = \left(1 + \frac{r}{k}\right)^k - 1$$

Using a similar argument, you can derive a formula for the effective interest rate when interest is compounded continuously. This derivation is left as a problem for you to do (see Problem 16). For reference, here is a summary of the situation.

Effective Interest Rate

If interest is compounded k times per year at an annual interest rate r, then

$$\text{Effective interest rate} = \left(1 + \frac{r}{k}\right)^k - 1$$

If interest is compounded continuously at an annual interest rate r, then

$$\text{Effective interest rate} = e^r - 1$$

EXAMPLE 5.4

One bank offers interest at an annual rate of 6.1 percent compounded quarterly, and a competing bank offers interest at an annual rate of 6 percent compounded continuously. Which bank offers the better deal?

Solution

The effective interest rate at the first bank is

$$\left(1 + \frac{0.061}{4}\right)^4 - 1 \simeq 0.0624$$

or 6.24 percent, and the effective interest rate at the second bank is only

$$e^{0.06} - 1 \simeq 0.0618$$

or 6.18 percent. Therefore, the first bank offers the better deal.

Present Value The **present value of B dollars payable t years from now** is the amount P that should be invested today so that it will be worth B dollars at the end of t years. To derive a formula for the present value, simply solve the compound interest formula for the principal P. For example, if interest is compounded k times per year, the balance after t years is

$$B = P\left(1 + \frac{r}{k}\right)^{kt}$$

and the present value of B dollars payable t years from now is

$$P = B\left(1 + \frac{r}{k}\right)^{-kt}$$

If interest is compounded continuously, the balance is

$$B = Pe^{rt}$$

and the present value is

$$P = Be^{-rt}$$

Present Value of Future Money

If interest is compounded k times per year at an annual interest rate r, the present value of B dollars payable t years from now is

$$P = B\left(1 + \frac{r}{k}\right)^{-kt} \quad \text{dollars}$$

If interest is compounded continuously at an annual interest rate r, the present value of B dollars payable t years from now is

$$P = Be^{-rt} \quad \text{dollars}$$

The concept of present value is illustrated in the next example.

EXAMPLE 5.5

How much should you invest now at an annual interest rate of 8 percent so that your balance 20 years from now will be $10,000 if the interest is compounded

(a) Quarterly   (b) Continuously

Solution

(a) The amount you should invest is the present value of $10,000 payable 20 years from now. Using the formula $P = B\left(1 + \frac{r}{k}\right)^{-kt}$ with $B = 10,000$, $r = 0.08$, $k = 4$, and $t = 20$, you find that this amount is

$$P = 10,000(1.02)^{-80} \simeq \$2,051.10$$

(b) Using the present-value formula $P = Be^{-rt}$ with $B = 10,000$, $r = 0.08$, and $t = 20$, you find that the amount you should invest is

$$P = 10,000e^{-1.6} \simeq \$2,018.97$$

The Amount of an Annuity   An **annuity** is a sequence of equal payments or deposits made at regular intervals over a specified period of time. Once a deposit is made, it earns interest at a fixed rate until the expiration or **term** of the annuity. The total amount of money (deposits plus interest) accumulated in this way is known as the **amount** of the annuity. The next example illustrates how to use the compound interest formula to calculate the amount of an annuity.

EXAMPLE 5.6

Each year on January first, you deposit $1,000 in an account that pays interest at an annual rate of 8 percent compounded continuously. How much will you have in the account at the end of 3 years (just before you make your 4th deposit)?

Solution

The three deposits of $1,000 and the value of each (at 8 percent compounded continuously) at the end of the 3-year term are shown in Figure

Figure 4.18 The annuity for Example 5.6.

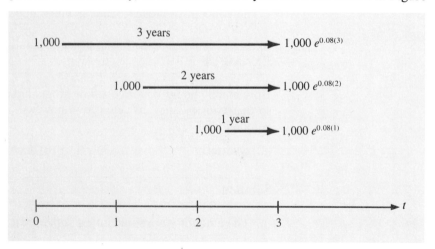

4.18. The amount of the annuity at the end of the term is the sum

$$\text{Amount} = 1,000e^{0.08(3)} + 1,000e^{0.08(2)} + 1,000e^{0.08(1)}$$

$$= 1,000(e^{0.24} + e^{0.16} + e^{0.08}) \simeq \$3,528.05$$

The Present Value of an Annuity   The **present value of an annuity** that consists of a periodic sequence of payments over a specified term is the amount of money that must be deposited today to permit periodic withdrawals generating the same sequence of payments over the specified term, after which nothing will be left in the account. The concept and its calculation are illustrated in the next example.

EXAMPLE 5.7

A bank pays interest at the annual rate of 8 percent compounded quarterly. How much should you deposit today so that you will be able to withdraw $500 at the end of each of the next 3 years, after which nothing will be left in the account? (That is, what is the present value of an annuity that pays $500 at the end of each year for a term of 3 years if the annual interest rate is 8 percent compounded quarterly?)

Solution

The present value of the annuity is the sum of the present values of the 3 withdrawals. Since $r = 0.08$ and $k = 4$, the appropriate present-value formula is

$$P(t) = B\left(1 + \frac{r}{k}\right)^{-kt} = B\left(1 + \frac{0.08}{4}\right)^{-4t} = B(1.02)^{-4t}$$

The withdrawals and corresponding present values are shown in Figure

Figure 4.19 The annuity for Example 5.7.

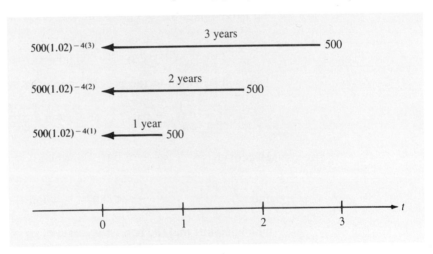

4.19. The present value of the annuity is the sum

$$\text{Present value} = 500(1.02)^{-4(1)} + 500(1.02)^{-4(2)} + 500(1.02)^{-4(3)}$$

$$= 500[(1.02)^{-4} + (1.02)^{-8} + (1.02)^{-12}] \approx \$1,282.91$$

Optimal Holding Time   Suppose you own an asset whose value increases with time. The longer you hold the asset, the more it will be worth. However, if there comes a time after which money invested at the prevailing interest rate grows more quickly than the value of the asset, you will be better off selling the asset and investing the proceeds. You determine the most advantageous time to sell the asset and invest the proceeds by comparing the prevailing interest rate with the percentage rate of growth of the value of the asset. The technique is illustrated in the next example.

EXAMPLE 5.8

Suppose you own a parcel of land whose market price t years from now will be $V(t) = 20{,}000e^{\sqrt{t}}$ dollars. If the prevailing interest rate remains constant at 7 percent compounded continuously, when will it be most advantageous to sell the land?

Solution

Since the derivative of the market price $V(t)$ is

$$V'(t) = 20{,}000e^{\sqrt{t}}\left(\frac{1}{2}\frac{1}{\sqrt{t}}\right) = \frac{10{,}000e^{\sqrt{t}}}{\sqrt{t}}$$

the percentage rate of change of $V(t)$ (expressed in decimal form) is

$$\frac{V'(t)}{V(t)} = \frac{10{,}000t^{-1/2}e^{\sqrt{t}}}{20{,}000e^{\sqrt{t}}} = \frac{1}{2\sqrt{t}}$$

This will be equal to the prevailing interest rate when

$$\frac{1}{2\sqrt{t}} = 0.07 \qquad \text{or} \qquad t = \left[\frac{1}{2(0.07)}\right]^2 \simeq 51.02$$

Moreover, $\dfrac{1}{2\sqrt{t}} > 0.07$ when $0 < t < 51.02$

and $\dfrac{1}{2\sqrt{t}} < 0.07$ when $t > 51.02$

which implies that the percentage rate of growth of the value of the land is greater than the prevailing interest rate when $0 < t < 51.02$ and less than the prevailing interest rate when $t > 51.02$. It follows that you should sell the land 51.02 years from now and invest the proceeds at the prevailing interest rate of 7 percent.

Comment: Economists use a different optimization criterion for problems like the one in Example 5.8. Instead of finding the time when the percentage rate of change of the value of the asset equals the prevailing interest rate, they determine the optimum time for selling the asset by finding when the **present value** of the selling price of the asset is greatest. In other words, they sell when today's dollar equivalent of the selling price is as large as possible. Actually, the two optimization criteria are the same (see Problem 32). In Example 5.9, we solve Example 5.8 as an economist would do it. Notice that the final answer is the same in both cases.

EXAMPLE 5.9

Suppose you own a parcel of land whose market price t years from now will be $V(t) = 20{,}000e^{\sqrt{t}}$ dollars. If the prevailing interest rate remains constant at 7 percent compounded continuously, when will the present value of the market price of the land be greatest?

Solution

In t years, the market price of the land will be $V(t) = 20{,}000e^{\sqrt{t}}$. The present value of this price is

$$P(t) = V(t)e^{-0.07t} = 20{,}000e^{\sqrt{t}}e^{-0.07t} = 20{,}000e^{\sqrt{t}-0.07t}$$

The goal is to maximize $P(t)$ for $t \geq 0$. The derivative of P is

$$P'(t) = 20{,}000e^{\sqrt{t}-0.07t}\left(\frac{1}{2\sqrt{t}} - 0.07\right)$$

which is zero if and only if

$$\frac{1}{2\sqrt{t}} - 0.07 = 0 \qquad \text{or} \qquad t = \left[\frac{1}{2(0.07)}\right]^2 \simeq 51.02$$

(and which is undefined when $t = 0$).

Since $P'(t)$ is positive if $0 < t < 51.02$ and negative if $t > 51.02$, it follows that the graph of P is increasing for $0 < t < 51.02$ and decreasing

Figure 4.20 Present value: $P(t) = 20{,}000e^{\sqrt{t}\,-0.07t}$.

for $t > 51.02$ as shown in Figure 4.20 and that $t = 51.02$ is the time at which the present value will be greatest.

Problems 4.5

1. Suppose $5,000 is invested at an annual interest rate of 5 percent. Compute the balance after 20 years in each of the following cases:
   (a) The interest is simple interest.
   (b) The interest is compounded semiannually.
   (c) The interest is compounded continuously.

2. Suppose $5,000 is invested at an annual interest rate of 6 percent. Compute the balance after 20 years in each of the following cases:
   (a) The interest is simple interest.
   (b) The interest is compounded semiannually.
   (c) The interest is compounded continuously.

3. How quickly will money double if it is invested at an annual interest rate of 12 percent and interest is compounded
   (a) Quarterly   (b) Continuously

4. How quickly will money double if it is invested at an annual simple interest rate of 8 percent?

5. Derive a formula for the amount of time it takes for money to double if the annual interest rate is r and interest is compounded k times per year.

6. Derive a formula for the amount of time it takes for money to double if the annual interest rate is r and interest is compounded continuously.

7. How quickly will money triple if it is invested at an annual interest rate of 6 percent compounded semiannually?

8. How quickly will money triple if it is invested at an annual interest rate of 6 percent compounded continuously?

9. Derive a formula for the amount of time it takes for money to triple if the annual interest rate is r and interest is compounded k times per year.

10. Derive a formula for the amount of time it takes for money to triple if the annual interest rate is r and interest is compounded continuously.

11. Determine how quickly $1,000 will grow to $2,500 if the annual interest rate is 6 percent and interest is compounded
    (a) Quarterly   (b) Continuously

12. Determine how quickly $600 will grow to $1,000 if the annual interest rate is 10 percent and interest is compounded
    (a) Semiannually   (b) Continuously

13. Find the effective interest rate if the nominal interest rate is 6 percent per year and interest is compounded
    (a) Quarterly   (b) Continuously

14. Which investment has the greater effective interest rate: 8.2 percent per year compounded quarterly or 8.1 percent per year compounded continuously?

15. Which investment has the greater effective interest rate: 10.25 percent per year compounded semiannually or 10.20 percent per year compounded continuously?

16. Derive the formula for the effective interest rate if the nominal rate is r and interest is compounded continuously.

17. A bank compounds interest quarterly. What (nominal) interest rate does it offer if $1,000 grows to $2,203.76 in 8 years?

18. A bank compounds interest continuously. What (nominal) interest rate does it offer if $1,000 grows to $2,054.44 in 12 years?

19. A certain bank offers an interest rate of 6 percent per year compounded annually. A competing bank compounds its interest continuously. What (nominal) interest rate should the competing bank offer so that the effective interest rates of the two banks will be equal?

20. How much should you invest now at an annual interest rate of 6 percent so that your balance 10 years from now will be $5,000 if interest is compounded
    (a) Quarterly   (b) Continuously

21. How much should you invest now at an annual interest rate of 8 percent so that your balance 15 years from now will be $20,000 if interest is compounded
    (a) Quarterly   (b) Continuously

22. Find the present value of $8,000 payable 10 years from now if the annual interest rate is 6.25 percent and interest is compounded
    (a) Semiannually   (b) Continuously

The Amount of an Annuity

23. At the end of each year you deposit $2,000 in an account that pays interest at an annual rate of 10 percent compounded continuously. How much will you have in your account immediately after you make your 3rd deposit?

The Amount of an Annuity

24. At the end of each interest period you deposit $100 in an account that pays interest at an annual rate of 8 percent compounded quarterly. How much will you have in the account just after you make your 4th deposit?

The Amount of an
Annuity

25. At the end of each interest period you deposit $50 in an account that pays interest at an annual rate of 6 percent compounded semi-annually. How much will you have in the account just after you make your 4th deposit?

The Present Value of
an Annuity

26. A bank offers interest at an annual rate of 5 percent compounded continuously. How much should you deposit today so that you will be able to make withdrawals of $2,000 at the end of each of the next 3 years, after which nothing will be left in the account?

The Present Value of
an Annuity

27. How much should you invest now (at the beginning of an interest period) at an annual interest rate of 6 percent compounded annually to enable you to make withdrawals of $500 at the end of each of the next 4 years, after which nothing will be left in the account?

The Present Value of
an Annuity

28. Find the present value of an annuity that makes 4 annual payments of $1,000 (starting today) if the prevailing annual interest rate is 9 percent compounded continuously.

Optimal Holding Time

29. Suppose you own a parcel of land whose value t years from now will be $V(t) = 8,000e^{\sqrt{t}}$ dollars. If the prevailing interest rate remains constant at 6 percent per year compounded continuously, when should you sell the land to maximize its present value?

Optimal Holding Time

30. Suppose your family owns a rare book whose value t years from now will be $V(t) = 200e^{\sqrt{2t}}$ dollars. If the prevailing interest rate remains constant at 6 percent per year compounded continuously, when will it be most advantageous for your family to sell the book and invest the proceeds?

Optimal Holding Time

31. Suppose you own a stamp collection that is currently worth $1,200 and whose value increases linearly at the rate of $200 per year. If the prevailing interest rate remains constant at 8 percent per year compounded continuously, when will it be most advantageous for you to sell the collection and invest the proceeds?

Optimal Holding Time

32. Let $V(t)$ be the value of an asset t years from now and assume that the prevailing interest rate remains fixed at r percent per year compounded continuously.
   (a) Show that the present value of the asset $P(t) = V(t)e^{-rt}$ has a first-order critical point where $V'(t) = V(t)r$. (Using economic arguments, it can be shown that this critical point corresponds to a maximum.)
   (b) Explain why the present value of $V(t)$ is maximized at a value of t where the percentage rate of change (expressed in decimal form) equals r.

Chapter Summary and Review Problems

Important Terms, Symbols, and Formulas

Exponential function: $f(x) = a^x$

Laws of exponents: $a^r a^s = a^{r+s}$ $\dfrac{a^r}{a^s} = a^{r-s}$ $(a^r)^s = a^{rs}$

The number e: $e = \lim\limits_{n \to \infty} \left(1 + \dfrac{1}{n}\right)^n \approx 2.718$

Exponential growth: $Q(t) = Q_0 e^{kt}$

$$Q'(t) = kQ(t)$$

$$\frac{\text{Percentage}}{\text{rate of change}} = 100k$$

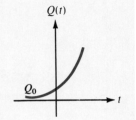

Exponential decay: $Q(t) = Q_0 e^{-kt}$

$$Q'(t) = -kQ(t)$$

$$\frac{\text{Percentage}}{\text{rate of change}} = -100k$$

Learning curve: $Q(t) = B - Ae^{-kt}$

Logistic curve: $Q(t) = \dfrac{B}{1 + Ae^{-Bkt}}$

The natural logarithm: $y = \ln x$ if and only if $x = e^y$

Inverse relationship: $e^{\ln x} = x$ and $\ln e^x = x$

Properties of logarithms:

$$\ln uv = \ln u + \ln v \qquad \ln \frac{u}{v} = \ln u - \ln v \qquad \ln u^v = v \ln u$$

Doubling time and half-life

Differentiation formulas:

$$\frac{d}{dx}(\ln x) = \frac{1}{x} \quad \text{and} \quad \frac{d}{dx}[\ln h(x)] = \frac{h'(x)}{h(x)}$$

$$\frac{d}{dx}(e^x) = e^x \quad \text{and} \quad \frac{d}{dx}[e^{h(x)}] = h'(x)e^{h(x)}$$

Logarithmic differentiation

Simple interest formula: $B(t) = P(1 + rt)$

Interest compounded k times per year:

$$\text{Balance} = B(t) = P\left(1 + \frac{r}{k}\right)^{kt}$$

$$\text{Effective rate} = \left(1 + \frac{r}{k}\right)^{k} - 1$$

$$\text{Present value} = B\left(1 + \frac{r}{k}\right)^{-kt}$$

Continuously compounded interest:

$$\text{Balance} = B(t) = Pe^{rt}$$

$$\text{Effective rate} = e^r - 1$$

$$\text{Present value} = Be^{-rt}$$

The amount of an annuity; the present value of an annuity

Optimal holding time

Review Problems   1. Sketch the graph of each of the following functions:

(a) $f(x) = 5e^{-x}$

(b) $f(x) = 5 - 2e^{-x}$

(c) $f(x) = 1 - \dfrac{6}{2 + e^{-3x}}$

(d) $f(x) = \dfrac{3 + 2e^{-2x}}{1 + e^{-2x}}$

2. (a) Find $f(4)$ if $f(x) = Ae^{-kx}$, and $f(0) = 10, f(1) = 25$.

(b) Find $f(3)$ if $f(x) = Ae^{kx}$ and $f(1) = 3, f(2) = 10$.

(c) Find $f(9)$ if $f(x) = 30 + Ae^{-kx}$ and $f(0) = 50, f(3) = 40$.

(d) Find $f(10)$ if $f(x) = \dfrac{6}{1 + Ae^{-kx}}$ and $f(0) = 3, f(5) = 2$.

3. The value of a certain industrial machine is decreasing exponentially. If the machine was originally worth $50,000 and was worth $20,000 when it was 5 years old, how much will it be worth when it is 10 years old?

4. It is estimated that if x thousand dollars are spent on advertising, approximately $Q(x) = 50 - 40e^{-0.1x}$ thousand units of a certain commodity will be sold.
   (a) Sketch the relevant portion of this sales function.
   (b) How many units will be sold if no money is spent on advertising?
   (c) How many units will be sold if $8,000 is spent on advertising?
   (d) How much should be spent on advertising to generate sales of 35,000 units?
   (e) According to this model, what is the most optimistic sales projection?

5. The daily output of a worker who has been on the job for t weeks is $Q(t) = 120 - Ae^{-kt}$ units. Initially the worker could produce 30 units per day, and after 8 weeks the worker could produce 80 units per day. How many units could the worker produce per day after 4 weeks?

6. It is estimated that t years from now the population of a certain country will be $P(t) = \dfrac{30}{1 + 2e^{-0.05t}}$ million.
   (a) Sketch the graph of $P(t)$.
   (b) What is the current population?
   (c) What will the population be 20 years from now?
   (d) What will happen to the population in the long run?

7. In 1626, Peter Minuit bought Manhattan Island from the Indians for $24. Suppose the same land was worth $375 billion in 1986. Assuming that the Indians invested their $24 in a bank that paid $100r$ percent per year compounded continuously and that r stayed fixed from 1626 to 1986, what would r have to be to generate the same appreciation?

8. Evaluate the following expressions without using tables or a calculator.
   (a) $\ln e^5$ (b) $e^{\ln 2}$ (c) $e^{3\ln 4 - \ln 2}$ (d) $\ln 9e^2 + \ln 3e^{-2}$

9. Solve for x.
   (a) $8 = 2e^{0.04x}$ (b) $5 = 1 + 4e^{-6x}$
   (c) $4 \ln x = 8$ (d) $5^x = e^3$

10. The number of bacteria in a certain culture grows exponentially. If 5,000 bacteria were initially present and 8,000 were present 10 minutes later, how long will it take for the number of bacteria to double?

11. Differentiate the following functions:
    (a) $f(x) = 2e^{3x+5}$ (b) $f(x) = x^2 e^{-x}$

(c) $f(x) = \ln \sqrt{x^2 + 4x + 1}$ (d) $f(x) = x \ln x^2$

(e) $f(x) = \dfrac{x}{\ln 2x}$

12. Use implicit differentiation to find $\dfrac{dy}{dx}$. Simplify your answers.

   (a) $e^{y/x} = 3xy^2$ (b) $\ln \dfrac{x}{y} = x^3 y^2$

13. An environmental study of a certain suburban community suggests that t years from now, the average level of carbon monoxide in the air will be $Q(t) = 4e^{0.03t}$ parts per million.
    (a) At what rate will the carbon monoxide level be changing with respect to time 2 years from now?
    (b) At what percentage rate will the carbon monoxide level be changing with respect to time t years from now? Does this percentage rate of change depend on t or is it constant?

14. A manufacturer can produce cameras at a cost of $40 apiece and estimates that if they are sold for p dollars apiece, consumers will buy approximately $D(p) = 800e^{-0.01p}$ cameras per week. At what price should the manufacturer sell the cameras to maximize profit?

15. Determine where the given function is increasing, decreasing, concave upward, and concave downward. Find the relative extrema and inflection points and draw the graph.

   (a) $f(x) = xe^{-2x}$ (b) $f(x) = e^x - e^{-x}$

   (c) $f(x) = \dfrac{4}{1 + e^{-x}}$ (d) $f(x) = \ln (x^2 + 1)$

   (e) $f(x) = \dfrac{\ln \sqrt[3]{x}}{x}$

16. Find $f'(x)$ by logarithmic differentiation.
   (a) $f(x) = \sqrt{(x^2 + 1)(x^2 + x + 2)}$

   (b) $f(x) = \dfrac{e^{-3x}\sqrt{1 - 2x}}{(x^2 + 2x - 3)^2}$

   (c) $f(x) = x^{2x+1}$ (for $x > 0$)

   (d) $f(x) = (\ln \sqrt{x})^x$ (for $x > 0$)

17. How quickly will $2,000 grow to $5,000 when invested at an annual interest rate of 8 percent if interest is compounded
    (a) Quarterly   (b) Continuously

18. Which investment has the greater effective interest rate: 8.25 percent per year compounded quarterly or 8.20 percent per year compounded continuously?

19. How much should you invest now at an annual interest rate of 6.25 percent so that your balance 10 years from now will be $2,000 if interest is compounded
    (a) Monthly   (b) Continuously

20. Find the present value of an annuity that makes 4 semiannual payments of $2,000 (starting today) if the prevailing annual interest rate is 10 percent compounded continuously.

21. Suppose you own a coin collection whose value t years from now will be $V(t) = 2,000e^{\sqrt{t}}$ dollars. If the prevailing interest rate remains constant at 7 percent per year compounded continuously, when will it be most advantageous for you to sell the collection and invest the proceeds?

CHAPTER 5
ANTIDIFFERENTIATION

322

1   Antidifferentiation: The Indefinite Integral

In many problems, the derivative of a function is known and the goal is to find the function itself. For example, a sociologist who knows the rate at which the population is growing may wish to use this information to predict future population levels; a physicist who knows the speed of a moving body may wish to calculate the future position of the body; an economist who knows the rate of inflation may wish to estimate future prices.

The process of obtaining a function from its derivative is called **antidifferentiation** or **integration.**

> **Antiderivative**
>
> A function $F(x)$ for which
>
> $$F'(x) = f(x)$$
>
> for every x in the domain of f is said to be an antiderivative (or indefinite integral) of f.

Later in this section you will learn techniques you can use to find antiderivatives. Once you have found what you believe to be an antiderivative of a function, you can always check your answer by differentiating. You should get the original function back. Here is an example.

EXAMPLE 1.1

Verify that $F(x) = \frac{1}{3}x^3 + 5x + 2$ is an antiderivative of $f(x) = x^2 + 5$.

Solution

$F(x)$ is an antiderivative of $f(x)$ if and only if $F'(x) = f(x)$. Differentiate F and you will find that

$$F'(x) = x^2 + 5 = f(x)$$

as required.

The General Antiderivative of a Function   A function has more than one antiderivative. For example, one antiderivative of the function $f(x) = 3x^2$ is $F(x) = x^3$, since

$$F'(x) = 3x^2 = f(x)$$

But so is $G(x) = x^3 + 12$, since the derivative of the constant 12 is zero and

$$G'(x) = 3x^2 = f(x)$$

In general, if F is one antiderivative of f, any function obtained by adding a constant to F is also an antiderivative of f. In fact, it turns out that all the antiderivatives of f can be obtained by adding constants to any one particular antiderivative of f.

The Antiderivatives of a Function

If F and G are antiderivatives of f, then there is a constant C such that

$$G(x) = F(x) + C$$

There is a simple geometric explanation for the fact that any two antiderivatives of the same function must differ by a constant. If F is an antiderivative of f, then $F'(x) = f(x)$. This says that for each value of x, $f(x)$ is the slope of the tangent to the graph of $F(x)$. If G is another antiderivative of f, the slope of its tangent is also $f(x)$. Hence, the graph of G is "parallel" to the graph of F and can be obtained by translating the graph of F vertically. That is, there is some constant C for which $G(x) = F(x) + C$. The situation is illustrated in Figure 5.1, which shows several antiderivatives of the function $f(x) = 3x^2$.

Figure 5.1 Some antiderivatives of $f(x) = 3x^2$.

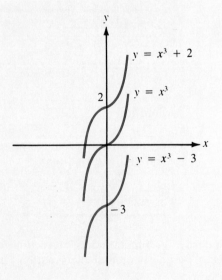

Integral Notation It is customary to write

$$\int f(x)\ dx = F(x) + C$$

to express the fact that every antiderivative of $f(x)$ is of the form $F(x) + C$. For example, you can express the fact that every antiderivative of $3x^2$ is of the form $x^3 + C$ by writing

$$\int 3x^2\ dx = x^3 + C$$

The symbol $\int$ is called an **integral sign** and indicates that you are to find the most general form of the antiderivative of the function following it. The integral sign resembles an elongated "s." The "s" stands for "sum." In Chapter 6 you will see a surprising connection between antiderivatives and sums which is so important that it is known as the **fundamental theorem of calculus.**

In the expression $\int f(x)\ dx = F(x) + C$, the function $f(x)$ that is to be integrated is called the **integrand.** The (unspecified) constant C that is added to $F(x)$ to give the most general form of the antiderivative is known as the **constant of integration.**

The symbol dx that appears after the integrand may seem mysterious. Its role is to indicate that x is the variable with respect to which the integration is to be performed. Analogous notation is used if the function is expressed in terms of a variable other than x. For example, $\int 3t^2\ dt = t^3 + C$. In the expression $\int 3px^2\ dx$, the dx tells you that x rather than p is the variable. Thus, $\int 3px^2\ dx = px^3 + C$. (Do you see how to evaluate $\int 3px^2\ dp$?)

Here is a restatement of the definition of the indefinite integral (or antiderivative) in integral notation.

The Indefinite Integral

$$\int f(x)\ dx = F(x) + C$$

if and only if $\qquad\qquad F'(x) = f(x)$

for every x in the domain of f.

Integration Rules Integration is the reverse of differentiation. Hence, many rules for integration can be obtained by stating the corresponding rules for differentiation in reverse.

The Power Rule for Integrals According to the power rule for derivatives, $\frac{d}{dx}(x^n) = nx^{n-1}$. That is, to differentiate a power function, reduce its power by 1 and multiply by the original power. Stated in reverse, this rule says that to integrate a power function, *raise* its power by 1 and *divide* by the *new* power. Here is a more precise statement of the rule.

The Power Rule for Integrals

For $n \neq -1$, $\qquad \displaystyle\int x^n \, dx = \frac{1}{n+1}x^{n+1} + C$

That is, to integrate x^n (for $n \neq -1$), increase the power of x by 1 and divide by the new power.

To verify this rule, simply observe that

$$\frac{d}{dx}\left(\frac{1}{n+1}x^{n+1}\right) = \frac{n+1}{n+1}x^n = x^n$$

The rule holds for all values of n except, of course, $n = -1$, in which case $\frac{1}{n+1}$ is undefined. This special case will be discussed shortly. In the meantime, see if you can think of an antiderivative of x^{-1} on your own.

The rule for integrating power functions is illustrated in the next example.

EXAMPLE 1.2

Find the following integrals:

(a) $\displaystyle\int x^{3/5} \, dx$ (b) $\displaystyle\int 1 \, dx$ (c) $\displaystyle\int \frac{1}{\sqrt{x}} \, dx$

Solution

(a) Increase the power of x by 1 and then divide by the new power to get

$$\int x^{3/5} \, dx = \frac{5}{8}x^{8/5} + C$$

(b) Since $1 = x^0$, it follows that

$$\int 1 \, dx = x + C$$

(c) Begin by rewriting $\dfrac{1}{\sqrt{x}}$ as $x^{-1/2}$. Then,

$$\int \frac{1}{\sqrt{x}} \, dx = \int x^{-1/2} \, dx = 2x^{1/2} + C = 2\sqrt{x} + C$$

The Integral of $\dfrac{1}{x}$ Were you able to discover an antiderivative for x^{-1}, that is, a function whose derivative is $\dfrac{1}{x}$? The natural logarithm $\ln x$ is such a function, and so it appears that $\int \dfrac{1}{x} \, dx = \ln x + C$. Actually, this is true only when x is positive, since $\ln x$ is not defined for nonpositive values of x. When x is negative, it turns out that the function $\ln |x|$ is an antiderivative of $\dfrac{1}{x}$, for if x is negative, $|x| = -x$ and

$$\frac{d}{dx}(\ln |x|) = \frac{d}{dx}[\ln (-x)] = \left(\frac{1}{-x}\right)(-1) = \frac{1}{x}$$

Since $|x| = x$ when x is positive, the integral of $\dfrac{1}{x}$ is given by the following single formula.

The Integral of $\dfrac{1}{x}$

$$\int \frac{1}{x} \, dx = \ln |x| + C$$

The Integral of e^x Integration of the exponential function e^x is trivial since e^x is its own derivative.

The Integral of e^x

$$\int e^x \, dx = e^x + C$$

The Constant Multiple Rule and Sum Rule It is easy to rewrite the constant multiple rule and the sum rule for differentiation as rules for integration.

The Constant Multiple Rule for Integrals

For any constant c,

$$\int cf(x)\, dx = c \int f(x)\, dx$$

That is, the integral of a constant times a function is equal to the constant times the integral of the function.

The Sum Rule for Integrals

$$\int [f(x) + g(x)]\, dx = \int f(x)\, dx + \int g(x)\, dx$$

That is, the integral of a sum is the sum of the individual integrals.

Here is an example illustrating the use of these rules.

EXAMPLE 1.3

Find $\int \left(3e^x + \dfrac{2}{x} - \dfrac{1}{2}x^2 \right) dx$.

Solution

$$\int \left(3e^x + \frac{2}{x} - \frac{1}{2}x^2 \right) dx = 3 \int e^x\, dx + 2 \int \frac{1}{x}\, dx - \frac{1}{2} \int x^2\, dx$$

$$= 3e^x + 2 \ln |x| - \frac{1}{6} x^3 + C$$

Notice that in place of separate constants of integration for each of the three antiderivatives computed in Example 1.3, a single constant C was added at the end.

Integration of Products and Quotients   You may have noticed that no general rules have been given for the integration of products and quotients. This is because there are no general rules. Occasionally, you will be able to rewrite a product or a quotient in a form in which you can integrate it using the techniques you have learned in this section. Here is an example.

EXAMPLE 1.4

Find $\displaystyle\int \frac{3x^5 + 2x - 5}{x^3}\, dx$.

Solution

Perform the indicated division to get

$$\frac{3x^5 + 2x - 5}{x^3} = 3x^2 + \frac{2}{x^2} - \frac{5}{x^3} = 3x^2 + 2x^{-2} - 5x^{-3}$$

and then integrate term by term:

$$\int \frac{3x^5 + 2x - 5}{x^3}\, dx = \int (3x^2 + 2x^{-2} - 5x^{-3})\, dx$$

$$= x^3 - 2x^{-1} + \frac{5}{2}x^{-2} + C = x^3 - \frac{2}{x} + \frac{5}{2x^2} + C$$

Practical Applications   Here are two problems in which the rate of change of a quantity is known and the goal is to find an expression for the quantity itself. Since the rate of change is the derivative of the quantity, you find the expression for the quantity itself by antidifferentiation.

EXAMPLE 1.5

It is estimated that x months from now the population of a certain town will be changing at the rate of $2 + 6\sqrt{x}$ people per month. The current population is 5,000. What will the population be 9 months from now?

Solution

Let $P(x)$ denote the population of the town x months from now. Then the rate of change of the population with respect to time is the derivative

$$\frac{dP}{dx} = 2 + 6\sqrt{x}$$

It follows that the population function $P(x)$ is an antiderivative of $2 + 6\sqrt{x}$. That is,

$$P(x) = \int \frac{dP}{dx}\, dx = \int (2 + 6\sqrt{x})\, dx = 2x + 4x^{3/2} + C$$

for some constant C. To determine C, use the fact that at present (when $x = 0$) the population is 5,000. That is,

$$5,000 = 2(0) + 4(0)^{3/2} + C \qquad \text{or} \qquad C = 5,000$$

Hence, $\qquad\qquad P(x) = 2x + 4x^{3/2} + 5,000$

and the population 9 months from now will be

$$P(9) = 2(9) + 4(27) + 5,000 = 5,126$$

EXAMPLE 1.6

A manufacturer has found that marginal cost is $3q^2 - 60q + 400$ dollars per unit when q units have been produced. The total cost of producing the first 2 units is \$900. What is the total cost of producing the first 5 units?

Solution

Recall that the marginal cost is the derivative of the total cost function $C(q)$. Thus,

$$C'(q) = 3q^2 - 60q + 400$$

and so $C(q)$ must be the antiderivative

$$C(q) = \int C'(q)\, dq = \int (3q^2 - 60q + 400)\, dq = q^3 - 30q^2 + 400q + K$$

for some constant K. (The letter K was used for the constant to avoid confusion with the cost function C.)

   The value of K is determined by the fact that $C(2) = 900$. In particular,

$$900 = (2)^3 - 30(2)^2 + 400(2) + K \qquad \text{or} \qquad K = 212$$

Hence, $\qquad\qquad C(q) = q^3 - 30q^2 + 400q + 212$

and the cost of producing the first 5 units is

$$C(5) = (5)^3 - 30(5)^2 + 400(5) + 212 = \$1,587$$

Geometric Application   The next example shows how integration can be used to find the equation of a function $f(x)$ when the slope of the tangent line is known at each point on the graph of $f(x)$.

EXAMPLE 1.7

Find the function $f(x)$ whose tangent has slope $3x^2 + 1$ for each value of x and whose graph passes through the point $(2, 6)$.

Solution

The slope of the tangent is the derivative of f. Thus

$$f'(x) = 3x^2 + 1$$

and so $f(x)$ is the antiderivative

$$f(x) = \int f'(x)\, dx = \int (3x^2 + 1)\, dx = x^3 + x + C$$

To find C, use the fact that the graph of f passes through $(2, 6)$. That is, substitute $x = 2$ and $f(2) = 6$ into the equation for $f(x)$ and solve for C to get

$$6 = (2)^3 + 2 + C \qquad \text{or} \qquad C = -4$$

That is, the desired function is $f(x) = x^3 + x - 4$.

Inventory Storage Costs   In the next example, you will see how to calculate the total cost resulting from the storage of unused inventory.

EXAMPLE 1.8

A retailer receives a shipment of 10,000 kilograms of rice that will be used up over a 5-month period at the constant rate of 2,000 kilograms per month. If storage costs are 1 cent per kilogram per month, how much will the retailer pay in storage costs over the next 5 months?

Solution

Let $S(t)$ denote the total storage cost (in dollars) over t months. Since the rice is used up at a constant rate of 2,000 kilograms per month, the number of kilograms of rice in storage after t months is 10,000–2,000t. Therefore, since storage costs are 1 cent per kilogram per month, the rate of change of the storage cost with respect to time is

$$\frac{dS}{dt} = \left(\begin{array}{c}\text{cost per}\\ \text{kilogram}\end{array}\right)\left(\begin{array}{c}\text{number of}\\ \text{kilograms}\end{array}\right) = 0.01(10{,}000 - 2{,}000t)$$

It follows that $S(t)$ is an antiderivative of

$$0.01(10{,}000 - 2{,}000t) = 100 - 20t$$

That is, $$S(t) = \int \frac{dS}{dt}\, dt = \int (100 - 20t)\, dt$$

$$= 100t - 10t^2 + C$$

for some constant C. To determine C, use the fact that at the time the shipment arrives (when $t = 0$) there is no cost, so that

$$0 = 100(0) - 10(0)^2 + C \quad \text{or} \quad C = 0$$

Hence, $$S(t) = 100t - 10t^2$$

and the total storage cost over the next 5 months will be

$$S(5) = 100(5) - 10(5)^2 = \$250$$

Motion along a Line   Recall from Section 3 of Chapter 2 that if an object moves along a straight line with displacement $s(t)$, then its velocity is given by $v = \dfrac{ds}{dt}$ and its acceleration by $a = \dfrac{dv}{dt}$. If the acceleration of the object is given, then its velocity and displacement can be found by integration. Here is an example.

EXAMPLE 1.9

After its brakes are applied, a certain car decelerates at the constant rate of 22 feet per second per second. If the car is traveling at 45 miles per hour (66 feet per second) when the brakes are applied, how far does it travel before coming to a complete stop?

Solution

Let $s(t)$ denote the displacement (distance) of the car t seconds after the brakes are applied. Since the car decelerates at 22 feet per second, it follows that $a(t) = -22$ (the negative sign indicates that the car is slowing down), and

$$\frac{dv}{dt} = a(t) = -22$$

Integrating, you find that the velocity at time t is given by

$$v(t) = \int -22 \, dt = -22t + C_1$$

To evaluate C_1, note that $v = 66$ when $t = 0$ so that

$$66 = v(0) = -22(0) + C_1$$

and $C_1 = 66$. Thus, the velocity at time t is $v(t) = -22t + 66$.
    Next, to find the displacement $s(t)$, begin with the fact that

$$\frac{ds}{dt} = v(t) = -22t + 66$$

and use integration to show that

$$s(t) = \int (-22t + 66)dt = -11t^2 + 66t + C_2$$

Since $s(0) = 0$ (do you see why?), it follows that $C_2 = 0$ and

$$s(t) = -11t^2 + 66t$$

Finally, to find the stopping distance, note that the car stops when $v(t) = 0$, and this occurs when

$$v(t) = -22t + 66 = 0$$

Solving this equation, you find that the car stops after 3 seconds of deceleration, and in that time it has traveled

$$s(3) = -11(3) + 66(3) = 165 \text{ feet}$$

Problems 5.1

In Problems 1 through 20, find the indicated integral. Check your answers by differentiation.

1. $\int x^5 \, dx$

2. $\int x^{3/4} \, dx$

3. $\int \frac{1}{x^2} \, dx$

4. $\int \sqrt{x} \, dx$

5. $\int 5 \, dx$

6. $\int 3e^x \, dx$

7. $\int (3x^2 - \sqrt{5x} + 2) \, dx$

8. $\int (x^{1/2} - 3x^{2/3} + 6) \, dx$

9. $\int \left(3\sqrt{x} - \frac{2}{x^3} + \frac{1}{x}\right) dx$

10. $\int \left(\frac{1}{2x} - \frac{2}{x^2} + \frac{3}{\sqrt{x}}\right) dx$

11. $\int \left(\frac{e^x}{2} + x\sqrt{x}\right) dx$

12. $\int \left(\sqrt{x^3} - \frac{1}{2\sqrt{x}} + \sqrt{2}\right) dx$

13. $\int \left(\frac{1}{3x} - \frac{3}{2x^2} + e^2 + \frac{\sqrt{x}}{2}\right) dx$

14. $\int \left(2e^x + \frac{6}{x} + \ln 2\right) dx$

15. $\int \frac{x^2 + 2x + 1}{x^2} \, dx$

16. $\int \frac{x^2 + 3x - 2}{\sqrt{x}} \, dx$

17. $\int (x^3 - 2x^2)\left(\frac{1}{x} - 5\right) dx$

18. $\int x^3 \left(2x + \frac{1}{x}\right) dx$

19. $\int \sqrt{x}(x^2 - 1) \, dx$

20. $\int x(2x + 1)^2 \, dx$

Population Growth   21. It is estimated that t months from now the population of a certain town will be changing at the rate of $4 + 5t^{2/3}$ people per month. If the current population is 10,000, what will the population be 8 months from now?

Distance and Speed   22. An object is moving so that its speed after t minutes is $v(t) = 1 + 4t + 3t^2$ meters per minute. How far does the object travel during the 3rd minute?

Distance and Speed   23. An object is moving so that its speed after t minutes is $v(t) = 3 + 2t + 6t^2$ meters per minute. How far does the object travel during the 2nd minute?

Land Values   24. It is estimated that t years from now the value of a certain parcel of land will be increasing at the rate of $r(t)$ dollars per year. Find an expression for the amount by which the value of the land will increase during the next 5 years.

Admission to Events   25. The promoters of a county fair estimate that t hours after the gates open at 9:00 A.M., visitors will be entering the fair at the rate of $r(t)$ people per hour. Find an expression for the number of people who will enter the fair between 11:00 A.M. and 1:00 P.M.

Sales Revenue   26. A bicycle manufacturer expects that x months from now consumers will be buying 5,000 bicycles per month at the price of $P(x) = 80 + 3\sqrt{x}$ dollars per bicycle. What is the total revenue the manufacturer can expect from the sale of the bicycles over the next 16 months?

Sales Revenue   27. A bicycle manufacturer expects that x months from now consumers will be buying $f(x) = 5,000 + 60\sqrt{x}$ bicycles per month at the price of $P(x) = 80 + 3\sqrt{x}$ dollars per bicycle. What is the total revenue the manufacturer can expect from the sale of the bicycles over the next 16 months?

Storage Cost   28. A retailer receives a shipment of 12,000 pounds of soybeans that will be used at a constant rate of 300 pounds per week. If the cost of storing the soybeans is 0.2 cent per pound per week, how much will the retailer have to pay in storage costs over the next 40 weeks?

Water Pollution   29. It is estimated that t years from now the population of a certain lakeside community will be changing at the rate of $0.6t^2 + 0.2t + 0.5$ thousand people per year. Environmentalists have found that the level of pollution in the lake increases at the rate of approximately 5 units per 1,000 people. By how much will the pollution in the lake increase during the next 2 years?

Air Pollution   30. An environmental study of a certain community suggests that t years from now the level of carbon monoxide in the air will be changing at the rate of $0.1t + 0.1$ parts per million per year. If the current level of carbon monoxide in the air is 3.4 parts per million, what will the level be 3 years from now?

Marginal Cost   31. A manufacturer has found that marginal cost is $6q + 1$ dollars per unit when q units have been produced. The total cost (including overhead) of producing the 1st unit is $130. What is the total cost of producing the first 10 units?

Marginal Profit   32. A manufacturer estimates marginal revenue to be $100q^{-1/2}$ dollars per unit when the level of production is q units. The corresponding marginal cost has been found to be $0.4q$ dollars per unit. Suppose the manufacturer's profit is $520 when the level of production is 16 units. What is the manufacturer's profit when the level of production is 25 units?

Marginal Profit   33. The marginal profit (the derivative of profit) of a certain company is $100 - 2q$ dollars per unit when q units are produced. If the company's profit is $700 when 10 units are produced, what is the company's maximum possible profit?

Geometry   34. Find the function whose tangent has slope $4x + 1$ for each value of x and whose graph passes through the point $(1, 2)$.

Geometry   35. Find the function whose tangent has slope $3x^2 + 6x - 2$ for each value of x and whose graph passes through the point $(0, 6)$.

Geometry   36. Find the function whose tangent has slope $x^3 - \dfrac{2}{x^2} + 2$ for each value of x and whose graph passes through the point $(1, 3)$.

Geometry   37. Find a function whose graph has a relative minimum when $x = 1$ and a relative maximum when $x = 4$.

In Problems 38 through 43, find the indicated integral. In each case, start with an educated guess, check your answer by differentiation, and then modify your guess if necessary.

38. $\displaystyle\int e^{3x}\, dx$ (*Hint:* Try $e^{3x} + C$.)   39. $\displaystyle\int 3e^{5x}\, dx$

40. $\displaystyle\int \dfrac{1}{2x + 1}\, dx$ (*Hint:* Try $\ln|2x + 1| + C$.)   41. $\displaystyle\int \dfrac{2}{3x + 2}\, dx$

42. $\displaystyle\int 5(3x + 1)^4\, dx$ 43. $\displaystyle\int (2x + 3)^5\, dx$

Storage Cost   44. A manufacturer receives N units of a certain raw material that are initially placed in storage and then withdrawn and used at a constant rate until the supply is exhausted 1 year later. Suppose storage costs remain fixed at p dollars per unit per year.

(a) Find an expression for the total storage cost the manufacturer will pay during the year.

(b) Show that the total storage cost in part (a) is the same as the cost of storing $\dfrac{N}{2}$ units for the entire year.

Sales Revenue   45. A manufacturer expects that x months from now consumers will be buying $n(x)$ lamps per month at the price of $p(x)$ dollars per lamp. Find an expression for the total revenue the manufacturer can expect from the sale of the lamps over the next 12 months.

Oil Production   46. Suppose that t months from now an oil well will be producing crude oil at the rate of $r(t)$ barrels per month and that the price of crude oil will be $p(t)$ dollars per barrel. Assuming that the oil is sold as soon as it is extracted from the ground, find an expression for the total revenue from the oil well over the next 2 years.

Motion along a Line   47. After its brakes are applied, a certain car decelerates at the constant rate of 25 feet per second per second. If the car is traveling at 60 miles per hour when the brakes are applied, how far does it travel before coming to a complete stop? (Note: 60 mph is the same as 88 feet per second.)

Motion along a Line   48. Suppose a certain car supplies a constant deceleration of A feet per second per second. If it is traveling at 60 miles per hour (88 feet per second) when the brakes are applied, its stopping distance is 180 feet.

(a) What is A?

(b) What would the stopping distance have been if the car had been traveling at only 30 miles per hour when the brakes were applied?

(c) At what speed is the car traveling when the brakes are applied if the stopping distance is 200 feet?

Spy Story   49. The spy (who defused the bomb in Problem 26 of Chapter 3, Section 4) is driving a sports car at 60 miles per hour (88 feet per second) in the remote republic of San Dimas. Suddenly he sees a camel in the road 199 feet in front of him. If he has a reaction time of 00.7 second before applying the brakes and the car decelerates at a constant rate of 28 feet per second per second, will he stop before hitting the camel?

2 Integration by Substitution

This section deals with the integral version of the chain rule. To refresh your memory, here is a typical application of the chain rule for differentiation.

According to the chain rule, the derivative of the function $(x^2 + 3x + 5)^9$ is

$$\frac{d}{dx}[(x^2 + 3x + 5)^9] = 9(x^2 + 3x + 5)^8(2x + 3)$$

Notice that the derivative is a product and that one of its factors, $2x + 3$, is the derivative of an expression, $x^2 + 3x + 5$, which occurs in the other factor. More precisely, the derivative is a product of the form

$$g(u)\frac{du}{dx}$$

where, in this case, $g(u) = 9u^8$ and $u = x^2 + 3x + 5$.

You can integrate many products of the form $g(u)\frac{du}{dx}$ by applying the chain rule in reverse. Specifically, if G is an antiderivative of g, then

$$\int g(u)\frac{du}{dx}\, dx = G(u) + C$$

since, by the chain rule,

$$\frac{d}{dx}[G(u)] = G'(u)\frac{du}{dx} = g(u)\frac{du}{dx}$$

To summarize:

Integral Form of the Chain Rule

$$\int g(u)\frac{du}{dx}\, dx = \int g(u)\, du$$

That is, to integrate a product of the form $g(u)\frac{du}{dx}$ in which one of the factors $\frac{du}{dx}$ is the derivative of an expression u that appears in the other factor:

1. Find the antiderivative $\int g(u)\, du$ of the factor $g(u)$ with respect to u.

2. Replace u in the answer by its expression in terms of x.

Here are two examples.

EXAMPLE 2.1

Find $\int 9(x^2 + 3x + 5)^8(2x + 3)\, dx.$

Solution

The integrand $9(x^2 + 3x + 5)^8(2x + 3)$ is a product in which one of the factors $2x + 3$ is the derivative of an expression $x^2 + 3x + 5$ that appears in the other factor. In particular,

$$9(x^2 + 3x + 5)^8(2x + 3) = g(u)\frac{du}{dx}$$

where $\quad\quad g(u) = 9u^8 \quad$ and $\quad u = x^2 + 3x + 5$

Hence, by the integral form of the chain rule,

$$\int 9(x^2 + 3x + 5)^8(2x + 3)\,dx = \int g(u)\,du = \int 9u^8\,du = u^9 + C$$

$$= (x^2 + 3x + 5)^9 + C$$

The product to be integrated in the next example is not exactly of the form $g(u)\dfrac{du}{dx}$. However, it is a constant multiple of such a product, and you can integrate it by combining the constant multiple rule with the integral form of the chain rule.

EXAMPLE 2.2

Find $\int x^3 e^{x^4 + 2}\,dx$.

Solution

First use the constant multiple rule to rewrite the integral as

$$\int x^3 e^{x^4 + 2}\,dx = \int \frac{1}{4}(4x^3 e^{x^4 + 2})\,dx = \frac{1}{4}\int 4x^3 e^{x^4 + 2}\,dx$$

so that the new integrand $4x^3 e^{x^4 + 2}$ is a product in which one of the factors $4x^3$ is the derivative of an expression $x^4 + 2$ that appears in the other factor. In particular,

$$4x^3 e^{x^4 + 2} = g(u)\frac{du}{dx}$$

where $\quad\quad g(u) = e^u \quad$ and $\quad u = x^4 + 2$

Hence, by the integral form of the chain rule,

$$\int x^3 e^{x^4 + 2}\,dx = \frac{1}{4}\int 4x^3 e^{x^4 + 2}\,dx = \frac{1}{4}\int g(u)\,du$$

$$\int x^3 e^{x^4+2} \, dx = \frac{1}{4} \int e^u \, du = \frac{1}{4} e^u + C$$

$$= \frac{1}{4} e^{x^4+2} + C$$

Change of Variables   The integral form of the chain rule may be thought of as a technique for simplifying an integral by changing the variable of integration. In particular, you start with an integral $\int g(u) \dfrac{du}{dx} \, dx$ in which the variable of integration is x and transform it into the simpler integral $\int g(u) \, du$ in which the variable of integration is u. In this transformation, the expression $\dfrac{du}{dx} \, dx$ in the original integral is replaced in the simplified integral by the symbol du. You can remember this relationship between $\dfrac{du}{dx} \, dx$ and du by pretending that $\dfrac{du}{dx}$ is a quotient and writing

$$\frac{du}{dx} \, dx = du$$

These observations suggest the following general integration technique, called **integration by substitution,** in which the variable u is formally substituted for an appropriate expression in x and the original integral is transformed into a simpler one in which the variable of integration is u.

Integration by Substitution

Step 1. Introduce the letter u to stand for some expression in x that is chosen with the simplification of the integral as the goal.

Step 2. Rewrite the integral in terms of u. To rewrite dx, compute $\dfrac{du}{dx}$ and solve algebraically as if the symbol $\dfrac{du}{dx}$ were a quotient.

Step 3. Evaluate the resulting integral and then replace u by its expression in terms of x in the answer.

A Word of Advice   If the integrand is a product or quotient of two terms and one term is a constant multiple of the derivative of an expression that appears in the other, then this expression is probably a good choice for u.

The method of integration by substitution is illustrated in the next example using the integral from Example 2.1.

EXAMPLE 2.3

Find $\displaystyle\int 9(x^2 + 3x + 5)^8(2x + 3)\ dx$.

Solution

The integrand is a product in which one of the factors $2x + 3$ is the derivative of an expression $x^2 + 3x + 5$ that appears in the other factor. This suggests that you let $u = x^2 + 3x + 5$. Then

$$\frac{du}{dx} = 2x + 3 \quad \text{and so} \quad du = (2x + 3)\ dx$$

Substituting $u = x^2 + 3x + 5$ and $du = (2x + 3)\ dx$, you get

$$\int 9(x^2 + 3x + 5)^8(2x + 3)\ dx = \int 9u^8\ du = u^9 + C$$

$$= (x^2 + 3x + 5)^9 + C$$

Here are some additional examples illustrating the method of integration by substitution.

EXAMPLE 2.4

Find $\displaystyle\int \frac{3x}{x^2 - 1}\ dx$.

Solution

Observe that

$$\frac{d}{dx}(x^2 - 1) = 2x = \frac{2}{3}(3x)$$

Thus the integrand is a quotient in which one term $3x$ is a constant multiple of the derivative of an expression $x^2 - 1$ that appears in the other term. This suggests that you let $u = x^2 - 1$. Then,

$$\frac{du}{dx} = 2x \qquad du = 2x\ dx \qquad \text{or} \qquad \frac{3}{2} du = 3x\ dx$$

Substituting $u = x^2 - 1$ and $\frac{3}{2} du = 3x\ dx$, you get

$$\int \frac{3x}{x^2 - 1} \, dx = \int \frac{1}{u} \left(\frac{3}{2}\right) du = \frac{3}{2} \int \frac{1}{u} \, du$$

$$\int \frac{3x}{x^2 - 1} \, dx = \frac{3}{2} \ln |u| + C$$

$$= \frac{3}{2} \ln |x^2 - 1| + C$$

EXAMPLE 2.5

Find $\int \dfrac{3x + 6}{\sqrt{2x^2 + 8x + 3}} \, dx.$

Solution

Observe that

$$\frac{d}{dx}(2x^2 + 8x + 3) = 4x + 8 = 4(x + 2) = \frac{4}{3}(3x + 6)$$

Thus, the integrand is a quotient in which one term $3x + 6$ is a constant multiple of the derivative of an expression $2x^2 + 8x + 3$ that appears in the other term. This suggests that you let $u = 2x^2 + 8x + 3$. Then,

$$\frac{du}{dx} = 4x + 8$$

$$du = (4x + 8) \, dx = 4(x + 2) \, dx = \frac{4}{3}(3x + 6) \, dx$$

or

$$\frac{3}{4} \, du = (3x + 6) \, dx$$

Substituting $u = 2x^2 + 8x + 3$ and $\frac{3}{4} du = (3x + 6) \, dx$, you get

$$\int \frac{3x + 6}{\sqrt{2x^2 + 8x + 3}} \, dx = \int \frac{3}{4} \frac{1}{\sqrt{u}} \, du = \frac{3}{4} \int u^{-1/2} \, du$$

$$= \frac{3}{4}(2u^{1/2}) + C = \frac{3}{2}\sqrt{2x^2 + 8x + 3} + C$$

EXAMPLE 2.6

Find $\int \dfrac{(\ln x)^2}{x} \, dx.$

Solution

Observe that

$$\frac{d}{dx}(\ln x) = \frac{1}{x}$$

Thus, the integrand

$$\frac{(\ln x)^2}{x} = (\ln x)^2\left(\frac{1}{x}\right)$$

is a product in which one factor $\dfrac{1}{x}$ is the derivative of an expression $\ln x$ that appears in the other factor. This suggests that you let $u = \ln x$. Then,

$$\frac{du}{dx} = \frac{1}{x} \qquad \text{or} \qquad du = \frac{1}{x}\,dx$$

Substituting $u = \ln x$ and $du = \dfrac{1}{x}\,dx$, you get

$$\int \frac{(\ln x)^2}{x}\,dx = \int u^2\,du = \frac{1}{3}u^3 + C = \frac{1}{3}(\ln x)^3 + C$$

The next example is designed to show you the versatility of the formal method of substitution. It deals with an integral that does not seem to be of the form $\displaystyle\int g(u)\frac{du}{dx}\,dx$ but that nevertheless can be simplified significantly by a clever change of variables.

EXAMPLE 2.7

Find $\displaystyle\int \frac{x}{x+1}\,dx$.

Solution

There seems to be no easy way to integrate this quotient as it stands. But watch what happens if you make the substitution $u = x + 1$. Then $du = dx$ and $x = u - 1$ and so

$$\int \frac{x}{x+1}\,dx = \int \frac{u-1}{u}\,du = \int 1\,du - \int \frac{1}{u}\,du$$

$$= u - \ln|u| + C = x + 1 - \ln|x+1| + C$$

A Comparison of the Two Techniques   The integral form of the chain rule, which was used in Examples 2.1 and 2.2, is attractive because it involves nothing more than a familiar rule for differentiation applied in reverse. With practice in using this method you should be able to find integrals like those in Examples 2.1 through 2.6 by inspection, without writing down any intermediate steps.

Nevertheless, many people prefer to use the method of formal substitution. They like the fact that it involves straightforward manipulation of symbols, and they appreciate the convenience of the notation. This method of substitution is also somewhat more versatile, as you saw in Example 2.7.

For many of the integrals you will encounter, both methods work well and you should feel free to use the one with which you are most comfortable.

Problems 5.2

In Problems 1 through 26, find the indicated integral and check your answer by differentiation.

1. $\int (2x + 6)^5 \, dx$

2. $\int e^{5x} \, dx$

3. $\int \sqrt{4x - 1} \, dx$

4. $\int \frac{1}{3x + 5} \, dx$

5. $\int e^{1-x} \, dx$

6. $\int [(x - 1)^5 + 3(x - 1)^2 + 5] \, dx$

7. $\int xe^{x^2} \, dx$

8. $\int 2xe^{x^2 - 1} \, dx$

9. $\int x(x^2 + 1)^5 \, dx$

10. $\int 3x\sqrt{x^2 + 8} \, dx$

11. $\int x^2(x^3 + 1)^{3/4} \, dx$

12. $\int x^5 e^{1 - x^6} \, dx$

13. $\int \frac{2x^4}{x^5 + 1} \, dx$

14. $\int \frac{x^2}{(x^3 + 5)^2} \, dx$

15. $\int (x + 1)(x^2 + 2x + 5)^{12} \, dx$

16. $\int (3x^2 - 1)e^{x^3 - x} \, dx$

17. $\int \frac{3x^4 + 12x^3 + 6}{x^5 + 5x^4 + 10x + 12} \, dx$

18. $\int \frac{10x^3 - 5x}{\sqrt{x^4 - x^2 + 6}} \, dx$

19. $\int \frac{3x - 3}{(x^2 - 2x + 6)^2} \, dx$

20. $\int \frac{6x - 3}{4x^2 - 4x + 1} \, dx$

21. $\int \dfrac{\ln 5x}{x} dx$

22. $\int \dfrac{1}{x \ln x} dx$

23. $\int \dfrac{1}{x(\ln x)^2} dx$

24. $\int \dfrac{\ln x^2}{x} dx$

25. $\int \dfrac{2x \ln (x^2 + 1)}{x^2 + 1} dx$

26. $\int \dfrac{e^{\sqrt{x}}}{\sqrt{x}} dx$

In Problems 27 through 33, use an appropriate change of variables to find the indicated integral.

27. $\int \dfrac{x}{x - 1} dx$

28. $\int x\sqrt{x + 1}\, dx$

29. $\int \dfrac{x}{(x - 5)^6} dx$

30. $\int (x + 1)(x - 2)^9\, dx$

31. $\int \dfrac{x + 3}{(x - 4)^2} dx$

32. $\int \dfrac{x}{2x + 1} dx$

33. $\int (2x + 3)\sqrt{2x - 1}\, dx$

34. Find the function whose tangent has slope $x\sqrt{x^2 + 5}$ for each value of x and whose graph passes through the point (2, 10).

35. Find the function whose tangent has slope $\dfrac{2x}{1 - 3x^2}$ for each value of x and whose graph passes through the point (0, 5).

Tree Growth   36. A tree has been transplanted and after x years is growing at the rate of $1 + \dfrac{1}{(x + 1)^2}$ meters per year. After 2 years it has reached a height of 5 meters. How tall was it when it was transplanted?

Depreciation   37. The resale value of a certain industrial machine decreases at a rate that changes with time. When the machine is t years old, the rate at which its value is changing is $-960e^{-t/5}$ dollars per year. If the machine was bought new for \$5,000, how much will it be worth 10 years later?

Population Growth   38. It is projected that t years from now the population of a certain country will be changing at the rate of $e^{0.02t}$ million per year. If the current population is 50 million, what will the population be 10 years from now?

Land Value   39. It is estimated that x years from now the value of an acre of farmland will be increasing at the rate of $\dfrac{0.4x^3}{\sqrt{0.2x^4 + 8,000}}$ dollars per year. If the land is currently worth \$500 per acre, how much will it be worth in 10 years?

3 Integration by Parts

In this section you will see a technique you can use to integrate certain products $f(x)g(x)$. The technique is called **integration by parts,** and as you will see, it is a restatement of the product rule for differentiation. Here is the formal statement of the technique.

> **Integration by Parts**
>
> $$\int f(x)g(x)\,dx = f(x)G(x) - \int f'(x)G(x)\,dx$$
>
> where G is an antiderivative of g.

Why Integration by Parts Works

To see how integration by parts is a restatement of what happens when the product rule is used to differentiate $f(x)G(x)$, where G is an antiderivative of g, note that,

$$\frac{d}{dx}[f(x)G(x)] = f'(x)G(x) + f(x)G'(x) = f'(x)G(x) + f(x)g(x)$$

Expressed in terms of integrals, this says

$$f(x)G(x) = \int f'(x)G(x)\,dx + \int f(x)g(x)\,dx$$

or

$$\int f(x)g(x)\,dx = f(x)G(x) - \int f'(x)G(x)\,dx$$

which is precisely the formula for integration by parts.

How and When to Use Integration by Parts

Integration by parts is a technique for integrating products $f(x)g(x)$, in which one of the factors, say $g(x)$, can be easily integrated and the other, $f(x)$, becomes simpler when differentiated. To evaluate such an integral, $\int f(x)g(x)\,dx$, using integration by parts, first integrate g and multiply the result by f to get

$$f(x)G(x)$$

where G is an antiderivative of g. Then multiply the antiderivative G by the derivative of f and subtract the integral of this product from the result of the first step to get

$$f(x)G(x) - \int f'(x)G(x)\,dx$$

This expression will be equal to the original integral $\int f(x)g(x)\,dx$, and if you are lucky, the new integral $\int f'(x)G(x)\,dx$ will be easier to find than the original one.

Here is an informal, step-by-step summary of the procedure.

How to Use Integration by Parts to Integrate a Product

Step 1. Select one of the factors of the product as the one to be integrated and the other as the one to be differentiated. The factor selected for integration should be easy to integrate, and the factor selected for differentiation should become simpler when differentiated.

Step 2. Integrate the designated factor and multiply it by the other factor.

Step 3. Differentiate the designated factor, multiply it by the integrated factor from Step 2, and subtract the integral of this product from the result of Step 2.

Step 4. Complete the procedure by finding the new integral that was formed in Step 3.

Here are some examples illustrating the procedure. In each example, $g(x)$ is used to denote the factor that is to be integrated and $f(x)$ is used to denote the factor that is to be differentiated. As reminders, the letters I (for integrate) and D (for differentiate) are placed above the appropriate factors in the integrand.

With practice, you will become familiar with the pattern and should find that you can do integration by parts without the intermediate step of writing down the functions $g(x)$, $f(x)$, $G(x)$, and $f'(x)$.

EXAMPLE 3.1

Find $\int xe^{2x}\,dx$.

Solution

In this case, both factors x and e^{2x} are easy to integrate. Both are also easy to differentiate, but the process of differentiation simplifies x while it makes e^{2x} slightly more complicated. This suggests that you should try integration by parts with

$$g(x) = e^{2x} \qquad \text{and} \qquad f(x) = x$$

Then, $$G(x) = \frac{1}{2}e^{2x} \quad \text{and} \quad f'(x) = 1$$

and so $$\int \overset{\text{D\,I}}{xe^{2x}}\,dx = \left(\frac{1}{2}e^{2x}\right)(x) - \int\left(\frac{1}{2}e^{2x}\right)(1)\,dx$$

$$= \frac{1}{2}xe^{2x} - \frac{1}{2}\int e^{2x}\,dx$$

$$= \frac{1}{2}xe^{2x} - \frac{1}{4}e^{2x} + C = \frac{1}{2}\left(x - \frac{1}{2}\right)e^{2x} + C$$

EXAMPLE 3.2

Find $\int x\sqrt{x + 5}\,dx$.

Solution

Again, both factors in the product are easy to integrate and differentiate. However, the factor x is simplified by differentiation, whereas the derivative of $\sqrt{x + 5}$ is even more complicated than $\sqrt{x + 5}$ itself. This suggests that you should try integration by parts with

$$g(x) = \sqrt{x + 5} \quad \text{and} \quad f(x) = x$$

Then, $$G(x) = \frac{2}{3}(x + 5)^{3/2} \quad \text{and} \quad f'(x) = 1$$

and so $$\int \overset{\text{D\quad I}}{x\sqrt{x + 5}}\,dx = \frac{2}{3}x(x + 5)^{3/2} - \frac{2}{3}\int (x + 5)^{3/2}\,dx$$

$$= \frac{2}{3}x(x + 5)^{3/2} - \frac{4}{15}(x + 5)^{5/2} + C$$

Comment: Some integrals can be evaluated by either substitution or integration by parts. For instance, the integral in Example 3.2 can be found by substituting as follows:

Let $u = x + 5$. Then $du = dx$ and $x = u - 5$, and

$$\int x\sqrt{x + 5}\,dx = \int (u - 5)\sqrt{u}\,du = \int (u^{3/2} - 5u^{1/2})du$$

$$= \frac{u^{5/2}}{5/2} - \frac{5u^{3/2}}{3/2} + C = \frac{2}{5}(x + 5)^{5/2} - \frac{10}{3}(x + 5)^{3/2} + C$$

This form of the integral is not the same as that found in Example 3.2. To show that the two forms are equivalent, note that the antiderivative in Example 3.2 can be expressed as

$$\frac{2x}{3}(x + 5)^{3/2} - \frac{4}{15}(x + 5)^{5/2} = (x + 5)^{3/2}\left[\frac{2x}{3} - \frac{4}{15}(x + 5)\right]$$

$$= (x + 5)^{3/2}\left[\frac{2x}{3} - \frac{4}{3}\right] = (x + 5)^{3/2}\left[\frac{2}{5}(x + 5) - \frac{10}{3}\right]$$

$$= \frac{2}{5}(x + 5)^{5/2} - \frac{10}{3}(x + 5)^{3/2}$$

which is the form of the antiderivative obtained by substitution. This example shows that it is quite possible for you to do everything right and still not get the answer given at the back of the book.

EXAMPLE 3.3

Find $\int x \ln x \, dx$.

Solution

In this case, the factor x is easy to integrate, while the factor $\ln x$ is simplified by differentiation. This suggests that you try integration by parts with

$$g(x) = x \quad \text{and} \quad f(x) = \ln x$$

Then, $\qquad G(x) = \frac{1}{2}x^2 \quad \text{and} \quad f'(x) = \frac{1}{x}$

and so

$$\int \overset{I}{x} \overset{D}{\ln x} \, dx = \frac{1}{2}x^2 \ln x - \frac{1}{2}\int x^2\left(\frac{1}{x}\right) dx = \frac{1}{2}x^2 \ln x - \frac{1}{2}\int x \, dx$$

$$= \frac{1}{2}x^2 \ln x - \frac{1}{4}x^2 + C = \frac{1}{2}x^2 \left(\ln x - \frac{1}{2}\right) + C$$

The Integral of ln x In the next example, you will see how to use integration by parts to integrate the natural logarithm $\ln x$.

EXAMPLE 3.4

Find $\int \ln x \, dx$.

Solution

The trick is to write $\ln x$ as the product $1(\ln x)$, in which the factor 1 is easy to integrate and the factor $\ln x$ is simplified by differentiation. This suggests that you use integration by parts with

$$g(x) = 1 \quad \text{and} \quad f(x) = \ln x$$

Then,

$$G(x) = x \quad \text{and} \quad f'(x) = \frac{1}{x}$$

and so

$$\int \ln x \, dx = \int \overset{\text{I}}{1}\overset{\text{D}}{(\ln x)} \, dx = x \ln x - \int x\left(\frac{1}{x}\right) dx = x \ln x - \int 1 \, dx$$

$$= x \ln x - x + C = x(\ln x - 1) + C$$

Repeated Applications of Integration by Parts   Sometimes integration by parts leads to a new integral that also must be integrated by parts. This situation is illustrated in the next example.

EXAMPLE 3.5

Find $\int x^2 e^x \, dx$.

Solution

Since the factor e^x is easy to integrate and the factor x^2 is simplified by differentiation, try integration by parts with

$$g(x) = e^x \quad \text{and} \quad f(x) = x^2$$

Then,

$$G(x) = e^x \quad \text{and} \quad f'(x) = 2x$$

and so

$$\int \overset{\text{D}}{x^2}\overset{\text{I}}{e^x} \, dx = x^2 e^x - 2 \int x e^x \, dx$$

To find $\int x e^x \, dx$, you have to integrate by parts again, this time with

$$g(x) = e^x \quad \text{and} \quad f(x) = x$$

Then,

$$G(x) = e^x \quad \text{and} \quad f'(x) = 1$$

and so

$$\int x^2 e^x \, dx = x^2 e^x - 2 \int \overset{\text{D}}{x}\overset{\text{I}}{e^x} \, dx$$

$$= x^2e^x - 2\left(xe^x - \int e^x \, dx\right)$$

$$= x^2e^x - 2(xe^x - e^x) + C = (x^2 - 2x + 2)e^x + C$$

Problems 5.3

In Problems 1 through 25, use integration by parts to find the given integral.

1. $\int xe^{-x} \, dx$

2. $\int xe^{x/2} \, dx$

3. $\int xe^{-x/5} \, dx$

4. $\int xe^{0.1x} \, dx$

5. $\int (1 - x)e^x \, dx$

6. $\int (3 - 2x)e^{-x} \, dx$

7. $\int x \ln 2x \, dx$

8. $\int x \ln x^2 \, dx$

9. $\int x\sqrt{x - 6} \, dx$

10. $\int x\sqrt{1 - x} \, dx$

11. $\int x(x + 1)^8 \, dx$

12. $\int (x + 1)(x + 2)^6 \, dx$

13. $\int \frac{x}{\sqrt{x + 2}} \, dx$

14. $\int \frac{x}{\sqrt{2x + 1}} \, dx$

15. $\int x^2e^{-x} \, dx$

16. $\int x^2e^{3x} \, dx$

17. $\int x^3e^x \, dx$

18. $\int x^3e^{2x} \, dx$

19. $\int x^2 \ln x \, dx$

20. $\int x(\ln x)^2 \, dx$

21. $\int \frac{\ln x}{x^2} \, dx$

22. $\int \frac{\ln x}{x^3} \, dx$

23. $\int x^3e^{x^2} \, dx$ [*Hint:* Rewrite the integrand as $x^2(xe^{x^2})$.]

24. $\int x^3(x^2 - 1)^{10} \, dx$

25. $\int x^7(x^4 + 5)^8 \, dx$

26. Find the function whose tangent has slope $(x + 1)e^{-x}$ for each value of x and whose graph passes through the point $(1, 5)$.

27. Find the function whose tangent has slope $x \ln\sqrt{x}$ for each value of $x > 0$ and whose graph passes through the point $(2, -3)$.

Distance

28. After t seconds, an object is moving at the speed of $te^{-t/2}$ meters per second. Express the distance the object travels as a function of time.

Efficiency

29. After t hours on the job, a factory worker can produce $100te^{-0.5t}$ units per hour. How many units does the worker produce during the first 3 hours?

Fund-Raising

30. After t weeks, contributions in response to a local fund-raising campaign were coming in at the rate of $2,000te^{-0.2t}$ dollars per week. How much money was raised during the first 5 weeks?

Marginal Cost

31. A manufacturer has found that marginal cost is $0.5(q + 1)e^{0.3q}$ dollars per unit when q units have been produced. The total cost of producing 100 units is $200. What is the total cost of producing the first 20 units?

Population Growth

32. It is projected that t years from now the population of a certain city will be changing at the rate of $t \ln\sqrt{t + 1}$ thousand people per year. If the current population is 2 million, what will the population be 5 years from now?

Reduction Formula

33. (a) Use integration by parts to derive the formula

$$\int x^n e^{ax}\, dx = \frac{1}{a}x^n e^{ax} - \frac{n}{a}\int x^{n-1}e^{ax}\, dx$$

(b) Use the formula in part (a) to find $\int x^3 e^{5x}\, dx$.

4 The Use of Integral Tables

Most of the integrals you will encounter in the social, managerial, and life sciences can be evaluated using the techniques you have learned so far in this chapter. From time to time, however, an integral will turn up that cannot be handled by these techniques. For such occasions, it is helpful to know how to use a **table of integrals.**

A table of integrals is a list of integration formulas. Extensive tables listing several hundred formulas can be found in most mathematics handbooks, and condensed versions appear in many calculus texts. Here is a tiny sampling of the formulas that appear in a table of integrals.

A Small Table of Integrals

1. $\displaystyle\int \frac{dx}{p^2 - x^2} = \frac{1}{2p} \ln\left|\frac{p + x}{p - x}\right|$

2. $\displaystyle\int \frac{dx}{x(ax + b)} = \frac{1}{b} \ln\left|\frac{x}{ax + b}\right|$

3. $\displaystyle \int \frac{dx}{\sqrt{x^2 \pm p^2}} = \ln \left| x + \sqrt{x^2 \pm p^2} \right|$

4. $\displaystyle \int x^n e^{ax}\, dx = \frac{1}{a}x^n e^{ax} - \frac{n}{a} \int x^{n-1} e^{ax}\, dx$

In these formulas, the letters a, b, p, and n denote constants. The term p^2 in the first and third formulas may be any *positive* constant (since any positive number is the square of its square root). The compact fractional notation $\dfrac{dx}{p^2 - x^2}$ in the first formula is an abbreviation for $\dfrac{1}{p^2 - x^2}\, dx$. Similar notation occurs in some of the other formulas. Also, to keep the formulas simple, the constant C is omitted from each of the integrals in the table. You add the C.

For convenience, most tables of integrals are divided into sections. Integrals containing similar expressions are grouped together in the same section. For example, the first formula would be found in a section entitled "Expressions Containing $p^2 - x^2$," the second in a section called "Expressions Containing $ax + b$," and the fourth in the section "Expressions Containing Exponential and Logarithmic Functions."

The use of the first three integration formulas is illustrated in the following examples.

EXAMPLE 4.1

Find $\displaystyle \int \frac{1}{x(3x - 6)}\, dx$.

Solution

Apply the second formula with $a = 3$ and $b = -6$ to get

$$\int \frac{1}{x(3x - 6)}\, dx = -\frac{1}{6} \ln \left| \frac{x}{3x - 6} \right| + C$$

EXAMPLE 4.2

Find $\displaystyle \int \frac{1}{6 - 3x^2}\, dx$.

Solution

If the coefficient of x^2 were 1 instead of 3, you could use the first formula. This suggests that you first rewrite the integrand as

$$\frac{1}{6 - 3x^2} = \frac{1}{3}\left(\frac{1}{2 - x^2}\right)$$

and then apply the first formula with $p = \sqrt{2}$:

$$\int \frac{1}{6 - 3x^2}\, dx = \frac{1}{3}\int \frac{1}{2 - x^2}\, dx$$

$$= \frac{1}{3}\left(\frac{1}{2\sqrt{2}}\right)\ln\left|\frac{\sqrt{2} + x}{\sqrt{2} - x}\right| + C$$

$$= \frac{\sqrt{2}}{12}\ln\left|\frac{\sqrt{2} + x}{\sqrt{2} - x}\right| + C$$

EXAMPLE 4.3

Find $\displaystyle\int \frac{1}{3x^2 - 6}\, dx$.

Solution

Since

$$\frac{1}{3x^2 - 6} = -\frac{1}{6 - 3x^2}$$

you can apply the first formula as in Example 4.2 to get

$$\int \frac{1}{3x^2 - 6}\, dx = -\int \frac{1}{6 - 3x^2}\, dx = -\frac{\sqrt{2}}{12}\ln\left|\frac{\sqrt{2} + x}{\sqrt{2} - x}\right| + C$$

EXAMPLE 4.4

Find $\displaystyle\int \frac{1}{x^2 + 2x}\, dx$.

Solution

Factor the integrand as $\displaystyle\frac{1}{x^2 + 2x} = \frac{1}{x(x + 2)}$

and apply the second formula with $a = 1$ and $b = 2$:

$$\int \frac{1}{x^2 + 2x}\, dx = \int \frac{1}{x(x + 2)}\, dx = \frac{1}{2}\ln\left|\frac{x}{x + 2}\right| + C$$

EXAMPLE 4.5

Find $\int \dfrac{1}{3x^2 + 6}\, dx$.

Solution

It is natural to try to match this integral to the one in the first formula by writing

$$\int \frac{1}{3x^2 + 6}\, dx = -\frac{1}{3}\int \frac{1}{-2 - x^2}\, dx$$

However, since -2 is negative, it cannot be written as the square p^2 of any real number p, and so the formula does not apply.

There is a formula for integrals of the form $\int \dfrac{1}{x^2 + p^2}\, dx$ that can be used in this case. You can find it in a table of integrals under a heading like "Expressions Containing $x^2 \pm p^2$." However, the antiderivative will be written in terms of inverse trigonometric functions and cannot be expressed in more elementary terms.

EXAMPLE 4.6

Find $\int \dfrac{1}{\sqrt{4x^2 - 9}}\, dx$.

Solution

To put this integral in the form of the third formula, rewrite the integrand as

$$\frac{1}{\sqrt{4x^2 - 9}} = \frac{1}{\sqrt{4(x^2 - 9/4)}} = \frac{1}{2\sqrt{x^2 - 9/4}}$$

Then apply the formula with $p^2 = \dfrac{9}{4}$, using minus signs in place of the symbol $\pm$, to get

$$\int \frac{1}{\sqrt{4x^2 - 9}} \, dx = \frac{1}{2} \int \frac{1}{\sqrt{x^2 - 9/4}} \, dx = \frac{1}{2} \ln |x + \sqrt{x^2 - 9/4}| + C$$

Reduction Formulas   The fourth formula expresses an integral in terms of a simpler one of the same type. If the formula is subsequently applied to the new integral, further simplification may occur. Successive applications of the formula usually lead to an integral that can be found by elementary methods. A formula of this type is called a **reduction formula.** The use of reduction formulas is illustrated in the next example.

EXAMPLE 4.7

Find $\int x^2 e^{5x} \, dx$.

Solution

Apply the fourth formula with $n = 2$ and $a = 5$ to get

$$\int x^2 e^{5x} \, dx = \frac{1}{5} x^2 e^{5x} - \frac{2}{5} \int x e^{5x} \, dx$$

Now apply the fourth formula again, this time with $n = 1$ and $a = 5$, to get

$$\int x e^{5x} \, dx = \frac{1}{5} x e^{5x} - \frac{1}{5} \int e^{5x} \, dx = \frac{1}{5} x e^{5x} - \frac{1}{25} e^{5x} + C$$

Combine these results to conclude that

$$\int x^2 e^{5x} \, dx = \frac{1}{5} x^2 e^{5x} - \frac{2}{5} \left(\frac{1}{5} x e^{5x} - \frac{1}{25} e^{5x} \right) + C$$

$$= \left(\frac{1}{5} x^2 - \frac{2}{25} x + \frac{2}{125} \right) e^{5x} + C$$

A Word of Advice   No special formula was really needed to find the integral in Example 4.7. You could have found this integral quite easily using integration by parts. Indeed, if the formula had not been so conveniently displayed at the beginning of this section, you would have been much better off integrating by parts directly than hunting through a table of integrals for the appropriate formula. Try not to succumb to the temptation to rely excessively on tables when computing integrals. Most of the integrals you will encounter can be found quite easily without the aid of formulas. Moreover, before you can use a formula, you must find it, and this can be time-consuming.

In general, it is good strategy to use a table of integrals only as a last resort.

Problems 5.4

In Problems 1 through 10, use one of the integration formulas listed in this section to find the given integral.

1. $\int \dfrac{1}{x(2x-3)}\,dx$

2. $\int \dfrac{3}{4x(x-5)}\,dx$

3. $\int \dfrac{1}{\sqrt{x^2+25}}\,dx$

4. $\int \dfrac{1}{\sqrt{9x^2-4}}\,dx$

5. $\int \dfrac{1}{4-x^2}\,dx$

6. $\int \dfrac{1}{3x^2-9}\,dx$

7. $\int \dfrac{1}{3x^2+2x}\,dx$

8. $\int \dfrac{4}{x^2-x}\,dx$

9. $\int x^2 e^{3x}\,dx$

10. $\int x^3 e^{-x}\,dx$

Locate a table of integrals and use it to find the integrals in Problems 11 through 18.

11. $\int \dfrac{x}{2-x^2}\,dx$

12. $\int \dfrac{x+3}{\sqrt{2x+4}}\,dx$

13. $\int (\ln 2x)^2\,dx$

14. $\int (x^2+1)^{3/2}\,dx$

15. $\int \dfrac{1}{3x\sqrt{2x+5}}\,dx$

16. $\int \dfrac{x}{\sqrt{4-x^2}}\,dx$

17. $\int \dfrac{1}{2-3e^{-x}}\,dx$

18. $\int \dfrac{1}{\sqrt{3x^2-6x+2}}\,dx$

19. One table of integrals lists the formula

$$\int \frac{dx}{\sqrt{x^2\pm p^2}} = \ln\left|\frac{x+\sqrt{x^2\pm p^2}}{p}\right|$$

while another table lists

$$\int \frac{dx}{\sqrt{x^2\pm p^2}} = \ln|x+\sqrt{x^2\pm p^2}|$$

Can you reconcile this apparent contradiction?

20. The following two formulas appear in a table of integrals:

$$\int \frac{dx}{p^2 - x^2} = \frac{1}{2p} \ln \left| \frac{p + x}{p - x} \right|$$

and

$$\int \frac{dx}{a + bx^2} = \frac{1}{2\sqrt{-ab}} \ln \left| \frac{a + x\sqrt{-ab}}{a - x\sqrt{-ab}} \right| \qquad (\text{for } -ab \geq 0)$$

(a) Use the second formula to derive the first.

(b) Apply both formulas to the integral $\int \frac{1}{9 - 4x^2} \, dx$. Which do you find easier to use in this problem?

In Problems 21 and 22, use the following integral formula to find the given integral.

$$\int (\ln x)^n dx = x(\ln x)^n - n \int (\ln x)^{n-1} \, dx$$

21. $\int (\ln x)^3 \, dx$

22. $\int x(\ln x)^2 \, dx$ (*Hint:* First use integration by parts.)

Chapter Summary and Review Problems

Important Terms, Symbols, and Formulas

Antiderivative; indefinite integral:

$$\int f(x) \, dx = F(x) + C \qquad \text{if and only if } F'(x) = f(x)$$

Power rule: $\int x^n \, dx = \frac{1}{n + 1} x^{n+1} + C$ (for $n \neq -1$)

The integral of $\frac{1}{x}$: $\int \frac{1}{x} \, dx = \ln |x| + C$

Constant multiple rule: $\int cf(x) \, dx = c \int f(x) \, dx$

Sum rule: $\int [f(x) + g(x)] \, dx = \int f(x) \, dx + \int g(x) \, dx$

The integral of e^x: $\int e^x \, dx = e^x + C$

Integration by substitution:

$$\int g(u) \frac{du}{dx} \, dx = G(u) + C \qquad \text{where } G \text{ is an antiderivative of } g$$

Integration by parts:

$$\int f(x)g(x) \, dx = f(x)G(x) - \int f'(x)G(x) \, dx \qquad \text{where } G \text{ is an antiderivative of } g$$

Table of integrals

Reduction formula

Review Problems   In Problems 1 through 20, find the indicated integral.

1. $\int \left(x^5 - 3x^2 + \dfrac{1}{x^2} \right) dx$

2. $\int \left(x^{2/3} - \dfrac{1}{x} + 5 + \sqrt{x} \right) dx$

3. $\int \sqrt{3x + 1} \; dx$

4. $\int (3x + 1)\sqrt{3x^2 + 2x + 5} \; dx$

5. $\int (x + 2)(x^2 + 4x + 2)^5 \; dx$

6. $\int \dfrac{x + 2}{x^2 + 4x + 2} \; dx$

7. $\int \dfrac{3x + 6}{(2x^2 + 8x + 3)^2} \; dx$

8. $\int (x - 5)^{12} \; dx$

9. $\int x(x - 5)^{12} \; dx$

10. $\int 5e^{3x} \; dx$

11. $\int 5xe^{3x} \; dx$

12. $\int xe^{-x/2} \; dx$

13. $\int x^5 e^{x^3} \; dx$

14. $\int (2x + 1)e^{0.1x} \; dx$

15. $\int x \ln 3x \; dx$

16. $\int \ln 3x \; dx$

17. $\int \dfrac{\ln 3x}{x} \; dx$

18. $\int \dfrac{\ln 3x}{x^2} \; dx$

19. $\int x^3(x^2 + 1)^8 \; dx$

20. $\int 2x \ln (x^2 + 1) \; dx$

21. Find the function whose tangent has slope $x(x^2 + 1)^3$ for each value of x and whose graph passes through the point $(1, 5)$.

22. It is estimated that x weeks from now, the number of commuters using a new subway line will be increasing at the rate of $18x^2 + 500$ per week. Currently, 8,000 commuters use the subway. How many will be using it 5 weeks from now?

23. Statistics compiled by the local department of corrections indicate that x years from now the number of inmates in county prisons will be increasing at the rate of $280e^{0.2x}$ per year. Currently 2,000 inmates are housed in county prisons. How many inmates should the county expect 10 years from now?

24. A manufacturer estimates marginal revenue to be $200q^{-1/2}$ dollars per unit when the level of production is q units. The corresponding

marginal cost has been found to be $0.4q$ dollars per unit. If the manufacturer's profit is $2,000 when the level of production is 25 units, what is the profit when the level of production is 36 units?

25. In a certain section of the country, the price of large Grade A eggs is currently $1.60 per dozen. Studies indicate that x weeks from now, the price will be changing at the rate of $0.2 + 0.003x^2$ cents per week. How much will eggs cost 10 weeks from now?

26. The resale value of a certain industrial machine decreases at a rate that changes with time. When the machine is t years old, the rate at which its value is changing is $220(t - 10)$ dollars per year. If the machine was bought new for $12,000, how much will it be worth 10 years later?

27. A tree has been transplanted and after x years is growing at the rate of $f(x) = 0.5 + \dfrac{1}{(x + 1)^2}$ meters per year. By how much does the tree grow during the 2nd year?

28. It is estimated that t days from now a farmer's crop will be increasing at the rate of $0.3t^2 + 0.6t + 1$ bushels per day. By how much will the value of the crop increase during the next 5 days if the market price remains fixed at $3 per bushel?

In Problems 29 through 32, use one of the formulas in the table of integrals given at the beginning of Section 4 to find the given integral.

29. $\displaystyle\int \frac{5}{8 - 2x^2}\, dx$

30. $\displaystyle\int \frac{2}{\sqrt{9x^2 + 16}}\, dx$

31. $\displaystyle\int x^2 e^{-x/2}\, dx$

32. $\displaystyle\int \frac{4}{x(9 + 5x)}\, dx$

CHAPTER 6
FURTHER TOPICS IN INTEGRATION

1   The Definite Integral

Suppose you know the rate $f(x) = \dfrac{dF}{dx}$ at which a certain quantity F is changing and wish to find the amount by which the quantity F will change between $x = a$ and $x = b$. You would first find F by antidifferentiation and then compute the difference

$$\text{Change in } F \text{ between } x = a \text{ and } x = b = F(b) - F(a)$$

The numerical result of such a computation is called a **definite integral** of the function f and is denoted by the symbol $\displaystyle\int_a^b f(x)\ dx$.

The Definite Integral

The definite integral of f from a to b is the difference

$$\int_a^b f(x)\ dx = F(b) - F(a)$$

where F is an antiderivative of f. That is, the definite integral is the net change in the antiderivative between $x = a$ and $x = b$.

The symbol $\displaystyle\int_a^b f(x)\ dx$ is read "the (definite) integral of f from a to b." The numbers a and b are called **limits of integration.** In computations involving definite integrals, it is often convenient to use the symbol $F(x)\Big|_a^b$ to stand for the difference $F(b) - F(a)$.

Applications   Here are three practical problems involving net change whose solutions are definite integrals.

EXAMPLE 1.1

A study indicates that x months from now the population of a certain town will be increasing at the rate of $2 + 6\sqrt{x}$ people per month. By how much will the population of the town increase during the next 4 months?

Solution

Let $P(x)$ denote the population of the town x months from now. Then the rate of change of the population with respect to time is $\dfrac{dP}{dx} = 2 + 6\sqrt{x}$, and the amount by which the population will increase during the next 4 months is the definite integral

$$P(4) - P(0) = \int_0^4 (2 + 6\sqrt{x})\, dx = (2x + 4x^{3/2} + C)\Big|_0^4$$

$$= [2(4) + 4(4)^{3/2} + C] - [2(0) + 4(0)^{3/2} + C]$$

$$= (40 + C) - (0 + C) = 40 \text{ people}$$

Comment: Notice what happens to the constant C in the evaluation of a definite integral. It appears in the expressions for both $F(b)$ and $F(a)$ and is eventually eliminated by the subtraction. You may therefore omit the constant C altogether when evaluating definite integrals.

EXAMPLE 1.2

At a certain factory, the marginal cost is $3(q - 4)^2$ dollars per unit when the level of production is q units. By how much will the total manufacturing cost increase if the level of production is raised from 6 units to 10 units?

Solution

Let $C(q)$ denote the total cost of producing q units. Then the marginal cost is the derivative $\dfrac{dC}{dq} = 3(q - 4)^2$, and the increase in cost if production is raised from 6 units to 10 units is the definite integral

$$C(10) - C(6) = \int_6^{10} 3(q - 4)^2\, dq = (q - 4)^3 \Big|_6^{10}$$

$$= (10 - 4)^3 - (6 - 4)^3 = 216 - 8 = \$208$$

EXAMPLE 1.3

In a certain community, the demand for gasoline is increasing exponentially at the rate of 5 percent per year. If the current demand is 4 million gallons per year, how much gasoline will be consumed in the community during the next 3 years?

Solution

Let $Q(t)$ denote the total consumption (in million-gallon units) of gasoline in the community over the next t years. Then the demand (million gallons per year) is the rate of change $\dfrac{dQ}{dt}$ of the total consumption with respect to time. The fact that this demand is increasing exponentially at the rate of 5 percent per year and is currently equal to 4 million gallons per year implies that

$$\frac{dQ}{dt} = 4e^{0.05t} \qquad \text{million gallons per year}$$

Hence, the total consumption during the next 3 years is the definite integral

$$Q(3) - Q(0) = \int_0^3 4e^{0.05t} \, dt = 80e^{0.05t} \Big|_0^3$$

$$= 80(e^{0.15} - 1) \approx 12.95 \text{ million gallons}$$

The Evaluation of Definite Integrals by Substitution

In the next example, you will see how the method of substitution can be used to evaluate a definite integral.

EXAMPLE 1.4

Evaluate $\displaystyle\int_0^1 8x(x^2 + 1)^3 \, dx$.

Solution

The integrand is a product in which one factor $8x$ is a constant multiple of the derivative of an expression $x^2 + 1$ that appears in the other factor. This suggests that you let $u = x^2 + 1$. Then $du = 2x \, dx$, and so

$$\int 8x(x^2 + 1)^3 \, dx = \int 4u^3 \, du = u^4$$

The limits of integration, 0 and 1, refer to the variable x and not to u. You may, therefore, proceed in one of two ways. Either you can rewrite the antiderivative in terms of x, or you can find the values of u that correspond to $x = 0$ and $x = 1$.

If you choose the first alternative, you find that

$$\int 8x(x^2 + 1)^3 \, dx = u^4 = (x^2 + 1)^4$$

and so $\displaystyle\int_0^1 8x(x^2 + 1)^3 \, dx = (x^2 + 1)^4 \Big|_0^1 = 16 - 1 = 15$

If you choose the second alternative, you use the fact that $u = x^2 + 1$ to conclude that $u = 1$ when $x = 0$ and $u = 2$ when $x = 1$. Hence,

$$\int_0^1 8x(x^2 + 1)^3 \, dx = u^4 \Big|_1^2 = 16 - 1 = 15$$

Probably the most efficient approach is to adopt the second alternative and write the solution compactly as follows:

$$\int_0^1 8x(x^2 + 1)^3 \, dx = \int_1^2 4u^3 \, du = u^4 \Big|_1^2 = 16 - 1 = 15$$

Here is a summary of the method of substitution for definite integrals.

Integration by Substitution

$$\int_a^b g[u(x)] \frac{du}{dx} \, dx = \int_{u(a)}^{u(b)} g(u) \, du$$

Here is one more example.

EXAMPLE 1.5

Evaluate $\displaystyle\int_1^e \frac{\ln x}{x} \, dx$.

Solution

The integrand can be written as the product $(\ln x)\left(\dfrac{1}{x}\right)$ in which one of the factors $\dfrac{1}{x}$ is the derivative of the other $\ln x$. This suggests that you let $u = \ln x$. Then $du = \dfrac{1}{x} \, dx$, $u(1) = 0$, and $u(e) = 1$. Hence,

$$\int_1^e \frac{\ln x}{x} \, dx = \int_0^1 u \, du = \frac{1}{2}u^2 \Big|_0^1 = \frac{1}{2} - 0 = \frac{1}{2}$$

The Evaluation of Definite Integrals by Parts

The formula for integration by parts can be rephrased for definite integrals as follows.

Integration by Parts

$$\int_a^b f(x)g(x)\ dx = f(x)G(x)\Big|_a^b - \int_a^b f'(x)G(x)\ dx$$

where G is an antiderivative of g.

Here are two examples.

EXAMPLE 1.6

Evaluate $\displaystyle\int_0^{\ln 2} xe^x\ dx$.

Solution

Both factors x and e^x are easy to integrate, while the factor x is simplified by differentiation. This suggests that you try integration by parts with

$$g(x) = e^x \quad \text{and} \quad f(x) = x$$

Then, $\qquad G(x) = e^x \quad \text{and} \quad f'(x) = 1$

and so

$$\int_0^{\ln 2} xe^x\ dx = xe^x\Big|_0^{\ln 2} - \int_0^{\ln 2} e^x\ dx = (xe^x - e^x)\Big|_0^{\ln 2}$$

$$= (2 \ln 2 - 2) - (0 - 1) = 2 \ln 2 - 1$$

EXAMPLE 1.7

Evaluate $\displaystyle\int_1^e x \ln x\ dx$.

Solution

Since the factor x is easy to integrate and the factor $\ln x$ is simplified by differentiation, try integration by parts with

$$g(x) = x \quad \text{and} \quad f(x) = \ln x$$

Then, $\qquad G(x) = \dfrac{1}{2}x^2 \quad \text{and} \quad f'(x) = \dfrac{1}{x}$

and so
$$\int_1^e x \ln x \, dx = \frac{1}{2}x^2 \ln x \Big|_1^e - \frac{1}{2} \int_1^e x^2 \left(\frac{1}{x}\right) dx$$

$$= \frac{1}{2}x^2 \ln x \Big|_1^e - \frac{1}{2} \int_1^e x \, dx$$

$$= \left(\frac{1}{2}x^2 \ln x - \frac{1}{4}x^2\right)\Big|_1^e$$

$$= \left(\frac{1}{2}e^2 \ln e - \frac{1}{4}e^2\right) - \left[\frac{1}{2}(1)^2 \ln 1 - \frac{1}{4}(1)^2\right]$$

$$= \frac{1}{2}e^2 - \frac{1}{4}e^2 + \frac{1}{4} = \frac{1}{4}(e^2 + 1)$$

Definite Integrals and Sums   In this text, we *define* the definite integral to be the net change in the antiderivative and then *prove* that this net change is equal to the "limiting value" of a certain sum. Some calculus texts take the opposite approach. In particular, they *define* the definite integral as the limiting value of a sum and then *prove* that it can be computed as the net change in the antiderivative. In either case, the important result is the connection between antiderivatives and sums. This relationship is known as the **fundamental theorem of calculus** and will be derived and applied in Section 4.

Problems 6.1

In Problems 1 through 24, evaluate the given definite integral.

1. $\int_0^1 (x^4 - 3x^3 + 1) \, dx$

2. $\int_{-1}^0 (3x^5 - 3x^2 + 2x - 1) \, dx$

3. $\int_2^5 (2 + 2t + 3t^2) \, dt$

4. $\int_1^9 \left(\sqrt{t} - \frac{1}{\sqrt{t}}\right) dt$

5. $\int_1^3 \left(1 + \frac{1}{x} + \frac{1}{x^2}\right) dx$

6. $\int_{\ln 1/2}^{\ln 2} (e^t - e^{-t}) \, dt$

7. $\int_{-3}^{-1} \frac{t + 1}{t^3} \, dt$

8. $\int_0^6 x^2(x - 1) \, dx$

9. $\int_1^2 (2x - 4)^5 \, dx$

10. $\int_{-3}^0 (2x + 6)^4 \, dx$

11. $\int_0^4 \frac{1}{\sqrt{6t + 1}} \, dt$

12. $\int_1^2 \frac{x^2}{(x^3 + 1)^2} \, dx$

13. $\int_0^1 (t^3 + t) \sqrt{t^4 + 2t^2 + 1} \, dt$

14. $\int_0^1 \frac{6x}{x^2 + 1} \, dx$

15. $\int_2^{e+1} \frac{x}{x-1}\, dx$

16. $\int_1^2 (t+1)(t-2)^9\, dt$

17. $\int_1^{e^2} \ln t\, dt$

18. $\int_{1/2}^{e/2} t \ln 2t\, dt$

19. $\int_{-2}^2 xe^{-x}\, dx$

20. $\int_0^1 x^2 e^{2x}\, dx$

21. $\int_1^{e^2} \frac{(\ln x)^2}{x}\, dx$

22. $\int_e^{e^2} \frac{1}{x \ln x}\, dx$

23. $\int_0^{10} (20+t)e^{-0.1t}\, dt$

24. $\int_0^5 te^{-(5-t)/20}\, dt$

Population Growth   25. A study indicates that x months from now the population of a certain town will be increasing at the rate of $5 + 3x^{2/3}$ people per month. By how much will the population of the town increase over the next 8 months?

Distance and Speed   26. An object is moving so that its speed after t minutes is $5 + 2t + 3t^2$ meters per minute. How far does the object travel during the 2nd minute?

Depreciation   27. The resale value of a certain industrial machine decreases over a 10-year period at a rate that changes with time. When the machine is x years old, the rate at which its value is changing is $220(x - 10)$ dollars per year. By how much does the machine depreciate during the 2nd year?

Admission to Events   28. The promoters of a county fair estimate that t hours after the gates open at 9:00 A.M. visitors will be entering the fair at the rate of $-4(t + 2)^3 + 54(t + 2)^2$ people per hour. How many people will enter the fair between 10:00 A.M. and noon?

Marginal Cost   29. At a certain factory, the marginal cost is $6(q - 5)^2$ dollars per unit when the level of production is q units. By how much will the total manufacturing cost increase if the level of production is raised from 10 to 13 units?

Oil Production   30. A certain oil well that yields 400 barrels of crude oil a month will run dry in 2 years. The price of crude oil is currently $18 per barrel and is expected to rise at a constant rate of 3 cents per barrel per month. If the oil is sold as soon as it is extracted from the ground, what will the total future revenue from the well be?

Farming   31. It is estimated that t days from now a farmer's crop will be increasing at the rate of $0.3t^2 + 0.6t + 1$ bushels per day. By how much will the value of the crop increase during the next 5 days if the market price remains fixed at $3 per bushel?

Energy Consumption   32. It is estimated that the demand for oil is increasing exponentially at the rate of 10 percent per year. If the demand for oil is currently 30 billion barrels per year, how much oil will be consumed during the next 10 years?

Sales Revenue   33. It is estimated that the demand for a manufacturer's product is increasing exponentially at the rate of 2 percent per year. If the current demand is 5,000 units per year and if the price remains fixed at $400 per unit, how much revenue will the manufacturer receive from the sale of the product over the next 2 years?

Efficiency   34. After t hours on the job, a factory worker can produce $100te^{-0.5t}$ units per hour. How many units does a worker who arrives on the job at 8:00 A.M. produce between 10:00 A.M. and noon?

35. (a) Show that $\displaystyle\int_a^b f(x) \, dx + \int_b^c f(x) \, dx = \int_a^c f(x) \, dx$.

(b) Use the formula in part (a) to evaluate $\displaystyle\int_{-1}^1 |x| \, dx$. (*Hint:* Evaluate $\displaystyle\int_{-1}^0 |x| \, dx$ and $\displaystyle\int_0^1 |x| \, dx$ and combine the results.)

(c) Evaluate $\displaystyle\int_0^4 (1 + |x - 3|)^2 \, dx$.

Even and Odd Functions   36. (a) Show that if F is an antiderivative of f, then

$$\int_a^b f(-x) \, dx = -F(-b) + F(-a)$$

(*Hint:* Use the method of substitution with $u = -x$.)

(b) A function f is said to be **even** if $f(-x) = f(x)$. [For example, $f(x) = x^2$ is even.] Use Problem 35 and part (a) to show that if f is even, then

$$\int_{-a}^a f(x) \, dx = 2 \int_0^a f(x) \, dx$$

(c) Use part (b) to evaluate $\displaystyle\int_{-1}^1 |x| \, dx$ and $\displaystyle\int_{-2}^2 x^2 \, dx$.

(d) A function f is said to be **odd** if $f(-x) = -f(x)$. Use Problem 35 and part (a) to show that if f is odd, then

$$\int_{-a}^a f(x) \, dx = 0$$

(e) Evaluate $\displaystyle\int_{-12}^{12} x^3 \, dx$.

2 Area and Integration

There is a surprising connection between definite integrals and the geometric concept of area. If $f(x)$ is continuous and nonnegative on the interval $a \leq x \leq b$, then the region R under the graph of f between $x = a$ and $x = b$ (see Figure 6.1) has area equal to the definite integral $\int_a^b f(x)\ dx$.

Figure 6.1 The area under the curve $y = f(x)$ between $x = a$ and $x = b$.

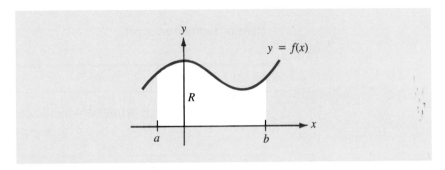

The Area under a Curve

Suppose $f(x)$ is continuous and $f(x) \geq 0$ on the interval $a \leq x \leq b$. Then the region R under the curve $y = f(x)$ and between $x = a$ and $x = b$ has area

$$\text{Area of } R = \int_a^b f(x)\ dx$$

A justification of this area formula will be given later in this section, but first we shall examine a few examples that illustrate the use of the formula. We begin by using the formula to find an area that you already know from elementary geometry.

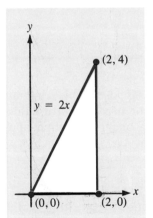

Figure 6.2 The region under $y = 2x$ between $x = 0$ and $x = 2$.

EXAMPLE 2.1

Find the area of the region bounded by the line $y = 2x$, the x axis, and the vertical line $x = 2$.

Solution

The region in question is the triangle in Figure 6.2, and its area is

$$\text{Area} = \frac{1}{2}(\text{base})(\text{height}) = \frac{1}{2}(2)(4) = 4$$

To compute this area using calculus, apply the integral formula with $f(x) = 2x$. Take $b = 2$ since the region is bounded on the right by the line $x = 2$, and take $a = 0$ since, on the left, the boundary consists of the single point $(0, 0)$, which is part of the vertical line $x = 0$. You will find, as expected, that

$$\text{Area} = \int_0^2 2x\, dx = x^2 \Big|_0^2 = 4$$

Here is another example.

EXAMPLE 2.2

Find the area of the region bounded by the curve $y = -x^2 + 4x - 3$ and the x axis.

Solution

From the factored form of the polynomial

$$y = -x^2 + 4x - 3 = -(x - 3)(x - 1)$$

you see that the x intercepts of the curve are $(1, 0)$ and $(3, 0)$. From the corresponding graph (Figure 6.3) you see that the region in question is below the curve $y = -x^2 + 4x - 3$, above the x axis, and extends from $x = 1$ to $x = 3$. Hence,

$$\text{Area} = \int_1^3 (-x^2 + 4x - 3)\, dx = \left(-\frac{1}{3}x^3 + 2x^2 - 3x\right)\Big|_1^3$$

$$= (-9 + 18 - 9) - \left(-\frac{1}{3} + 2 - 3\right) = \frac{4}{3}$$

Figure 6.3 The region bounded by $y = -x^2 + 4x - 3$ and the x axis.

Why the Integral Formula for Area Works   To see why the integral formula for area works, suppose $f(x)$ is continuous and nonnegative on the interval $a \le x \le b$. For any value of x in this interval, let $A(x)$ denote the area of the region under the graph of f between a and x as shown in Figure 6.4. Note that $A(x)$ is a function of x on the interval $a \le x \le b$.

Your goal is to show that $A(b) = \int_a^b f(x)\, dx$. The key step is to establish that the derivative $A'(x)$ of the area function is equal to $f(x)$. To do this, consider the difference quotient

$$\frac{A(x + \Delta x) - A(x)}{\Delta x}$$

points

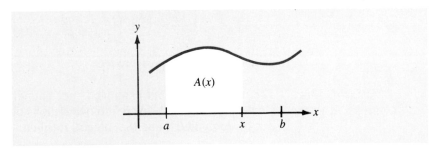

Figure 6.4 The area $A(x)$.

The expression $A(x + \Delta x) - A(x)$ in the numerator is just the area under the curve between x and $x + \Delta x$. If Δx is small, this area is approximately the same as the area of the rectangle whose height is $f(x)$ and whose width is Δx, as indicated in Figure 6.5. That is,

$$A(x + \Delta x) - A(x) \simeq f(x)\,\Delta x$$

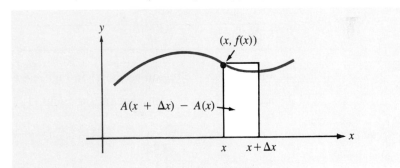

Figure 6.5 The area $A(x + \Delta x) - A(x)$ and an approximating rectangle.

or, equivalently, $$\dfrac{A(x + \Delta x) - A(x)}{\Delta x} \simeq f(x)$$

As Δx approaches zero, the error in the approximation approaches zero and it follows that

$$\lim_{\Delta x \to 0} \frac{A(x + \Delta x) - A(x)}{\Delta x} = f(x)$$

But by the definition of the derivative,

$$\lim_{\Delta x \to 0} \frac{A(x + \Delta x) - A(x)}{\Delta x} = A'(x)$$

Hence, $$A'(x) = f(x)$$

and so

$$\int_a^b f(x)\,dx = \int_a^b A'(x)\,dx = A(b) - A(a)$$

But $A(a)$ is the area under the curve between $x = a$ and $x = a$, which is clearly zero. Hence,

$$\int_a^b f(x)\ dx\ =\ A(b)$$

and the area formula is proved.

Areas of More Complex Regions   The region in the next example is not bounded above by a single curve. However, it can be broken into two regions that are, and the area of each can be computed using the integral formula.

EXAMPLE 2.3

Find the area of the region R in the first quadrant that lies under the curve $y = \dfrac{1}{x}$ and is bounded by this curve and the lines $y = x$, $y = 0$, and $x = 2$.

Solution

First sketch the region as shown in Figure 6.6. Note that the curve $y = \dfrac{1}{x}$ and the line $y = x$ intersect at the point $(1, 1)$.

Figure 6.6 The region in the first quadrant bounded by $y = \dfrac{1}{x}$, $y = x$, $y = 0$, and $x = 2$.

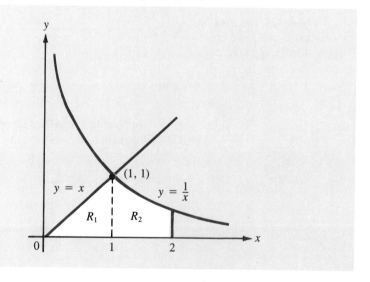

Observe that to the left of $x = 1$, R is bounded above by the line $y = x$, while to the right of $x = 1$, it is bounded above by the curve $y = \dfrac{1}{x}$. This suggests that you break R into two subregions, R_1 and R_2, as shown in Figure 6.6, and apply the integral formula for area to each subregion separately. In particular,

$$\text{Area of } R_1 = \int_0^1 x \, dx = \frac{1}{2}x^2 \bigg|_0^1 = \frac{1}{2}$$

and $$\text{Area of } R_2 = \int_1^2 \frac{1}{x} \, dx = \ln |x| \bigg|_1^2 = \ln 2 - \ln 1 = \ln 2$$

The area of R is the sum of these two areas. That is,

$$\text{Area of } R = \text{area of } R_1 + \text{area of } R_2 = \frac{1}{2} + \ln 2 \simeq 1.19$$

Probability Density Functions

One of the most important applications of the integral formula for area in the social, managerial, and life sciences is the computation of probabilities. Here, by way of a very brief introduction, is a simplified outline of a typical situation. A more detailed discussion of the topic will be given in Chapter 8, Section 3.

The life span x of an electronic component selected at random from a manufacturer's stock is a quantity that cannot be predicted with certainty. The **probability** that the life span x lies in some interval $a \leq x \leq b$ is denoted by $P(a \leq x \leq b)$ and is defined to be the fraction of all the components manufactured by the company that can be expected to have life spans in this range. A probability of the form $P(a \leq x \leq b)$ can be interpreted as the area under a certain curve and evaluated as a definite integral. The curve is the graph of a function $f(x)$ known as the **probability density function** for x. More precisely, a probability density function for the variable x is a positive continuous function $f(x)$ with the property that $P(a \leq x \leq b)$ is the area under its graph between $x = a$ and $x = b$. The situation is illustrated in Figure 6.7.

Probability density functions are constructed by statisticians from experimental data and theoretical considerations using techniques explained in most probability and statistics texts. The purpose of the discussion in

Figure 6.7 A probability density function.

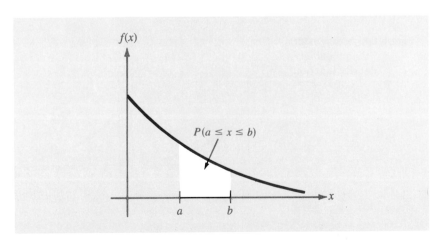

this section is to show you how to use integral calculus to compute probabilities once the appropriate probability density function is known. Here is an example.

EXAMPLE 2.4

Suppose the probability density function for the life span of electronic components produced by a certain company is $f(x) = 0.02e^{-0.02x}$, where x denotes the life span (in months) of a randomly selected component.

(a) What is the probability that the life span of a component selected at random will be between 20 and 30 months?
(b) What is the probability that the life span of a component selected at random will be less than or equal to 20 months?
(c) What is the probability that the life span of a component selected at random will be greater than 20 months?

Solution

(a) The desired probability $P(20 \leq x \leq 30)$ is the area (Figure 6.8a) under the graph of the density function between $x = 20$ and $x = 30$. Using the integral formula to compute this area, you get

$$P(20 \leq x \leq 30) = \int_{20}^{30} 0.02e^{-0.02x} \, dx = -e^{-0.02x}\Big|_{20}^{30}$$

$$\approx -0.5488 + 0.6703 \approx 0.1215$$

That is, approximately 12.15 percent of the components manufactured by the company will have life spans of between 20 and 30 months.

(b) The desired probability is $P(0 \leq x \leq 20)$, which is the area (Figure 6.8b) under the density function between $x = 0$ and $x = 20$. That is,

Figure 6.8 Areas under the probability density function $f(x) = 0.02e^{-0.02x}$.

$f(x)$ $f(x)$ $f(x)$

20 30   20   20

$P(20 \leq x \leq 30)$ $P(0 \leq x \leq 20)$ $P(x > 20)$
(a)   (b)   (c)

$$P(0 \le x \le 20) = \int_0^{20} 0.02e^{-0.02x}\, dx = -e^{-0.02x}\Big|_0^{20}$$

$$\approx -0.6703 + 1 \approx 0.3297$$

Thus, roughly $\frac{1}{3}$ of the components will fail during the first 20 months.

(c) The fraction of components whose life span is greater than 20 months is 1 (the fraction representing 100 percent of all the components) minus the fraction whose life span is less than or equal to 20 months. Hence,

$$P(x > 20) = 1 - P(0 \le x \le 20) \approx 1 - 0.3297 \approx 0.6703$$

In geometric terms, this is the area (Figure 6.8c) under the density function to the right of $x = 20$.

The Area between Two Curves   In some practical problems you may have to compute the area between two curves. Suppose that $f(x)$ and $g(x)$ are nonnegative functions and that $f(x) \ge g(x)$ on the interval $a \le x \le b$ as shown in Figure 6.9a.

To find the area of the region R between the curves from $x = a$ to $x = b$, you simply subtract the area under the lower curve $y = g(x)$ (Figure 6.9c) from the area under the upper curve $y = f(x)$ (Figure 6.9b). That is,

$$\text{Area of } R = \int_a^b f(x)\, dx - \int_a^b g(x)\, dx = \int_a^b [f(x) - g(x)]\, dx$$

It can be shown that this formula remains valid even if the functions f and g are not assumed to be nonnegative.

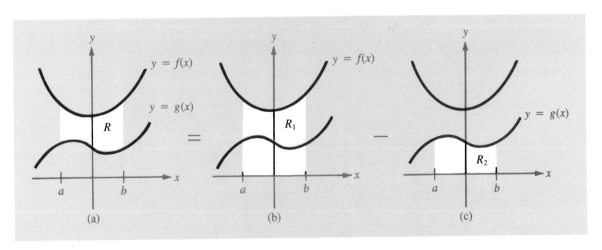

Figure 6.9 Area of R = area of R_1 − area of R_2.

The Area between Two Curves

If $f(x)$ and $g(x)$ are continuous on the interval $a \le x \le b$ with $f(x) \ge g(x)$, and if R is the region bounded by the graphs of f and g and the vertical lines $x = a$ and $x = b$, then

$$\text{Area of } R = \int_a^b [f(x) - g(x)] \, dx$$

Here are two examples.

EXAMPLE 2.5

Find the area of the region bounded by the curves $y = x^2 + 1$ and $y = 2x - 2$ between $x = -1$ and $x = 2$.

Solution

So that you can visualize the situation, begin by sketching the region as shown in Figure 6.10. Then apply the integral formula with $f(x) = x^2 + 1$, $g(x) = 2x - 2$, $a = -1$, and $b = 2$ to get

$$\text{Area} = \int_{-1}^{2} [(x^2 + 1) - (2x - 2)] \, dx = \int_{-1}^{2} (x^2 - 2x + 3) \, dx$$

$$= \left(\frac{1}{3}x^3 - x^2 + 3x \right)\Big|_{-1}^{2} = \frac{14}{3} - \left(-\frac{13}{3} \right) = 9$$

Figure 6.10 The region bounded by $y = x^2 + 1$ and $y = 2x - 2$ between $x = -1$ and $x = 2$.

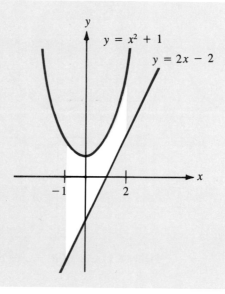

EXAMPLE 2.6

Find the area of the region bounded by the curves $y = x^3$ and $y = x^2$.

Solution

Sketch the curves as shown in Figure 6.11a.

Find the points of intersection by solving the equations of the two curves simultaneously to get

$$x^3 = x^2 \qquad x^3 - x^2 = 0 \qquad x^2(x - 1) = 0$$

or

$$x = 0 \qquad \text{and} \qquad x = 1$$

The corresponding points $(0, 0)$ and $(1, 1)$ are the points of intersection.

Notice that for $0 \leq x \leq 1$, the graph of $y = x^2$ lies above that of $y = x^3$ (since the square of a fraction between 0 and 1 is greater than its cube). Hence, the region in question is bounded above by the curve $y = x^2$ and below by the curve $y = x^3$ and extends from $x = 0$ to $x = 1$. Hence,

$$\text{Area} = \int_0^1 (x^2 - x^3)\, dx = \left(\frac{1}{3}x^3 - \frac{1}{4}x^4\right)\Big|_0^1 = \frac{1}{12}$$

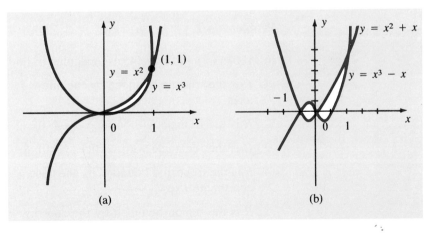

Figure 6.11 The area between intersecting curves. (a) The region bounded by $y = x^3$ and $y = x^2$. (b) The region bounded by $y = x^3 - x$ and $y = x^2 + x$ from $x = -1$ to $x = 2$.

EXAMPLE 2.7

Find the area of the region bounded by the curves $y = x^3 - x$ and $y = x^2 + x$.

Solution

First make a rough sketch of the two curves as shown in Figure 6.11b. You find the points of intersection by solving the equations of the two curves simultaneously to get

$$x^3 - x = x^2 + x \qquad x^3 - x^2 - 2x = 0 \qquad x(x + 1)(x - 2) = 0$$

or $x = 0$, $x = -1$, and $x = 2$. The region whose area you wish to compute lies between $x = -1$ and $x = 2$, but since the two curves cross at $x = 0$, neither curve is always above the other between $x = -1$ and $x = 2$. However, since the curve $y = x^3 - x$ is above $y = x^2 + x$ between $x = -1$ and $x = 0$, and since $y = x^2 + x$ is above $y = x^3 - x$ for $x = 0$ to $x = 2$, it follows that

$$\begin{bmatrix} \text{Area from} \\ x = -1 \text{ to } x = 0 \end{bmatrix} = \int_{-1}^{0} [(x^3 - x) - (x^2 + x)]dx = \int_{-1}^{0} (x^3 - x^2 - 2x)\, dx = \frac{5}{12}$$

and $\quad \begin{bmatrix} \text{Area from} \\ x = 0 \text{ to } x = 2 \end{bmatrix} = \int_{0}^{2} [(x^2 + x) - (x^3 - x)]dx = \int_{0}^{2} (-x^3 + x^2 + 2x)\, dx = \frac{8}{3}$

Thus, the *total area* between the two curves (between $x = -1$ and $x = 2$) is the sum

$$\begin{bmatrix} \text{Area between} \\ \text{the curves} \end{bmatrix} = \begin{bmatrix} \text{Area from} \\ x = -1 \text{ to } x = 0 \end{bmatrix} + \begin{bmatrix} \text{Area from} \\ x = 0 \text{ to } x = 2 \end{bmatrix}$$

$$= \frac{5}{12} + \frac{8}{3} = \frac{37}{12}$$

Problems 6.2

In Problems 1 through 14, use calculus to find the area of the region R.

1. R is the triangle bounded by the line $y = 4 - 3x$ and the coordinate axes.

2. R is the triangle with vertices $(-4, 0)$, $(2, 0)$, and $(2, 6)$.

3. R is the rectangle with vertices $(1, 0)$, $(-2, 0)$, $(-2, 5)$, and $(1, 5)$.

4. R is the trapezoid bounded by the lines $y = x + 6$ and $x = 2$ and the coordinate axes.

5. R is the region bounded by the curve $y = \sqrt{x}$, the lines $x = 4$ and $x = 9$, and the x axis.

6. R is the region bounded by the curve $y = 4x^3$, the line $x = 2$, and the x axis.

7. R is the region bounded by the curve $y = 1 - x^2$ and the x axis.

8. R is the region bounded by the curve $y = -x^2 - 6x - 5$ and the x axis.

9. R is the region bounded by the curve $y = e^x$, the lines $x = 0$ and $x = \ln\frac{1}{2}$, and the x axis.

10. R is the region in the first quadrant bounded by the curve $y = 4 - x^2$ and the lines $y = 3x$ and $y = 0$.

11. R is the region bounded by the curve $y = \sqrt{x}$ and the lines $y = 2 - x$ and $y = 0$.

12. R is the region in the first quadrant that lies under the curve $y = \frac{16}{x}$ and that is bounded by this curve and the lines $y = x$, $y = 0$, and $x = 8$.

13. R is the region that lies below the curve $y = x^2 + 4$ and is bounded by this curve, the line $y = -x + 10$, and the coordinate axes.

14. R is the region bounded by the curve $y = x^2 - 2x$ and the x axis. (*Hint*: Reflect the region across the x axis and integrate the corresponding function.)

Product Reliability   15. The probability density function for the life span of the light bulbs manufactured by a certain company is $f(x) = 0.01e^{-0.01x}$, where x denotes the life span (in hours) of a randomly selected bulb.
    (a) What is the probability that the life span of a randomly selected bulb will be between 50 and 60 hours?
    (b) What is the probability that the life span of a randomly selected bulb will be less than or equal to 60 hours?
    (c) What is the probability that the life span of a randomly selected bulb will be greater than 60 hours?

Product Reliability   16. The probability density function for the life of an appliance is $f(x) = 0.2e^{-0.2x}$, where x denotes the life (in months) of a randomly selected appliance.
    (a) What is the probability that a randomly selected appliance will last between 10 and 15 months?
    (b) What is the probability that a randomly selected appliance will last less than 8 months?
    (c) What is the probability that a randomly selected appliance will last longer than 1 year?

Duration of Telephone   17. The probability density function for the duration of telephone calls in
Calls   a certain city is $f(x) = 0.5e^{-0.5x}$, where x denotes the duration (in minutes) of a randomly selected call.

(a) What percentage of the calls can be expected to last between 2 and 3 minutes?

(b) What percentage of the calls can be expected to last 2 minutes or less?

(c) What percentage of the calls can be expected to last more than 2 minutes?

Airplane Arrivals 18. The probability density function for the time interval between the arrivals of successive planes at a certain airport is $f(x) = 0.2e^{-0.2x}$, where x is the time (in minutes) between the arrivals of a randomly selected pair of successive planes.

(a) What is the probability that two successive planes selected at random will arrive within 5 minutes of each other?

(b) What is the probability that two successive planes selected at random will arrive more than 6 minutes apart?

In Problems 19 through 33, find the area of the region R.

19. R is the region bounded by the curves $y = x^2 + 3$ and $y = 1 - x^2$ between $x = -2$ and $x = 1$.

20. R is the region bounded by the curves $y = x^2 + 5$ and $y = -x^2$, the line $x = 3$, and the y axis.

21. R is the region bounded by the curve $y = e^x$ and the lines $y = 1$ and $x = 1$.

22. R is the region bounded by the curve $y = x^2$ and the line $y = x$.

23. R is the region bounded by the curve $y = x^2$ and the line $y = 4$.

24. R is the region bounded by the curves $y = \sqrt{x}$ and $y = x^2$.

25. (a) R is the region to the right of the y axis that is bounded above by the curve $y = 4 - x^2$ and below by the line $y = 3$.

(b) R is the region to the right of the y axis that lies below the line $y = 3$ and is bounded by the curve $y = 4 - x^2$, the line $y = 3$, and the coordinate axes.

26. R is the region bounded by the curves $y = x^3 - 6x^2$ and $y = -x^2$.

27. R is the region bounded by the line $y = x$ and the curve $y = x^3$.

28. R is the region bounded by the curve $y = \dfrac{1}{x^2}$ and the lines $y = x$ and $y = \dfrac{x}{8}$.

29. R is the region in the first quadrant bounded by the curve $y = x^2 + 2$ and the lines $y = 11 - 8x$ and $y = 11$.

30. R is the region bounded by the curves $y = x^2 - 2x$ and $y = -x^2 + 4$.

31. R is the region bounded by the curves $y = x^2 - 3x + 1$ and $y = -x^2 + 2x + 2$.

32. R is the region bounded by the curves $y = x^3 - x$ and $y = -x^2 + x$.

33. R is the region bounded by the curves $y = x^3 - 2x^2 + 5$ and $y = x^2 + 4x - 7$.

skip ③ **Applications to Business and Economics**

In this section you will see four examples from business and economics in which quantities can be expressed as definite integrals and represented geometrically as areas between or under curves. The first two examples deal with the general concept of **net gain,** while the last two involve consumers' willingness to spend and a related concept known as **consumers' surplus.**

Net Excess Profit

Suppose that x years from now, two investment plans will be generating profit at the rates of $R_1(x)$ and $R_2(x)$ dollars per year, respectively, and that for the next N years the rate $R_2(x)$ will be greater than the rate $R_1(x)$, as illustrated in Figure 6.12.

Figure 6.12 Net excess profit as the area between two curves.

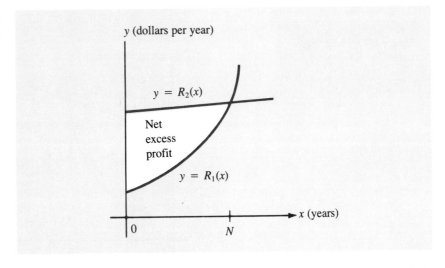

The functions $R_2(x)$ and $R_1(x)$ represent the rates at which profit is generated by the second and first plans, respectively. Hence, their difference, $R_2(x) - R_1(x)$, represents the rate at which the profit generated by the second plan exceeds that generated by the first plan, and the net *excess* profit generated by the second plan during the next N years is the definite integral of this rate of change from $x = 0$ to $x = N$. That is,

$$\text{Net excess profit} = \int_0^N [R_2(x) - R_1(x)] \, dx$$

which can be interpreted geometrically as the area between the curves $y = R_2(x)$ and $y = R_1(x)$ from $x = 0$ to $x = N$. Here is an example.

EXAMPLE 3.1

Suppose that x years from now, one investment plan will be generating profit at the rate of $R_1(x) = 50 + x^2$ dollars per year, while a second plan will be generating profit at the rate of $R_2(x) = 200 + 5x$ dollars per year.

(a) For how many years will the second plan be the more profitable one?
(b) Compute your net excess profit if you invest in the second plan instead of the first for the period of time in part (a).
(c) Interpret the net excess profit from part (b) as the area between two curves.

Solution

To help you visualize the situation, begin by sketching the curves $y = R_1(x)$ and $y = R_2(x)$ as shown in Figure 6.13.

Figure 6.13 Net excess profit for Example 3.1.

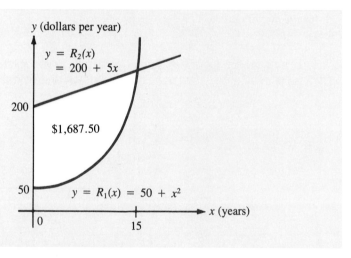

(a) As the graph indicates, the rate $R_2(x)$ at which the second plan generates profit is initially greater than the rate $R_1(x)$ at which the first plan generates profit. The second plan will be the more profitable one until $R_1(x) = R_2(x)$, that is, until

$$50 + x^2 = 200 + 5x \qquad x^2 - 5x - 150 = 0$$

$$(x - 15)(x + 10) = 0 \qquad \text{or} \qquad x = 15 \text{ years}$$

(b) For $0 \leq x \leq 15$, the rate at which the profit generated by the second plan exceeds that of the first plan is $R_2(x) - R_1(x)$ dollars per year. Hence, the net excess profit generated over the 15-year period by the second plan is the definite integral

$$\int_0^{15} [R_2(x) - R_1(x)] \, dx = \int_0^{15} [(200 + 5x) - (50 + x^2)] \, dx$$

$$= \int_0^{15} (150 + 5x - x^2) \, dx$$

$$= \left(150x + \frac{5}{2}x^2 - \frac{1}{3}x^3 \right) \Big|_0^{15} = \$1,687.50$$

(c) In geometric terms, the definite integral giving the net excess profit in part (b) is the area of the shaded region in Figure 6.13 between the curves $y = R_2(x)$ and $y = R_1(x)$ from $x = 0$ to $x = 15$.

Net Earnings from Industrial Equipment

The net earnings generated by an industrial machine over a period of time is the difference between the total revenue generated by the machine and the total cost of operating and servicing the machine. In the next example, the net earnings of a machine are calculated as a definite integral and interpreted as the area between two curves.

EXAMPLE 3.2

Suppose that when it is x years old, an industrial machine generates revenue at the rate of $R(x) = 5,000 - 20x^2$ dollars per year and results in costs that accumulate at the rate of $C(x) = 2,000 + 10x^2$ dollars per year.

(a) For how many years is the use of the machine profitable?
(b) What are the net earnings generated by the machine during the period of time in part (a)?
(c) Interpret the net earnings in part (b) as the area between two curves.

Solution

To help you visualize the situation, begin by sketching the curves $y = R(x)$ and $y = C(x)$ as shown in Figure 6.14.

(a) Use of the machine will be profitable as long as the rate at which revenue is generated is greater than the rate at which costs accumulate, that is, until $R(x) = C(x)$ or

$$5,000 - 20x^2 = 2,000 + 10x^2 \qquad 30x^2 = 3,000 \qquad x^2 = 100$$

or

$$x = 10 \text{ years}$$

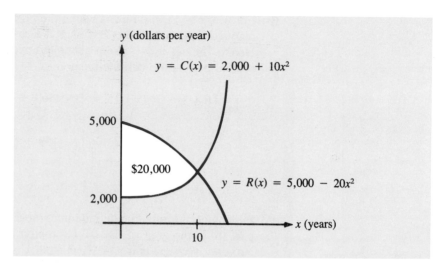

Figure 6.14 The net earnings from an industrial machine.

(b) The functions $R(x)$ and $C(x)$ represent the rates of change of total revenue and total cost, respectively, and hence their difference, $R(x) - C(x)$, represents the rate of change of the net earnings generated by the machine. It follows that the net earnings for the next 10 years are represented by the definite integral

$$\int_0^{10} [R(x) - C(x)] \, dx = \int_0^{10} [(5{,}000 - 20x^2) - (2{,}000 + 10x^2)] \, dx$$

$$= \int_0^{10} (3{,}000 - 30x^2) \, dx$$

$$= (3{,}000x - 10x^3)\Big|_0^{10} = \$20{,}000$$

(c) In geometric terms, the definite integral representing the net earnings is the area of the shaded region in Figure 6.14 between the curves $y = R(x)$ and $y = C(x)$ from $x = 0$ to $x = 10$.

The Consumers' Demand Curve and Willingness to Spend

In studying consumer behavior, economists often assume that the price a consumer or group of consumers is willing to pay to buy an additional unit of a commodity is a function of the number of units of the commodity that the consumer or group has already bought.

For example, a young couple on a limited budget might be willing to spend up to $500 to own one television set. For the convenience of having two sets (to eliminate the conflict between *Monday Night Football* and the network news, for example), the couple might be willing to spend an additional $300 to buy a second set. Since there would be very little use for more than two sets, the couple might be willing to spend no more than $50 for a third set.

A function $p = D(q)$ giving the price per unit that consumers are willing to pay to get the qth unit of a commodity is known in economics as the **consumers' demand function** for the commodity. As in Figure 6.15, the consumers' demand function is usually a decreasing function of q. That is, the price that consumers are willing to pay to get one additional unit usually decreases as the number of units already bought increases. A typical consumers' demand function is sketched in Figure 6.15a.

Figure 6.15 A consumers' demand function and the total amount consumers are willing to spend.

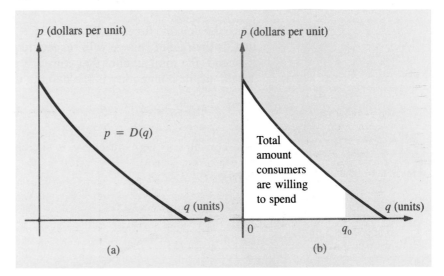

(a)

(b)

The consumers' demand function $p = D(q)$ giving the price per unit that consumers are willing to pay when the level of consumption is q units can be thought of as the rate of change with respect to q of the *total* amount consumers are willing to spend for q units. [In economic terminology, $D(q)$ is the **marginal willingness to spend**.] If $A(q)$ is the total amount (in dollars) that consumers are willing to spend to get q units of the commodity [and if $A(q)$ is differentiable], then $D(q) = \dfrac{dA}{dq}$. Hence, the total amount that consumers are willing to spend to get q_0 units of the commodity is the definite integral

$$A(q_0) = A(q_0) - A(0) = \int_0^{q_0} \frac{dA}{dq}\, dq = \int_0^{q_0} D(q)\, dq$$

In geometric terms, this total willingness to spend is the area of the region in Figure 6.15b under the demand curve $p = D(q)$ between $q = 0$ and $q = q_0$.

EXAMPLE 3.3

Suppose the consumers' demand function for a certain commodity is $D(q) = 4(25 - q^2)$ dollars per unit.

(a) Find the total amount of money consumers are willing to spend to get 3 units of the commodity.

(b) Sketch the demand curve and interpret the answer to part (a) as an area.

Solution

(a) Since the demand function $D(q) = 4(25 - q^2)$, measured in dollars per unit, is the rate of change with respect to q of consumers' willingness to spend, the total amount that consumers are willing to spend to get 3 units of the commodity is the definite integral

$$\int_0^3 D(q)\, dq = 4 \int_0^3 (25 - q^2)\, dq = 4\left(25q - \frac{1}{3}q^3\right)\Big|_0^3 = \$264$$

(b) The consumers' demand curve is sketched in Figure 6.16. In geo-

Figure 6.16 The consumers' demand curve $p = 4(25 - q^2)$ and consumers' willingness to spend for 3 units.

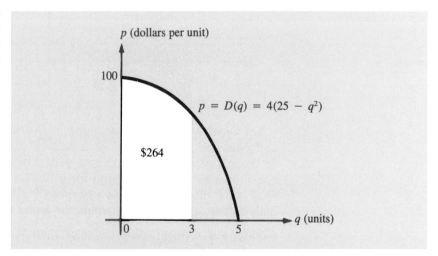

metric terms, the total amount, \$264, that consumers are willing to spend to get 3 units of the commodity is the area under the demand curve from $q = 0$ to $q = 3$.

Consumers' Surplus   In a competitive economy, the total amount that consumers actually spend on a commodity is usually less than the total amount they would have been willing to spend. The difference between the two amounts can be thought of as a savings realized by consumers and is known in economics as the **consumers' surplus.** That is,

$$\begin{bmatrix} \text{Consumers'} \\ \text{surplus} \end{bmatrix} = \begin{pmatrix} \text{total amount consumers} \\ \text{would be willing to spend} \end{pmatrix} - \begin{pmatrix} \text{actual consumer} \\ \text{expenditure} \end{pmatrix}$$

Market conditions determine the price per unit at which a commodity is sold. Once the price, say p_0, is known, the demand equation $p = D(q)$ determines the number of units q_0 that consumers will buy. The actual consumer expenditure for q_0 units of the commodity at the price of p_0 dollars per unit is $p_0 q_0$ dollars. The consumers' surplus is calculated by subtracting this amount from the total amount consumers would have been willing to spend to get q_0 units of the commodity.

To get a better feel for the concept of consumers' surplus, consider once again the example of the couple that was willing to spend $500 for its first television set, $300 for a second set, and $50 for a third set. Suppose the market price for television sets is $300 per set. Then the couple would buy only two sets and would spend a total of $2 \times \$300 = \$600$. This is less than the $\$500 + \$300 = \$800$ that the couple would have been willing to spend to get the two sets. The savings of $\$800 - \$600 = \$200$ is the couple's consumers' surplus.

Consumers' surplus has a simple geometric interpretation, which is illustrated in Figure 6.17. The symbols p_0 and q_0 denote the market price

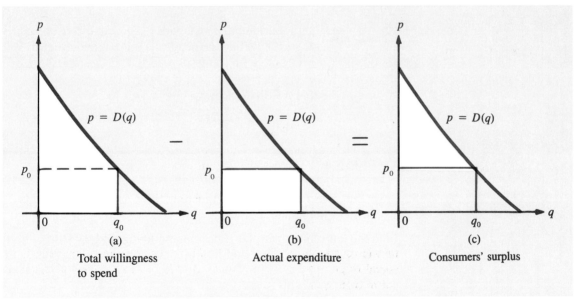

Figure 6.17 Geometric interpretation of consumers' surplus.

and corresponding demand, respectively. Figure 6.17a shows the region under the demand curve from $q = 0$ to $q = q_0$. Its area, as we have seen, represents the total amount that consumers are willing to spend to get q_0 units of the commodity. The rectangle in Figure 6.17b has an area of $p_0 q_0$ and hence represents the actual consumer expenditure for q_0 units at p_0 dollars per unit. The difference between these two areas (Figure 6.17c) represents the consumers' surplus. That is, consumers' surplus is the area of the region between the demand curve $p = D(q)$ and the horizontal line $p = p_0$ and hence is equal to the definite integral

$$\int_0^{q_0} [D(q) - p_0]\, dq = \int_0^{q_0} D(q)\, dq - \int_0^{q_0} p_0\, dq$$

$$= \int_0^{q_0} D(q)\, dq - p_0 q \Big|_0^{q_0}$$

$$= \int_0^{q_0} D(q)\, dq - p_0 q_0$$

Here is a summary of the situation.

Consumers' Surplus

If q_0 units of a commodity are sold at a price of p_0 dollars per unit and if $p = D(q)$ is the consumers' demand function for the commodity, then

$$\begin{bmatrix} \text{Consumers'} \\ \text{surplus} \end{bmatrix} = \begin{bmatrix} \text{total amount consumers} \\ \text{are willing to spend} \\ \text{for } q_0 \text{ units} \end{bmatrix} - \begin{bmatrix} \text{actual consumer} \\ \text{expenditure for} \\ q_0 \text{ units} \end{bmatrix}$$

$$= \int_0^{q_0} D(q)\, dq - p_0 q_0$$

EXAMPLE 3.4

Suppose the consumers' demand function for a certain commodity is $D(q) = 4(25 - q^2)$ dollars per unit.

(a) Find the consumers' surplus if the commodity is sold for $64 per unit.
(b) Sketch the demand curve and interpret the consumers' surplus as an area.

Solution

(a) First find the number of units that will be bought by solving the demand equation $p = D(q)$ for q when $p = \$64$ to get

$$64 = 4(25 - q^2) \qquad 16 = 25 - q^2 \qquad q^2 = 9 \qquad \text{or} \qquad q = 3$$

That is, 3 units will be bought when the price is $64 per unit. The corresponding consumers' surplus is

$$\int_0^3 D(q) \, dq - 64(3) = 4 \int_0^3 (25 - q^2) \, dq - 192$$

$$= 4 \left(25q - \frac{1}{3}q^3 \right) \Big|_0^3 - 192 = 264 - 192 = \$72$$

(b) The consumers' demand curve is sketched in Figure 6.18. The con-

Figure 6.18 The consumers' demand curve $p = 4(25 - q^2)$ and corresponding consumers' surplus when the price is $64 per unit.

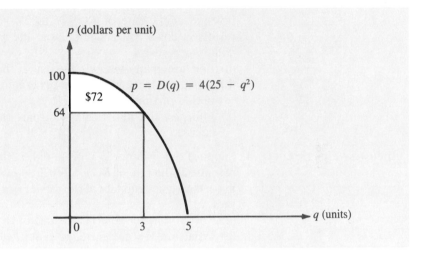

sumers' surplus from part (a) is equal to the area of the region between the demand curve and the horizontal line $p = 64$.

Problems 6.3

Investment

1. Suppose that x years from now, one investment plan will be generating profit at the rate of $R_1(x) = 100 + x^2$ dollars per year, while a second plan will be generating profit at the rate of $R_2(x) = 220 + 2x$ dollars per year.
   (a) For how many years will the second plan be the more profitable one?
   (b) How much excess profit will you earn if you invest in the second plan instead of the first for the period of time in part (a)?
   (c) Interpret the excess profit from part (b) as the area between two curves.

Investment

2. Suppose that x years from now, one investment plan will be generating profit at the rate of $R_1(x) = 60e^{0.12x}$ dollars per year, while a second plan will be generating profit at the rate of $R_2(x) = 160e^{0.08x}$ dollars per year.

    (a) For how many years will the second plan be the more profitable one?

    (b) How much excess profit will you earn if you invest in the second plan instead of the first for the period of time in part (a)?

    (c) Interpret the excess profit from part (b) as the area between two curves.

Industrial Equipment   3. Suppose that when it is x years old, an industrial machine generates revenue at the rate of $R(x) = 6{,}025 - 10x^2$ dollars per year and results in costs that accumulate at the rate of $C(x) = 4{,}000 + 15x^2$ dollars per year.

    (a) For how many years is the use of the machine profitable?

    (b) What are the net earnings generated by the machine during the period of time in part (a)?

    (c) Interpret the net earnings in part (b) as the area between two curves.

Industrial Equipment   4. Suppose that when it is x years old, an industrial machine generates revenue at the rate of $R(x) = 6{,}025 - 8x^2$ dollars per year and results in costs that accumulate at the rate of $C(x) = 4{,}681 + 13x^2$ dollars per year.

    (a) For how many years is the use of the machine profitable?

    (b) What are the net earnings generated by the machine during its period of profitability?

    (c) Interpret the net earnings in part (a) as the area between two curves.

Efficiency   5. After t hours on the job, one factory worker is producing $Q_1(t) = 60 - 2(t - 1)^2$ units per hour, while a second worker is producing $Q_2(t) = 50 - 5t$ units per hour.

    (a) If both arrive on the job at 8:00 A.M., how many more units will the first worker have produced by noon than the second worker?

    (b) Interpret the answer in part (a) as the area between two curves.

Fund-Raising   6. It is estimated that t weeks from now, contributions in response to a fund-raising campaign will be coming in at the rate of $R(t) = 5{,}000e^{-0.2t}$ dollars per week, while campaign expenses are expected to accumulate at the constant rate of \$676 per week.

    (a) For how many weeks will the campaign be profitable?

    (b) What net earnings will be generated by the campaign during the period of time in part (a)?

    (c) Interpret the net earnings in part (b) as the area between two curves.

Fund-Raising   7. It is estimated that t weeks from now, contributions to a charity will be coming in at the rate of $R(t) = 6{,}537e^{-0.3t}$ dollars per week, while expenses are expected to accumulate at the constant rate of \$593 per week.

(a) For how many weeks will the fund-raising campaign be profitable?

(b) What net earnings will be generated by the campaign during the period of time in part (a)?

(c) Interpret the net earnings in part (a) as the area between two curves.

Consumers' Willingness to Spend   For the consumers' demand functions $D(q)$ in Problems 8 through 13

(a) Find the total amount of money consumers are willing to spend to get q_0 units of the commodity.

(b) Sketch the demand curve and interpret the consumers' willingness to spend in part (a) as an area.

8. $D(q) = 2(64 - q^2)$ dollars per unit; $q_0 = 6$ units

9. $D(q) = \dfrac{300}{(0.1q + 1)^2}$ dollars per unit; $q_0 = 5$ units

10. $D(q) = \dfrac{400}{0.5q + 2}$ dollars per unit; $q_0 = 12$ units

11. $D(q) = \dfrac{300}{4q + 3}$ dollars per unit; $q_0 = 10$ units.

12. $D(q) = 40e^{-0.05q}$ dollars per unit; $q_0 = 10$ units

13. $D(q) = 50e^{-0.04q}$ dollars per unit; $q_0 = 15$ units.

Consumers' Surplus   For the consumers' demand functions $D(q)$ in Problems 14 through 19

(a) Find the consumers' surplus if the market price of the commodity is p_0 dollars per unit.

(b) Sketch the demand curve and interpret the consumers' surplus in part (a) as an area.

14. $D(q) = 2(64 - q^2)$ dollars per unit; $p_0 = \$110$ per unit

15. $D(q) = 150 - 2q - 3q^2$ dollars per unit; $p_0 = \$117$ per unit

16. $D(q) = \dfrac{300}{(0.1q + 1)^2}$ dollars per unit; $p_0 = \$12$ per unit

17. $D(q) = \dfrac{400}{0.5q + 2}$ dollars per unit; $p_0 = \$20$ per unit

18. $D(q) = 40e^{-0.05q}$ dollars per unit; $p_0 = \$11.46$ per unit

19. $D(q) = 30e^{-0.04q}$ dollars per unit; $p_0 = \$10$ per unit

Consumers' Surplus   20. Parts for a piece of heavy machinery are sold by the manufacturer in units of 1,000. The dollar price per unit q is given by $p = 110 - q$, and the total cost of producing those q units is $C(q) = q^3 - 25q^2 + 2q + 3,000$.

(a) For what value of q is the manufacturer's profit maximized?

(b) Find the consumers' surplus at the price that corresponds to maximum profit.

Consumers' Surplus   21. Repeat Problem 20 with $p = 124 - 2q$ and $C(q) = 2q^3 - 59q^2 + 4q + 7,600$.

Consumers' Surplus   22. Suppose that the demand function for a certain commodity is $D(q) = 32 - q^2$ and that the supply function for the same commodity is $S(q) = \frac{1}{3}q^2 + 2q + 5$ (this is the price at which q units will be supplied to the marketplace). If the quantity sold and the corresponding price are determined so that supply equals demand, find the corresponding consumers' surplus.

Consumers' Surplus   23. Repeat Problem 22 for a commodity whose demand and supply functions are $D(q) = \dfrac{16}{q + 2} - 3$ and $S(q) = \frac{1}{3}(q + 1)$.

4 The Definite Integral as the Limit of a Sum

In Section 1, we defined the definite integral of a function to be a certain net change in its antiderivative. In particular,

$$\int_a^b f(x)\,dx = F(b) - F(a)$$

where F is an antiderivative of f. In this section we shall introduce an important relationship between antiderivatives and limits of sums known as the **fundamental theorem of calculus.** By combining this theorem with our definition of the definite integral, we shall be able to interpret certain limits of sums arising in practical problems as definite integrals, which we can then evaluate by antidifferentiation.

The discussion begins with a geometric derivation of the theorem based on the interpretation of definite integrals as areas.

Geometric Derivation of the Fundamental Theorem of Calculus   Suppose that $f(x)$ is nonnegative and continuous on the interval $a \leq x \leq b$. You can approximate the area under the graph of f between $x = a$ and $x = b$ as follows: First divide the interval $a \leq x \leq b$ into n equal subintervals of width Δx and let x_j denote the beginning of the jth subinterval. Then draw n rectangles such that the base of the jth rectangle is the jth subinterval and the height of the jth rectangle is $f(x_j)$. The situation is illustrated in Figure 6.19.

The area of the jth rectangle is $f(x_j)\,\Delta x$ and is an approximation to the area under the curve from $x = x_j$ to $x = x_{j+1}$. The sum of the areas of all n rectangles is

$$f(x_1)\,\Delta x + f(x_2)\,\Delta x + \cdots + f(x_n)\,\Delta x$$

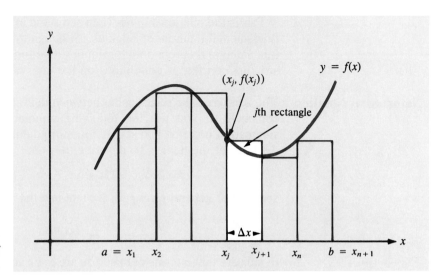

Figure 6.19 An approximation by rectangles of the area under a curve.

This sum is an approximation to the total area under the curve between $x = a$ and $x = b$ and hence is an approximation to the corresponding definite integral

$$\int_a^b f(x)\ dx$$

That is, $f(x_1)\ \Delta x + f(x_2)\ \Delta x + \cdots + f(x_n)\ \Delta x \simeq \int_a^b f(x)\ dx$

As Figure 6.20 suggests, the sum of the areas of the rectangles ap-

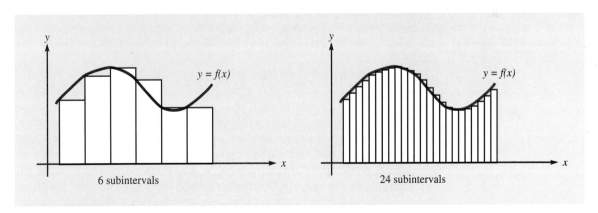

6 subintervals

24 subintervals

Figure 6.20 The approximation improves as the number of subintervals increases.

proaches the actual area under the curve as the number of rectangles increases without bound. That is, as n increases without bound,

$$f(x_1)\ \Delta x + f(x_2)\ \Delta x + \cdots + f(x_n)\ \Delta x \to \int_a^b f(x)\ dx = F(b) - F(a)$$

where F is an antiderivative of f.

This is the relationship between sums and integrals that is known as the fundamental theorem of calculus. Although this argument establishes the fundamental theorem only for nonnegative functions, it actually holds for any function that is continuous on the interval $a \leq x \leq b$.

Summation Notation

You can write the relationship between definite integrals and sums more compactly if you use the following **summation notation.** (The use of summation notation is discussed in more detail in the Algebra Review at the back of the book.) To describe the sum

$$f(x_1) \, \Delta x + f(x_2) \, \Delta x + \cdots + f(x_n) \, \Delta x$$

specify the general term $f(x_j) \, \Delta x$ and use the symbol

$$\sum_{j=1}^{n} f(x_j) \, \Delta x$$

to indicate that n terms of this form are to be added together, starting with the term in which $j = 1$ and ending with the term in which $j = n$. Thus,

$$\sum_{j=1}^{n} f(x_j) \, \Delta x = f(x_1) \, \Delta x + f(x_2) \, \Delta x + \cdots + f(x_n) \, \Delta x$$

and the fundamental theorem of calculus can be stated compactly as follows.

The Fundamental Theorem of Calculus

Suppose f is continuous on the interval $a \leq x \leq b$, which is divided into n equal subintervals of length Δx by $x_1, x_2, \ldots, x_n$, where x_j is the left-hand endpoint of the jth subinterval. Then,

$$\lim_{n \to \infty} \sum_{j=1}^{n} f(x_j) \, \Delta x = \int_{a}^{b} f(x) \, dx = F(b) - F(a)$$

where F is any antiderivative of f.

Perspective: Actually, this is a somewhat restricted version of a more general characterization of definite integrals. The relationship between definite integrals and limits of sums is still valid if $f(x_j)$ in the jth term of the sum is replaced by $f(x_j^*)$, where x_j^* is any point whatsoever in the jth subinterval. Moreover, the n subintervals need not have equal width as long as the width of the largest eventually approaches zero as n increases. For most applications, however, the restricted characterization is sufficient, and you will have no need to use the more general result.

Applications Involving the Definite Integral as the Limit of a Sum

Next, we shall examine several applications that involve the characterization of the definite integral as the limit of a sum. The first two applications involve an integral formula for the amount of an income stream, and the others come from the social sciences, biology, and geometry. Here is a summary of the general procedure used in these applications.

Applications Involving Limits of Sums

The following steps often lead to the numerical value of a quantity associated with an interval $a \leq x \leq b$.

Step 1. Divide the interval $a \leq x \leq b$ into n equal subintervals of length Δx and let x_j denote the left-hand endpoint of the jth subinterval.

Step 2. Approximate the contribution to the total quantity that comes from the jth subinterval.

Step 3. Add the approximations from Step 2 to approximate the total quantity.

Step 4. Take the limit as n increases without bound to pass from the approximation to the exact value of the quantity.

Step 5. Use the fact that

$$\lim_{n \to \infty} \sum_{j=1}^{n} f(x_j) \, \Delta x = \int_{a}^{b} f(x) \, dx$$

to transform the limit of the sum in Step 4 to a definite integral.

Step 6. Evaluate the definite integral using antidifferentiation.

The Amount of an Income Stream

Suppose that a stream of income is transferred continuously into an account in which it earns interest over a specified period of time (or **term**). The **amount** (or **future value**) of the income stream is the total amount (money transferred into the account plus interest) that is accumulated in this way during the specified term.

The calculation of the amount of an income stream is illustrated in the next example. The strategy is to approximate the continuous income stream by a sequence of discrete deposits (i.e., by an **annuity**). The amount of the approximating annuity is a certain sum whose limit (a definite integral) is the amount of the income stream. (The amount of an annuity was calculated in Example 5.6 of Chapter 4, and before proceeding, you may wish to reread that example.)

EXAMPLE 4.1

Money is transferred continuously into an account at the constant rate of $1,200 per year. The account earns interest at the annual rate of 8 percent compounded continuously. How much will be in the account at the end of 2 years?

Solution

Recall from Chapter 4 that P dollars invested at 8 percent compounded continuously will be worth $Pe^{0.08t}$ dollars t years later.

To approximate the future value of the income stream, divide the 2-year time interval $0 \leq t \leq 2$ into n equal subintervals of length Δt years and let t_j denote the beginning of the jth subinterval. Then, during the jth subinterval (of length Δt years),

$$\text{Money deposited} = (\text{dollars per year})(\text{number of years}) = 1{,}200\Delta t$$

If all this money were deposited at the beginning of the subinterval (at time t_j), it would remain in the account for $2 - t_j$ years and hence would grow to $(1{,}200 \, \Delta t)e^{0.08(2-t_j)}$ dollars. Thus,

$$\begin{array}{c}\text{Future value of money deposited} \\ \text{during } j\text{th subinterval}\end{array} \simeq 1{,}200e^{0.08(2-t_j)} \, \Delta t$$

The situation is illustrated in Figure 6.21.

Figure 6.21 The (approximate) future value of the money deposited during the jth subinterval.

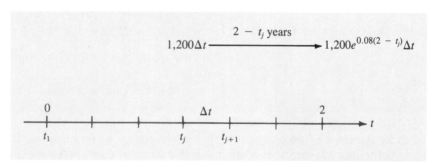

The future value of the entire income stream is the sum of the future values of the money deposited during each of the n subintervals. Hence,

$$\text{Future value of income stream} \simeq \sum_{j=1}^{n} 1{,}200e^{0.08(2-t_j)} \, \Delta t$$

(Note that this is only an approximation because it is based on the assumption that all $1{,}200\Delta t$ dollars are deposited at time t_j rather than continuously throughout the jth subinterval.)

As n increases without bound, the length of each subinterval approaches zero and the approximation approaches the true future value of the income stream. Hence,

$$\begin{array}{c}\text{Future value of} \\ \text{income stream}\end{array} = \lim_{n \to \infty} \sum_{j=1}^{n} 1{,}200e^{0.08(2-t_j)} \, \Delta t$$

$$= \int_{0}^{2} 1{,}200e^{0.08(2-t)} \, dt$$

$$= 1,200e^{0.16} \int_0^2 e^{-0.08t}\, dt$$

$$= -\frac{1,200}{0.08} e^{0.16} (e^{-0.08t}) \Big|_0^2$$

$$= -15,000e^{0.16}(e^{-0.16} - 1)$$

$$= -15,000 + 15,000e^{0.16}$$

$$= -15,000(1 - e^{0.16})$$

$$\simeq \$2,602.66$$

The Present Value of an Income Stream

The **present value** of an investment or business venture that generates income continuously at a certain rate over a specified period of time is the amount of money that must be deposited today at the prevailing interest rate to generate the same income stream over the same term.

The calculation of the present value of a continuous income stream is illustrated in the next example. Once again, the strategy is to approximate the continuous income stream by a sequence of discrete payments (i.e., by an annuity). The present value of the approximating annuity is a certain sum whose limit (a definite integral) is the present value of the income stream. (The present value of an annuity was calculated in Example 5.7 of Chapter 4, and before proceeding, you may wish to reread that example.)

EXAMPLE 4.2

The management of a national chain of ice-cream parlors is selling a 5-year franchise to operate its newest outlet in Madison, Wisconsin. Past experience in similar localities suggests that t years from now the franchise will be generating profit at the rate of $f(t) = 14,000 + 490t$ dollars per year. If the prevailing annual interest rate remains fixed during the next 5 years at 7 percent compounded continuously, what is the present value of the franchise?

Solution

Recall from Chapter 4 that if the prevailing annual interest rate is 7 percent compounded continuously, the present value of B dollars payable t years from now is $Be^{-0.07t}$.

To approximate the present value of the franchise, divide the 5-year time interval $0 \le t \le 5$ into n equal subintervals of length Δt years and let t_j denote the beginning of the jth subinterval. Then,

$$\text{Profit from the } j\text{th interval} = (\text{dollars per year})(\text{number of years}) \approx f(t_j)\,\Delta t$$

Hence $$\text{Present value of the profit from the } j\text{th subinterval} \approx f(t_j)e^{-0.07t_j}\,\Delta t$$

and $$\text{Present value of franchise} \approx \sum_{j=1}^{n} f(t_j)e^{-0.07t_j}\,\Delta t$$

The situation is illustrated in Figure 6.22.

Figure 6.22 The (approximate) present value of the profit generated during the jth subinterval.

The approximation approaches the true present value as n increases without bound. Thus,

$$\begin{aligned}
\text{Present value of franchise} &= \lim_{n \to \infty} \sum_{j=1}^{n} f(t_j)e^{-0.07t_j}\,\Delta t \\
&= \int_0^5 f(t)e^{-0.07t}\,dt \\
&= \int_0^5 (14{,}000 + 490t)e^{-0.07t}\,dt \\
&= -\frac{1}{0.07}(14{,}000 + 490t)e^{-0.07t}\Big|_0^5 + 7{,}000\int_0^5 e^{-0.07t}\,dt \qquad \text{integration by parts} \\
&= \left[-1{,}000(200 + 7t)e^{-0.07t} - 100{,}000e^{-0.07t}\right]\Big|_0^5 \\
&= -1{,}000(300 + 7t)e^{-0.07t}\Big|_0^5 \\
&= -1{,}000[335e^{-0.35} - 300] \approx \$63{,}929.49
\end{aligned}$$

Survival and Renewal Functions   In the next example, a **survival function** gives the fraction of individuals in a group or population that can be expected to remain in the group for any specified period of time. A **renewal function** giving the rate at which new members arrive is also known, and the goal is to predict the size of the group at some future time. Problems of this type arise in many fields, including sociology, demography, and ecology.

EXAMPLE 4.3

A new county mental health clinic has just opened. Statistics compiled at similar facilities suggest that the fraction of patients who will still be receiving treatment at the clinic t months after their initial visit is given by the function $f(t) = e^{-t/20}$. The clinic initially accepts 300 people for treatment and plans to accept new patients at the rate of 10 per month. Approximately how many people will be receiving treatment at the clinic 15 months from now?

Solution

Since $f(15)$ is the fraction of patients whose treatment continues at least 15 months, it follows that of the current 300 patients, only $300f(15)$ will still be receiving treatment 15 months from now.

To approximate the number of *new* patients who will be receiving treatment 15 months from now, divide the 15-month time interval $0 \leq t \leq 15$ into n equal subintervals of length Δt months and let t_j denote the beginning of the jth subinterval. Since new patients are accepted at the rate of 10 per month, the number of new patients accepted during the jth subinterval is $10\Delta t$. Fifteen months from now, approximately $15 - t_j$ months will have elapsed since these $10\Delta t$ new patients had their initial visits, and so approximately $(10\Delta t)f(15 - t_j)$ of them will still be receiving treatment at that time. (See Figure 6.23.) It follows that the total number

Figure 6.23 New members arriving during the jth subinterval.

of new patients still receiving treatment 15 months from now can be approximated by the sum

$$\sum_{j=1}^{n} 10f(15 - t_j)\,\Delta t$$

Adding this to the number of current patients who will still be receiving treatment in 15 months, you get

$$P \simeq 300f(15) + \sum_{j=1}^{n} 10f(15 - t_j) \, \Delta t$$

where P is the total number of patients who will be receiving treatment 15 months from now.

As n increases without bound, the approximation improves and approaches the true value of P. That is,

$$P = 300f(15) + \lim_{n \to \infty} \sum_{j=1}^{n} 10f(15 - t_j) \, \Delta t$$

$$= 300f(15) + \int_{0}^{15} 10f(15 - t) \, dt$$

Since $f(t) = e^{-t/20}$, it follows that $f(15) = e^{-3/4}$ and $f(15 - t) = e^{-(15-t)/20} = e^{-3/4} e^{t/20}$. Hence,

$$P = 300e^{-3/4} + 10e^{-3/4} \int_{0}^{15} e^{t/20} \, dt$$

$$= 300e^{-3/4} + \left(200e^{-3/4} e^{t/20} \Big|_{0}^{15} \right)$$

$$= 300e^{-3/4} + 200(1 - e^{-3/4})$$

$$= 100e^{-3/4} + 200 \simeq 247.24$$

That is, 15 months from now, the clinic will be treating approximately 247 patients.

The Flow of Blood through an Artery

Biologists have found that the speed of blood in an artery is a function of the distance of the blood from the artery's central axis. According to Poiseuille's law, the speed (in centimeters per second) of blood that is r centimeters from the central axis of the artery is $S(r) = k(R^2 - r^2)$, where R is the radius of the artery and k is a constant. In the next example, you will see how to use this information to compute the rate (in cubic centimeters per second) at which blood passes through the artery.

EXAMPLE 4.4

Find an expression for the rate (in cubic centimeters per second) at which blood flows through an artery of radius R if the speed of the blood r centimeters from the central axis is $S(r) = k(R^2 - r^2)$, where k is a constant.

Solution

To approximate the volume of blood that flows through a cross section of the artery per second, divide the interval $0 \le r \le R$ into n equal subintervals of width Δr centimeters and let r_j denote the beginning of the jth subinterval. These subintervals determine n concentric rings as illustrated in Figure 6.24.

Figure 6.24 The division of the cross section into concentric rings.

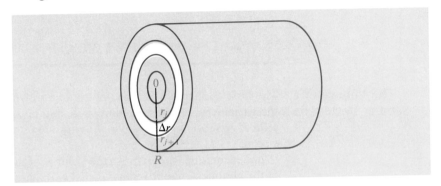

If Δr is small, the area of the jth ring is approximately the area of a rectangle whose length is the circumference of the (inner) boundary of the ring and whose width is Δr. That is,

$$\text{Area of } j\text{th ring} \simeq 2\pi r_j\, \Delta r$$

If you multiply the area of the jth ring (square centimeters) by the speed (centimeters per second) of the blood flowing through this ring, you get the rate (cubic centimeters per second) at which blood flows through the jth ring. Since the speed of blood flowing through the jth ring is approximately $S(r_j)$ centimeters per second, it follows that

$$\begin{pmatrix}\text{Rate of flow} \\ \text{through } j\text{th ring}\end{pmatrix} \simeq \begin{pmatrix}\text{area of} \\ j\text{th ring}\end{pmatrix}\begin{pmatrix}\text{speed of blood} \\ \text{through } j\text{th ring}\end{pmatrix}$$

$$\simeq (2\pi r_j\, \Delta r)S(r_j)$$

$$= (2\pi r_j\, \Delta r)[k(R^2 - r_j^2)]$$

$$= 2\pi k(R^2 r_j - r_j^3)\, \Delta r$$

The rate of flow of blood through the entire cross section is the sum of n such terms, one for each of the n concentric rings. That is,

$$\text{Rate of flow} \simeq \sum_{j=1}^{n} 2\pi k(R^2 r_j - r_j^3)\, \Delta r$$

As n increases without bound, this approximation approaches the true value of the rate of flow. That is,

$$\text{Rate of flow} = \lim_{n \to \infty} \sum_{j=1}^{n} 2\pi k (R^2 r_j - r_j^3) \, \Delta r$$

$$= \int_{0}^{R} 2\pi k (R^2 r - r^3) \, dr$$

$$= 2\pi k \left(\frac{R^2}{2} r^2 - \frac{1}{4} r^4 \right) \Big|_{0}^{R}$$

$$= \frac{\pi k R^4}{2} \quad \text{cubic centimeters per second}$$

The Volume of a Solid of Revolution   The next application is a geometric one in which the characterization of the definite integral as the limit of a sum is used to find the volume of a **solid of revolution** formed by revolving a region R in the xy plane about the x axis.

The technique is to express the volume of the solid as the limit of a sum of the volumes of approximating disks. In particular, suppose that S is the solid (Figure 6.25b) formed by rotating about the x axis the region R (Figure 6.25a) under the curve $y = f(x)$ from $x = a$ to $x = b$.

Divide the interval $a \leq x \leq b$ into n equal subintervals of length Δx, and let x_j denote the beginning of the jth subinterval. Then approximate the region R by the n rectangles shown in Figure 6.26a and the solid S by the corresponding n cylindrical disks (Figure 6.26b) formed by revolving these rectangles about the x axis.

The radius r_j of the jth disk (Figure 6.27b) is the height $f(x_j)$ of the jth rectangle (Figure 6.27a), and so

Figure 6.25 The region under the curve $y = f(x)$ from $x = a$ to $x = b$ and the corresponding solid of revolution.

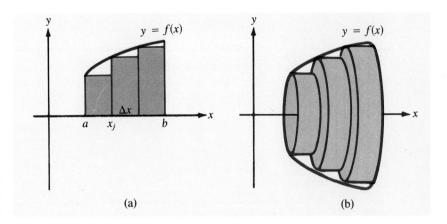

Figure 6.26 The approximation of R by rectangles and the corresponding approximation of S by disks.

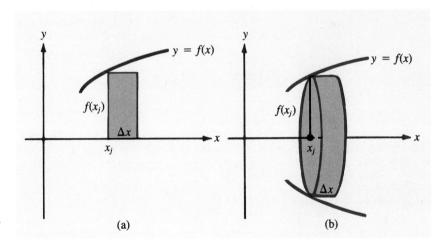

Figure 6.27 The jth rectangle and the jth disk.

Volume of jth disk = (area of circular cross section)(width)

$$= \pi r_j^2(\text{width}) = \pi[f(x_j)]^2 \, \Delta x$$

The total volume of S is approximately the sum of the volumes of the n disks. That is,

$$\text{Volume of } S \simeq \sum_{j=1}^{n} \pi[f(x_j)]^2 \, \Delta x$$

The approximation improves as n increases without bound and

$$\text{Volume of } S = \lim_{n \to \infty} \sum_{j=1}^{n} \pi[f(x_j)]^2 \, \Delta x = \pi \int_{a}^{b} [f(x)]^2 \, dx$$

To summarize:

Volume Formula

Suppose $f(x)$ is continuous and nonnegative on $a \leq x \leq b$ and let R be the region under the curve $y = f(x)$ between $x = a$ and $x = b$. Then the solid S formed by revolving R about the x axis has volume

$$\text{Volume of } S = \pi \int_a^b [f(x)]^2 \, dx$$

Here is an example.

EXAMPLE 4.5

Find the volume of the solid S formed by revolving the region under the curve $y = x^2 + 1$ from $x = 0$ to $x = 2$ about the x axis.

Solution

The region, the solid of revolution, and the jth disk are shown in Figure 6.28.

Figure 6.28 Region, solid of revolution, and jth disk for Example 4.5.

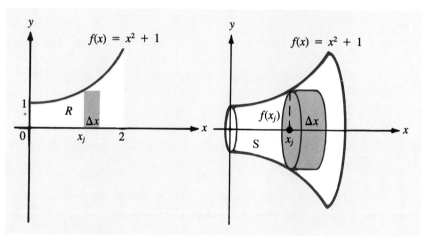

The radius of the jth disk is $f(x_j) = x_j^2 + 1$. Hence,

Volume of jth disk $= \pi[f(x_j)]^2 \, \Delta x = \pi(x_j^2 + 1)^2 \, \Delta x$

and

$$\text{Volume of } S = \lim_{n \to \infty} \sum_{j=1}^{n} \pi(x_j^2 + 1)^2 \, \Delta x$$

$$= \pi \int_0^2 (x^2 + 1)^2 \, dx$$

$$= \pi \int_0^2 (x^4 + 2x^2 + 1) \, dx$$

$$= \pi \left(\frac{1}{5}x^5 + \frac{2}{3}x^3 + x \right)\Big|_0^2 = \frac{206}{15} \pi \simeq 43.14$$

The Average Value of a Function

In many practical situations one is interested in the **average value** of a continuous function over an interval, such as the average level of air pollution over a 24-hour period, the average speed of a truck during a 3-hour trip, the average productivity of a worker during a production run, and the average blood pressure of a patient during an operation. Here is a simple formula involving a definite integral that you can use to compute averages of this type.

The Average Value of a Function

The average value of a continuous function $f(x)$ over the interval $a \leq x \leq b$ is given by the formula

$$\text{Average value} = \frac{1}{b - a} \int_a^b f(x) \, dx$$

Derivation of the Average-Value Formula

To see why the average-value formula is valid, divide the interval $a \leq x \leq b$ into n equal subintervals of length Δx and let x_j denote the beginning of the jth subinterval. The numerical average of the corresponding n function values $f(x_1), f(x_2), \ldots, f(x_n)$ is

$$\frac{f(x_1) + f(x_2) + \cdots + f(x_n)}{n} = \frac{1}{n} \sum_{j=1}^n f(x_j)$$

As n increases without bound, this numerical average becomes increasingly sensitive to fluctuations in f and therefore approximates with increasing accuracy the average value of f over the entire interval $a \leq x \leq b$.

That is, $$\begin{array}{c} \text{Average value of } f \\ \text{over } a \leq x \leq b \end{array} = \lim_{n \to \infty} \frac{1}{n} \sum_{j=1}^n f(x_j)$$

To write this limit of a sum as a definite integral, observe that if the interval $a \leq x \leq b$ is divided into n equal subintervals of length Δx, then $\Delta x = \dfrac{b - a}{n}$. Hence, $\dfrac{1}{n} = \dfrac{\Delta x}{b - a}$ and so

$$\frac{1}{n} \sum_{j=1}^n f(x_j) = \frac{\Delta x}{b - a} \sum_{j=1}^n f(x_j) = \frac{1}{b - a} \sum_{j=1}^n f(x_j) \, \Delta x$$

It follows from the characterization of the definite integral as the limit of a sum that

$$\text{Average value of } f \text{ over } a \le x \le b = \lim_{n \to \infty} \frac{1}{n} \sum_{j=1}^{n} f(x_j)$$

$$= \lim_{n \to \infty} \frac{1}{b - a} \sum_{j=1}^{n} f(x_j) \, \Delta x$$

$$= \frac{1}{b - a} \int_{a}^{b} f(x) \, dx$$

and the average-value formula is proved.

Application of the Average-Value Formula   The use of the integral formula for average value is illustrated in the next example.

EXAMPLE 4.6

For several weeks, the highway department has been recording the speed of freeway traffic flowing past a certain downtown exit. The data suggest that between the hours of 1:00 and 6:00 P.M. on a normal weekday, the speed of traffic at the exit is approximately $S(t) = t^3 - 10.5t^2 + 30t + 20$ miles per hour, where t is the number of hours past noon. Compute the average speed of the traffic between the hours of 1:00 and 6:00 P.M.

Solution

The goal is to find the average value of $S(t)$ on the interval $1 \le t \le 6$. From the integral formula,

$$\text{Average speed} = \frac{1}{6 - 1} \int_{1}^{6} (t^3 - 10.5t^2 + 30t + 20) \, dt$$

$$= \frac{1}{5}\left(\frac{1}{4}t^4 - \frac{10.5}{3}t^3 + 15t^2 + 20t \right)\Big|_{1}^{6}$$

$$= \frac{1}{5}(228 - 31.75) = 39.25 \text{ miles per hour}$$

Geometric Interpretation of Average Value   The integral formula for average value has an interesting geometric interpretation. To see this, multiply both sides of the equation by $b - a$ to get

$$(b - a)(\text{average value}) = \int_{a}^{b} f(x) \, dx$$

If $f(x)$ is nonnegative, the integral on the right-hand side of this equation is equal to the area under the graph of f from $x = a$ to $x = b$. The product on the left-hand side is the area of a rectangle whose width is $b - a$ and whose height is the average value of f over the interval $a \leq x \leq b$. It follows that the average value of $f(x)$ over the interval $a \leq x \leq b$ is equal to the height of the rectangle whose base is this interval and whose area is the same as the area under the graph of f from $x = a$ to $x = b$. The situation is illustrated in Figure 6.29.

Figure 6.29 Geometric interpretation of average value: Area of rectangle = area under curve.

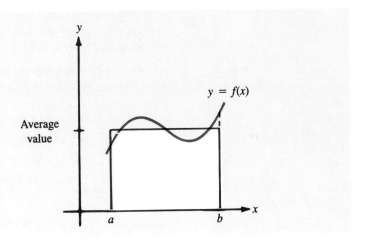

Problems 6.4

Use the characterization of the definite integral as the limit of a sum to solve the following problems.

The Amount of an Income Stream

1. Money is transferred continuously into an account at the constant rate of $2,400 per year. The account earns interest at the annual rate of 6 percent compounded continuously. How much will be in the account at the end of 5 years?

The Amount of an Income Stream

2. Money is transferred continuously into an account at the constant rate of $1,000 per year. The account earns interest at the annual rate of 10 percent compounded continuously. How much will be in the account at the end of 10 years?

The Amount of an Income Stream

3. Money is transferred continuously into an account at the constant rate of $8,000 per year. The account earns interest at the annual rate of 8 percent compounded continuously. How much will be in the account at the end of 3 years?

The Amount of an Income Stream

4. Money is transferred continuously into an account at the constant rate of $6,000 per year. The account earns interest at the annual rate

of 6 percent compounded continuously. How much will be in the account at the end of 10 years?

5. An investment will generate income continuously at the constant rate of $2,400 per year for 5 years. If the prevailing annual interest rate remains fixed at 6 percent compounded continuously, what is the present value of the investment?

6. An investment will generate income continuously at the constant rate of $6,000 per year for 6 years. If the prevailing annual interest rate remains fixed at 8 percent compounded continuously, what is the present value of the investment?

7. An investment will generate income continuously at the constant rate of $1,200 per year for 5 years. If the prevailing annual interest rate remains fixed at 12 percent compounded continuously, what is the present value of the investment?

8. The management of a national chain of pizza parlors is selling a 6-year franchise to operate its newest outlet in Orlando, Florida. Past experience in similar localities suggests that t years from now the franchise will be generating profit at the rate of $f(t) = 10,000 + 500t$ dollars per year. The prevailing rate of interest remains fixed during the next 6 years at 6 percent compounded continuously. What is the present value of the franchise?

9. The management of a national chain of fast-food outlets is selling a 10-year franchise in Cleveland, Ohio. Past experience in similar localities suggests that t years from now the franchise will be generating profit at the rate of $f(t) = 10,000 + 500t$ dollars per year. If the prevailing annual interest rate remains fixed at 10 percent compounded continuously, what is the present value of the franchise?

10. Having been left partially disabled after a head-on collision with a camel (see Problem 49 of Chapter 5, Section 1), the spy is contemplating retirement from the service. As compensation for many years of dedicated public service, the government offers the spy a choice between a 10-year pension of 5,000 pounds sterling per year and a flat sum of 35,000 pounds sterling to be paid immediately. Assuming that an annual interest rate of 10 percent compounded continuously will be available at banks throughout this period, decide which offer the spy should accept. (*Hint:* Compare the flat sum of 35,000 pounds with the present value of the pension. Assume that the pension is paid continuously.)

11. A national consumers' association has compiled statistics suggesting that the fraction of its members who are still active t months after joining is given by the function $f(t) = e^{-0.2t}$. A new local chapter has

200 charter members and expects to attract new members at the rate of 10 per month. How many members can the chapter expect to have at the end of 8 months?

Survival and Renewal   12. The administrators of a town estimate that the number of people who will still be residing in the town t years after they arrive is given by the function $f(t) = e^{-0.04t}$. If the current population is 20,000 people and new townspeople arrive at the rate of 500 per year, what will the population be 10 years from now?

Computer Dating   13. The operators of a new computer dating service estimate that the fraction of people who will retain their memberships in the service for at least t months is given by the function $f(t) = e^{-t/10}$. There are 8,000 charter members, and the operators expect to attract 200 new members per month. How many members will the service have 10 months from now?

Population Density   14. The population density r miles from the center of a certain city is $D(r) = 25,000e^{-0.05r}$ people per square mile. How many people live between 1 and 2 miles from the city center?

Population Density   15. The population density r miles from the center of a certain city is $D(r) = 5,000e^{-0.1r}$ people per square mile. How many people live within 3 miles of the center of the city? (*Hint:* Divide a circular disk of radius 3 into concentric rings.)

Poiseuille's Law   16. Calculate the rate (in cubic centimeters per second) at which blood flows through an artery of radius 0.1 centimeter if the speed of the blood r centimeters from the central axis is $8 - 800r^2$ centimeters per second.

Poiseuille's Law   17. Blood flows through an artery of radius R. At a distance r centimeters from the central axis of the artery, the speed of the blood is given by $S(r) = k(R^2 - r^2)$. Show that the average velocity of the blood is one-half the maximum speed.

Volume of a Solid of   In Problems 18 through 25, find the volume of the solid of revolution
Revolution   formed by revolving the region R about the x axis.

18. R is the region under the line $y = 3x + 1$ from $x = 0$ to $x = 1$.

19. R is the region under the curve $y = \sqrt{x}$ from $x = 1$ to $x = 4$.

20. R is the region under the curve $y = x^2 + 2$ from $x = -1$ to $x = 3$.

21. R is the region under the curve $y = 4 - x^2$ from $x = -2$ to $x = 2$.

22. R is the region under the curve $y = \sqrt{4 - x^2}$ from $x = -2$ to $x = 2$.

23. R is the region under the curve $y = \dfrac{1}{x}$ from $x = 1$ to $x = 10$.

24. R is the region under the curve $y = \dfrac{1}{\sqrt{x}}$ from $x = 1$ to $x = e^2$.

25. R is the region under the curve $y = e^{-0.1x}$ from $x = 0$ to $x = 10$.

Volume of a Sphere

26. Use integral calculus to prove that the volume of a sphere of radius r is $\dfrac{4}{3}\pi r^3$. (*Hint:* Think of the sphere as the solid formed by revolving the region under the semicircle $y = \sqrt{r^2 - x^2}$ in the accompanying figure from $x = -r$ to $x = r$ about the x axis.)

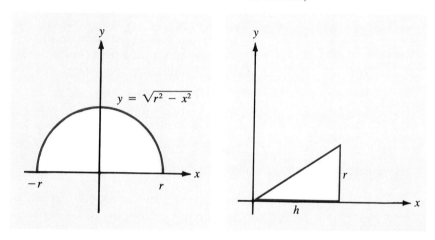

Volume of a Cone

27. Use integral calculus to prove that the volume of a right circular cone of height h and radius r is $\dfrac{1}{3}\pi r^2 h$. (*Hint:* Think of the cone as the solid formed by revolving the triangle shown in the accompanying figure about the x axis.)

In Problems 28 through 31, find the average value of the given function over the specified interval. In each case, sketch the graph of the function along with the rectangle whose base is the given interval and whose height is the average value of the function.

28. $f(x) = x;\ 0 \le x \le 4$

29. $f(x) = 2x - x^2;\ 0 \le x \le 2$

30. $f(x) = (x + 2)^2;\ -4 \le x \le 0$

31. $f(x) = \dfrac{1}{x};\ 1 \le x \le 2$

Temperature

32. Records indicate that t hours past midnight, the temperature at the local airport was $f(t) = -0.3t^2 + 4t + 10$ degrees Celsius. What was the average temperature at the airport between 9:00 A.M. and noon?

Food Prices

33. Records indicate that t months after the beginning of the year, the price of ground beef in local supermarkets was $P(t) = 0.09t^2 - 0.2t +$

1.6 dollars per pound. What was the average price of ground beef during the first 3 months of the year?

Efficiency   34. After t months on the job, a postal clerk can sort mail at the rate of $Q(t) = 700 - 400e^{-0.5t}$ letters per hour. What is the average rate at which the clerk sorts mail during the first 3 months on the job?

Bacterial Growth   35. The number of bacteria present in a certain culture after t minutes of an experiment was $Q(t) = 2,000e^{0.05t}$. What was the average number of bacteria present during the first 5 minutes of the experiment?

Speed and Distance   36. A car is driven so that after t hours its speed is $S(t)$ miles per hour.
   (a) Write down a definite integral that gives the average speed of the car during the first N hours.
   (b) Write down a definite integral that gives the total distance the car travels during the first N hours.
   (c) Discuss the relationship between the integrals in parts (a) and (b).

Inventory   37. An inventory of 60,000 kilograms of a certain commodity is used at a constant rate and is exhausted after 1 year. What is the average inventory for the year?

Fluid Flow   38. Find an expression for the rate (in cubic centimeters per second) at which a fluid flows through a cylindrical pipe of radius R if the speed of the fluid r centimeters from the central axis of the pipe is $S(r)$ centimeters per second.

The Area of a Disk   39. Use the characterization of the definite integral as the limit of a sum to show that the area of a circular disk of radius R is πR^2. (*Hint:* Divide the disk into n concentric rings as in Example 4.4.)

Present Value   40. A certain investment generates income continuously over a period of N years. After t years, the investment will be generating income at the rate of $f(t)$ dollars per year. Derive an expression for the present value of this investment if the prevailing annual interest remains fixed at $100r$ percent compounded continuously.

Group Membership   41. Let $f(t)$ denote the fraction of the membership of a certain group that will remain in the group for at least t years. Suppose that the group has just been formed with an initial membership of P_0 and that t years from now new members will be added to the group at the rate of $r(t)$ per year. Find an expression for the size of the group N years from now.

Nuclear Waste   42. A certain nuclear power plant produces radioactive waste in the form of strontium 90 at the constant rate of 500 pounds per year. The waste decays exponentially with a half-life of 28 years. How much of the radioactive waste from the nuclear plant will be present after 140 years? (*Hint:* Think of this as a survival and renewal problem.)

Total Revenue   43. Consider the following problem: "A certain oil well that yields 300 barrels of crude oil a month will run dry in 3 years. It is estimated that t months from now the price of crude oil will be $P(t) = 18 + 0.3\sqrt{t}$ dollars per barrel. If the oil is sold as soon as it is extracted from the ground, what will the total future revenue from the well be?"
   (a) Solve this problem using antidifferentiation, as in Chapter 5.
   (b) Solve the problem using definite integration. [Hint: Divide the 3-year (36 month) time interval $0 \le t \le 36$ into n equal subintervals of length Δt and let t_j denote the beginning of the jth subinterval. Find an expression that estimates the revenue $R(t_j)$ obtained during the jth subinterval. Then express the total revenue as the limit of a sum.]

Inventory Storage   44. Problem 44 of Section 1, Chapter 5, says:
Costs   "A manufacturer receives N units of a certain raw material that are initially placed in storage and then withdrawn and used at a constant rate until the supply is exhausted 1 year later. Suppose storage costs remain fixed at p dollars per unit per year. Find an expression for the total storage cost the manufacturer will pay during the year."

Solve this problem using definite integration. [Hint: Let $Q(t)$ denote the number of units in storage after t months and find an expression for $Q(t)$. Then subdivide the interval $0 \le t \le 1$ into n equal subintervals and express the total storage cost as the limit of a sum.]

5   Numerical Integration

In this section you will see some techniques you can use to approximate definite integrals. Numerical methods such as these are needed when the function to be integrated does not have an elementary antiderivative. For instance, neither $\sqrt{x^3 + 1}$ nor $\dfrac{e^x}{x}$ has an elementary antiderivative.

Approximation by Rectangles   If $f(x)$ is positive on the interval $a \le x \le b$, the definite integral $\displaystyle\int_a^b f(x)\, dx$ is equal to the area under the graph of f between $x = a$ and $x = b$. As you saw in Section 4 of this chapter, one way to approximate this area is to use n rectangles, as shown in Figure 6.30. In particular, you divide the interval $a \le x \le b$ into n equal subintervals of width $\Delta x = \dfrac{b - a}{n}$ and let x_j denote the beginning of the jth subinterval. The base of the jth rectangle is the jth subinterval, and its height is $f(x_j)$. Hence, the area of the jth rectangle is $f(x_j)\,\Delta x$. The sum of the areas of all n rectangles is an approximation to the area under the curve and hence an approximation to the corresponding definite integral. Thus,

$$\int_a^b f(x)\, dx \simeq f(x_1)\,\Delta x + f(x_2)\,\Delta x + \cdots + f(x_n)\,\Delta x$$

Figure 6.30 Approximation by rectangles.

This approximation improves as the number of rectangles increases, and you can estimate the integral to any desired degree of accuracy by making n large enough. However, since fairly large values of n are usually required to achieve reasonable accuracy, approximation by rectangles is rarely used in practice.

Approximation by Trapezoids

The accuracy of the approximation improves significantly if trapezoids are used instead of rectangles. Figure 6.31 shows the area from Figure 6.30 approximated by n trapezoids. Notice how much better the approximation is in this case.

Figure 6.31 Approximation by trapezoids.

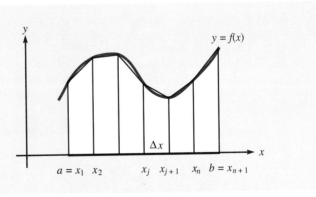

The jth trapezoid is shown in greater detail in Figure 6.32. Notice that it consists of a rectangle with a right triangle on top of it. Since,

$$\text{Area of rectangle} = f(x_{j+1}) \, \Delta x$$

and

$$\text{Area of triangle} = \frac{1}{2}[f(x_j) - f(x_{j+1})] \, \Delta x$$

it follows that

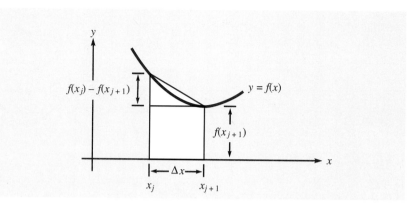

Figure 6.32 The jth trapezoid.

$$\text{Area of trapezoid} = f(x_{j+1})\, \Delta x + \frac{1}{2}[f(x_j) - f(x_{j+1})]\, \Delta x$$

$$= \frac{1}{2}[f(x_j) + f(x_{j+1})]\, \Delta x$$

The sum of the areas of all n trapezoids is an approximation to the area under the curve and hence an approximation to the corresponding definite integral. Thus,

$$\int_a^b f(x)\, dx \simeq \frac{1}{2}[f(x_1) + f(x_2)]\, \Delta x + \frac{1}{2}[f(x_2) + f(x_3)]\, \Delta x + \cdots$$

$$+ \frac{1}{2}[f(x_n) + f(x_{n+1})]\, \Delta x$$

$$= \frac{\Delta x}{2}[f(x_1) + 2f(x_2) + \cdots + 2f(x_n) + f(x_{n+1})]$$

This approximation formula is known as the **trapezoidal rule** and applies even if the function f is not positive.

The Trapezoidal Rule

$$\int_a^b f(x)\, dx \simeq \frac{\Delta x}{2}[f(x_1) + 2f(x_2) + \cdots + 2f(x_n) + f(x_{n+1})]$$

The use of the trapezoidal rule is shown in the following example.

EXAMPLE 5.1

Use the trapezoidal rule with $n = 10$ to approximate $\int_1^2 \frac{1}{x}\, dx$.

Solution

Since
$$\Delta x = \frac{2-1}{10} = 0.1$$

the interval $1 \le x \le 2$ is divided into 10 subintervals, as shown in Figure 6.33, by

$$x_1 = 1, x_2 = 1.1, x_3 = 1.2, \ldots, x_{10} = 1.9, x_{11} = 2$$

Figure 6.33 Division of the interval $1 \le x \le 2$ into 10 subintervals.

Then, by the trapezoidal rule,

$$\int_1^2 \frac{1}{x}\, dx \approx \frac{0.1}{2}\left(\frac{1}{1} + \frac{2}{1.1} + \frac{2}{1.2} + \frac{2}{1.3} + \frac{2}{1.4} + \frac{2}{1.5} + \frac{2}{1.6} + \frac{2}{1.7}\right.$$
$$\left. + \frac{2}{1.8} + \frac{2}{1.9} + \frac{1}{2}\right)$$

$$\approx 0.693771$$

The definite integral in Example 5.1 can be evaluated directly. In particular,

$$\int_1^2 \frac{1}{x}\, dx = \ln |x|\,\Big|_1^2 = \ln 2 \approx 0.693147$$

Thus the approximation of this particular integral by the trapezoidal rule with $n = 10$ is accurate (after round off) to two decimal places.

The Accuracy of the Trapezoidal Rule   The difference between the true value of the integral $\int_a^b f(x)\, dx$ and the approximation generated by the trapezoidal rule when n subintervals are used is denoted by E_n. The following estimate for the absolute value of E_n is proved in more advanced courses.

Error Estimate for the Trapezoidal Rule

If M is the maximum value of $|f''(x)|$ on the interval $a \le x \le b$, then

$$|E_n| \le \frac{M(b-a)^3}{12n^2}$$

The use of this formula is illustrated in the next example.

EXAMPLE 5.2

Estimate the accuracy of the approximation of $\int_1^2 \dfrac{1}{x} \, dx$ by the trapezoidal rule with $n = 10$.

Solution

Starting with $f(x) = \dfrac{1}{x}$, compute the derivatives

$$f'(x) = -\frac{1}{x^2} \quad \text{and} \quad f''(x) = \frac{2}{x^3}$$

and observe that the largest value of $|f''(x)|$ for $1 \le x \le 2$ is $|f''(1)| = 2$. Apply the error formula with

$$M = 2 \quad a = 1 \quad b = 2 \quad \text{and} \quad n = 10$$

to get

$$|E_{10}| \le \frac{2(2-1)^3}{12(10)^2} \approx 0.00167$$

That is, the error in the approximation in Example 5.1 is guaranteed to be no greater than 0.00167. (In fact, to five decimal places, the error is 0.00062, as you can see by comparing the approximation obtained in Example 5.1 with the decimal representation of ln 2.)

How to Choose the Number of Subintervals With the aid of the error estimate you can decide in advance how many subintervals to use to achieve a desired degree of accuracy. Here is an example.

EXAMPLE 5.3

How many subintervals are required to guarantee that the error will be less than 0.00005 in the approximation of $\int_1^2 \dfrac{1}{x} \, dx$ using the trapezoidal rule?

Solution

From Example 5.2 you know that $M = 2$, $a = 1$, and $b = 2$, so that

$$|E_n| \leq \frac{2(2-1)^3}{12n^2} = \frac{1}{6n^2}$$

The goal is to find the smallest positive integer n for which

$$\frac{1}{6n^2} < 0.00005$$

or, equivalently, $$n^2 > \frac{1}{6(0.00005)}$$

or $$n > \sqrt{\frac{1}{6(0.00005)}} \simeq 57.74$$

The smallest such integer is $n = 58$, and so 58 subintervals are required to ensure the desired accuracy.

Approximation Using Parabolas: Simpson's Rule

The relatively large number of subintervals required in Example 5.3 to ensure accuracy to within 0.00005 suggests that approximation by trapezoids may not be efficient enough for some applications. There is another approximation technique, called **Simpson's rule,** which is no harder to use than the trapezoidal rule but which often requires substantially fewer calculations to achieve a given degree of accuracy. Like the trapezoidal rule, it is based on the approximation of the area under a curve by columns, but unlike the trapezoidal rule, it uses parabolas rather than lines as the tops of the columns.

To be more specific, the approximation of a definite integral using parabolas is based on the following construction (illustrated in Figure 6.34 for $n = 6$). Divide the interval $a \leq x \leq b$ into an even number of

Figure 6.34 Approximation using parabolas.

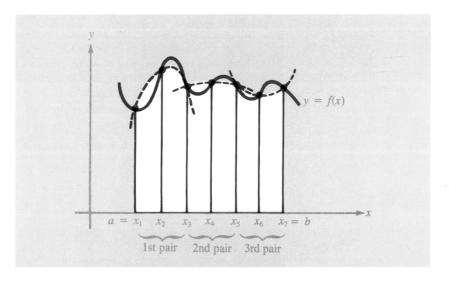

subintervals so that adjacent subintervals can be paired with none left over. Approximate the portion of the graph that lies above the first pair of subintervals by the (unique) parabola that passes through the three points $(x_1, f(x_1))$, $(x_2, f(x_2))$, and $(x_3, f(x_3))$ and use the area under this parabola between x_1 and x_3 to approximate the corresponding area under the curve. Do the same for the remaining pairs of subintervals and use the sum of the resulting areas to approximate the total area under the graph. It can be shown that this construction leads to the following approximation formula.

Simpson's Rule

$$\int_a^b f(x)\, dx \approx \frac{\Delta x}{3} [f(x_1) + 4f(x_2) + 2f(x_3) + 4f(x_4) + 2f(x_5)$$

$$+ \cdots + 2f(x_{n-1}) + 4f(x_n) + f(x_{n+1})]$$

Notice that the first and last function values in the approximating sum in Simpson's rule are multiplied by 1, while the others are multiplied alternately by 4 and 2.

The proof of Simpson's rule is based on the fact that the equation of a parabola is a polynomial of the form $y = Ax^2 + Bx + C$. For each pair of subintervals, the three given points are used to find the coefficients A, B, and C, and the resulting polynomial is then integrated to get the corresponding area. The details of the proof are straightforward but tedious and will be omitted.

EXAMPLE 5.4

Use Simpson's rule with $n = 10$ to approximate $\int_1^2 \frac{1}{x}\, dx$.

Solution

As in Example 5.1, $\Delta x = 0.1$, and hence the interval $1 \le x \le 2$ is divided into the 10 subintervals by

$$x_1 = 1,\ x_2 = 1.1,\ x_3 = 1.2,\ \ldots,\ x_{10} = 1.9,\ x_{11} = 2$$

Then, by Simpson's rule,

$$\int_1^2 \frac{1}{x}\, dx \approx \frac{0.1}{3} \left(\frac{1}{1} + \frac{4}{1.1} + \frac{2}{1.2} + \frac{4}{1.3} + \frac{2}{1.4} + \frac{4}{1.5} + \frac{2}{1.6} + \frac{4}{1.7} + \frac{2}{1.8} + \frac{4}{1.9} + \frac{1}{2} \right)$$

$$\approx 0.693150$$

Notice that this is an excellent approximation to the true value

$$\ln 2 = 0.693147$$

The Accuracy of Simpson's Rule The error estimate for Simpson's rule turns out to involve the fourth derivative $f^{(4)}(x)$.

> **Error Estimate for Simpson's Rule**
>
> If M is the maximum value of $|f^{(4)}(x)|$ on the interval $a \leq x \leq b$, then
>
> $$|E_n| \leq \frac{M(b - a)^5}{180n^4}$$

Here is an application of the formula.

EXAMPLE 5.5

Estimate the accuracy of the approximation of $\int_1^2 \frac{1}{x} \, dx$ by Simpson's rule with $n = 10$.

Solution

Starting with $f(x) = \frac{1}{x}$, compute the derivatives

$$f'(x) = -\frac{1}{x^2} \qquad f''(x) = \frac{2}{x^3} \qquad f^{(3)}(x) = -\frac{6}{x^4} \qquad f^{(4)}(x) = \frac{24}{x^5}$$

and observe that the largest value of $|f^{(4)}(x)|$ on the interval $1 \leq x \leq 2$ is $|f^{(4)}(1)| = 24$.

Now apply the error formula with $M = 24$, $a = 1$, $b = 2$, and $n = 10$ to get

$$|E_{10}| \leq \frac{24(2 - 1)^5}{180(10)^4} \approx 0.000013$$

That is, the error in the approximation in Example 5.4 is guaranteed to be no greater than 0.000013.

How to Choose the Number of Subintervals   In the next example, the error estimate is used to determine the number of subintervals that are required to ensure a specified degree of accuracy.

EXAMPLE 5.6

How many subintervals are required to ensure accuracy to within 0.00005 in the approximation of $\int_1^2 \frac{1}{x}\, dx$ by Simpson's rule?

Solution

From Example 5.5 you know that $M = 24$, $a = 1$, and $b = 2$. Hence,

$$|E_n| \le \frac{24(2 - 1)^5}{180n^4} = \frac{2}{15n^4}$$

The goal is to find the smallest positive (even) integer for which

$$\frac{2}{15n^4} < 0.00005$$

or, equivalently, $$n^4 > \frac{2}{15(0.00005)}$$

or $$n > \left[\frac{2}{15(0.00005)}\right]^{1/4} \simeq 7.19$$

The smallest such (even) integer is $n = 8$, and so eight subintervals are required to ensure the desired accuracy.

Problems 5.5

In Problems 1 through 8, approximate the given integral using (a) the trapezoidal rule and (b) Simpson's rule with the specified number of subintervals.

1. $\int_1^2 x^2\, dx$; $n = 4$

2. $\int_4^6 \frac{1}{\sqrt{x}}\, dx$; $n = 10$

3. $\int_0^1 \frac{1}{1 + x^2}\, dx$; $n = 4$

4. $\int_2^3 \frac{1}{x^2 - 1}\, dx$; $n = 4$

5. $\int_{-1}^0 \sqrt{1 + x^2}\, dx$; $n = 4$

6. $\int_0^3 \sqrt{9 - x^2}\, dx$; $n = 6$

7. $\int_0^1 e^{-x^2}\, dx$; $n = 4$

8. $\int_0^2 e^{x^2}\, dx$; $n = 10$

In Problems 9 through 14, approximate the given integral and estimate the error $|E_n|$ using (a) the trapezoidal rule and (b) Simpson's rule with the specified number of subintervals.

9. $\displaystyle\int_1^2 \frac{1}{x^2}\, dx;\ n = 4$

10. $\displaystyle\int_0^2 x^3\, dx;\ n = 8$

11. $\displaystyle\int_1^3 \sqrt{x}\, dx;\ n = 10$

12. $\displaystyle\int_1^2 \ln x\, dx;\ n = 4$

13. $\displaystyle\int_0^1 e^{x^2}\, dx;\ n = 4$

14. $\displaystyle\int_0^{0.6} e^{x^3}\, dx;\ n = 6$

In Problems 15 through 20, determine how many subintervals are required to guarantee accuracy to within 0.00005 in the approximation of the given integral by (a) the trapezoidal rule and (b) Simpson's rule.

15. $\displaystyle\int_1^3 \frac{1}{x}\, dx$

16. $\displaystyle\int_0^4 (x^4 + 2x^2 + 1)\, dx$

17. $\displaystyle\int_1^2 \frac{1}{\sqrt{x}}\, dx$

18. $\displaystyle\int_1^2 \ln(1 + x)\, dx$

19. $\displaystyle\int_{1.2}^{2.4} e^x\, dx$

20. $\displaystyle\int_0^2 e^{x^2}\, dx$

21. A quarter circle of radius 1 has the equation $y = \sqrt{1 - x^2}$ for $0 \le x \le 1$ and has area $\frac{\pi}{4}$. It follows that $\displaystyle\int_0^1 \sqrt{1 - x^2}\, dx = \frac{\pi}{4}$. Use this formula to estimate π to 3 decimal places by applying:

    (a) The trapezoidal rule   (b) Simpson's rule

22. Suppose that after t months on the job, a clerk can file letters at the rate of $Q(t) = \dfrac{e^{-0.4t}}{t}$ letters per hour. Use the trapezoidal rule with $n = 10$ to estimate the average rate at which the clerk files letters from the first until the sixth month on the job (a five month period).

23. Use the trapezoidal rule to estimate the area bounded by the curve $y = \sqrt{x^3 + 1}$, the x axis, and the lines $x = 0$ and $x = 1$.

24. The management of a national chain of fast-food restaurants is selling a 5-year franchise to operate its newest outlet in Tulare, California. Past experience in similar localities suggests that t years from now the franchise will be generating profit at the rate of $f(t) = 12,000\sqrt{t}$ dollars per year. Suppose the prevailing annual interest rate remains fixed during the next 5 years at 7 percent compounded continuously. Use Simpson's rule with $n = 10$ to estimate the present value of the franchise.

Chapter Summary and Review Problems

**Important Terms,
Symbols, and
Formulas**

Definite integral: $\int_a^b f(x)\, dx = F(b) - F(a)$

Integration by substitution:

$$\int_a^b g[u(x)]\frac{du}{dx}\, dx = \int_{u(a)}^{u(b)} g(u)\, du$$

Integration by parts:

$$\int_a^b f(x)g(x)\, dx = f(x)G(x)\Big|_a^b - \int_a^b f'(x)G(x)\, dx$$

where G is an antiderivative of g

Area under a curve (left-hand figure):

$$\text{Area of } R = \int_a^b f(x)\, dx$$

 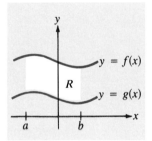

Area between two curves (right-hand figure):

$$\text{Area of } R = \int_a^b [f(x) - g(x)]\, dx$$

Probability density function

Net gain

Demand function: $p = D(q)$

Consumers' willingness to spend

Consumers' surplus

Fundamental theorem of calculus:

$$\lim_{n\to\infty} \sum_{j=1}^{n} f(x_j)\, \Delta x = \int_a^b f(x)\, dx = F(b) - F(a)$$

Amount of an income stream

Present value of an income stream

Survival and renewal functions

Volume of a solid of revolution:

$$\text{Volume} = \pi \int_a^b [f(x)]^2 \, dx$$

$$\text{Average value} = \frac{1}{b-a} \int_a^b f(x) \, dx$$

Trapezoidal rule:

$$\int_a^b f(x) \, dx \approx \frac{\Delta x}{2} [f(x_1) + 2f(x_2) + \cdots + 2f(x_n) + f(x_{n+1})]$$

Error estimate: $$|E_n| \leq \frac{M(b-a)^3}{12n^2}$$

where M is the maximum value of $|f''(x)|$ for $a \leq x \leq b$.

Simpson's rule:

$$\int_a^b f(x) \, dx \approx \frac{\Delta x}{3} [f(x_1) + 4f(x_2) + 2f(x_3) + 4f(x_4) + 2f(x_5) + \cdots$$

$$+ 2f(x_{n-1}) + 4f(x_n) + f(x_{n+1})]$$

Error estimate: $$|E_n| \leq \frac{M(b-a)^5}{180n^4}$$

where M is the maximum value of $|f^{(4)}(x)|$ for $a \leq x \leq b$.

Review Problems   In Problems 1 through 10, evaluate the given definite integral.

1. $\int_0^1 (5x^4 - 8x^3 + 1) \, dx$ 2. $\int_1^4 (\sqrt{x} + x^{-3/2}) \, dx$

3. $\int_{-1}^2 30(5x - 2)^2 \, dx$ 4. $\int_0^1 2xe^{x^2 - 1} \, dx$

5. $\int_0^1 (x - 3)(x^2 - 6x + 2)^3 \, dx$ 6. $\int_{-1}^1 \frac{(3x + 6) \, dx}{(x^2 + 4x + 5)^2}$

7. $\int_{-1}^1 xe^x \, dx$ 8. $\int_e^{e^2} \frac{dx}{x(\ln x)^2}$

9. $\int_1^e x^2 \ln x \, dx$ 10. $\int_0^{10} (2x + 1)e^{0.2x} \, dx$

11. A study indicates that x months from now, the population of a certain town will be increasing at the rate of $10 + 2\sqrt{x}$ people per month. By how much will the population of the town increase over the next 9 months?

12. It is estimated that t days from now a farmer's crop will be increasing

at the rate of $0.3t^2 + 0.6t + 1$ bushels per day. By how much will the value of the crop increase during the next 6 days if the market price remains fixed at $2 per bushel?

13. It is estimated that the demand for oil is increasing exponentially at the rate of 10 percent per year. If the demand for oil is currently 40 billion barrels per year, how much oil will be consumed during the next 5 years?

In Problems 14 through 19, find the area of the region R.

14. R is the region bounded by the curve $y = 3x^2 + 2$, the lines $x = -1$ and $x = 3$, and the x axis.

15. R is the region bounded by the curve $y = \dfrac{1}{x^2}$, the x axis, and the vertical lines $x = 1$ and $x = 4$.

16. R is the region bounded by the curve $y = 2 + x - x^2$ and the x axis.

17. R is the region bounded by the curves $y = \dfrac{8}{x}$ and $y = \sqrt{x}$ and the lines $y = 0$ and $x = 8$.

18. R is the region bounded by the curve $y = x^4$ and the line $y = x$.

19. R is the region in the first quadrant bounded by the line $y = 7x$ and the curves $y = 8 - x^2$ and $y = x^2$.

20. The probability density function for the duration of telephone calls in a certain city is $f(x) = 0.4e^{-0.4x}$, where x denotes the duration (in minutes) of a randomly selected call.
    (a) What percentage of the calls can be expected to last between 1 and 2 minutes?
    (b) What percentage of the calls can be expected to last 2 minutes or less?
    (c) What percentage of the calls can be expected to last more than 2 minutes?

21. When it is x years old, a certain industrial machine generates revenue at the rate of $R(x) = 4{,}575 - 5x^2$ dollars per year and results in costs that accumulate at the rate of $C(x) = 1{,}200 + 10x^2$ dollars per year.
    (a) For how many years is the use of the machine profitable?
    (b) What are the net earnings generated by the machine during the period of time in part (a)?
    (c) Interpret the net earnings in part (b) as the area between two curves.

22. Suppose the consumers' demand function for a certain commodity is $D(q) = 30 - 2q - q^2$ dollars per unit.
    (a) Find the number of units that will be bought if the market price is $15 per unit.

(b) Compute the consumers' willingness to spend to get the number of units in part (a).

(c) Compute the consumers' surplus when the market price is $15 per unit.

(d) Sketch the demand curve and interpret the consumers' willingness to spend and the consumers' surplus as areas.

23. Money is transferred continuously into an account at the constant rate of $1,200 per year. The account earns interest at the annual rate of 8 percent compounded continuously. How much will be in the account at the end of 5 years?

24. What is the present value of an investment that will generate income continuously at a constant rate of $1,000 per year for 10 years if the prevailing annual interest rate remains fixed at 7 percent compounded continuously?

25. In a certain community the fraction of the homes placed on the market that remain unsold for at least t weeks is approximately $f(t) = e^{-0.2t}$. If 200 homes are currently on the market and if additional homes are placed on the market at the rate of 8 per week, approximately how many homes will be on the market 10 weeks from now?

In Problems 26 and 27, find the volume of the solid of revolution formed by revolving the region R about the x axis.

26. R is the region under the curve $y = x^2 + 1$ from $x = -1$ to $x = 2$.

27. R is the region under the curve $y = e^{-x/20}$ from $x = 0$ to $x = 10$.

28. Records indicate that t months after the beginning of the year, the price of chicken in local supermarkets was $P(t) = 0.06t^2 - 0.2t + 1.2$ dollars per pound. What was the average price of chicken during the first 6 months of the year?

In Problems 29 and 30, approximate the given integral and estimate the error $|E_n|$ using (a) the trapezoidal rule and (b) Simpson's rule with the specified number of subintervals.

29. $\int_1^3 \frac{1}{x}\, dx;\ n = 10$ 　　　　　　 30. $\int_0^2 e^{x^2}\, dx;\ n = 8$

In Problems 31 and 32, determine how many subintervals are required to guarantee accuracy to within 0.00005 in the approximation of the given integral by (a) the trapezoidal rule and (b) Simpson's rule.

31. $\int_1^3 \sqrt{x}\, dx$ 　　　　　　 32. $\int_{0.5}^1 e^{x^2}\, dx$

CHAPTER 7
DIFFERENTIAL EQUATIONS

1   Introduction; Separable Differential Equations

Any equation that contains a derivative is called a **differential equation.** For example, the equations

$$\frac{dy}{dx} = 3x^2 + 5 \qquad \frac{dP}{dt} = kP \qquad \text{and} \qquad \left(\frac{dy}{dx}\right)^2 + 3\frac{dy}{dx} + 2y = e^x$$

are all differential equations.

Applications of Differential Equations

Many practical situations, especially those involving rates of change, can be described mathematically by differential equations. For example, the assumption that population grows at a rate proportional to its size can be expressed by the differential equation $\frac{dP}{dt} = kP$, where P denotes the population size at time t and k is the constant of proportionality. In economics, statements about marginal cost and marginal revenue can be formulated as differential equations.

Here are two examples.

EXAMPLE 1.1

Write a differential equation describing the fact that the rate at which people hear about a new increase in postal rates is proportional to the number of people in the country who have not heard about it.

Solution

Let $Q(t)$ denote the number of people who have heard about the rate increase at time t and B the total population of the country. Then

$$\text{Rate at which people hear about the increase} = \frac{dQ}{dt}$$

and

$$\text{Number of people who have not heard about the increase} = B - Q$$

Hence, the desired differential equation is

$$\frac{dQ}{dt} = k(B - Q)$$

where k is the constant of proportionality. Notice that the constant k must be positive because $\frac{dQ}{dt} > 0$ (since Q is an increasing function of t) and $B - Q > 0$ (since $B > Q$).

EXAMPLE 1.2

Write a differential equation describing the fact that when environmental factors impose an upper bound on its size, population grows at a rate that is jointly proportional to its current size and the difference between its upper bound and current size.

Solution

Let $P(t)$ denote the size of the population at time t and let B denote the upper bound imposed on the population by the environment. Then,

$$\text{Rate of population growth } = \frac{dP}{dt}$$

and $$\begin{array}{l}\text{Difference between upper} \\ \text{bound and population}\end{array} = B - P$$

Since "jointly proportional" means "proportional to the product" (as you saw in Chapter 1, Section 4), it follows that the desired differential equation is

$$\frac{dP}{dt} = kP(B - P)$$

where k is the constant of proportionality. (Convince yourself that the constant k must be positive in the context of this problem.)

General and Particular Solutions Any function that satisfies a differential equation is said to be a **solution** of that equation. Here is an example to illustrate this concept.

EXAMPLE 1.3

Verify that the function $y = e^x - x$ is a solution of the differential equation $\frac{dy}{dx} - y = x - 1$.

Solution

Substitute

$$y = e^x - x \quad \text{and} \quad \frac{dy}{dx} = e^x - 1$$

in the left-hand side of the differential equation to get

$$\frac{dy}{dx} - y = (e^x - 1) - (e^x - x) = x - 1$$

as required.

You can easily check that the function $y = 3e^x - x$ is also a solution of the differential equation in Example 1.3. In fact, every function of the form $y = Ce^x - x$, where C is a constant, is a solution of this equation. Moreover, it can be shown that *every* solution of this differential equation is of this form. For this reason, the function $y = Ce^x - x$ is said to be the **general solution** of this differential equation. A solution obtained by replacing C by a specific number is sometimes called a **particular solution** of the differential equation.

EXAMPLE 1.4

Find the particular solution of the differential equation $\frac{dy}{dx} - y = x - 1$ that satisfies the condition that $y = 4$ when $x = 0$.

Solution

Use the given condition to determine the numerical value of the constant C in the general solution $y = Ce^x - x$. In particular, substitute $y = 4$ and $x = 0$ into the general solution to get $C = 4$ and conclude that the desired particular solution is $y = 4e^x - x$.

Differential Equations of the Form
$$\frac{dy}{dx} = g(x)$$

The simplest type of differential equation is one of the form

$$\frac{dy}{dx} = g(x)$$

in which the derivative of the quantity in question is given explicitly as a function of the independent variable. The differential equations

$$\frac{dy}{dx} = 3x^2 + 5 \quad \text{and} \quad \frac{dQ}{dt} = 5te^{-2t}$$

are of this form. On the other hand, the differential equation

$$\frac{dP}{dt} = kP$$

describing population growth is not of this form, because the derivative of P is expressed in terms of P itself rather than the independent variable t.

A differential equation of the form $\frac{dy}{dx} = g(x)$ is particularly easy to solve. Its general solution is simply $y = G(x) + C$, where G is an antiderivative of g.

Differential Equations of the Form $\frac{dy}{dx} = g(x)$

The general solution of the differential equation

$$\frac{dy}{dx} = g(x)$$

is

$$y = \int g(x)\, dx + C$$

You have already solved differential equations of the form $\frac{dy}{dx} = g(x)$ in Chapter 5. Here, for purposes of review, are two practical problems involving differential equations of this form.

EXAMPLE 1.5

The resale value of a certain industrial machine decreases over a 10-year period at a rate that depends on the age of the machine. When the machine is x years old, the rate at which its value is changing is $220(x - 10)$ dollars per year. Express the value of the machine as a function of its age and initial value. If the machine was originally worth $12,000, how much will it be worth when it is 10 years old?

Solution

Let $V(x)$ denote the value of the machine when it is x years old. The derivative $\frac{dV}{dx}$ is equal to the rate $220(x - 10)$ at which the value of the machine is changing. Hence, you begin with the differential equation

$$\frac{dV}{dx} = 220(x - 10) = 220x - 2{,}200$$

To find V, solve this differential equation by integration.

$$V(x) = \int (220x - 2{,}200)\, dx = 110x^2 - 2{,}200x + C$$

Notice that C is equal to $V(0)$, the initial value of the machine. A more descriptive symbol for this constant is V_0. Using this notation, you can write the general solution as

$$V(x) = 110x^2 - 2{,}200x + V_0$$

If $V_0 = 12{,}000$, the corresponding particular solution is

$$V(x) = 110x^2 - 2{,}200x + 12{,}000$$

and the value of the machine after 10 years is

$$V(10) = 11{,}000 - 22{,}000 + 12{,}000 = \$1{,}000$$

Graphs showing the rate of depreciation and the resale value of the machine are sketched in Figure 7.1. Note that the rate of *depreciation* is

Figure 7.1 Depreciation of industrial machinery.

Rate of depreciation:
$R(x) = -220(x - 10)$

Value of machine:
$V(x) = 110x^2 - 2{,}200x + 12{,}000$

the negative of the rate of change $\dfrac{dV}{dx} = 220(x - 10)$. (Do you see why?)

EXAMPLE 1.6

An oil well that yields 300 barrels of crude oil a month will run dry in 3 years. It is estimated that t months from now the price of crude oil will be $P(t) = 18 + 0.3\sqrt{t}$ dollars per barrel. If the oil is sold as soon as it is extracted from the ground, what will the total future revenue from the well be?

Solution

Let $R(t)$ denote the revenue generated during the next t months. To construct the relevant differential equation, use the relationship

$$\text{Rate of change of revenue} \atop \text{with respect to time} = \frac{\text{dollars}}{\text{per month}} = \left(\frac{\text{dollars}}{\text{per barrel}}\right)\left(\frac{\text{barrels}}{\text{per month}}\right)$$

Since
$$\text{Rate of change of revenue} \atop \text{with respect to time} = \frac{dR}{dt}$$

$$\text{Dollars per barrel} = P(t) = 18 + 0.3\sqrt{t}$$

and
$$\text{Barrels per month} = 300$$

it follows that

$$\frac{dR}{dt} = P(t)\,(300) \qquad \text{or} \qquad \frac{dR}{dt} = 300(18 + 0.3\sqrt{t}) = 5{,}400 + 90\sqrt{t}$$

The general solution of this differential equation is

$$R(t) = \int (5{,}400 + 90\sqrt{t})\,dt = 5{,}400t + 60t^{3/2} + C$$

Since $R(0) = 0$, it follows that $C = 0$, and so the appropriate particular solution is

$$R(t) = 5{,}400t + 60t^{3/2}$$

Since the well will run dry in 36 months, the total future revenue will be

$$R(36) = 5{,}400(36) + 60(216) = \$207{,}360$$

Separable Differential Equations

Many useful differential equations can be formally rewritten so that all the terms containing the independent variable appear on one side of the equation and all the terms containing the dependent variable appear on the other. Differential equations with this special property are said to be **separable** and can be solved by the following procedure involving two integrations.

> **Separable Differential Equations**
>
> A differential equation that can be written in the form
>
> $$g(y)\,dy = h(x)\,dx$$
>
> is said to be separable. Its general solution is obtained by integrating both sides of this equation. That is,
>
> $$\int g(y)\,dy = \int h(x)\,dx + C$$

A proof that this procedure works will be given later in this section. First, here are some examples to illustrate how the procedure is used.

EXAMPLE 1.7

Find the general solution of the differential equation $\dfrac{dy}{dx} = \dfrac{2x}{y^2}$.

Solution

To separate the variables, pretend that the derivative $\dfrac{dy}{dx}$ is actually a quotient and write

$$y^2 \, dy = 2x \, dx$$

Now integrate both sides of this equation to get

$$\int y^2 \, dy = \int 2x \, dx$$

or $\dfrac{1}{3}y^3 = x^2 + C_1$ (where C_1 is an arbitrary constant)

and solve for y to conclude that

$$y = (3x^2 + 3C_1)^{1/3}$$

or

$$y = (3x^2 + C)^{1/3}$$ (where $C = 3C_1$ is an arbitrary constant)

Exponential Models   In the next three examples, you will see how separable differential equations lead to some of the exponential models that were introduced in Chapter 4.

Exponential Growth and Decay   In Chapter 4, Section 4, you saw that if a quantity grows exponentially, its rate of change is proportional to its size. Using separable differential equations, we can now establish the converse of this result.

EXAMPLE 1.8

Show that a quantity that grows at a rate proportional to its size grows exponentially.

Solution

Let $Q(t)$ denote the size of the quantity at time t and begin with the differential equation

$$\frac{dQ}{dt} = kQ$$

where k is the constant of proportionality. Separate the variables by writing

$$\frac{1}{Q} \, dQ = k \, dt$$

and integrate both sides of this equation to get

$$\int \frac{1}{Q} \, dQ = \int k \, dt$$

or
$$\ln |Q| = kt + C$$

Solve this equation for $|Q|$ by applying the exponential function to each side to get

$$|Q| = e^{kt+C} = e^C e^{kt}$$

Since $|Q| = \pm Q$, depending on whether Q is positive or negative, you can drop the absolute value sign and write

$$Q = \pm e^C e^{kt}$$

Finally, since the constant $\pm e^C$ is the value of Q when $t = 0$, you can introduce Q_0 to stand for this constant and, using functional notation, write

$$Q(t) = Q_0 e^{kt} \qquad \text{(where } Q_0 = \pm e^C\text{)}$$

which is precisely the equation describing exponential growth.

You can use a similar argument to show that a quantity that decreases at a rate proportional to its size decreases exponentially. For practice, work out the details.

Exponential Growth and Decay

A quantity grows (or decays) exponentially if and only if its rate of change is proportional to its size.

A Technical Detail　The solution of Example 1.8 is not quite correct as stated. In particular, the argument that the equation

$$|Q| = e^C e^{kt}$$

can be rewritten as
$$Q = \pm e^C e^{kt}$$

where $\pm e^C$ is a *constant* is valid only if Q never changes sign. If Q were to change sign, the expression $\pm e^C$ would be positive part of the time and negative part of the time and hence would not be constant. Fortunately, it turns out that the continuity of Q guarantees that this cannot happen and that the expression $\pm e^C$ is really a constant. Feel free to ignore this technical point when solving similar separable differential equations and simply assume that the function does not change sign.

Learning Curves   As you saw in Chapter 4, Section 2, the graphs of functions of the form $Q(t) = B - Ae^{-kt}$ are called learning curves because functions of this form often describe the relationship between the efficiency with which an individual performs a task and the amount of training or experience the individual has had. In general, any quantity that grows at a rate that is proportional to the difference between its size and a fixed upper bound can be represented by a function of this form. This is illustrated in the next example.

EXAMPLE 1.9

The rate at which people hear about a new increase in postal rates is proportional to the number of people in the country who have not heard about it. Express the number of people who have heard about the increase as a function of time.

Solution

Let $Q(t)$ denote the number of people who have heard about the increase by time t and B the total population of the country. Then (as you saw in Example 1.1 of this chapter),

$$\frac{dQ}{dt} = k(B - Q)$$

where k is the constant of proportionality. Separate the variables by writing

$$\frac{1}{B - Q}\, dQ = k\, dt$$

and integrate to get

$$\int \frac{1}{B - Q}\, dQ = \int k\, dt$$

or

$$-\ln|B - Q| = kt + C$$

(Be sure you see where the minus sign came from.) This time you can drop the absolute value sign immediately since $B - Q$ cannot be negative in this context. Hence,

$$-\ln (B - Q) = kt + C$$

$$\ln (B - Q) = -kt - C$$

$$B - Q = e^{-kt-C} = e^{-kt}e^{-C}$$

or

$$Q = B - e^{-C}e^{-kt}$$

Denoting the constant e^{-C} by A and using functional notation, you can conclude that

$$Q(t) = B - Ae^{-kt}$$

which is precisely the general equation of a learning curve. For reference, the graph of Q is sketched in Figure 7.2.

Figure 7.2 A learning curve: $Q(t) = B - Ae^{-kt}$.

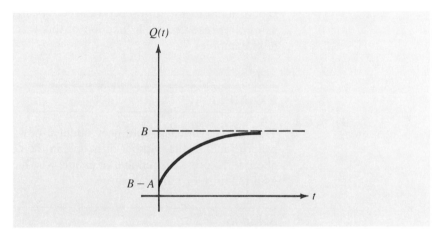

Logistic Curves

In Chapter 4, Section 2, you learned that the graphs of functions of the form $Q(t) = \dfrac{B}{1 + Ae^{-Bkt}}$ are called **logistic curves.** They are used as models of population growth when environmental factors impose an upper bound on the possible size of the population, and they also describe such phenomena as the spread of rumors and epidemics. The calculation in the next example can be used to show that any quantity that satisfies a differential equation of the form $\dfrac{dQ}{dt} = kQ(B - Q)$ can be represented by a logistic curve.

EXAMPLE 1.10

The rate at which an epidemic spreads through a community is jointly proportional to the number of residents who have been infected and the number of susceptible residents who have not. Express the number of residents who have been infected as a function of time.

Solution

Let $Q(t)$ denote the number of residents who have been infected by time t and B the total number of susceptible residents. Then the number of susceptible residents who have not been infected is $B - Q$, and the differential equation describing the spread of the epidemic is

$$\frac{dQ}{dt} = kQ(B - Q)$$

where k is the constant of proportionality. This is a separable differential equation whose solution is

$$\int \frac{1}{Q(B - Q)} \, dQ = \int k \, dt$$

The trick to finding the integral on the left-hand side is to observe that

$$\frac{1}{Q(B - Q)} = \frac{1}{B} \left[\frac{B}{Q(B - Q)} \right] = \frac{1}{B} \left(\frac{1}{Q} + \frac{1}{B - Q} \right)$$

(which you can verify by adding the fractions in the expression on the right). Hence,

$$\frac{1}{B} \int \frac{1}{Q} \, dQ + \frac{1}{B} \int \frac{1}{B - Q} \, dQ = \int k \, dt$$

$$\frac{1}{B} \ln |Q| - \frac{1}{B} \ln |B - Q| = kt + C$$

$$\frac{1}{B} \ln \left| \frac{Q}{B - Q} \right| = kt + C \qquad \text{since } \ln \frac{u}{v} = \ln u - \ln v$$

or

$$\frac{1}{B} \ln \frac{Q}{B - Q} = kt + C \qquad \text{since } Q > 0 \text{ and } B > Q$$

Solve for Q to get

$$\ln \frac{Q}{B - Q} = Bkt + BC$$

$$\frac{Q}{B - Q} = e^{Bkt + BC} = e^{Bkt} e^{BC} = A_1 e^{Bkt} \qquad \text{where } A_1 = e^{BC}$$

$$Q = (B - Q)A_1 e^{Bkt} = BA_1 e^{Bkt} - QA_1 e^{Bkt}$$

$$Q + QA_1 e^{Bkt} = A_1 B e^{Bkt}$$

$$Q(1 + A_1 e^{Bkt}) = A_1 B e^{Bkt}$$

$$Q = \frac{A_1 B e^{Bkt}}{1 + A_1 e^{Bkt}}$$

To make this formula more attractive, divide the numerator and denominator by $A_1 e^{Bkt}$ to get

$$Q = \frac{B}{(1/A_1)\, e^{-Bkt} + 1}$$

Finally, let A denote the constant $\dfrac{1}{A_1}$ and use functional notation to get

$$Q(t) = \frac{B}{1 + Ae^{-Bkt}}$$

which is, as promised, the general equation of a logistic curve.

For reference, the graph of Q is sketched in Figure 7.3. It is not hard to show that the inflection point occurs when $Q(t) = \dfrac{B}{2}$. (See Problem 60

Figure 7.3 A logistic curve:
$$Q(t) = \frac{B}{1 + Ae^{-Bkt}}.$$

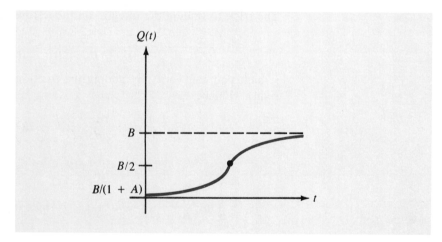

at the end of this section.) This corresponds to the fact that the epidemic is spreading most rapidly when half the susceptible residents have been infected.

Dilution Problem   The next example is typical of an important class of dilution problems that lead to separable differential equations.

EXAMPLE 1.11

The residents of a certain community have voted to discontinue the fluoridation of their water supply. The local reservoir currently holds 200 million gallons of fluoridated water that contains 1,600 pounds of fluoride. The fluoridated water is flowing out of the reservoir at the rate of 4 million gallons per day and is being replaced at the same rate by unfluoridated water. At all times, the remaining fluoride is evenly distributed in the reservoir. Express the amount of fluoride in the reservoir as a function of time.

Solution

Begin with the following relationship:

$$\text{Rate of change of fluoride} \atop \text{with respect to time} = \left(\text{concentration of} \atop \text{fluoride in water} \right) \left(\text{rate of flow of} \atop \text{fluoridated water} \right)$$

or

$$\text{Pounds} \atop \text{per day} = \left(\text{pounds per} \atop \text{million gallons} \right) \left(\text{million gallons} \atop \text{per day} \right)$$

Let $Q(t)$ denote the number of pounds of fluoride in the reservoir after t days. Then,

$$\text{Rate of change of fluoride} \atop \text{with respect to time} = \frac{dQ}{dt} \quad \text{pounds per day}$$

$$\text{Concentration of} \atop \text{fluoride in water} = \frac{\text{number of pounds of fluoride in reservoir}}{\text{number of million gallons of water in reservoir}}$$

$$= \frac{Q}{200} \quad \text{pounds per million gallons}$$

and

$$\text{Rate of flow of} \atop \text{fluoridated water} = -4 \text{ million gallons per day}$$

where the minus sign indicates that the water is leaving the reservoir. These observations are summarized in Figure 7.4.

Figure 7.4 Diagram showing the amount of fluoride flowing into and out of the reservoir at time t.

Since the rate of change of the fluoride in the reservoir equals the inflow rate minus the outflow rate, it follows that

$$\frac{dQ}{dt} = 0 - \frac{Q}{200}(4) = \frac{-Q}{50}$$

Solving this differential equation by separation of variables, you get

$$\int \frac{1}{Q} \, dQ = - \int \frac{1}{50} \, dt$$

$$\ln Q = - \frac{t}{50} + C$$

or $\qquad Q = Q_0 e^{-t/50} \qquad$ where $Q_0 = e^C$

Initially, 1,600 pounds of fluoride were in the reservoir. Replacing Q_0 by 1,600 and using functional notation, you can conclude that

$$Q(t) = 1,600 e^{-t/50}$$

That is, the amount of fluoride in the reservoir decreases exponentially.

Figure 7.5 The amount of fluoride: $Q(t) = 1,600 e^{-t/50}$.

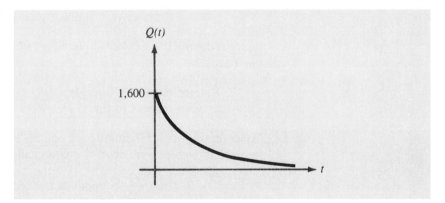

The situation is illustrated in Figure 7.5.

A Price Adjustment Model

Let p denote the price of a particular commodity and let $S(p)$ and $D(p)$ denote the supply and demand functions of that commodity, respectively. In Section 4 of Chapter 1, you saw that market equilibrium occurs at the price for which demand equals supply.

In a certain dynamic model of price adjustment, price, supply, and demand are all regarded as functions of time t, and the time rate of change of price $\frac{dp}{dt}$ is assumed to be proportional to the shortage $D - S$; that is,

$$\frac{dp}{dt} = k(D - S)$$

where k is a positive constant. Here is an example involving this model.

EXAMPLE 1.12

Suppose the price $p(t)$ of a particular commodity varies in such a way that its rate of change with respect to time is proportional to the shortage $D - S$, where $D(p)$ and $S(p)$ are the linear demand and supply functions $D = 8 - 2p$ and $S = 2 + p$.

(a) If the price is \$5 when $t = 0$ and \$3 when $t = 2$, find $p(t)$.

(b) Determine what happens to $p(t)$ in the "long run" (as t increases without bound).

Solution

(a) The rate of change of $p(t)$ is given by the separable differential equation

$$\frac{dp}{dt} = k(D - S) = k[(8 - 2p) - (2 + p)]$$

$$= k(6 - 3p)$$

Separating the variables, integrating, and solving for p, you get

$$\frac{-1}{3} \ln |6 - 3p| = kt + C_1$$

$$\ln |6 - 3p| = -3kt - 3C_1$$

$$6 - 3p = e^{-3kt - 3C_1}$$

$$= e^{-3kt}e^{-3C_1} = Ce^{-3kt} \qquad \text{where } C = e^{-3C_1}$$

$$p(t) = 2 - \frac{1}{3}Ce^{-3kt}$$

To evaluate the constant C, use the fact that $p(0) = 5$, so that

$$5 = p(0) = 2 - \frac{1}{3}Ce^0 = 2 - \frac{1}{3}C$$

$$C = -9$$

and $$p(t) = 2 + 3e^{-3kt}$$

You still need to find k. Since $p = 3$ when $t = 2$, it follows that

$$3 = p(2) = 2 + 3e^{-3k(2)} = 2 + 3e^{-6k}$$

and by solving this equation for e^{-6k} and then taking logarithms, you get

$$e^{-6k} = \frac{3 - 2}{3} = \frac{1}{3}$$

$$-6k = \ln\left(\frac{1}{3}\right) = -1.0986$$

$$k = \frac{-1.0986}{-6} = 0.1831$$

Thus, the price at time t is

$$p(t) = 2 + 3e^{-6(0.1831t)} = 2 + 3e^{-1.0986t}$$

(b) As t increases without bound, $e^{-1.0986t}$ approaches zero and $p(t)$ approaches 2, which is the price at which supply equals demand. That is, in the "long run," the price $p(t)$ approaches the equilibrium price of the commodity (see Figure 7.6).

Figure 7.6 The price $p(t)$ approaches the equilibrium price $p = 2$.

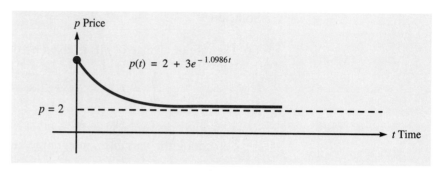

Why the Method of Separation of Variables Works

It is not hard to see why the method of separation of variables works. Before the variables were separated, the differential equation

$$g(y)\, dy = h(x)\, dx$$

was

$$\frac{dy}{dx} = \frac{h(x)}{g(y)}$$

or, equivalently,

$$g(y)\frac{dy}{dx} - h(x) = 0$$

The left-hand side of this equation can be rewritten in terms of the antiderivatives of g and h. In particular, if G is an antiderivative of g and H an antiderivative of h, it follows from the chain rule that

$$\frac{d}{dx}[G(y) - H(x)] = G'(y)\frac{dy}{dx} - H'(x) = g(y)\frac{dy}{dx} - h(x)$$

Hence, the differential equation $g(y)\dfrac{dy}{dx} - h(x) = 0$ says that

$$\frac{d}{dx}[G(y) - H(x)] = 0$$

But constants are the only functions whose derivatives are identically zero, and so

$$G(y) - H(x) = C$$

for some constant C. That is,

$$G(y) = H(x) + C$$

or, equivalently, $$\int g(y)\,dy = \int h(x)\,dx + C$$

and the proof is complete.

Problems 7.1

In Problems 1 through 12, write a differential equation describing the given situation. Define all variables you introduce. (Do not try to solve the differential equation at this time.)

Growth of Bacteria   1. The number of bacteria in a culture grows at a rate that is proportional to the number present.

Radioactive Decay   2. A sample of radium decays at a rate that is proportional to its size.

Investment Growth   3. An investment grows at a rate equal to 7 percent of its size.

Concentration of Drugs   4. The rate at which the concentration of a drug in the bloodstream decreases is proportional to the concentration.

Population Growth   5. The population of a certain town increases at the constant rate of 500 people per year.

Marginal Cost   6. A manufacturer's marginal cost is $60 per unit.

Temperature Change   7. The rate at which the temperature of an object changes is proportional to the difference between its own temperature and the temperature of the surrounding medium.

Dissolution of Sugar   8. After being placed in a container of water, sugar dissolves at a rate proportional to the amount of undissolved sugar remaining in the container.

Recall from Memory   9. When a person is asked to recall a set of facts, the rate at which the facts are recalled is proportional to the number of relevant facts in the person's memory that have not yet been recalled.

The Spread of an Epidemic   10. The rate at which an epidemic spreads through a community is jointly proportional to the number of people who have caught the disease and the number who have not.

Corruption in Government   11. The rate at which people are implicated in a government scandal is jointly proportional to the number of people already implicated and the number of people involved who have not yet been implicated.

The Spread of a Rumor   12. The rate at which a rumor spreads through a community is jointly proportional to the number of people in the community who have heard the rumor and the number who have not.

13. Verify that the function $y = Ce^{kx}$ is a solution of the differential equation $\dfrac{dy}{dx} = ky$.

14. Verify that the function $Q = B - Ce^{-kt}$ is a solution of the differential equation $\dfrac{dQ}{dt} = k(B - Q)$.

15. Verify that the function $y = C_1 e^x + C_2 x e^x$ is a solution of the differential equation $\dfrac{d^2 y}{dx^2} - 2\dfrac{dy}{dx} + y = 0$.

16. Verify that the function $y = \dfrac{1}{20}x^4 - \dfrac{C_1}{x} + C_2$ is a solution of the differential equation $x\dfrac{d^2 y}{dx^2} + 2\dfrac{dy}{dx} = x^3$.

In Problems 17 through 31, find the general solution of the given differential equation.

17. $\dfrac{dy}{dx} = 3x^2 + 5x - 6$

18. $\dfrac{dP}{dt} = \sqrt{t} + e^{-t}$

19. $\dfrac{dV}{dx} = \dfrac{2}{x + 1}$

20. $\dfrac{dA}{dt} = 2te^{t^2 + 5}$

21. $\dfrac{d^2 p}{dt^2} = 50$ (*Hint:* Integrate twice.)

22. $\dfrac{d^2 y}{dx^2} = 3x^2 + 5x - 6$

23. $\dfrac{dy}{dx} = 3y$

24. $\dfrac{dy}{dx} = y^2$

25. $\dfrac{dy}{dx} = e^y$

26. $\dfrac{dy}{dx} = e^{x+y}$

27. $\dfrac{dy}{dx} = \dfrac{x}{y}$

28. $\dfrac{dy}{dx} = \dfrac{y}{x}$

29. $\dfrac{dy}{dx} = y + 10$

30. $\dfrac{dy}{dx} = 80 - y$

31. $\dfrac{dy}{dx} = y(1 - 2y)$ $\left[Hint: \dfrac{1}{y(1 - 2y)} = \dfrac{1}{y} + \dfrac{2}{1 - 2y} \right]$

In Problems 32 through 39, find the particular solution of the given differential equation that satisfies the given condition.

32. $\dfrac{dy}{dx} = e^{5x}$; $y = 1$ when $x = 0$

33. $\dfrac{dy}{dx} = 5x^4 - 3x^2 - 2$; $y = 4$ when $x = 1$

34. $\dfrac{d^2y}{dt^2} = 3t^2 + 2t - 1$; $y = 3$ and $\dfrac{dy}{dt} = 0$ when $t = 1$ (*Hint:* Integrate twice.)

35. $\dfrac{d^2A}{dt^2} = e^{-t/2}$; $A = 2$ and $\dfrac{dA}{dt} = 1$ when $t = 0$

36. $\dfrac{dy}{dx} = \dfrac{x}{y^2}$; $y = 3$ when $x = 2$ 37. $\dfrac{dy}{dx} = 4x^3y^2$; $y = 2$ when $x = 1$

38. $\dfrac{dy}{dx} = 5(8 - y)$; $y = 6$ when $x = 0$

39. $\dfrac{dy}{dx} = y(1 - y)$; $y = \dfrac{1}{3}$ when $x = 0$

Depreciation   40. The resale value of a certain industrial machine decreases at a rate that depends on its age. When the machine is t years old, the rate at which its value is changing is $-960e^{-t/5}$ dollars per year.
    (a) Express the value of the machine in terms of its age and initial value.
    (b) If the machine was originally worth $5,200, how much will it be worth when it is 10 years old?

Marginal Cost   41. At a certain factory, the marginal cost is $3(q - 4)^2$ dollars per unit when the level of output is q units.
    (a) Express the total production cost in terms of the overhead (the cost of producing 0 units) and the number of units produced.
    (b) What is the cost of producing 14 units if the overhead is $436?

Retail Prices   42. In a certain section of the country, the price of chicken is currently $3 per kilogram. It is estimated that x weeks from now the price will be increasing at the rate of $3\sqrt{x} + 1$ cents per week. How much will chicken cost 8 weeks from now?

Air Pollution   43. In a certain Los Angeles suburb, a reading of air pollution levels taken at 7:00 A.M. shows the ozone level to be 0.25 part per million. A 12-hour forecast of air conditions predicts that t hours later the ozone level will be changing at the rate of $\dfrac{0.24 - 0.03t}{\sqrt{36 + 16t - t^2}}$ parts per million per hour.
    (a) Express the ozone level as a function of t.
    (b) At what time will the peak ozone level occur? What will the ozone level be at this time?

Oil Production   44. A certain oil well that yields 400 barrels of crude oil per month will run dry in 2 years. The price of crude oil is currently $20 per barrel and is expected to rise at the constant rate of 4 cents per barrel per month. If the oil is sold as soon as it is extracted from the ground, what will the total future revenue from the well be?

Water Pollution   45. It is estimated that t years from now the population of a certain lakeside community will be changing at the rate of $0.6t^2 + 0.2t + 0.5$ thousand per year. Environmentalists have found that the level of pollution in the lake increases at the rate of approximately 5 units per 1,000 people. If the level of pollution in the lake is currently 60 units, what will the pollution level be 2 years from now?

Investment   46. A $1,000 investment grows at a rate equal to 7 percent of its size. Express the value of the investment as a function of time.

Drug Concentration   47. The rate at which the concentration of a drug in the bloodstream decreases is proportional to the concentration. Express the concentration of the drug in the bloodstream as a function of time.

Exponential Decay   48. Show that a quantity that decays at a rate proportional to its size decays exponentially.

Exponential Functions   49. Show that if a differentiable function is equal to its own derivative, then the function must be of the form $y = Ce^x$.

Recall from Memory   50. Some psychologists believe that when a person is asked to recall a set of facts, the rate at which the facts are recalled is proportional to the number of relevant facts in the subject's memory that have not yet been recalled. Express the number of facts that have been recalled as a function of time and draw the graph.

Dissolution of Sugar   51. After being placed in a container of water, sugar dissolves at a rate proportional to the amount of undissolved sugar remaining in the container. Express the amount of sugar that has been dissolved as a function of time and draw the graph.

Newton's Law of Cooling   52. The rate at which the temperature of an object changes is proportional to the difference between its own temperature and that of the surrounding medium. A cold drink is removed from a refrigerator on a hot summer day and placed in a room where the temperature is 80° F. Express the temperature of the drink as a function of time (minutes) if the temperature of the drink was 40° F when it left the refrigerator and was 50° F after 20 minutes in the room.

Newton's Law of Cooling   53. The rate at which the temperature of an object changes is proportional to the difference between its own temperature and that of the surrounding medium. Express the temperature of the object as a function of time and draw the graph if the temperature of the object is greater than that of the surrounding medium.

Dilution

54. A tank contains 200 gallons of clear water. Brine (salt water) containing 2 pounds of salt per gallon flows into the tank at the rate of 5 gallons per minute, and the mixture, which is stirred so that the salt is evenly distributed at all times, runs out of the tank at the same rate. Express the amount of salt in the tank as a function of time and draw the graph. [*Hint:* Let $Q(t)$ denote the number of pounds of salt in the tank after t minutes. Then

$$\frac{dQ}{dt} = \left(\begin{array}{c}\text{rate at which}\\\text{salt enters tank}\end{array}\right) - \left(\begin{array}{c}\text{rate at which}\\\text{salt leaves tank}\end{array}\right)]$$

Dilution

55. A tank currently holds 200 gallons of brine that contains 3 pounds of salt per gallon. Brine containing 2 pounds of salt per gallon flows into the tank at the rate of 5 gallons per minute, while the mixture, which is kept uniform, runs out of the tank at the same rate. Express the amount of salt in the tank as a function of time and draw the graph.

Air Purification

56. A 2,400-cubic-foot room contains an activated charcoal air filter through which air passes at the rate of 400 cubic feet per minute. The ozone in the air is absorbed by the charcoal as the air flows through the filter, and the purified air is recirculated in the room. Assuming that the remaining ozone is evenly distributed throughout the room at all times, determine how long it takes the filter to remove 50 percent of the ozone from the room.

The Spread of an Epidemic

57. The rate at which an epidemic spreads through a community is jointly proportional to the number of residents who have been infected and the number of susceptible residents who have not. Express the number of residents who have been infected as a function of time (in weeks) if the community has 2,000 susceptible residents, if 500 residents had the disease initially, and if 855 residents had been infected by the end of the 1st week.

The Spread of an Epidemic

58. The rate at which an epidemic spreads through a community is jointly proportional to the number of residents who have been infected and the number of susceptible residents who have not. Show that the epidemic is spreading most rapidly when one-half of the susceptible residents have been infected. (*Hint:* You do not have to solve a differential equation to do this. Just start with a formula for the *rate* at which the epidemic is spreading and use calculus to maximize this rate.)

Corruption in Government

59. The number of people implicated in a certain major government scandal increases at a rate jointly proportional to the number of people already implicated and the number involved who have not yet been implicated. Suppose that 7 people were implicated when a Washington newspaper first made the scandal public, that 9 more

were implicated over the next 3 months, and that another 12 were implicated during the following 3 months. Approximately how many people are involved in the scandal? (*Warning:* This problem will test your algebraic ingenuity!)

Logistic Curves 60. Show that if a quantity Q satisfies the differential equation $\dfrac{dQ}{dt} = kQ(B - Q)$, where k and B are positive constants, then the rate of change $\dfrac{dQ}{dt}$ is greatest when $Q(t) = \dfrac{B}{2}$. What does this result tell you about the inflection point of a logistic curve? Explain. (*Hint:* See the hint for Problem 58.)

Price Adjustment 61. Suppose the price $p(t)$ of a particular commodity varies in such a way that its rate of change $\dfrac{dp}{dt}$ is proportional to the shortage $D - S$, where $D = 7 - p$ and $S = 1 + p$ are the demand and supply functions for the commodity.
(a) If the price is \$6 when $t = 0$ and \$4 when $t = 4$, find $p(t)$.
(b) Show that as t increases without bound, $p(t)$ approaches the price where supply equals demand.

Price Adjustment 62. Suppose a particular commodity has linear demand and supply functions $D(p) = a - bp$ and $S(p) = r + sp$ for positive constants a, b, r, s. Further assume that the price p is a function of time and that its time rate of change $\dfrac{dp}{dt}$ is proportional to the shortage $D - S$. Find $p(t)$ and show that as t increases without bound, $p(t)$ approaches the price where supply equals demand.

Fick's Law 63. When a cell is placed in a liquid containing a solute, the solute passes through the cell wall by diffusion. As a result, the concentration of the solute inside the cell changes, increasing if the concentration of the solute outside the cell is greater than the concentration inside and decreasing if the opposite is true. In biology, Fick's law asserts that the concentration of the solute inside the cell changes at a rate that is jointly proportional to the area of the cell wall and the difference between the concentrations of the solute inside and outside the cell. Assuming that the concentration of the solute outside the cell is constant and greater than the concentration inside, derive a formula for the concentration of the solute inside the cell.

2 First-Order Linear Differential Equations

A **first-order linear** differential equation is one with the general form

$$\frac{dy}{dx} + p(x)y = q(x)$$

For instance, $\dfrac{dy}{dx} + \dfrac{y}{x} = e^{-x}$ (for $x > 0$)

is a first-order linear differential equation with $p(x) = \dfrac{1}{x}$ and $q(x) = e^{-x}$. In its present form, this equation cannot be solved by simple integration or by separation of variables. However, if you multiply both sides of the equation by x, it becomes

$$x\frac{dy}{dx} + y = xe^{-x}$$

Notice that the left side of this equation is the derivative (with respect to x) of the product xy since

$$\frac{d}{dx}(xy) = x\frac{dy}{dx} + y\frac{dx}{dx} = x\frac{dy}{dx} + y$$

Thus, the given differential equation now reads

$$\frac{d}{dx}(xy) = xe^{-x}$$

which you can solve by integrating both sides with respect to x:

$$\int \frac{d}{dx}(xy)\, dx = \int xe^{-x}dx$$

$$xy = \int xe^{-x}dx = -(x+1)e^{-x} + C \qquad \text{(using integration by parts)}$$

Solving for y, you conclude that the given first-order linear differential equation has the general solution

$$y = \frac{1}{x}[e^{-x}(-x-1) + C]$$

In this example, multiplication by the function x converted the given first-order linear differential equation into one that could be solved by simple integration. Such a function is called an **integrating factor** of the differential equation. In more advanced texts, it is shown that every first-order linear differential equation $\dfrac{dy}{dx} + p(x)y = q(x)$ has an integrating factor equal to $I(x) = e^{\int p(x)dx}$. For instance, in the introductory example, where $p(x) = \dfrac{1}{x}$, the integrating factor is

$$I(x) = e^{\int 1/x \ dx} = e^{\ln x} = x \qquad (\text{remember, } x > 0)$$

To summarize:

Solution of a First-Order Linear Differential Equation

The first-order linear differential equation $\dfrac{dy}{dx} + p(x)y = q(x)$ has the general solution

$$y = \frac{1}{I(x)}\left[\int I(x)q(x) \ dx + C\right]$$

where C is an arbitrary constant and $I(x)$ is the integrating factor

$$I(x) = e^{\int p(x)dx}$$

EXAMPLE 2.1

Find the general solution of the differential equation

$$\frac{dy}{dx} + 2y = 2x$$

Solution

This is a first-order linear differential equation with $p(x) = 2$ and $q(x) = 2x$. The integrating factor is

$$I(x) = e^{\int 2 dx} = e^{2x}$$

and the general solution of the first-order linear equation is

$$y = \frac{1}{e^{2x}}\left[\int 2x \ e^{2x} \ dx + C\right]$$

The integral on the right can be found using integration by parts:

$$\int 2x \ e^{2x} \ dx = \left(x - \frac{1}{2}\right)e^{2x}$$

(Verify the details.) Thus, the general solution is

$$y = \frac{1}{e^{2x}}\left[xe^{2x} - \frac{1}{2} e^{2x} + C\right] = x - \frac{1}{2} + Ce^{-2x}$$

In Example 1.11 of Section 7.1, you saw a dilution problem that was modeled by a separable differential equation. The next example deals with a similar dilution problem, but this one is modeled by a first-order linear differential equation.

EXAMPLE 2.2

A tank contains 20 pounds of salt dissolved in 50 gallons of water. Suppose that 3 gallons of brine containing 2 pounds of dissolved salt per gallon runs into the tank every minute and that the mixture (kept uniform by stirring) runs out of the tank at the rate of 2 gallons per minute. Find an equation for the amount of salt in the tank after t minutes.

Solution

Let $S(t)$ denote the amount of salt in the tank at time t. Since 3 gallons of brine flows into the tank each minute and each gallon contains 2 pounds of salt, then $(3)(2) = 6$ pounds of salt flow into the tank each minute. To find the number of pounds of salt flowing out of the tank each minute, note that at time t, there are $S(t)$ pounds of salt and $50 + (3 - 2)t = 50 + t$ gallons of solution in the tank (because there is a net increase of $3 - 2 = 1$ gallon of solution every minute). Thus, the concentration of salt in solution at time t is $\dfrac{S(t)}{50 + t}$ pounds per gallon, and salt is leaving the tank at the rate

$$\left[\frac{S(t)}{50 + t} \text{ pounds/gallon} \right] [2 \text{ gallons/minute}] = \frac{2S(t)}{50 + t} \text{ pounds/minute}$$

It follows that the net rate of change $\dfrac{dS}{dt}$ of salt in the tank is given by

$$\underbrace{\frac{dS}{dt}}_{\substack{\text{net rate} \\ \text{of change}}} = \underbrace{6}_{\text{inflow}} - \underbrace{\frac{2S}{50 + t}}_{\text{outflow}}$$

(see Figure 7.7). This equation can be written as

$$\frac{dS}{dt} + \frac{2S}{50 + t} = 6$$

which is a first-order linear differential equation with $p(t) = \dfrac{2}{50 + t}$ and $q(t) = 6$. The integrating factor is

Figure 7.7 Rate of change of salt in solution equals inflow rate minus outflow rate.

$$I(t) = e^{\int P(t)dt} = e^{\int 2/(50+t)\, dt}$$

$$= e^{2\ln|50+t|} = \cdot(50 + t)^2$$

Hence, the general solution is

$$S(t) = \frac{1}{(50 + t)^2}\left[\int (6)(50 + t)^2\, dt + C\right]$$

$$= \frac{1}{(50 + t)^2}[2(50 + t)^3 + C]$$

$$= 2(50 + t) + \frac{C}{(50 + t)^2}$$

To evaluate C, note that there are 20 pounds of salt initially in the tank. Thus,

$$20 = S(0) = 2(50 + 0) + \frac{C}{(50 + 0)^2}$$

so that $C = -80(50)^2$. It follows that the amount of salt at time t is

$$S(t) = 2(50 + t) - \frac{80(50)^2}{(50 + t)^2}$$

Problems 7.2

In Problems 1 through 8, find the general solution of the given first-order linear differential equation.

1. $\dfrac{dy}{dx} + \dfrac{3y}{x} = x$

2. $\dfrac{dy}{dx} + \dfrac{2y}{x} = \sqrt{x} + 1$

3. $\dfrac{dy}{dx} + \dfrac{y}{2x} = \sqrt{x}\, e^x$

4. $x^4 \dfrac{dy}{dx} + 2x^3 y = 5$

5. $x^2 \dfrac{dy}{dx} + xy = 2$

6. $x \dfrac{dy}{dx} + 2y = x\, e^{x^3}$

7. $\dfrac{dy}{dx} + \left(\dfrac{2x + 1}{x}\right) y = e^{-2x}$

8. $\dfrac{dy}{dx} + \dfrac{xy}{x^2 + 1} = x$

In Problems 9 through 12, find the particular solution of the given differential equation that satisfies the given condition.

9. $x \dfrac{dy}{dx} - 2y = 2x^3$; $y = 2$ when $x = 1$

10. $\dfrac{dy}{dx} + xy = x + e^{-x^2/2}$; $y = -1$ when $x = 0$

11. $\dfrac{dy}{dx} + y = x$; $y = 4$ when $x = 0$

12. $\dfrac{dy}{dx} + y = x^2$; $y = 2$ when $x = 0$

Differential equations can often be solved in several different ways. In Problems 13 and 14, solve the given differential equation by regarding it as:

(a) Separable

(b) First-order linear

13. $\dfrac{dy}{dx} + 3y = 5$

14. $\dfrac{dy}{dx} + \dfrac{y}{x^2} = 0$

Geometry 15. Find a function $f(x)$ whose graph passes through the point $(-1, 2)$ and has the property that at each point (x, y) on the graph, the slope of the tangent line equals the sum $x + y$ of the coordinates.

Geometry 16. Find a function $f(x)$ whose graph passes through the point $(1, 1)$ and has the property that at each point (x, y) on the graph, the slope of the tangent line equals $x^2 + 2y$.

Real Estate 17. The price of a certain house is currently $200,000. Suppose it is estimated that after t months, the price $P(t)$ will be increasing at the rate of $0.01P(t) + 1,000t$ dollars per month. How much will the house cost 9 months from now?

Retail Prices 18. The price of a certain commodity is currently $3 per unit. It is estimated that t weeks from now, the price $P(t)$ will be increasing at the rate of $0.02P(t) + e^{0.1t}$ cents per week. How much will the commodity cost in 10 weeks?

Savings 19. Chris has a starting salary of $27,000 per year and figures that with salary increases and bonuses, her compensation will increase at an average annual rate of 9 percent. She regularly deposits 5 percent of her salary in a savings account that earns interest at the annual rate of 8 percent compounded continuously.

(a) Set up a first-order linear differential equation for the amount of money in her savings account t years from now. Assume her deposits are made continuously.

(b) How much money will be in her savings account in 20 years?

Savings   20. Chuck is saving to take an $8,000 tour of Europe in 6 years. His parents want to help by making a lump sum deposit of A dollars in a savings account that earns 9 percent interest per year compounded continuously. Chuck plans to add to this amount by making frequent deposits into the account totaling $800 per year. Set up and solve a differential equation for the amount of money $Q(t)$ in his account after t years and determine what A must be for him to meet his goal.

Dilution   21. A tank contains 10 pounds of salt dissolved in 30 gallons of water. Suppose that 2 gallons of brine containing 1 pound of dissolved salt per gallon runs into the tank every minute and that the mixture (kept uniform by stirring) runs out at the rate of 1 gallon per minute.

(a) Find the amount of salt in the tank at time t.

(b) Suppose the tank has a capacity of 50 gallons. How much salt is in the tank when it is full?

Dilution   22. A tank contains 5 pounds of salt dissolved in 40 gallons of water. Pure water runs into the tank at the rate of 1 gallon per minute, and the mixture (kept uniform by stirring) runs out at the rate of 3 gallon per minute.

(a) Find the amount of salt in the tank at time t.

(b) How long does it take for the amount of salt in solution to reach 4 pounds?

(c) How much salt is in solution at the time that the tank is completely drained?

Reduction of Air Pollutant   23. The air in a room with volume 2,500 cubic feet initially contains 0.25 percent carbon dioxide. Suppose the room receives fresh air containing 0.03 percent carbon dioxide at the rate of 1,000 cubic feet per minute.

(a) What percentage of carbon dioxide is in the room after t minutes?

(b) How long does it take before the room contains only 0.08 percent carbon dioxide?

(c) What percentage of carbon dioxide is left in the room in the "long run" (as time t increases without bound)?

Reduction of Water Pollutant   24. A lake has a volume of 6 billion cubic feet, and initially, its pollutant concentration is 0.22 percent. A river whose waters contain only 0.06 percent pollutants flows into the lake at the rate of 350 million cubic feet per day, and another river flows out of the lake carrying 400 million cubic feet per day. Assume that the water in the two rivers and the lake is always well mixed.

(a) Find the concentration of pollutants in the lake as a function of time.

(b) How long does it take to reduce the concentration of pollutants in the lake to 0.1 percent?

Absorption of a Drug   25. The rate at which a certain drug is absorbed into the blood system is given by $\frac{dx}{dt} = r - sx$, where $x(t)$ is the concentration of the drug in the bloodstream at time t and r and s are positive constants. Assume that initially there is no trace of the drug in the blood.

(a) Find $x(t)$.

(b) What happens to $x(t)$ in the "long run" (as t increases without bound)?

Falling Body with Air Resistance   26. In a medium for which air resistance is proportional to the velocity of a falling body, it can be shown that the velocity $v(t)$ of a body with mass m satisfies the differential equation

$$m\frac{dv}{dt} = mg - kv$$

where k is a positive constant and g is the constant acceleration due to gravity. Find $v(t)$ if the initial velocity of the body is v_0 feet per second.

Population Growth with Immigration   27. Let $P(t)$ be the population of a certain country at time t. Suppose that the birth rate r and the death rate s of the country are constant and that there is a constant rate of immigration m.

(a) Explain why the population satisfies the differential equation

$$\frac{dP}{dt} = (r - s)P(t) + m$$

(b) Find $P(t)$.

(c) If the population of the country is 100 million in 1990, with a growth rate (birth rate minus death rate) of 2 percent, and if immigration is allowed at the rate of 300,000 people per year, what will the population be in the year 2000?

Spy Story   28. While the spy is preoccupied with analyzing his retirement options (Problem 10 in Chapter 6, Section 4), he is captured by the archfiend André Scélérat. He is tied to a chair and placed in a closet into which poison gas is being slowly pumped. The closet has volume 400 cubic feet. The poison gas enters slowly at the rate of 0.2 cubic foot per minute, and the mixed air (gas and fresh air) escapes at the same rate through a crack under the door. The gas becomes deadly when the air in the room contains 0.8 cubic foot of poison gas. If the spy takes 3 minutes to get untied and another 45 seconds to kick open the door, will he survive?

Chapter Summary and Review Problems

Important Terms, Symbols, and Formulas

Differential equation

General solution; particular solution

Separable differential equation:

If $g(y)\ dy = h(x)\ dx$, then $\int g(y)\ dy = \int h(x)\ dx + C$

First-order linear differential equation:

If $\dfrac{dy}{dt} + p(x)y = q(x)$, then the integrating factor is $I(x) = e^{\int p(x)dx}$ and the general solution is $y = \dfrac{1}{I(x)}\left[\int I(x)q(x)dx + C\right]$

Exponential models:

Model	Differential Equation	General Solution	Graph
Exponential growth	$\dfrac{dQ}{dt} = kQ$ $k > 0$	$Q(t) = Q_0 e^{kt}$	
Exponential decay	$\dfrac{dQ}{dt} = -kQ$ $k > 0$	$Q(t) = Q_0 e^{-kt}$	
Learning curve	$\dfrac{dQ}{dt} = k(B - Q)$	$Q(t) = B - Ae^{-kt}$	

Model	Differential Equation	General Solution	Graph
Logistic curve	$\dfrac{dQ}{dt} = kQ(B - Q)$	$Q(t) = \dfrac{B}{1 + Ae^{-Bkt}}$	

Review Problems In Problems 1 through 10, find the general solution of the given differential equation.

1. $\dfrac{dy}{dx} = x^3 - 3x^2 + 5$

2. $\dfrac{dy}{dx} = 0.02y$

3. $\dfrac{dy}{dx} = k(80 - y)$

4. $\dfrac{dy}{dx} = y(1 - y)$

5. $\dfrac{dy}{dx} = e^{x+y}$ (*Hint:* $e^{x+y} = e^x e^y$.)

6. $\dfrac{dy}{dx} = ye^{-2x}$

7. $\dfrac{dy}{dx} + \dfrac{4y}{x} = e^{-x}$

8. $\dfrac{dy}{dx} + \dfrac{y}{x} = \dfrac{2}{x + 1}$

9. $\dfrac{dy}{dx} + \dfrac{2y}{x + 1} = \dfrac{4}{x + 2}$

10. $x^3 \dfrac{dy}{dx} + xy = 5$

In Problems 11 through 18, find the particular solution of the given differential equation that satisfies the given condition.

11. $\dfrac{dy}{dx} = 5x^4 - 3x^2 - 2$; $y = 4$ when $x = 1$

12. $\dfrac{dy}{dx} = 0.06y$; $y = 100$ when $x = 0$

13. $\dfrac{dy}{dx} = 3 - y$; $y = 2$ when $x = 0$

14. $\dfrac{d^2y}{dx^2} = 2$; $y = 5$ and $\dfrac{dy}{dx} = 3$ when $x = 0$

15. $\dfrac{dy}{dx} - \dfrac{5y}{x} = x^2$; $y = 4$ when $x = -1$

16. $\dfrac{dy}{dx} + \dfrac{y}{x + 1} = x$; $y = 0$ when $x = 3$

17. $\dfrac{dy}{dx} - xy = e^{x^2/2}$; $y = 4$ when $x = 0$

18. $\dfrac{dy}{dx} + 3x^2y = x^2$; $y = 2$ when $x = 0$

19. The resale value of a certain industrial machine decreases at a rate proportional to the difference between its current value and its scrap value of $5,000. The machine was bought new for $40,000 and was worth $30,000 after 4 years. How much will it be worth when it is 8 years old?

20. A certain oil well that yields 600 barrels of crude oil per month will run dry in 3 years. The price of crude oil is currently $24 per barrel and is expected to rise at the constant rate of 8 cents per barrel per month. If the oil is sold as soon as it is extracted from the ground, what will the total future revenue from the well be?

21. A tank currently holds 200 gallons of brine that contains 3 pounds of salt per gallon. Clear water flows into the tank at the rate of 4 gallons per minute, while the mixture, which is kept uniform, runs out of the tank at the rate of 5 gallons per minute. How much salt is in the tank at the end of 100 minutes?

22. The rate at which the population of a certain country is growing is jointly proportional to the upper bound of 10 million imposed by environmental factors and the difference between the upper bound and the size of the population. Express the population (in millions) of the country as a function of time (in years measured from 1985) if the population in 1985 was 4 million and the population in 1990 was 4.74 million.

23. A nation has 5 billion dollars in currency. Each day, about 18 million dollars comes into the banks and the same amount is paid out. Suppose the country decides that whenever a dollar bill comes into a bank, it is destroyed and replaced by a new style of currency. How long will it take for 90 percent of the currency in circulation to be the new style? [*Hint:* Think of this situation as if it were a dilution problem like Example 1.11, Section 7.1.]

24. A tank initially contains 30 pounds of salt dissolved in 50 gallons of water. Brine containing 2 pounds of salt per gallon runs into the tank at the rate of 2 gallons per minute, and the mixture, kept uniform by stirring, flows out at the rate of 1 gallon per minute.
    (a) Find the amount of salt in solution in the tank at time t.
    (b) How many gallons of solution are in the tank at the time when there are 40 pounds of salt?

25. **Pareto's law** in economics says that the rate of change (decrease) of

the number of people P in a stable economy who have an income of at least x dollars is directly proportional to the number of such people and inversely proportional to their income. Express this law as a differential equation and solve for P in terms of x.

26. A **Gompertz function** is used by demographers to predict populations. Let P_0 be the initial population of a region (a country, the world, etc.) and let β be the growth rate of the population. Then the Gompertz function $P(t)$ represents the population at time t and satisfies the differential equation

$$\frac{dP}{dt} = P(\ln P_0)(\ln \beta)\beta^t$$

Solve this equation to find P as a function of t. Assume that P_0 and β are constant.

CHAPTER 8
LIMITS AT INFINITY AND IMPROPER INTEGRALS

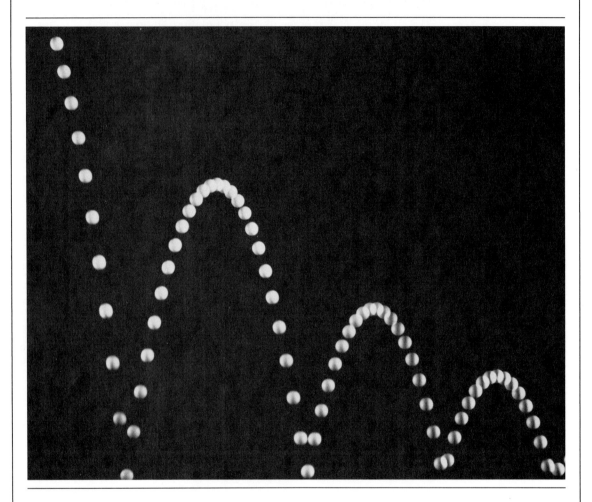

1   Limits at Infinity and L'Hôpital's Rule

The limit of a function as its variable increases without bound (sometimes called the **limit at infinity**) may give useful information in practical situations. For example, if the variable represents time, such a limit describes what will happen to the function "in the long run." This interpretation is illustrated in Figure 8.1. The curve in Figure 8.1a represents the value of an industrial machine that decreases exponentially and approaches $1,000 (scrap value) in the long run. The curve in Figure 8.1b is a logistic curve for bounded population growth. It shows a population that will eventually approach a fixed upper bound B.

Figure 8.1 (a) Depreciation of machinery. (b) Population growth.

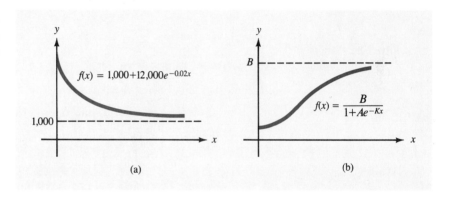

Later in this chapter, you will see some important consequences of the application of this limit concept to definite integrals. Limits at infinity will also play a central role in the discussion of infinite series in Chapter 11. The purpose of this introductory section is to show you some of the computational techniques you can use to find the limit of a function as its variable increases without bound.

Notation and Properties for Limits at Infinity

The symbol $\lim_{x \to \infty} f(x)$ is read "the limit of $f(x)$ as x approaches infinity" and denotes the behavior of $f(x)$ as x increases without bound. If $f(x)$ approaches a finite number L, you write

$$\lim_{x \to \infty} f(x) = L$$

If $f(x)$ increases or decreases without bound, you write

$$\lim_{x \to \infty} f(x) = \infty \qquad \text{or} \qquad \lim_{x \to \infty} f(x) = -\infty$$

The situation is illustrated in Figure 8.2. The limit properties stated in Section 5 of Chapter 1 all hold for limits at infinity. Here is a list of those properties.

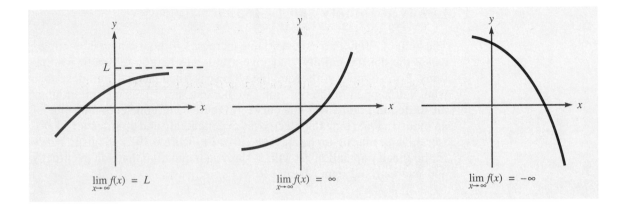

Figure 8.2 Limits at infinity.

Properties of Limits at Infinity

If $\lim_{x \to \infty} f(x)$ and $\lim_{x \to \infty} g(x)$ exist, then

$$\lim_{x \to \infty} [f(x) + g(x)] = \lim_{x \to \infty} f(x) + \lim_{x \to \infty} g(x)$$

$$\lim_{x \to \infty} [f(x) - g(x)] = \lim_{x \to \infty} f(x) - \lim_{x \to \infty} g(x)$$

$$\lim_{x \to \infty} [kf(x)] = k \lim_{x \to \infty} f(x) \quad \text{for any constant } k$$

$$\lim_{x \to \infty} [f(x)g(x)] = [\lim_{x \to \infty} f(x)][\lim_{x \to \infty} g(x)]$$

$$\lim_{x \to \infty} \left[\frac{f(x)}{g(x)} \right] = \frac{\lim_{x \to \infty} f(x)}{\lim_{x \to \infty} g(x)} \quad \text{if } \lim_{x \to \infty} g(x) \neq 0$$

$$\lim_{x \to \infty} [f(x)]^p = [\lim_{x \to \infty} f(x)]^p \quad \text{for any number } p$$
$$\text{if } [\lim_{x \to \infty} f(x)]^p \text{ is defined}$$

Review of Some Basic Limits

For reference, here is a summary of some important limits that you should already know.

Limits of Basic Functions

Powers of x: If $n > 0$,

$$\lim_{x \to \infty} x^n = \infty \quad \text{and} \quad \lim_{x \to \infty} \frac{1}{x^n} = 0$$

Exponential functions: If $k > 0$,

$$\lim_{x \to \infty} e^{kx} = \infty \quad \text{and} \quad \lim_{x \to \infty} e^{-kx} = 0$$

Natural logarithm: $\displaystyle\lim_{x \to \infty} \ln x = \infty$

The graphs of these basic functions are sketched in Figure 8.3.

Figure 8.3 Limits of some basic functions.

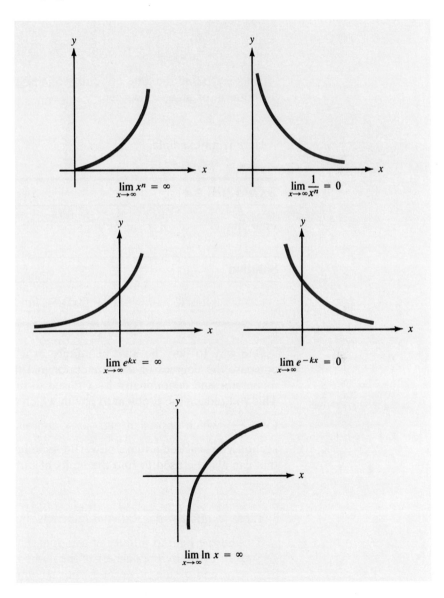

$$\lim_{x \to \infty} x^n = \infty$$

$$\lim_{x \to \infty} \frac{1}{x^n} = 0$$

$$\lim_{x \to \infty} e^{kx} = \infty$$

$$\lim_{x \to \infty} e^{-kx} = 0$$

$$\lim_{x \to \infty} \ln x = \infty$$

Elementary Calculations   The following examples illustrate elementary techniques you can use to find the limits of functions that are built from one or more of the five basic functions. The first example involves a polynomial. In general, the limit of a polynomial as x increases without bound is determined by its term of highest degree, which increases (or decreases) more rapidly than the other terms of lower degree.

Limit at Infinity of a Polynomial

If $a_n \neq 0$,

$$\lim_{x \to \infty} (a_0 + a_1 x + \cdots + a_n x^n) = \lim_{x \to \infty} a_n x^n$$

That is, to find the limit at infinity of a polynomial, take the limit of the term of highest degree.

Here is an example.

EXAMPLE 1.1

Find $\lim_{x \to \infty} (1 - x^2 + x^3 - 3x^4)$.

Solution

$$\lim_{x \to \infty} (1 - x^2 + x^3 - 3x^4) = \lim_{x \to \infty} (-3x^4) = -\infty$$

One way to find the limit at infinity of a rational function is to first compare the degrees of the numerator and the denominator and divide numerator and denominator by x raised to the smaller of these degrees. This will reduce the problem to one in which most of the terms are of the form $\dfrac{a}{x^n}$, which approach zero as x approaches infinity. Later in this section you will see a more powerful technique involving differentiation that can also be used to find the limits of rational functions.

Limit at Infinity of a Rational Function

To find the limit at infinity of a rational function:
Step 1. Compare the degrees of the numerator and the denominator and divide numerator and denominator by x raised to the smaller of these degrees.
Step 2. Take the limits of the new numerator and denominator.

Here are two examples illustrating the technique.

EXAMPLE 1.2

Find $\lim\limits_{x \to \infty} \dfrac{2x^2 + 3x + 1}{3x^2 - 5x + 2}$.

Solution

Divide numerator and denominator by x^2 to get

$$\lim_{x \to \infty} \frac{2x^2 + 3x + 1}{3x^2 - 5x + 2} = \lim_{x \to \infty} \frac{2 + 3/x + 1/x^2}{3 - 5/x + 2/x^2} = \frac{2 + 0 + 0}{3 - 0 + 0} = \frac{2}{3}$$

EXAMPLE 1.3

Find $\lim\limits_{x \to \infty} \dfrac{-x^3 + 2x + 1}{x - 3}$

Solution

Divide numerator and denominator by x to get

$$\lim_{x \to \infty} \frac{-x^3 + 2x + 1}{x - 3} = \lim_{x \to \infty} \frac{-x^2 + 2 + 1/x}{1 - 3/x}$$

Since

$$\lim_{x \to \infty} \left(-x^2 + 2 + \frac{1}{x} \right) = -\infty \quad \text{and} \quad \lim_{x \to \infty} \left(1 - \frac{3}{x} \right) = 1$$

it follows that

$$\lim_{x \to \infty} \frac{-x^3 + 2x + 1}{x - 3} = -\infty$$

In the next two examples, the elementary limit properties of the exponential function are used to find the limits of more complicated functions.

EXAMPLE 1.4

Find $\lim\limits_{x \to \infty} \dfrac{e^{-x}}{1 + e^{2x}}$.

Solution

The limits of the numerator and denominator are, respectively,

$$\lim_{x \to \infty} e^{-x} = 0 \quad \text{and} \quad \lim_{x \to \infty} (1 + e^{2x}) = \infty$$

The behavior of the numerator and that of the denominator reinforce each other. Both tend to drive the quotient to zero as x increases without bound. It follows that

$$\lim_{x \to \infty} \frac{e^{-x}}{1 + e^{2x}} = 0$$

EXAMPLE 1.5

Find $\lim\limits_{x \to \infty} \dfrac{x^2 + 1}{-e^{-x}}$.

Solution

The limits of the numerator and the denominator are, respectively,

$$\lim_{x \to \infty} (x^2 + 1) = \infty \quad \text{and} \quad \lim_{x \to \infty} (-e^{-x}) = 0$$

Again, the behavior of the numerator and that of the denominator reinforce each other. This time, both tend to drive the absolute value of the quotient to infinity as x increases without bound. Moreover, the sign of the quotient is negative, since the numerator is positive and the denominator negative. It follows that

$$\lim_{x \to \infty} \frac{x^2 + 1}{-e^{-x}} = -\infty$$

L'Hôpital's Rule: $\dfrac{0}{0}$ and $\dfrac{\infty}{\infty}$ forms

In each of the four preceding examples, you were able to determine the limit of a quotient easily from the limits of its numerator and denominator. This determination is not always so straightforward. Consider, for example, the quotient $\dfrac{x}{e^x}$. As x increases without bound, the numerator approaches infinity, which tends to drive the quotient to infinity, while the denominator also approaches infinity, which tends to drive the quotient to zero. Without further analysis, it is not clear whether the numerator or the denominator will dominate. The quotient might approach zero or infinity or any number in between. In this case one says that the limit of the quotient is of the indeterminate form $\dfrac{\infty}{\infty}$.

Similar uncertainty arises if the limits of the numerator and the denominator are both zero. In this case one says that the limit of the quotient is of the indeterminate form $\frac{0}{0}$.

There is a powerful technique, known as L'Hôpital's rule, which you can use to analyze indeterminate forms. The rule says, in effect, that if your attempt to find the limit of a quotient leads to one of the indeterminate forms $\frac{0}{0}$ or $\frac{\infty}{\infty}$, take the derivatives of the numerator and the denominator and try again. Here is a summary of the procedure.

L'Hôpital's Rule

If $\lim_{x \to \infty} f(x) = 0$ and $\lim_{x \to \infty} g(x) = 0$, then

$$\lim_{x \to \infty} \frac{f(x)}{g(x)} = \lim_{x \to \infty} \frac{f'(x)}{g'(x)}$$

If $\lim_{x \to \infty} f(x) = \infty$ and $\lim_{x \to \infty} g(x) = \infty$, then

$$\lim_{x \to \infty} \frac{f(x)}{g(x)} = \lim_{x \to \infty} \frac{f'(x)}{g'(x)}$$

The proof of L'Hôpital's rule is technical and beyond the scope of this book. However, the fact that the rule involves the ratio of the derivatives of f and g should not come as a complete surprise. The value of an indeterminate form depends on the relative rates with which the functions in the numerator and the denominator approach their limits, and these rates are given by the derivatives of the functions.

Advice on Using L'Hôpital's Rule

The use of L'Hôpital's rule is illustrated in the following examples. As you read these examples, pay particular attention to the following two points:

1. L'Hôpital's rule involves differentiation of the numerator and the denominator *separately*. A common mistake is to differentiate the entire quotient using the quotient rule.
2. L'Hôpital's rule applies only to quotients whose limits are indeterminate forms $\frac{0}{0}$ or $\frac{\infty}{\infty}$. Limits of the form $\frac{0}{\infty}$ or $\frac{\infty}{0}$, for example, are *not* indeterminate, as you saw in Examples 1.4 and 1.5, and L'Hôpital's rule does not apply to such limits.

EXAMPLE 1.6

Find $\lim_{x \to \infty} \dfrac{x}{e^x}$.

Solution

This limit is of the indeterminate form $\dfrac{\infty}{\infty}$. Applying L'Hôpital's rule, you get

$$\lim_{x \to \infty} \frac{x}{e^x} = \lim_{x \to \infty} \frac{1}{e^x} = 0$$

EXAMPLE 1.7

Find $\displaystyle\lim_{x \to \infty} \frac{e^{1/x} - 1}{1/x}$.

Solution

This limit is of the indeterminate form $\dfrac{0}{0}$. Applying L'Hôpital's rule and then simplifying the resulting quotient, you get

$$\lim_{x \to \infty} \frac{e^{1/x} - 1}{1/x} = \lim_{x \to \infty} \frac{(-1/x^2)e^{1/x}}{-1/x^2} \quad \text{(L'Hôpital's rule)}$$

$$= \lim_{x \to \infty} e^{1/x} \quad \text{(algebraic simplification)}$$

$$= e^0 = 1$$

EXAMPLE 1.8

Find $\displaystyle\lim_{x \to \infty} \frac{e^{-x}}{1 + e^{-x}}$.

Solution

This limit is of the form $\dfrac{0}{1}$, which is *not* indeterminate and can be calculated directly as follows:

$$\lim_{x \to \infty} \frac{e^{-x}}{1 + e^{-x}} = \frac{0}{1} = 0$$

To convince yourself that L'Hôpital's rule does not apply in this case, differentiate the numerator and the denominator. The resulting quotient is equal to 1, and its limit is certainly not zero.

Sometimes one application of L'Hôpital's rule is not enough. In the next example, the use of L'Hôpital's rule leads to a new quotient whose limit is also of an indeterminate form. A second application of L'Hôpital's rule is needed to determine the limit.

EXAMPLE 1.9

Find $\lim\limits_{x \to \infty} \dfrac{x \ln x}{e^{2x}}$.

Solution

This limit is of the indeterminate form $\dfrac{\infty}{\infty}$. Applying L'Hôpital's rule (and using the product rule to differentiate the numerator), you get

$$\lim_{x \to \infty} \frac{x \ln x}{e^{2x}} = \lim_{x \to \infty} \frac{1 + \ln x}{2e^{2x}}$$

Since the new limit is also of the form $\dfrac{\infty}{\infty}$, you apply L'Hôpital's rule again to get

$$\lim_{x \to \infty} \frac{1 + \ln x}{2e^{2x}} = \lim_{x \to \infty} \frac{1/x}{4e^{2x}} = 0$$

and you conclude that

$$\lim_{x \to \infty} \frac{x \ln x}{e^{2x}} = 0$$

Comment: In problems involving repeated application of L'Hôpital's rule, do not forget to verify that each new limit is of an indeterminate form before you apply the rule again. In Example 1.9, for instance, the limit of the quotient $\dfrac{1/x}{4e^{2x}}$ was of the form $\dfrac{0}{\infty}$, and further application of L'Hôpital's rule would have been invalid.

L'Hôpital's rule gives you an alternative way to find limits of rational functions. Here is the calculation for the rational functions from Example 1.2.

EXAMPLE 1.10

Find $\lim\limits_{x \to \infty} \dfrac{2x^2 + 3x + 1}{3x^2 - 5x + 2}$.

Solution

This limit is of the indeterminate form $\dfrac{\infty}{\infty}$. Repeated application of L'Hôpital's rule gives

$$\lim_{x\to\infty} \frac{2x^2 + 3x + 1}{3x^2 - 5x + 2} = \lim_{x\to\infty} \frac{4x + 3}{6x - 5} = \lim_{x\to\infty} \frac{4}{6} = \frac{2}{3}$$

The Indeterminate Form $0 \cdot \infty$

The limit of a product in which one of the factors approaches zero and the other approaches infinity is said to be of the indeterminate form $0 \cdot \infty$, since it is not clear which factor will dominate. To evaluate such a limit, try writing the product as a quotient whose limit is of the form $\dfrac{0}{0}$ or $\dfrac{\infty}{\infty}$ and apply L'Hôpital's rule. Here is an example.

EXAMPLE 1.11

Find $\lim\limits_{x\to\infty} e^{-x} \ln x$.

Solution

This limit is of the indeterminate form $0 \cdot \infty$ and can be rewritten as

$$\lim_{x\to\infty} \frac{e^{-x}}{1/\ln x} \qquad \left(\text{of the form } \frac{0}{0}\right)$$

or as

$$\lim_{x\to\infty} \frac{\ln x}{e^x} \qquad \left(\text{of the form } \frac{\infty}{\infty}\right)$$

Applying L'Hôpital's rule to the simpler second quotient, you get

$$\lim_{x\to\infty} e^{-x} \ln x = \lim_{x\to\infty} \frac{\ln x}{e^x} = \lim_{x\to\infty} \frac{1/x}{e^x} = 0$$

Other Indeterminate Forms

The limit of an expression of the form $f(x)^{g(x)}$ may be one of the indeterminate forms 1^{∞}, ∞^0, or 0^0. To evaluate such a limit, try using logarithms to simplify the situation. The technique is illustrated in the next example.

EXAMPLE 1.12

Find $\lim\limits_{x\to\infty} x^{1/x}$.

Solution

This limit is of the indeterminate form ∞^0. To simplify the problem, let

$$y = x^{1/x}$$

and take the logarithm of both sides of this equation to get

$$\ln y = \frac{1}{x} \ln x$$

As a preliminary calculation find the limit of $\ln y$ as follows:

$$\lim_{x \to \infty} \ln y = \lim_{x \to \infty} \frac{1}{x} \ln x \quad (0 \cdot \infty)$$

$$= \lim_{x \to \infty} \frac{\ln x}{x} \quad (\infty/\infty)$$

$$= \lim_{x \to \infty} \frac{1/x}{1} \quad \text{(L'Hôpital's rule)}$$

$$= 0$$

Having completed this preliminary calculation, you know the limit of the natural logarithm of y but still have to find the limit of the original function y itself. To proceed, observe that since

$$\ln y \to 0$$

it follows from the identity $y = e^{\ln y}$ that

$$y \to e^0 = 1$$

That is, $\displaystyle\lim_{x \to \infty} x^{1/x} = 1$

As a final illustration of this technique, here is the limit that was used in Chapter 4, Section 1, to define the number e.

EXAMPLE 1.13

Find $\displaystyle\lim_{x \to \infty} \left(1 + \frac{1}{x} \right)^x$.

Solution

This limit is of the indeterminate form 1^∞. To simplify the problem let

$$y = \left(1 + \frac{1}{x} \right)^x$$

Then
$$\ln y = x \ln \left(1 + \frac{1}{x}\right)$$

and

$$\lim_{x \to \infty} \ln y = \lim_{x \to \infty} x \ln \left(1 + \frac{1}{x}\right) \qquad (\infty \cdot 0)$$

$$\lim_{x \to \infty} \ln y = \lim_{x \to \infty} \frac{\ln (1 + 1/x)}{1/x} \qquad \left(\frac{0}{0}\right)$$

$$= \lim_{x \to \infty} \frac{\dfrac{(-1/x^2)}{(1 + 1/x)}}{-1/x^2} \qquad \text{(L'Hôpital's rule)}$$

$$= \lim_{x \to \infty} \frac{1}{1 + 1/x} \qquad \text{(algebraic simplification)}$$

$$= 1$$

Since $\ln y \to 1$, it follows that $y \to e^1 = e$. That is,

$$\lim_{x \to \infty} \left(1 + \frac{1}{x}\right)^x = e$$

Problems 8.1

For Problems 1 through 38, find $\lim_{x \to \infty} f(x)$.

1. $f(x) = x^3 - 4x^2 - 4$

2. $f(x) = 1 - x + 2x^2 - 3x^3$

3. $f(x) = (1 - 2x)(x + 5)$

4. $f(x) = (1 + x^2)^3$

5. $f(x) = \dfrac{x^2 - 2x + 3}{2x^2 + 5x + 1}$

6. $f(x) = \dfrac{1 - 3x^3}{2x^3 - 6x + 2}$

7. $f(x) = \dfrac{2x + 1}{3x^2 + 2x - 7}$

8. $f(x) = \dfrac{x^2 + x - 5}{1 - 2x - x^3}$

9. $f(x) = \dfrac{3x^2 - 6x + 2}{2x - 9}$

10. $f(x) = \dfrac{1 - 2x^3}{x + 1}$

11. $f(x) = 2 - 3e^x$

12. $f(x) = 5 + e^{-3x}$

13. $f(x) = \dfrac{3}{2 + 5e^{-8x}}$

14. $f(x) = \dfrac{e^x}{2 + e^{-x}}$

15. $f(x) = \dfrac{e^{2x}}{3x}$

16. $f(x) = \dfrac{e^{-2x}}{3x}$

17. $f(x) = \dfrac{2x}{e^{-3x}}$

18. $f(x) = \dfrac{2x}{e^{3x}}$

19. $f(x) = \dfrac{x^2}{e^x}$

20. $f(x) = \dfrac{e^{2x}}{x^2}$

21. $f(x) = \dfrac{\sqrt{x}}{e^x}$

22. $f(x) = \dfrac{e^x}{x^{3/2}}$

23. $f(x) = \dfrac{\ln x}{x}$

24. $f(x) = \dfrac{\sqrt{x}}{\ln x}$

25. $f(x) = \dfrac{\ln (x^2 + 1)}{x}$

26. $f(x) = \dfrac{(\ln x)^2}{x}$

27. $f(x) = \dfrac{\ln (2x + 1)}{\ln (3x - 1)}$

28. $f(x) = \dfrac{\ln (x^2 + 4)}{\ln (x - 1)}$

29. $f(x) = \dfrac{e^x}{x \ln x}$

30. $f(x) = \dfrac{x \ln x}{x + \ln x}$

31. $f(x) = \dfrac{x}{e^{\sqrt{x}}}$

32. $f(x) = xe^{-x^2}$

33. $f(x) = e^{-2x} \ln x$

34. $f(x) = x(e^{1/x} - 1)$

35. $f(x) = x^{2/x}$

36. $f(x) = (\ln x)^{1/x}$

37. $f(x) = \left(1 + \dfrac{2}{x}\right)^x$

38. $f(x) = \left(1 + \dfrac{1}{x}\right)^{x^2}$

39. In the next section you will encounter several limits of the form $\lim\limits_{x \to \infty} x^P e^{-kx}$. Show that for any constant P and any positive constant k, this limit is equal to zero. (*Hint:* Write $x^P e^{-kx}$ as a quotient and show that after a sufficient number of applications of L'Hôpital's rule, the limit can be rewritten as

$$\lim_{x \to \infty} \frac{C}{e^{kx}} \quad \text{or} \quad \lim_{x \to \infty} \frac{C}{x^n e^{kx}}$$

where C and n are constants and $n > 0$.)

40. Evaluate the limit

$$\lim_{x \to \infty} \frac{a_n x^n + a_{n-1} x^{n-1} + \cdots + a_1 x + a_0}{b_m x^m + b_{m-1} x^{m-1} + \cdots + b_1 x + b_0}$$

for constants $a_0, a_1, \ldots, a_n$ and $b_0, b_1, \ldots, b_m$ in each of the following cases:

(a) $n < m$

(b) $n = m$

(c) $n > m$ (Note: There are two possible answers, depending on the signs of a_n and b_m.)

2 Improper Integrals

This section extends the concept of the definite integral to integrals of the form

$$\int_a^\infty f(x)\, dx$$

in which the upper limit of integration is not a finite number. Such integrals are known as **improper integrals** and arise in a variety of practical situations.

Geometric Interpretation   If f is nonnegative, the improper integral $\int_a^\infty f(x)\, dx$ may be interpreted as the area of the region under the graph of f to the right of $x = a$ (Figure 8.4a). Although this region has infinite extent, its area may be either finite or infinite, depending on how quickly $f(x)$ approaches zero as x increases.

A reasonable strategy for finding the area of such a region is to first use a definite integral to compute the area from $x = a$ to some finite number $x = N$ and then to let N approach infinity in the resulting expression. That is,

$$\text{Total area} = \lim_{N \to \infty} (\text{area from } a \text{ to } N) = \lim_{N \to \infty} \int_a^N f(x)\, dx$$

This strategy is illustrated in Figure 8.4 and motivates the following definition of the improper integral.

Figure 8.4

$$\text{Area} = \int_a^\infty f(x)\, dx$$
$$= \lim_{N \to \infty} \int_a^N f(x)\, dx.$$

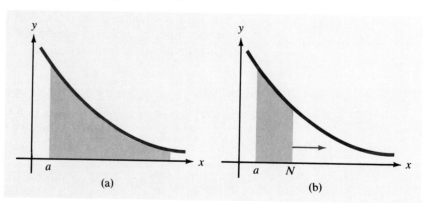

(a)   (b)

The Improper Integral

$$\int_a^\infty f(x)\, dx = \lim_{N \to \infty} \int_a^N f(x)\, dx$$

If the limit defining the improper integral is a finite number, the integral is said to **converge**. Otherwise the integral is said to **diverge**. Here are some examples.

EXAMPLE 2.1

Evaluate $\displaystyle\int_1^\infty \frac{1}{x^2}\, dx$.

Solution

First compute the integral from 1 to N and then let N approach infinity. Arrange your work as follows:

$$\int_1^\infty \frac{1}{x^2}\, dx = \lim_{N\to\infty} \int_1^N \frac{1}{x^2}\, dx = \lim_{N\to\infty} \left(-\frac{1}{x}\Big|_1^N\right) = \lim_{N\to\infty} \left(-\frac{1}{N} + 1\right) = 1$$

EXAMPLE 2.2

Evaluate $\displaystyle\int_1^\infty \frac{1}{x}\, dx$.

Solution

$$\int_1^\infty \frac{1}{x}\, dx = \lim_{N\to\infty} \int_1^N \frac{1}{x}\, dx = \lim_{N\to\infty} \left(\ln |x| \Big|_1^N\right) = \lim_{N\to\infty} \ln N = \infty$$

Notice that the improper integral of the function $f(x) = \dfrac{1}{x^2}$ in Example 2.1 converged while that of the function $f(x) = \dfrac{1}{x}$ in Example 2.2 diverged. In geometric terms, this says that the area to the right of $x = 1$ under the curve $y = \dfrac{1}{x^2}$ is finite while the corresponding area under the curve $y = \dfrac{1}{x}$ is infinite. The reason for the difference is that, as x increases, $\dfrac{1}{x^2}$ approaches zero more quickly than does $\dfrac{1}{x}$. (See Figure 8.5.)

The evaluation of improper integrals arising from practical problems often involves limits of the form

$$\lim_{N\to\infty} N^P e^{-kN} \qquad \text{(for } k > 0)$$

Figure 8.5 The areas under $y = \dfrac{1}{x}$ and $y = \dfrac{1}{x^2}$

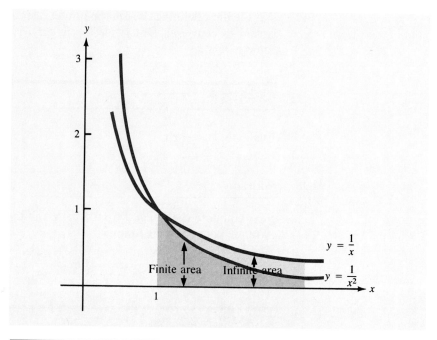

These limits are of the indeterminate form $\infty \cdot 0$ and can be found using L'Hôpital's rule. (See Problem 39 of the preceding section.)

A Useful Limit for Improper Integrals

For any power p and positive number k,

$$\lim_{N \to \infty} N^p e^{-kN} = 0$$

The next example involves a limit of this type and also requires the use of integration by parts.

EXAMPLE 2.3

Evaluate $\displaystyle\int_0^\infty xe^{-2x}\, dx$.

Solution

$$\int_0^\infty xe^{-2x}\, dx = \lim_{N \to \infty} \int_0^N xe^{-2x}\, dx$$

$$= \lim_{N \to \infty} \left(-\frac{1}{2} xe^{-2x} \Big|_0^N + \frac{1}{2} \int_0^N e^{-2x} \, dx \right) \quad \text{(integration by parts)}$$

$$= \lim_{N \to \infty} \left(-\frac{1}{2} xe^{-2x} - \frac{1}{4} e^{-2x} \right) \Big|_0^N$$

$$= \lim_{N \to \infty} \left(-\frac{1}{2} Ne^{-2N} - \frac{1}{4} e^{-2N} + 0 + \frac{1}{4} \right)$$

$$= \frac{1}{4} \quad \text{(since } Ne^{-2N} \to 0)$$

Applications of the Improper Integral

The following applications of the improper integral generalize applications of the definite integral that you saw in Chapter 6, Section 4. In each, the strategy is to use the characterization of the definite integral as the limit of a sum to construct an appropriate definite integral and then to let the upper limit of integration increase without bound. As you read these examples, you may want to refer back to the corresponding examples in Chapter 6.

Present Value

In Example 4.2 of Chapter 6 you saw that the present value of an investment that generates income over a finite period of time is given by a definite integral. The present value of an investment that generates income in perpetuity is given by an improper integral. Here is an example.

EXAMPLE 2.4

A donor wishes to make a gift to a private college from which the college will draw $7,000 per year in perpetuity to support the operation of its computer center. Assuming that the prevailing annual interest rate will remain fixed at 10 percent compounded continuously, how much should the donor give the college? That is, what is the present value of the endowment?

Solution

To find the present value of a gift that generates $7,000 per year for N years, divide the N-year time interval $0 \leq t \leq N$ into n subintervals of length Δt years and let t_j denote the beginning of the jth subinterval (Figure 8.6).

Then

$$\begin{array}{c} \text{Amount generated during} \\ j\text{th subinterval} \end{array} \simeq 7,000 \, \Delta t$$

Figure 8.6 The present value of the money generated during the jth subinterval.

and

$$
\begin{array}{c}
\text{Present value of amount}\\
\text{generated during}\\
j\text{th subinterval}
\end{array} \approx 7{,}000e^{-0.1t_j}\,\Delta t
$$

Hence,

$$
\begin{array}{c}
\text{Present value of}\\
N\text{-year gift}
\end{array} = \lim_{n \to \infty} \sum_{j=1}^{n} 7{,}000e^{-0.1t_j}\,\Delta t
$$

$$
= \int_0^N 7{,}000e^{-0.1t_j}\,dt
$$

To find the present value of the total gift, take the limit of this integral as N approaches infinity. That is,

$$
\begin{array}{c}
\text{Present value of}\\
\text{total gift}
\end{array} = \lim_{N \to \infty} \int_0^N 7{,}000e^{-0.1t}\,dt
$$

$$
= \lim_{N \to \infty} \left(-70{,}000e^{-0.1t} \,\Big|_0^N \right)
$$

$$
= \lim_{N \to \infty} -70{,}000(e^{-0.1N} - 1)
$$

$$
= \$70{,}000
$$

Nuclear Waste   The next example is similar in structure to the problem involving survival and renewal functions in Example 4.3 of Chapter 6.

EXAMPLE 2.5

It is estimated that t years from now, a certain nuclear power plant will be producing radioactive waste at the rate of $f(t) = 400t$ pounds per year. The waste decays exponentially at the rate of 2 percent per year. What will happen to the accumulation of radioactive waste from the plant in the long run?

Solution

To find the amount of radioactive waste present after N years, divide the N-year interval $0 \leq t \leq N$ into n equal subintervals of length Δt years and let t_j denote the beginning of the jth subinterval (Figure 8.7). Then,

Figure 8.7 Radioactive waste generated during the jth subinterval.

$$\begin{array}{l}\text{Amount of waste produced}\\\text{during }j\text{th subinterval}\end{array} \approx 400t_j\,\Delta t$$

Since the waste decays exponentially at the rate of 2 percent per year, and since there are $N - t_j$ years between times $t = t_j$ and $t = N$, it follows that

$$\begin{array}{l}\text{Amount of waste produced}\\\text{during }j\text{th subinterval}\\\text{still present at }t = N\end{array} \approx 400t_j e^{-0.02(N-t_j)}\,\Delta t$$

Thus,

$$\begin{aligned}\begin{array}{l}\text{Amount of waste present}\\\text{after }N\text{ years}\end{array} &= \lim_{n\to\infty}\sum_{j=1}^{n} 400t_j e^{-0.02(N-t_j)}\,\Delta t\\[2mm]
&= \int_0^N 400te^{-0.02(N-t)}\,dt\\[2mm]
&= 400e^{-0.02N}\int_0^N te^{0.02t}\,dt\end{aligned}$$

The amount of radioactive waste present in the long run is the limit of this expression as N approaches infinity. That is,

$$\begin{aligned}\begin{array}{l}\text{Amount of waste}\\\text{present in the}\\\text{long run}\end{array} &= \lim_{N\to\infty} 400e^{-0.02N}\int_0^N te^{0.02t}\,dt\\[2mm]
&= \lim_{N\to\infty} 400e^{-0.02N}(50te^{0.02t} - 2{,}500e^{0.02t})\Big|_0^N\\[2mm]
&= \lim_{N\to\infty} 400e^{-0.02N}(50Ne^{0.02N} - 2{,}500e^{0.02N} + 2{,}500)\\[2mm]
&= \lim_{N\to\infty} 400(50N - 2{,}500 + 2{,}500e^{-0.02N})\\[2mm]
&= \infty\end{aligned}$$

That is, in the long run the accumulation of radioactive waste from the plant will increase without bound.

Other Types of Improper Integrals   In the next section you will see applications of improper integrals to probability. In the course of the discussion, you will encounter improper integrals of the form $\displaystyle\int_{-\infty}^{\infty} f(x)\,dx$. Here, for future reference, is the definition of such an integral.

The Improper Integral from $-\infty$ to ∞

$$\int_{-\infty}^{\infty} f(x)\ dx = \lim_{N\to\infty} \int_{-N}^{0} f(x)\ dx + \lim_{N\to\infty} \int_{0}^{N} f(x)\ dx$$

If both limits are finite, the improper integral is said to converge to their sum. Otherwise, the improper integral is said to diverge.

If f is nonnegative, the improper integral $\int_{-\infty}^{\infty} f(x)\ dx$ may be interpreted as the total area under the graph of f (Figure 8.8).

Figure 8.8 Area $=$ $\int_{-\infty}^{\infty} f(x)\ dx$.

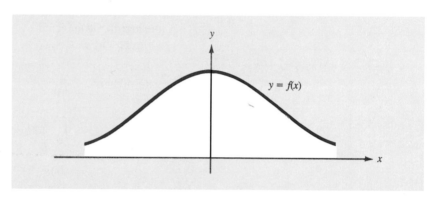

Problems 8.2

In Problems 1 through 24, evaluate the given improper integral.

1. $\displaystyle\int_{1}^{\infty} \frac{1}{x^3}\ dx$

2. $\displaystyle\int_{1}^{\infty} x^{-3/2}\ dx$

3. $\displaystyle\int_{1}^{\infty} \frac{1}{\sqrt{x}}\ dx$

4. $\displaystyle\int_{1}^{\infty} x^{-2/3}\ dx$

5. $\displaystyle\int_{3}^{\infty} \frac{1}{2x-1}\ dx$

6. $\displaystyle\int_{3}^{\infty} \frac{1}{\sqrt[3]{2x-1}}\ dx$

7. $\displaystyle\int_{3}^{\infty} \frac{1}{(2x-1)^2}\ dx$

8. $\displaystyle\int_{0}^{\infty} e^{-x}\ dx$

9. $\displaystyle\int_{0}^{\infty} 5e^{-2x}\ dx$

10. $\displaystyle\int_{1}^{\infty} e^{1-x}\ dx$

11. $\displaystyle\int_{1}^{\infty} \frac{x^2}{(x^3+2)^2}\ dx$

12. $\displaystyle\int_{1}^{\infty} \frac{x^2}{x^3+2}\ dx$

13. $\int_1^\infty \dfrac{x^2}{\sqrt{x^3 + 2}}\, dx$ 14. $\int_0^\infty xe^{-x^2}\, dx$

15. $\int_1^\infty \dfrac{e^{-\sqrt{x}}}{\sqrt{x}}\, dx$ 16. $\int_0^\infty xe^{-x}\, dx$

17. $\int_0^\infty 2xe^{-3x}\, dx$ 18. $\int_0^\infty xe^{1-x}\, dx$

19. $\int_0^\infty 5xe^{10-x}\, dx$ 20. $\int_1^\infty \dfrac{\ln x}{x}\, dx$

21. $\int_2^\infty \dfrac{1}{x \ln x}\, dx$ 22. $\int_2^\infty \dfrac{1}{x\sqrt{\ln x}}\, dx$

23. $\int_0^\infty x^2 e^{-x}\, dx$ 24. $\int_0^\infty x^3 e^{-x^2}\, dx$

Present Value of an Investment

25. An investment will generate \$2,400 per year in perpetuity. If the money is dispensed continuously throughout the year and if the prevailing annual interest rate remains fixed at 12 percent compounded continuously, what is the present value of the investment?

Present Value

26. An investment will generate income continuously at the constant rate of Q dollars per year in perpetuity. Assuming a fixed annual interest rate of r compounded continuously, use an improper integral to show that the present value of the investment is Q/r dollars.

Present Value of Rental Property

27. It is estimated that t years from now an apartment complex will be generating profit for its owner at the rate of $f(t) = 10,000 + 500t$ dollars per year. If the profit is generated in perpetuity and the prevailing annual interest rate remains fixed at 10 percent compounded continuously, what is the present value of the apartment complex?

Present Value of a Franchise

28. The management of a national chain of fast-food outlets is selling a permanent franchise in Seattle, Washington. Past experience in similar localities suggests that t years from now, the franchise will be generating profit at the rate of $f(t) = 12,000 + 900t$ dollars per year. If the prevailing interest rate remains fixed at 10 percent compounded continuously, what is the present value of the franchise?

Present Value

29. In t years, an investment will be generating $f(t) = A + Bt$ dollars per year, where A and B are constants. If the income is generated in perpetuity and the prevailing annual interest rate of r compounded continuously does not change, show that the present value of this investment is $\dfrac{A}{r} + \dfrac{B}{r^2}$.

Nuclear Waste

30. A certain nuclear power plant produces radioactive waste at the rate

of 600 pounds per year. The waste decays exponentially at the rate of 2 percent per year. How much radioactive waste from the plant will be present in the long run?

Health Care

31. The fraction of patients who will still be receiving treatment at a certain health clinic t months after their initial visit is $f(t) = e^{-t/20}$. If the clinic accepts new patients at the rate of 10 per month, approximately how many patients will be receiving treatment at the clinic in the long run?

Population Growth

32. Demographic studies conducted in a certain city indicate that the fraction of the residents that will remain in the city for at least t years is $f(t) = e^{-t/20}$. The current population of the city is 200,000, and it is estimated that new residents will be arriving at the rate of 100 people per year. If this estimate is correct, what will happen to the population of the city in the long run?

Medicine

33. A hospital patient receives intravenously 5 units of a certain drug per hour. The drug is eliminated exponentially, so that the fraction that remains in the patient's body for t hours is $f(t) = e^{-t/10}$. If the treatment is continued indefinitely, approximately how many units of the drug will be in the patient's body in the long run?

3 Probability Density Functions

Some of the most important applications of integration to the social, managerial, and life sciences are in the areas of probability and statistics. In Chapter 6, Section 2, you were introduced to the technique of integrating probability density functions to compute probabilities. The purpose of this section is to explore in more detail the relationship between integration and probability. Improper integrals will play an important role in the discussion.

Random Variables

The life span of a light bulb selected at random from a manufacturer's stock is a quantity that cannot be predicted with certainty. In the terminology of probability and statistics, the process of selecting the bulb at random is called a **random experiment** and the life span of the bulb is said to be a **random variable.** In general, a random variable is a number associated with the outcome of a random experiment.

A random variable that can take on only integer values is said to be **discrete.** The face value of a randomly selected playing card and the number of times heads comes up in three tosses of a coin are discrete random variables. So is the IQ of a randomly selected university student, since IQs are measured in whole numbers.

A random variable that can take on any value in some interval is said to be **continuous.** Some continuous random variables are the time a randomly selected motorist spends waiting at a traffic light, the time interval between the arrivals of randomly selected successive planes at an airport, and the time it takes a randomly selected subject to learn a particular task. Integral calculus is used in the study of continuous random variables.

Probability   The **probability** of an event that can result from a random experiment is a number between 0 and 1 that specifies the likelihood of the event. In particular, the probability is the fraction of the time the event can be expected to occur if the experiment is repeated a large number of times. For example, the probability that an evenly balanced tossed coin will come up heads is $\frac{1}{2}$ since this event can be expected approximately $\frac{1}{2}$ of the time if the coin is tossed repeatedly. In a group containing 13 men and 10 women, the probability is $\frac{10}{23}$ that a person selected at random is a woman. The probability of an event that is certain to occur is 1, while the probability of an event that cannot possibly occur is zero. For example, if you roll an ordinary die, the probability is 1 that you will get a number between 1 and 6, inclusive, while the probability is zero that you will get a 7.

Consider again the random experiment in which a light bulb is selected from a manufacturer's stock. A possible event resulting from this experiment is that the life span of the selected bulb is between 20 and 35 hours. If x is the random variable denoting the life span of a randomly selected bulb, this event can be described by the inequality $20 \le x \le 35$ and its probability denoted by the symbol $P(20 \le x \le 35)$. Similarly, the probability that the bulb will burn for at least 50 hours is denoted by $P(x \ge 50)$ or $P(50 \le x < \infty)$.

Probability Density
Functions   A **probability density function** for a continuous random variable x is a nonnegative function f with the property that $P(a \le x \le b)$ is the area under the graph of f from $x = a$ to $x = b$. A possible probability density function for the life span of a light bulb is sketched in Figure 8.9. Its shape

Figure 8.9 Probability density function for the life span of a light bulb.

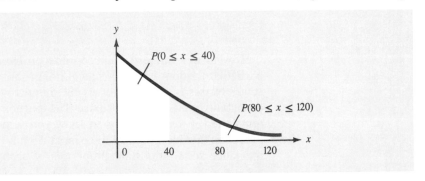

reflects the fact that most bulbs burn out relatively quickly. For example, the probability that a bulb will fail within the first 40 hours is represented by the area under the curve between $x = 0$ and $x = 40$. This is much greater than the area under the curve between $x = 80$ and $x = 120$, which represents the probability that the bulb will fail between its 80th hour and 120th hour of use.

The basic property of probability density functions can be restated in terms of the integrals you would use to compute the appropriate areas.

Probability Density Functions

A probability density function for a continuous random variable x is a nonnegative function f such that

$$P(a \le x \le b) = \int_a^b f(x)\, dx$$

The values of a and b in this formula need not be finite. If either is infinite, the corresponding probability is given by an improper integral. For example, the probability that x is greater than or equal to a is

$$P(x \ge a) = P(a \le x < \infty) = \int_a^\infty f(x)\, dx$$

The total area under the graph of a probability density function must be equal to 1. This is because the total area represents the probability that x is between $-\infty$ and ∞, which is an event that is certain to occur. This observation can be restated in terms of improper integrals.

A Property of Probability Density Functions

If f is a probability density function for a continuous random variable x,

$$\int_{-\infty}^{\infty} f(x)\, dx = 1$$

How to determine the appropriate probability density function for a particular random variable is a central problem in probability beyond the scope of this book. It involves techniques which can be found in most probability and statistics texts. The purpose of the discussion in the remainder of this chapter is to show you some of the probability density functions that have proved to be most useful, to illustrate their use, and to introduce some important properties of probability density functions in general.

Uniform Density Functions

A **uniform density function** (Figure 8.10) is constant over a bounded interval $A \le x \le B$ and zero outside the interval. A random variable that has a uniform density function is said to be **uniformly distributed.** Roughly speaking, a uniformly distributed random variable is one for which all the values in some bounded interval are "equally likely." More precisely, a continuous random variable is uniformly distributed if the probability that its value will be in a particular subinterval of the bounded interval is equal to the probability that it will be in any other subinterval that has the same length. An example of a uniformly distributed random variable is the waiting time of a motorist at a traffic light that remains red for, say, 40 seconds at a time. This random variable has a uniform distribution because all waiting times between 0 and 40 seconds are equally likely.

Figure 8.10 A uniform density function.

If k is the constant value of a uniform density function $f(x)$ on the interval $A \le x \le B$, the value of k is determined by the requirement that the total area under the graph of f be equal to 1. In particular,

$$1 = \int_{-\infty}^{\infty} f(x)\ dx = \int_{A}^{B} f(x)\ dx \qquad \begin{array}{l} \text{[since } f(x) = 0 \text{ outside}\\ \text{the interval } A \le x \le B] \end{array}$$

$$= \int_{A}^{B} k\ dx = kx \ \Big|_{A}^{B} = k(B - A)$$

and so

$$k = \frac{1}{B - A}$$

This observation leads to the following formula for a uniform density function.

Uniform Density Function

$$f(x) = \begin{cases} \dfrac{1}{B - A} & \text{if } A \le x \le B \\[2mm] 0 & \text{otherwise} \end{cases}$$

Here is a typical application involving a uniform density function.

EXAMPLE 3.1

A certain traffic light remains red for 40 seconds at a time. You arrive (at random) at the light and find it red. Use an appropriate uniform density function to find the probability that you will have to wait at least 15 seconds for the light to turn green.

Solution

Let x denote the time (in seconds) that you must wait. Since all waiting times between 0 and 40 are "equally likely," x is uniformly distributed over the interval $0 \leq x \leq 40$. The corresponding uniform density function is

$$f(x) = \begin{cases} \dfrac{1}{40} & \text{if } 0 \leq x \leq 40 \\ 0 & \text{otherwise} \end{cases}$$

and the desired probability is

$$P(15 \leq x \leq 40) = \int_{15}^{40} \frac{1}{40}\,dx = \frac{x}{40}\Big|_{15}^{40} = \frac{40 - 15}{40} = \frac{5}{8}$$

Exponential Density Functions

An **exponential density function** is a function $f(x)$ that is zero for $x < 0$ and that decreases exponentially for $x \geq 0$. This is, for $x \geq 0$,

$$f(x) = Ae^{-kx}$$

where A and k are positive constants.

The value of A is determined by the requirement that the total area under the graph of f be equal to 1. Thus,

$$1 = \int_{-\infty}^{\infty} f(x)\,dx = \int_{0}^{\infty} Ae^{-kx}\,dx = \lim_{N \to \infty} \int_{0}^{N} Ae^{-kx}\,dx$$

$$= \lim_{N \to \infty} \left(-\frac{A}{k} e^{-kx} \Big|_{0}^{N} \right) = \lim_{N \to \infty} \left(-\frac{A}{k} e^{-kN} + \frac{A}{k} \right) = \frac{A}{k}$$

and so

$$A = k$$

This calculation leads to the following general formula for an exponential density function. The corresponding graph is shown in Figure 8.11.

Exponential Density Function

$$f(x) = \begin{cases} ke^{-kx} & \text{if } x \geq 0 \\ 0 & \text{if } x < 0 \end{cases}$$

A random variable that has an exponential density function is said to be **exponentially distributed.** As you can see from the graph in Figure 8.11,

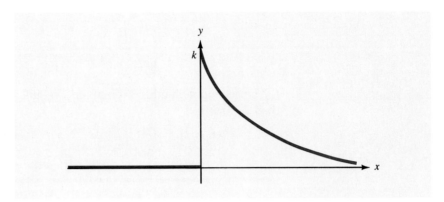

Figure 8.11 An exponential density function.

the value of an exponentially distributed random variable is much more likely to be small than large. Such random variables include the life span of electronic components, the duration of telephone calls, and the interval between the arrivals of successive planes at an airport. Here is an example.

EXAMPLE 3.2

The probability density function for the duration of telephone calls in a certain city is

$$f(x) = \begin{cases} 0.5e^{-0.5x} & \text{if } x \geq 0 \\ 0 & \text{if } x < 0 \end{cases}$$

where x denotes the duration (in minutes) of a randomly selected call.

(a) Find the probability that a randomly selected call will last between 2 and 3 minutes.
(b) Find the probability that a randomly selected call will last at least 2 minutes.

Solution

(a) $P(2 \leq x \leq 3) = \int_2^3 0.5e^{-0.5x}\, dx = -e^{-0.5x} \Big|_2^3$

$$= -e^{-1.5} + e^{-1} = 0.1447$$

(b) There are two ways to compute this probability. The first method is to evaluate an improper integral.

$$P(x \geq 2) = P(2 \leq x < \infty) = \int_2^\infty 0.5e^{-0.5x}\, dx$$

$$= \lim_{N \to \infty} \int_2^N 0.5e^{-0.5x} \, dx = \lim_{N \to \infty} \left(-e^{-0.5x} \,\Big|_2^N \right)$$

$$= \lim_{N \to \infty} (-e^{-0.5N} + e^{-1}) = e^{-1} = 0.3679$$

The second method is to compute 1 minus the probability that x is less than 2. That is,

$$P(x \geq 2) = 1 - \int_0^2 0.5e^{-0.5x} \, dx = 1 - \left(-e^{-0.5x} \,\Big|_0^2 \right)$$

$$= 1 - (-e^{-1} + 1) = e^{-1} = 0.3679$$

Normal Density Functions

The most widely used probability density functions are the **normal density functions.** You are probably already familiar with their famous "bell-shaped" graphs like the one in Figure 8.12. Detailed discussion of this important topic involves specialized ideas that are outside the scope of this text but may be found in any thorough statistics text.

Figure 8.12 A normal density function.

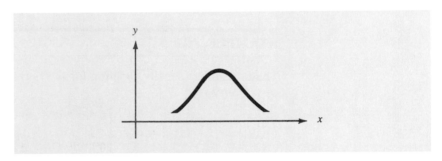

The Expected Value of a Random Variable

An important characteristic of a random variable is its "average" value. Familiar averages of random variables include the average highway mileage for a particular car model, the average waiting time at the check-in counter of a certain airline, and the average life span of workers in a hazardous profession. The average value of a random variable x is called its **expected value** or **mean** and is denoted by the symbol $E(x)$. For continuous random variables, you can calculate the expected value from the probability density function using the following formula.

Expected Value

If x is a continuous random variable with probability density function f, the expected value (or mean) of x is

$$E(x) = \int_{-\infty}^{\infty} xf(x) \, dx$$

Why the Expected Value Is the Average

To see why this integral gives the average value of a continuous random variable, first consider the simpler case in which x is a discrete random variable that takes on the values $x_1, x_2, \ldots, x_n$. If each of these values occurs with equal frequency, the probability of each is $\dfrac{1}{n}$ and the average value of x is

$$\frac{x_1 + x_2 + \cdots + x_n}{n} = x_1 \left(\frac{1}{n}\right) + x_2 \left(\frac{1}{n}\right) + \cdots + x_n \left(\frac{1}{n}\right)$$

More generally, if the values $x_1, x_2, \ldots, x_n$ occur with probabilities $p_1, p_2, \ldots, p_n$, respectively, the average value of x is the weighted sum

$$x_1 p_1 + x_2 p_2 + \cdots + x_n p_n = \sum_{j=1}^{n} x_j p_j$$

Now consider the continuous case. For simplicity, restrict your attention to a bounded interval $A \leq x \leq B$. Imagine that this interval is divided into n subintervals of width Δx, and let x_j denote the beginning of the jth subinterval. Then,

$$p_j = \begin{array}{c} \text{probability that } x \text{ is} \\ \text{in } j\text{th subinterval} \end{array} = \begin{array}{c} \text{area under graph of} \\ f \text{ from } x_j \text{ to } x_{j+1} \end{array} \approx f(x_j)\, \Delta x$$

where $f(x_j)\,\Delta x$ is the area of an approximating rectangle (Figure 8.13).

Figure 8.13 The jth subinterval with an approximating rectangle.

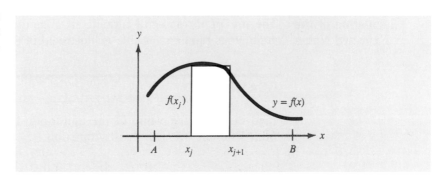

To approximate the average value of x over the interval $A \leq x \leq B$, treat x as if it were a discrete random variable that takes on the values $x_1, x_2, \ldots, x_n$ with probabilities $p_1, p_2, \ldots, p_n$, respectively. Then,

$$\begin{array}{c} \text{Average value of } x \\ \text{on } A \leq x \leq B \end{array} \approx \sum_{j=1}^{n} x_j p_j \approx \sum_{j=1}^{n} x_j f(x_j)\, \Delta x$$

The actual average value of x on the interval $A \leq x \leq B$ is the limit of this approximating sum as n increases without bound. That is,

$$\begin{array}{c} \text{Average value of } x \\ \text{on } A \leq x \leq B \end{array} = \lim_{n \to \infty} \sum_{j=1}^{n} x_j f(x_j)\, \Delta x = \int_{A}^{B} x f(x)\, dx$$

An extension of this argument shows that

$$\begin{array}{c}\text{Average value of } x \\ \text{on } -\infty < x < \infty\end{array} = \int_{-\infty}^{\infty} xf(x) \, dx$$

which is the formula for the expected value of a continuous random variable.

Geometric Interpretation of the Expected Value

To get a better feel for the expected value of a continuous random variable x, think of the probability density function for x as describing the distribution of mass on a beam lying on the x axis. Then the expected value of x is the point at which the beam will balance. If the graph of the density function is symmetric, the expected value is the point of symmetry, as illustrated in Figure 8.14.

Figure 8.14 The expected value of a symmetric random variable.

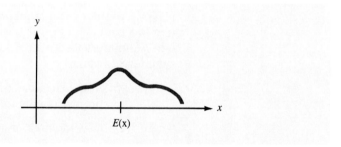

Calculation of the Expected Value

The use of the integral formula to compute the expected value of a continuous random variable is illustrated in the next two examples.

EXAMPLE 3.3

Find the expected value of the uniformly distributed random variable from Example 3.1 with density function

$$f(x) = \begin{cases} \dfrac{1}{40} & \text{if } 0 \le x \le 40 \\ \\ 0 & \text{otherwise} \end{cases}$$

Solution

$$E(x) = \int_{-\infty}^{\infty} xf(x) \, dx = \int_{0}^{40} \frac{x}{40} \, dx = \frac{x^2}{80} \bigg|_{0}^{40} = \frac{1,600}{80} = 20$$

In the context of Example 3.1, this says that the average waiting time at the red light is 20 seconds, a conclusion that should come as no surprise since the random variable is uniformly distributed between 0 and 40.

EXAMPLE 3.4

Find the expected value of the exponentially distributed random variable from Example 3.2 with density function

$$f(x) = \begin{cases} 0.5e^{-0.5x} & \text{if } x \geq 0 \\ 0 & \text{if } x < 0 \end{cases}$$

Solution

$$E(x) = \int_{-\infty}^{\infty} xf(x) \, dx = \int_{0}^{\infty} 0.5xe^{-0.5x} \, dx$$

$$= \lim_{N \to \infty} \int_{0}^{N} 0.5xe^{-0.5x} \, dx$$

$$= \lim_{N \to \infty} \left(-xe^{-0.5x} \Big|_{0}^{N} + \int_{0}^{N} e^{-0.5x} \, dx \right)$$

$$= \lim_{N \to \infty} \left(-xe^{-0.5x} - 2e^{-0.5x} \right) \Big|_{0}^{N}$$

$$= \lim_{N \to \infty} \left(-Ne^{-0.5N} - 2e^{-0.5N} + 2 \right)$$

$$= 2$$

That is, the average duration of telephone calls in the city in Example 3.2 is 2 minutes.

The Variance of a Random Variable

The expected value or mean of a random variable tells you the center of its distribution. Another concept that is useful in describing the distribution of a random variable is the **variance,** which tells you how spread out the distribution is. That is, the variance measures the tendency of the values of a random variable to cluster about their mean. Here is the definition.

Variance

If x is a continuous random variable with probability density function f, the variance of x is

$$\text{Var}(x) = \int_{-\infty}^{\infty} [x - E(x)]^2 f(x) \, dx$$

The definition of the variance is not as mysterious as it may seem at first glance. Notice that the formula for the variance is the same as that for the expected value, except that x has been replaced by the expression $[x - E(x)]^2$, which represents the square of the deviation of x from its mean $E(x)$. The variance, therefore, is simply the expected value or average of the squared deviations of the values of x from the mean. If the values of x tend to cluster about the mean as in Figure 8.15a, most of the deviations from the mean will be small and the variance, which is the average of the squares of these deviations, will also be small. On the other hand, if the values of the random variable are widely scattered as in Figure 8.15b, there will be many large deviations from the mean and the variance will be large.

Figure 8.15 The variance as a measure of the spread of a distribution.

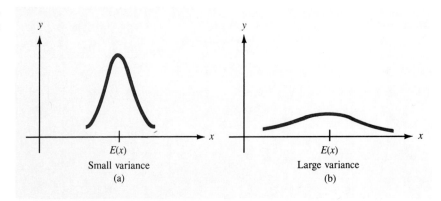

Small variance
(a)

Large variance
(b)

Calculation of the Variance   As you just saw, the formula

$$\text{Var}(x) = \int_{-\infty}^{\infty} [x - E(x)]^2 f(x) \, dx$$

defining the variance is fairly easily interpreted as a measure of the spread of the distribution of x. However, for all but the simplest density functions, this formula is cumbersome to use for the actual calculation of the variance. This is because the term $x - E(x)$ must be squared and multiplied by $f(x)$ before the integration can be performed. Here is an equivalent formula for the variance, which is easier to use for computational purposes.

Equivalent Formula for Variance

$$\text{Var}(x) = \int_{-\infty}^{\infty} x^2 f(x) \, dx - [E(x)]^2$$

The derivation of this formula is straightforward. It involves expanding the integrand in the definition of the variance and rearranging the resulting terms until the new formula emerges. As you read the steps, keep in mind that the expected value $E(x)$ is a constant and can be brought outside integrals by the constant multiple rule.

$$\text{Var}(x) = \int_{-\infty}^{\infty} [x - E(x)]^2 f(x)\, dx \qquad \text{(definition of variance)}$$

$$= \int_{-\infty}^{\infty} \{x^2 - 2xE(x) + [E(x)]^2\} f(x)\, dx \qquad \text{(integrand expanded)}$$

$$= \int_{-\infty}^{\infty} x^2 f(x)\, dx - 2E(x) \int_{-\infty}^{\infty} xf(x)\, dx + [E(x)]^2 \int_{-\infty}^{\infty} f(x)\, dx$$

$$\text{(sum rule and constant multiple rule)}$$

$$= \int_{-\infty}^{\infty} x^2 f(x)\, dx - 2E(x) \cdot E(x) + [E(x)]^2 (1)$$

$$\left[\text{definition of } E(x) \text{ and fact that } \int_{-\infty}^{\infty} f(x)\, dx = 1\right]$$

$$= \int_{-\infty}^{\infty} x^2 f(x)\, dx - [E(x)]^2$$

The use of this formula to compute the variance is illustrated in the next two examples.

EXAMPLE 3.5

Find the variance of the uniformly distributed random variable from Example 3.1 with density function

$$f(x) = \begin{cases} \dfrac{1}{40} & \text{if } 0 \le x \le 40 \\[2mm] 0 & \text{otherwise} \end{cases}$$

Solution

The first step is to compute $E(x)$. This was done in Example 3.3, where you found that $E(x) = 20$. Using this value in the variance formula, you get

$$\text{Var}(x) = \int_{-\infty}^{\infty} x^2 f(x)\, dx - [E(x)]^2$$

$$= \int_0^{40} \frac{x^2}{40}\, dx - 400 = \frac{x^3}{120}\bigg|_0^{40} - 400$$

$$= \frac{64{,}000}{120} - 400 = \frac{16{,}000}{120} = \frac{400}{3}$$

EXAMPLE 3.6

Find the variance of the exponentially distributed random variable from Example 3.2 with density function

$$f(x) = \begin{cases} 0.5e^{-0.5x} & \text{if } x \geq 0 \\ 0 & \text{if } x < 0 \end{cases}$$

Solution

Using the value $E(x) = 2$ obtained in Example 3.4 and integrating by parts twice, you get

$$\begin{aligned}
\text{Var}(x) &= \int_{-\infty}^{\infty} x^2 f(x)\, dx - [E(x)]^2 \\
&= \int_{0}^{\infty} 0.5x^2 e^{-0.5x}\, dx - 4 \\
&= \lim_{N \to \infty} \int_{0}^{N} 0.5x^2 e^{-0.5x}\, dx - 4 \\
&= \lim_{N \to \infty} \left(-x^2 e^{-0.5x} \Big|_{0}^{N} + 2\int_{0}^{N} xe^{-0.5x}\, dx \right) - 4 \\
&= \lim_{N \to \infty} \left[(-x^2 e^{-0.5x} - 4xe^{-0.5x}) \Big|_{0}^{N} + 4\int_{0}^{N} e^{-0.5x}\, dx \right] - 4 \\
&= \lim_{N \to \infty} (-x^2 - 4x - 8)e^{-0.5x} \Big|_{0}^{N} - 4 \\
&= \lim_{N \to \infty} (-N^2 - 4N - 8)e^{-0.5N} + 8 - 4 \\
&= 4
\end{aligned}$$

Problems 8.3

In Problems 1 through 8, integrate the given probability density function to find the indicated probabilities.

1. $f(x) = \begin{cases} \dfrac{1}{3} & \text{if } 2 \leq x \leq 5 \\ 0 & \text{otherwise} \end{cases}$

   (a) $P(2 \leq x \leq 5)$ (b) $P(3 \leq x \leq 4)$ (c) $P(x \geq 4)$

2. $f(x) = \begin{cases} \dfrac{x}{2} & \text{if } 0 \le x \le 2 \\ \\ 0 & \text{otherwise} \end{cases}$

    (a) $P(0 \le x \le 2)$ (b) $P(1 \le x \le 2)$ (c) $P(x \le 1)$

3. $f(x) = \begin{cases} \dfrac{1}{8}(4 - x) & \text{if } 0 \le x \le 4 \\ \\ 0 & \text{otherwise} \end{cases}$

    (a) $P(0 \le x \le 4)$ (b) $P(2 \le x \le 3)$ (c) $P(x \ge 1)$

4. $f(x) = \begin{cases} \dfrac{3}{32}(4x - x^2) & \text{if } 0 \le x \le 4 \\ \\ 0 & \text{otherwise} \end{cases}$

    (a) $P(0 \le x \le 4)$ (b) $P(1 \le x \le 2)$ (c) $P(x \le 1)$

5. $f(x) = \begin{cases} \dfrac{3}{x^4} & \text{if } 1 \le x < \infty \\ \\ 0 & \text{if } x < 1 \end{cases}$

    (a) $P(1 \le x < \infty)$ (b) $P(1 \le x \le 2)$ (c) $P(x \ge 2)$

6. $f(x) = \begin{cases} \dfrac{1}{10}e^{-x/10} & \text{if } x \ge 0 \\ \\ 0 & \text{if } x < 0 \end{cases}$

    (a) $P(0 \le x < \infty)$ (b) $P(x \le 2)$ (c) $P(x \ge 5)$

7. $f(x) = \begin{cases} 2xe^{-x^2} & \text{if } x \ge 0 \\ \\ 0 & \text{if } x < 0 \end{cases}$

    (a) $P(x \ge 0)$ (b) $P(1 \le x \le 2)$ (c) $P(x \le 2)$

8. $f(x) = \begin{cases} \dfrac{1}{4}xe^{-x/2} & \text{if } x \ge 0 \\ \\ 0 & \text{if } x < 0 \end{cases}$

    (a) $P(0 \le x < \infty)$ (b) $P(2 \le x \le 4)$ (c) $P(x \ge 6)$

Traffic Flow   9. A certain traffic light remains red for 45 seconds at a time. You arrive (at random) at the light and find it red. Use an appropriate uniform density function to find the probability that the light will turn green within 15 seconds.

Commuting   10. During the morning rush hour, commuter trains from Long Island to Manhattan run every 20 minutes. You arrive (at random) at the station during the rush hour and find no train at the platform. Assuming that the trains are running on schedule, use an appropriate uniform density function to find the probability that you will have to wait at least 8 minutes for your train.

Movie Theaters   11. A 2-hour movie runs continuously at a local theater. You leave for the theater without first checking the show times. Use an appropriate uniform density function to find the probability that you will arrive at the theater within 10 minutes of (before or after) the start of the film.

Experimental Psychology   12. Suppose x is the length of time (in minutes) that it takes a laboratory rat to traverse a certain maze. If x is exponentially distributed with density function

$$f(x) = \begin{cases} \dfrac{1}{3}\, e^{-x/3} & \text{if } x \geq 0 \\[2mm] 0 & \text{if } x < 0 \end{cases}$$

find the probability that a randomly selected rat will take more than 3 minutes to traverse the maze.

Customer Service   13. The time x (in minutes) a customer must spend waiting in line at a certain bank is exponentially distributed with density function

$$f(x) = \begin{cases} \dfrac{1}{4}\, e^{-x/4} & \text{if } x \geq 0 \\[2mm] 0 & \text{if } x < 0 \end{cases}$$

Find the probability that a randomly selected customer at the bank will have to stand in line at least 8 minutes.

Warranty Protection   14. The life span x (in months) of a certain electrical appliance is exponentially distributed with density function

$$f(x) = \begin{cases} 0.08e^{-0.08x} & \text{if } x \geq 0 \\[2mm] 0 & \text{if } x < 0 \end{cases}$$

The appliance carries a 1-year warranty from the manufacturer. Suppose you purchase one of these appliances, selected at random from the manufacturer's stock. Find the probability that the warranty will expire before your appliance becomes unusable.

15. Find the expected value and variance for the random variable in Problem 1.

16. Find the expected value and variance for the random variable in Problem 2.

17. Find the expected value and variance for the random variable in Problem 3.

18. Find the expected value and variance for the random variable in Problem 5.

19. Find the expected value and variance for the random variable in Problem 6.

Traffic Flow   20. Find the average waiting time for cars arriving on red at the traffic light in Problem 9.

Commuting   21. Find the average wait for rush-hour commuters arriving at the station in Problem 10 when no train is at the platform.

Experimental Psychology   22. Find the average time required for laboratory rats to traverse the maze in Problem 12.

Customer Service   23. Find the average waiting time for customers at the bank in Problem 13.

24. Show that the expected value of a uniformly distributed random variable with density function

$$f(x) = \begin{cases} \dfrac{1}{B - A} & \text{if } A \leq x \leq B \\ 0 & \text{otherwise} \end{cases}$$

is $\dfrac{A + B}{2}$.

25. Show that the variance of the uniformly distributed random variable in Problem 24 is $\dfrac{(B - A)^2}{12}$.

26. Show that the expected value of an exponentially distributed random variable with density function

$$f(x) = \begin{cases} ke^{-kx} & \text{if } x \geq 0 \\ 0 & \text{if } x < 0 \end{cases}$$

is $\dfrac{1}{k}$.

27. Show that the variance of the exponentially distributed random variable in Problem 26 is $\dfrac{1}{k^2}$.

Chapter Summary and Review Problems

Important Terms, Symbols, and Formulas

Limits at infinity: $\lim\limits_{x \to \infty} f(x)$

Limits of powers of x: If $n > 0$,

$$\lim\limits_{x \to \infty} x^n = \infty \qquad \text{and} \qquad \lim\limits_{x \to \infty} \frac{1}{x^n} = 0$$

Limits of exponential functions: If $k > 0$.

$$\lim\limits_{x \to \infty} e^{kx} = \infty \qquad \text{and} \qquad \lim\limits_{x \to \infty} e^{-kx} = 0$$

Limit of the natural logarithm: $\lim\limits_{x \to \infty} \ln x = \infty$

Limits of powers of x times exponential functions: If $k > 0$,

$$\lim\limits_{x \to \infty} x^n e^{-kx} = 0$$

Limits of polynomials: If $a_n \neq 0$,

$$\lim\limits_{x \to \infty} (a_0 + a_1 x + \cdots + a_n x^n) = \lim\limits_{x \to \infty} a_n x^n$$

Limits of rational functions: Divide numerator and denominator by x^k, where k is the smaller of the degrees of the numerator and denominator (or use L'Hôpital's rule).

Indeterminate forms: $\dfrac{0}{0}, \dfrac{\infty}{\infty}, 0 \cdot \infty, 1^\infty, \infty^0, 0^0$

L'Hôpital's rule: If $\lim\limits_{x \to \infty} f(x) = 0$ and $\lim\limits_{x \to \infty} g(x) = 0$, or if $\lim\limits_{x \to \infty} f(x) = \infty$ and $\lim\limits_{x \to \infty} g(x) = \infty$, then

$$\lim\limits_{x \to \infty} \frac{f(x)}{g(x)} = \lim\limits_{x \to \infty} \frac{f'(x)}{g'(x)}$$

Improper integrals:

$$\int_a^\infty f(x)\, dx = \lim\limits_{N \to \infty} \int_a^N f(x)\, dx$$

$$\int_{-\infty}^\infty f(x)\, dx = \lim\limits_{N \to \infty} \int_{-N}^0 f(x)\, dx + \lim\limits_{N \to \infty} \int_0^N f(x)\, dx$$

Discrete random variable; continuous random variable

Probability density function: $P(a \leq x \leq b) = \displaystyle\int_a^b f(x)\, dx$

Uniform density function: $f(x) = \begin{cases} \dfrac{1}{B - A} & \text{if } A \leq x \leq B \\ 0 & \text{otherwise} \end{cases}$

Exponential density function: $f(x) = \begin{cases} ke^{-kx} & \text{if } x \geq 0 \\ 0 & \text{if } x < 0 \end{cases}$

Expected value (mean): $E(x) = \displaystyle\int_{-\infty}^{\infty} xf(x) \, dx$

Variance: $\text{Var}(x) = \displaystyle\int_{-\infty}^{\infty} [x - E(x)]^2 \, f(x) \, dx = \int_{-\infty}^{\infty} x^2 f(x) \, dx - [E(x)]^2$

Review Problems   In Problems 1 through 12, find $\displaystyle\lim_{x \to \infty} f(x)$.

1. $f(x) = (x^2 + 1)(2 - x^4)$

2. $f(x) = \dfrac{1 + x^2 - 3x^3}{2x^3 + 5}$

3. $f(x) = \dfrac{(x + 1)(2 - x)}{x + 3}$

4. $f(x) = \dfrac{2}{3 + 4e^{-5x}}$

5. $f(x) = \dfrac{x^2 + 1}{e^{2x}}$

6. $f(x) = \dfrac{x^2 - 1}{e^{-2x}}$

7. $f(x) = \dfrac{1 + e^{-x}}{e^{-x}}$

8. $f(x) = \dfrac{\ln (3x + 1)}{\ln (x^2 - 4)}$

9. $f(x) = \dfrac{e^{\sqrt{x}}}{2x}$

10. $f(x) = x^2(e^{1/x} - 1)$

11. $f(x) = x^{1/x^2}$

12. $f(x) = \left(1 + \dfrac{2}{x}\right)^{3x}$

In Problems 13 through 22, evaluate the given improper integral.

13. $\displaystyle\int_0^{\infty} \dfrac{1}{\sqrt[3]{1 + 2x}} \, dx$

14. $\displaystyle\int_0^{\infty} (1 + 2x)^{-3/2} \, dx$

15. $\displaystyle\int_0^{\infty} \dfrac{3x}{x^2 + 1} \, dx$

16. $\displaystyle\int_0^{\infty} 3e^{-5x} \, dx$

17. $\displaystyle\int_0^{\infty} xe^{-2x} \, dx$

18. $\displaystyle\int_0^{\infty} 2x^2 e^{-x^3} \, dx$

19. $\displaystyle\int_0^{\infty} x^2 e^{-2x} \, dx$

20. $\displaystyle\int_2^{\infty} \dfrac{1}{x (\ln x)^2} \, dx$

21. $\displaystyle\int_0^{\infty} \dfrac{x - 1}{x + 2} \, dx$

22. $\displaystyle\int_0^{\infty} x^5 e^{-x^3} \, dx$

23. The publishers of a national magazine have found that the fraction of subscribers who remain subscribers for at least t years is $f(t) = e^{-t/10}$. Currently the magazine has 20,000 subscribers and estimates that new subscriptions will be sold at the rate of 1,000 per year.

Approximately how many subscribers will the magazine have in the long run?

24. It is estimated that t years from now, a certain investment will be generating income at the rate of $f(t) = 8,000 + 400t$ dollars per year. If the income is generated in perpetuity and the prevailing annual interest rate remains fixed at 10 percent compounded continuously, find the present value of the investment.

25. Demographic studies conducted in a certain city indicate that the fraction of the residents that will remain in the city for at least t years is $f(t) = e^{-t/20}$. The current population of the city is 100,000, and it is estimated that t years from now, new people will be arriving at the rate of $100t$ people per year. If this estimate is correct, what will happen to the population of the city in the long run?

In Problems 26 through 28, integrate the given probability density functions to find the indicated probabilities.

26. $f(x) = \begin{cases} \dfrac{1}{3} & \text{if } 1 \le x \le 4 \\ 0 & \text{otherwise} \end{cases}$

   (a) $P(1 \le x \le 4)$ (b) $P(2 \le x \le 3)$ (c) $P(x \le 2)$

27. $f(x) = \begin{cases} \dfrac{2}{9}(3 - x) & \text{if } 0 \le x \le 3 \\ 0 & \text{otherwise} \end{cases}$

   (a) $P(0 \le x \le 3)$ (b) $P(1 \le x \le 2)$

28. $f(x) = \begin{cases} 0.2e^{-0.2x} & \text{if } x \ge 0 \\ 0 & \text{if } x < 0 \end{cases}$

   (a) $P(x \ge 0)$ (b) $P(1 \le x \le 4)$ (c) $P(x \ge 5)$

29. Find the expected value and variance for the random variable in Problem 26.

30. Find the expected value and variance for the random variable in Problem 27.

31. Find the expected value and variance for the random variable in Problem 28.

32. A bakery turns out a fresh batch of chocolate chip cookies every 45 minutes. You arrive (at random) at the bakery, hoping to buy a fresh cookie. Use an appropriate uniform density function to find the

probability that you arrive within 5 minutes of (before or after) the time that the cookies come out of the oven.

33. The time x (in minutes) between the arrivals of successive cars at a toll booth is exponentially distributed with the density function

$$f(x) = \begin{cases} 0.5e^{-0.5x} & \text{if } x \geq 0 \\ 0 & \text{if } x < 0 \end{cases}$$

(a) Find the probability that a randomly selected pair of successive cars will arrive at the toll booth at least 6 minutes apart.

(b) Find the average time interval between the arrivals of successive cars at the toll booth.

CHAPTER 9
FUNCTIONS OF
TWO VARIABLES

1   Functions of Two Variables; Surfaces and Level Curves

In many practical situations, the value of one quantity may depend on the values of two or more others. For example, the amount of water in a reservoir may depend on the amount of rainfall and on the amount of water consumed by local residents. The demand for butter may depend on the price of butter and on the price of margarine. The output at a factory may depend on the amount of capital invested in the plant and on the size of the labor force. Relationships of this sort often can be represented mathematically by functions having more than one independent variable.

In this chapter, we shall restrict our attention to functions of two variables. As you shall see, such functions can be represented geometrically as surfaces in three-dimensional space. We shall use geometry to gain insight into the new concepts. The theory of functions of three or more variables is similar to that of functions of two variables and is discussed in advanced calculus texts.

Function of Two Variables

A function f of the two variables x and y is a rule that assigns to each ordered pair (x, y) of real numbers in some set one and only one real number denoted by $f(x, y)$.

The Domain of a Function of Two Variables

The domain of the function $f(x, y)$ is the set of all ordered pairs (x, y) of real numbers for which $f(x, y)$ can be evaluated.

EXAMPLE 1.1

Suppose $f(x, y) = \dfrac{3x^2 + 5y}{x - y}$.

(a) Find the domain of f.
(b) Compute $f(1, -2)$.

Solution

(a) Since division by any real number except zero is possible, the only ordered pairs (x, y) for which f cannot be evaluated are those for which $x = y$. Hence, the domain of f consists of all ordered pairs (x, y)

of real numbers for which $x \neq y$.

(b) $f(1, -2) = \dfrac{3(1)^2 + 5(-2)}{1 - (-2)} = \dfrac{3 - 10}{1 + 2} = -\dfrac{7}{3}$

EXAMPLE 1.2

Suppose $f(x, y) = xe^y + \ln x$.

(a) Find the domain of f.
(b) Compute $f(e^2, \ln 2)$

Solution

(a) Since xe^y is defined for all real numbers x and y and since $\ln x$ is defined only for $x > 0$, the domain of f consists of all ordered pairs (x, y) of real numbers for which $x > 0$.

(b) $f(e^2, \ln 2) = e^2 e^{\ln 2} + \ln e^2 = 2e^2 + 2 = 2(e^2 + 1) \simeq 16.78$

Applications   Here are two elementary applications of functions of two variables to business and economics.

EXAMPLE 1.3

A sports store in St. Louis carries two kinds of tennis rackets, the Michael Chang and the Boris Becker autograph brands. The consumer demand for each brand depends not only on its own price but also on the price of the competing brand. Sales figures indicate that if the Chang brand sells for x dollars per racket and the Becker brand for y dollars per racket, the demand for Chang rackets will be

$$D_1 = 300 - 20x + 30y \qquad \text{rackets per year}$$

and the demand for the Becker rackets will be

$$D_2 = 200 + 40x - 10y \qquad \text{rackets per year}$$

Express the sports store's total annual revenue from the sale of these rackets as a function of the prices x and y.

Solution

Let R denote the total monthly revenue. Then

R = (number of Chang rackets sold)(price per Chang racket)
+ (number of Becker rackets sold)(price per Becker racket)

Hence, $R(x, y) = (300 - 20x + 30y)(x) + (200 + 40x - 10y)(y)$
$= 300x + 200y + 70xy - 20x^2 - 10y^2$

Cobb-Douglas Production Functions   Output Q at a factory is often regarded as a function of the amount K of capital investment and the size L of the labor force. Output functions of the form

$$Q(K, L) = AK^\alpha L^{1-\alpha}$$

where A and α are positive constants and $0 < \alpha < 1$ have proved to be especially useful in economic analysis. Such functions are known as **Cobb-Douglas production functions.**

EXAMPLE 1.4

Suppose that at a certain factory, output is given by the Cobb-Douglas production function $Q(K, L) = 60K^{1/3}L^{2/3}$ units, where K is the capital investment measured in units of $1,000 and L the size of the labor force measured in worker-hours.

(a) Compute the output if the capital investment is $512,000 and 1,000 worker-hours of labor are used.
(b) Show that the output in part (a) will double if both the capital investment and the size of the labor force are doubled.

Solution

(a) Evaluate $Q(K, L)$ with $K = 512$ (thousand) and $L = 1,000$ to get

$$Q(512, 1,000) = 60(512)^{1/3}(1,000)^{2/3}$$

$$= 60(8)(100) = 48,000 \text{ units}$$

(b) Evaluate $Q(K, L)$ with $K = 2(512)$ and $L = 2(1,000)$ as follows to get

$$Q[2(512), 2(1,000)] = 60[2(512)]^{1/3}[2(1,000)]^{2/3}$$

$$= 60(2)^{1/3}(512)^{1/3}(2)^{2/3}(1,000)^{2/3}$$

$$= (2)^{1/3 + 2/3}[60(512)^{1/3}(1,000)^{2/3}]$$

$$= 2Q(512, 1,000) = 2(48,000) = 96,000 \text{ units}$$

That is, the output Q when $K = 2(512)$ and $L = 2(1,000)$ is twice the output when $K = 512$ and $L = 1,000$.

Using a calculation similar to the one in part (b) of the preceding example, you can show that if output is related to capital and labor by a Cobb-Douglas production function and if both capital and labor are multiplied by some positive number m, then output will also be multiplied by m. (The details are left as an exercise for you to do.) In economics, production functions with this property are said to have **constant returns to scale.**

Geometric Representation of Functions

Functions of one variable can be represented graphically as curves drawn on a two-dimensional coordinate system. As you will now see, functions of two variables can be represented graphically as **surfaces** drawn on a three-dimensional coordinate system. Unfortunately, there is no analogous way to visualize functions of more than two variables.

A Three-Dimensional Coordinate System

To construct a three-dimensional coordinate system, we adjoin a third axis (the z axis) to the familiar xy coordinate plane as shown in Figure 9.1. Notice that the xy plane is taken as horizontal. The z axis is perpendicular to the xy plane, and the upward direction is chosen to be the positive z direction. (For simplicity, only the positive coordinate axes are drawn in Figure 9.1.)

Figure 9.1 (left) A three-dimensional coordinate system.

Figure 9.2 (right) The point (1, 2, 4).

You can describe the location of a point in three-dimensional space by specifying three coordinates. For example, the point in Figure 9.2 that is 4 units above the xy plane and that lies directly over the point $(x, y) = (1, 2)$ is represented by the coordinates $x = 1$, $y = 2$, and $z = 4$ and is denoted by the ordered triple $(x, y, z) = (1, 2, 4)$. Similarly, the ordered triple $(2, -1, -3)$ represents the point that is 3 units below the xy plane and that lies directly under the point $(x, y) = (2, -1)$.

The Graph of $f(x, y)$

To graph a function $f(x, y)$ of the two independent variables x and y, it is customary to introduce the letter z to stand for the dependent variable and to write $z = f(x, y)$. The ordered pairs (x, y) in the domain of f are thought of as points in the xy plane, and the function f assigns a height z to each such point. Thus, if $f(1, 2) = 4$, we would express this fact geometrically by plotting the point $(1, 2, 4)$ on a three-dimensional coordinate system. The graph of f consists of all points (x, y, z) for which $z = f(x, y)$. The function may assign different heights to different points in its domain, and in general, its graph will be a surface in three-dimensional space. The situation is illustrated in Figure 9.3.

Figure 9.3 The graph of
$z = f(x, y).$

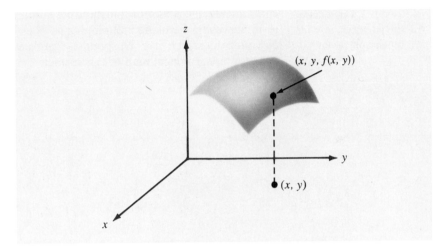

Level Curves   It is usually not easy to sketch the graph of a function of two variables. One way to visualize a surface is illustrated in Figure 9.4. Notice that when the plane $z = C$ intersects the surface $z = f(x, y)$, the result is a curve in space. The corresponding set of points (x, y) in the xy plane that

Figure 9.4 A level curve
of the surface $z = f(x, y).$

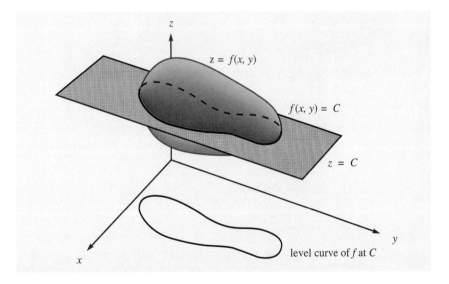

satisfy $f(x, y) = C$ is called the **level curve** of f at C, and an entire family of level curves is generated as C varies over a set of numbers. By sketching members of this family in the xy plane, you can obtain a useful representation of the surface $z = f(x, y)$.

For instance, imagine that the surface $z = f(x, y)$ is a "mountain" whose "elevation" at the point (x, y) is given by $f(x, y)$, as shown in Figure 9.5a. The level curve $f(x, y) = C$ lies directly below a path on the mountain where the elevation is always C. To graph the mountain, you can indicate the paths of constant elevation by sketching the family of level curves in the plane and pinning a "flag" to each curve to show the elevation to which it corresponds (Figure 9.5b). This "flat" figure is called a **topographical map** of the surface $z = f(x, y)$.

Figure 9.5 (a) The surface $z = f(x, y)$ as a mountain. (b) Level curves provide a topographical map of $z = f(x, y)$.

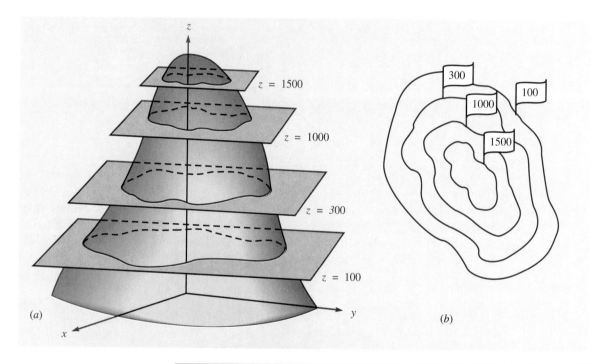

(a)

(b)

EXAMPLE 1.5 Fig. 9.5

Discuss the level curves of the function $f(x, y) = x^2 + y^2$.

Solution

The level curve $f(x, y) = C$ has the equation $x^2 + y^2 = C$. If $C = 0$, this is the point $(0, 0)$, and if $C > 0$, it is a circle of radius $\sqrt{C}$. If $C < 0$, there are no points that satisfy $x^2 + y^2 = C$.

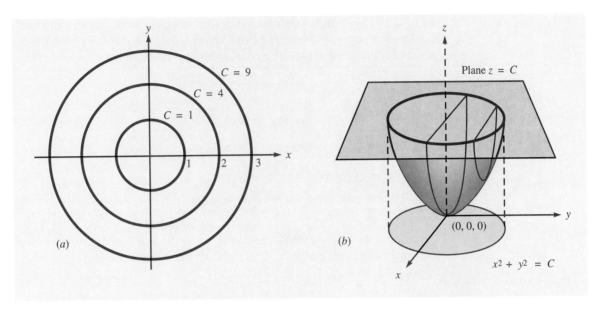

Figure 9.6 (a) Level curves of $f(x, y) = x^2 + y^2$ are the circles $x^2 + y^2 = C$. (b) The surface $z = x^2 + y^2$ is a bowl.

The graph of the surface $z = x^2 + y^2$ is shown in Figure 9.6. The level curves you have just found correspond to cross sections perpendicular to the z axis. It can be shown that cross sections perpendicular to the x axis and y axis are parabolas. (Try to see why this is true.) For this reason, the surface is shaped like a bowl. It is called a **circular paraboloid** or a **paraboloid of revolution.**

Level Curves in Economics: Isoquants and Indifference Curves

Level curves appear in many different applications. For instance, in economics, if the output $Q(x, y)$ of a production process is determined by two inputs x and y (say, hours of labor and capital investment), then the level curve $Q(x, y) = C$ is called the **curve of constant product C** or, more simply, an **isoquant.**

Another application of level curves in economics involves the concept of **indifference curves.** A consumer who is considering the purchase of a number of units of each of two commodities is associated with a **utility function** $U(x, y)$, which measures the total satisfaction (or **utility**) the consumer derives from having x units of the first commodity and y units of the second. A level curve $U(x, y) = C$ of the utility function is called an **indifference curve** and gives all the combinations of x and y that lead to the same level of consumer satisfaction. These terms are illustrated in Example 1.6.

EXAMPLE 1.6

Suppose the utility derived by a consumer from x units of one commodity and y units of a second commodity is given by the utility function

$U(x, y) = x^{3/2}y$. If the consumer currently owns $x = 16$ units of the first commodity and $y = 20$ units of the second, find the consumer's current level of utility and sketch the corresponding indifference curve.

Solution

The current level of utility is

$$U(16, 20) = (16)^{3/2}(20) = 1{,}280$$

and the corresponding indifference curve is

$$x^{3/2}y = 1{,}280$$

or $y = 1{,}280x^{-3/2}$. This curve consists of all points (x, y) where the level of utility $U(x, y)$ is 1,280. The curve $x^{3/2}y = 1{,}280$ and several other curves of the family $x^{3/2}y = C$ are shown in Figure 9.7.

Figure 9.7 Indifference curves for the utility function $U(x, y) = x^{3/2}y$.

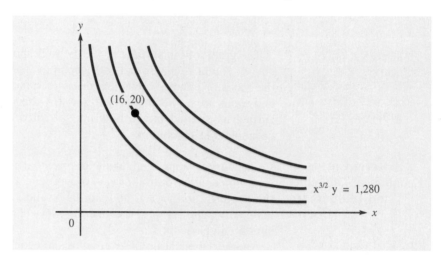

Computer Graphics   In practical work in the social, managerial, or life sciences, you will rarely, if ever, have to graph a function of two variables. Hence, we will spend no more time developing graphing procedures for such functions.

Computer software is now available for graphing functions of two variables. Such software often allows you to choose different scales along each coordinate axis and may also allow you to visualize a given surface from different viewpoints. These features permit you to obtain a detailed picture of the graph. A variety of computer-generated graphs are displayed in Figure 9.8.

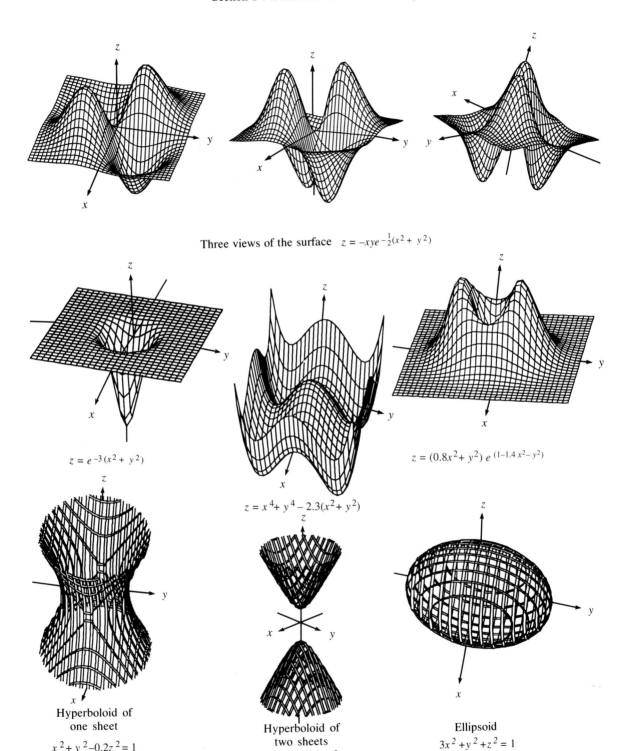

Three views of the surface $z = -xye^{-\frac{1}{2}(x^2 + y^2)}$

$z = e^{-3(x^2 + y^2)}$

$z = x^4 + y^4 - 2.3(x^2 + y^2)$

$z = (0.8x^2 + y^2)\, e^{\,(1-1.4\,x^2 - y^2)}$

Hyperboloid of
one sheet

$x^2 + y^2 - 0.2z^2 = 1$

Hyperboloid of
two sheets

$-10x^2 - 10y^2 + 5z^2 = 1$

Ellipsoid

$3x^2 + y^2 + z^2 = 1$

Problems 9.1

In Problems 1 through 6, specify the domain of f and compute the indicated function values.

1. $f(x, y) = (x - 1)^2 + 2xy^3; f(2, -1), f(1, 2)$

2. $f(x, y) = \dfrac{3x + 2y}{2x + 3y}; f(1, 2), f(-4, 6)$

3. $f(x, y) = \sqrt{y^2 - x^2}; f(4, 5), f(-1, 2)$

4. $f(x, y) = 10x^{1/2}y^{2/3}; f(16, 27), f(4, -1,331)$

5. $f(x, y) = \dfrac{x}{\ln y}; f(-1, e^3), f(\ln 9, e^2)$

6. $f(x, y) = \dfrac{e^{xy}}{2 - e^{xy}}; f(1, 0), f(\ln 2, 2)$

In Problems 7 through 14, sketch the indicated level curve $f(x, y) = C$ for each choice of constant C.

7. $f(x, y) = x + 2y; C = 1, C = 2, C = -3$

8. $f(x, y) = x^2 + y; C = 0, C = 4, C = 9$

9. $f(x, y) = x^2 - 4x - y; C = -4, C = 5$

10. $f(x, y) = \dfrac{x}{y}; C = -2, C = 2$

11. $f(x, y) = xy; C = 1, C = -1, C = 2, C = -2$

12. $f(x, y) = ye^x; C = 0, C = 1$

13. $f(x, y) = xe^y; C = 1, C = e$

14. $f(x, y) = \ln(x^2 + y^2); C = 4, C = \ln 4$

Production
15. Using x skilled workers and y unskilled workers, a manufacturer can produce $Q(x, y) = 10x^2y$ units per day. Currently there are 20 skilled workers and 40 unskilled workers on the job.
    (a) How many units are currently being produced each day?
    (b) By how much will the daily production level change if 1 more skilled worker is added to the current work force?
    (c) By how much will the daily production level change if 1 more unskilled worker is added to the current work force?
    (d) By how much will the daily production level change if 1 more skilled worker *and* 1 more unskilled worker are added to the current work force?

Production Cost
16. A manufacturer can produce electric typewriters at a cost of $80 apiece and manual typewriters at a cost of $20 apiece.

(a) Express the manufacturer's total monthly production cost as a function of the number of electric typewriters and the number of manual typewriters produced.

(b) Compute the total monthly cost if 500 electric and 800 manual typewriters are produced.

(c) The manufacturer wants to increase the output of electric typewriters by 50 a month from the level in part (b). What corresponding change should be made in the monthly output of manual typewriters so that the total monthly cost will not change?

Retail Sales 17. A paint store carries two brands of latex paint. Sales figures indicate that if the first brand is sold for x_1 dollars per gallon and the second for x_2 dollars per gallon, the demand for the first brand will be $D_1(x_1, x_2) = 200 - 10x_1 + 20x_2$ gallons per month and the demand for the second brand will be $D_2(x_1, x_2) = 100 + 5x_1 - 10x_2$ gallons per month.

(a) Express the paint store's total monthly revenue from the sale of the paint as a function of the prices x_1 and x_2.

(b) Compute the revenue in part (a) if the first brand is sold for $6 per gallon and the second for $5 per gallon.

Production 18. The output at a certain factory is $Q(K, L) = 120K^{2/3}L^{1/3}$ units, where K is the capital investment measured in units of $1,000 and L the size of the labor force measured in worker-hours.

(a) Compute the output if the capital investment is $125,000 and the size of the labor force is 1,331 worker-hours.

(b) What will happen to the output in part (a) if both the level of capital investment and the size of the labor force are cut in half?

Production 19. The Easy-Gro agricultural company estimates that when $100x$ worker-hours of labor are employed on y acres of land, the number of bushels of wheat produced is $f(x, y) = Ax^a y^b$, where A, a, and b are positive constants. Suppose the company decides to double the production factors x and y. Determine how this decision affects the production of wheat in each of these cases:
(a) $a + b > 1$ (b) $a + b < 1$ (c) $a + b = 1$

Production 20. Suppose that when x machines and y worker-hours are used each day, a certain factory will produce $Q(x, y) = 10xy$ transistor radios. Describe the relationship between the inputs x and y that results in an output of 1,000 radios each day. (Note that you are finding a level curve of Q.)

Retail Sales 21. A manufacturer with exclusive rights to a sophisticated new industrial machine is planning to sell a limited number of the machines to both foreign and domestic firms. The price the manufacturer can expect to receive for the machines will depend on the number of machines made available. It is estimated that if the manufacturer supplies x

machines to the domestic market and y machines to the foreign market, the machines will sell for $60 - \frac{x}{5} + \frac{y}{20}$ thousand dollars apiece at home and $50 - \frac{y}{10} + \frac{x}{20}$ thousand dollars apiece abroad.

Express the manufacturer's revenue R as a function of x and y.

Retail Sales   22. A manufacturer is planning to sell a new product at the price of A dollars per unit and estimates that if x thousand dollars is spent on development and y thousand dollars on promotion, consumers will buy approximately $\frac{320y}{y + 2} + \frac{160x}{x + 4}$ units of the product. If manufacturing costs are \$50 per unit, express the total profit as a function of x and y.

Constant Returns   23. Suppose output Q is given by the Cobb-Douglas production function to Scale $Q(K, L) = AK^\alpha L^{1-\alpha}$, where A and α are positive constants and $0 < \alpha < 1$. Prove that if K and L are both multiplied by the same positive number m, then the output Q will also be multiplied by m. That is, show that $Q(mK, mL) = mQ(K, L)$.

Constant Production   24. Using x skilled workers and y unskilled workers, a manufacturer can Curves   produce $Q(x, y) = 3x + 2y$ units per day. Currently the work force consists of 10 skilled workers and 20 unskilled workers.
(a) Compute the current daily output.
(b) Find an equation relating the levels of skilled and unskilled labor if the daily output is to remain at its current level.
(c) On a two-dimensional coordinate system, draw the isoquant (constant production curve) that corresponds to the current level of output.
(d) What change should be made in the level of unskilled labor y to offset an increase in skilled labor x of two workers so that the output will remain at its current level?

Indifference Curves   25. Suppose the utility derived by a consumer from x units of one commodity and y units of a second commodity is given by the utility function $U(x, y) = 2x^3y^2$. The consumer currently owns $x = 5$ units of the first commodity and $y = 4$ units of the second. Find the consumer's current level of utility and sketch the corresponding indifference curve.

Indifference Curves   26. Suppose the utility derived by a consumer from x units of one commodity and y units of a second commodity is given by the utility function $U(x, y) = (x + 1)(y + 2)$. The consumer currently owns $x = 25$ units of the first commodity and $y = 8$ units of the second. Find the consumer's current level of utility and sketch the corresponding indifference curve.

2 Partial Derivatives

In many problems involving functions of two variables, the goal is to find the rate of change of the function with respect to one of its variables when the other is held constant. That is, the goal is to differentiate the function with respect to the particular variable in question while keeping the other variable fixed. This process is known as **partial differentiation,** and the resulting derivative is said to be a **partial derivative** of the function.

Partial Derivatives

Suppose $z = f(x, y)$. The partial derivative of f with respect to x is denoted by

$$\frac{\partial z}{\partial x} \quad \text{or} \quad f_x(x, y)$$

and is the function obtained by differentiating f with respect to x, treating y as a constant. The partial derivative of f with respect to y is denoted by

$$\frac{\partial z}{\partial y} \quad \text{or} \quad f_y(x, y)$$

and is the function obtained by differentiating f with respect to y, treating x as a constant.

Computation of Partial Derivatives   No new rules are needed for the computation of partial derivatives. To compute f_x, simply differentiate f with respect to the single variable x, pretending that y is a constant. To compute f_y, simply differentiate f with respect to y, pretending that x is a constant. Here are some examples.

EXAMPLE 2.1

Find the partial derivatives f_x and f_y if $f(x, y) = x^2 + 2xy^2 + \dfrac{2y}{3x}$.

Solution

To simplify the computation, begin by rewriting the function as

$$f(x, y) = x^2 + 2xy^2 + \frac{2}{3}yx^{-1}$$

To compute f_x, think of f as a function of x and differentiate the sum term by term, treating y as a constant to get

$$f_x(x, y) = 2x + 2y^2 - \frac{2}{3}yx^{-2} = 2x + 2y^2 - \frac{2y}{3x^2}$$

To compute f_y, think of f as a function of y and differentiate term by term, treating x as a constant to get

$$f_y(x, y) = 0 + 4xy + \frac{2}{3}x^{-1} = 4xy + \frac{2}{3x}$$

EXAMPLE 2.2

Find the partial derivatives $\frac{\partial z}{\partial x}$ and $\frac{\partial z}{\partial y}$ if $z = (x^2 + xy + y)^5$.

Solution

Holding y fixed and using the chain rule to differentiate z with respect to x, you get

$$\frac{\partial z}{\partial x} = 5(x^2 + xy + y)^4(2x + y)$$

Holding x fixed and using the chain rule to differentiate z with respect to y, you get

$$\frac{\partial z}{\partial y} = 5(x^2 + xy + y)^4(x + 1)$$

EXAMPLE 2.3

Find the partial derivatives f_x and f_y if $f(x, y) = xe^{-2xy}$.

Solution

From the product rule,

$$f_x(x, y) = x(-2ye^{-2xy}) + e^{-2xy} = (-2xy + 1)e^{-2xy}$$

and from the constant multiple rule,

$$f_y(x, y) = x(-2xe^{-2xy}) = -2x^2e^{-2xy}$$

Geometric Interpretation of Partial Derivatives   As you saw in Section 1, functions of two variables can be represented graphically as surfaces drawn on three-dimensional coordinate systems. In particular, if $z = f(x, y)$, an ordered pair (x, y) in the domain of f can be identified with a point in the xy plane and the corresponding function value $z = f(x, y)$ can be thought of as assigning a height to this point. The graph of f is the surface consisting of all points (x, y, z) in three-dimensional space whose height z is equal to $f(x, y)$.

The partial derivatives of a function of two variables can be interpreted geometrically as follows. For each fixed number y_0, the points (x, y_0, z) form a vertical plane whose equation is $y = y_0$. If $z = f(x, y)$ and if y is kept fixed at $y = y_0$, then the corresponding points $(x, y_0, f(x, y_0))$ form a curve in three-dimensional space that is the intersection of the surface $z = f(x, y)$ with the plane $y = y_0$. At each point on this curve, the partial derivative $\dfrac{\partial z}{\partial x}$ is simply the slope of the line in the plane $y = y_0$ that is tangent to the curve at the point in question. That is, $\dfrac{\partial z}{\partial x}$ is the slope of the tangent "in the x direction." The situation is illustrated in Figure 9.9a.

Similarly, if x is kept fixed at $x = x_0$, the corresponding points $(x_0, y, f(x_0, y))$ form a curve that is the intersection of the surface $z = f(x, y)$ with the vertical plane $x = x_0$. At each point on this curve, the partial derivative $\dfrac{\partial z}{\partial y}$ is the slope of the tangent in the plane $x = x_0$. That is,

Figure 9.9 Geometric interpretation of partial derivatives.

$\dfrac{\partial z}{\partial y}$ is the slope of the tangent "in the y direction". The situation is illustrated in Figure 9.9b.

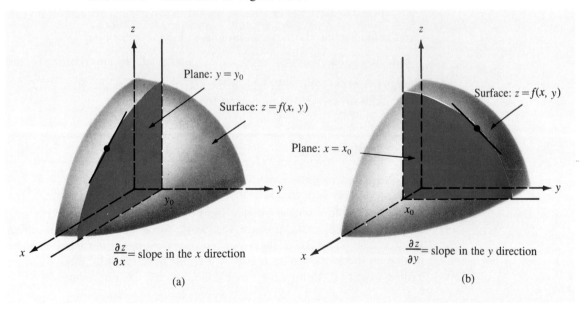

$\dfrac{\partial z}{\partial x}$ = slope in the x direction

(a)

$\dfrac{\partial z}{\partial y}$ = slope in the y direction

(b)

Marginal Analysis In economics, the term **marginal analysis** refers to the practice of using a derivative to estimate the change in the value of a function resulting from a 1-unit increase in one of its variables. In Chapter 2, Section 3, you saw some examples of marginal analysis involving ordinary derivatives of functions of one variable. Here are two examples illustrating marginal analysis for functions of two variables.

Marginal Products of Capital and Labor

EXAMPLE 2.4

Suppose the daily output Q at a factory depends on the amount K of capital (measured in units of $1,000) invested in the plant and equipment and also on the size L of the labor force (measured in worker-hours). In economics, the partial derivatives $\frac{\partial Q}{\partial K}$ and $\frac{\partial Q}{\partial L}$ are known as the **marginal products** (or marginal productivities) of capital and labor, respectively. Give economic interpretations of these two marginal products.

Solution

The marginal product of labor $\frac{\partial Q}{\partial L}$ is the rate at which output Q changes with respect to labor L for a fixed level K of capital investment. Hence, $\frac{\partial Q}{\partial L}$ is approximately the change in output that will result if capital investment is held fixed and labor is increased by 1 worker-hour.

Similarly, the marginal product of capital $\frac{\partial Q}{\partial K}$ is approximately the change in output that will result if the size of the labor force is held fixed and capital investment is increased by $1,000.

EXAMPLE 2.5

It is estimated that the weekly output at a certain plant is given by the function $Q(x, y) = 1{,}200x + 500y + x^2y - x^3 - y^2$ units, where x is the number of skilled workers and y the number of unskilled workers employed at the plant. Currently the work force consists of 30 skilled workers and 60 unskilled workers. Use marginal analysis to estimate the change in the weekly output that will result from the addition of 1 more skilled worker if the number of unskilled workers is not changed.

Solution

The partial derivative

$$Q_x(x, y) = 1{,}200 + 2xy - 3x^2$$

is the rate of change of output with respect to the number of skilled workers. For any values of x and y, this is an approximation of the number of additional units that will be produced each week if the number of skilled workers is increased from x to $x + 1$ while the number of unskilled workers is kept fixed at y. In particular, if the work force is increased from 30 skilled and 60 unskilled workers to 31 skilled and 60 unskilled workers, the resulting change in output is approximately

$$Q_x(30, 60) = 1{,}200 + 2(30)(60) - 3(30)^2 = 2{,}100 \text{ units}$$

For practice, compute the change in output exactly by subtracting appropriate values of Q. Is the approximation a good one?

Second-Order Partial Derivatives

Partial derivatives can themselves be differentiated. The resulting functions are called **second-order partial derivatives.** Here is a summary of the definition and notation for the four possible second-order partial derivatives of a function of two variables.

Second-Order Partial Derivatives

If $z = f(x, y)$, the partial derivative of f_x with respect to x is

$$f_{xx} = (f_x)_x \qquad \text{or} \qquad \frac{\partial^2 z}{\partial x^2} = \frac{\partial}{\partial x}\left(\frac{\partial z}{\partial x}\right)$$

The partial derivative of f_x with respect to y is

$$f_{xy} = (f_x)_y \qquad \text{or} \qquad \frac{\partial^2 z}{\partial y \partial x} = \frac{\partial}{\partial y}\left(\frac{\partial z}{\partial x}\right)$$

The partial derivative of f_y with respect to x is

$$f_{yx} = (f_y)_x \qquad \text{or} \qquad \frac{\partial^2 z}{\partial x \partial y} = \frac{\partial}{\partial x}\left(\frac{\partial z}{\partial y}\right)$$

The partial derivative of f_y with respect to y is

$$f_{yy} = (f_y)_y \qquad \text{or} \qquad \frac{\partial^2 z}{\partial y^2} = \frac{\partial}{\partial y}\left(\frac{\partial z}{\partial y}\right)$$

The computation of second-order partial derivatives is illustrated in the next example.

EXAMPLE 2.6

Compute the four second-order partial derivatives of the function $f(x, y) = xy^3 + 5xy^2 + 2x + 1$.

Solution

Since
$$f_x = y^3 + 5y^2 + 2$$
it follows that
$$f_{xx} = 0 \quad \text{and} \quad f_{xy} = 3y^2 + 10y$$
Since
$$f_y = 3xy^2 + 10xy$$
it follows that
$$f_{yx} = 3y^2 + 10y \quad \text{and} \quad f_{yy} = 6xy + 10x$$

Mixed Second-Order Partial Derivatives   The two partial derivatives f_{xy} and f_{yx} are sometimes called the **mixed second-order partial derivatives** of f. Notice that the mixed partial derivatives in Example 2.6 were equal. This is not an accident. It turns out that for virtually all the functions $f(x, y)$ you will encounter in practical work, the mixed partial derivatives will be equal; that is,

$$f_{xy} = f_{yx}$$

This means you will get the same answer if you first differentiate $f(x, y)$ with respect to x and then differentiate the resulting function with respect to y as you would if you performed the differentiation in the reverse order.

An Application to Economics   In the next example, you will see how a second-order partial derivative can convey useful information in a practical situation.

EXAMPLE 2.7

Suppose the output Q at a factory depends on the amount K of capital invested in the plant and equipment and also on the size L of the labor force, measured in worker-hours. Give an economic interpretation of the sign of the second-order partial derivative $\dfrac{\partial^2 Q}{\partial L^2}$.

Solution

If $\dfrac{\partial^2 Q}{\partial L^2}$ is negative, the marginal product of labor $\dfrac{\partial Q}{\partial L}$ decreases as L increases. This implies that for a fixed level of capital investment, the effect on output of the addition of 1 worker-hour of labor is greater when the work force is small than when the work force is large.

Similarly, if $\dfrac{\partial^2 Q}{\partial L^2}$ is positive, it follows that for a fixed level of capital investment, the effect on output of the addition of 1 worker-hour of labor is greater when the work force is large than when it is small.

For most factories operating with adequate work forces, the derivative $\frac{\partial^2 Q}{\partial L^2}$ is generally negative. Can you give an economic explanation for this fact?

Problems 9.2

In Problems 1 through 16, compute all the first-order partial derivatives of the given function.

1. $f(x, y) = 2xy^5 + 3x^2y + x^2$ 2. $z = 5x^2y + 2xy^3 + 3y^2$

3. $z = (3x + 2y)^5$ 4. $f(x, y) = (x + xy + y)^3$

5. $f(x, y) = \frac{3y}{2x}$ 6. $z = \frac{x^2}{y^3}$

7. $z = xe^{xy}$ 8. $f(x, y) = xye^x$

9. $f(x, y) = \frac{e^{2-x}}{y^2}$ 10. $f(x, y) = xe^{x+2y}$

11. $f(x, y) = \frac{2x + 3y}{y - x}$ 12. $z = \frac{xy^2}{x^2y^3 + 1}$

13. $z = x \ln y$ 14. $f(x, y) = x \ln xy$

15. $f(x, y) = \frac{\ln (x + 2y)}{y^2}$ 16. $z = \ln \left(\frac{x}{y} + \frac{y}{x}\right)$

Marginal Analysis   17. At a certain factory, the daily output is $Q(K, L) = 60K^{1/2}L^{1/3}$ units, where K denotes the capital investment measured in units of $1,000 and L the size of the labor force measured in worker-hours. Suppose that the current capital investment is $900,000 and that 1,000 worker-hours of labor are used each day. Use marginal analysis to estimate the effect of an additional capital investment of $1,000 on the daily output if the size of the labor force is not changed.

Consumer Demand   18. The monthly demand for a certain brand of toasters is given by a function $f(x, y)$, where x is the amount of money (measured in units of $1,000) spent on advertising and y is the selling price (in dollars) of the toasters. Give economic interpretations of the partial derivatives f_x and f_y. Under normal economic conditions, what will the sign of each of these partial derivatives be?

Marginal Analysis   19. A grocer's daily profit from the sale of two brands of orange juice is

$$P(x, y) = (x - 30)(70 - 5x + 4y) + (y - 40)(80 + 6x - 7y)$$

cents, where x is the price per can of the first brand and y is the price per can of the second. Currently the first brand sells for 50 cents per can and the second for 52 cents per can. Use marginal analysis to estimate the change in the daily profit that will result if the grocer raises the price of the second brand by 1 cent per can but keeps the price of the first brand unchanged.

Consumer Demand

20. Two competing brands of power lawnmowers are sold in the same town. The price of the first brand is x dollars per mower, and the price of the second brand is y dollars per mower. The local demand for the first brand of mower is given by a function $D(x, y)$.
    (a) How would you expect the demand for the first brand of mower to be affected by an increase in x? By an increase in y?
    (b) Translate your answers in part (a) into conditions on the signs of the partial derivatives of D.
    (c) If $D(x, y) = a + bx + cy$, what can you say about the signs of the coefficients b and c if your conclusions in parts (a) and (b) are to hold?

Substitute Commodities

21. In economics, two commodities are said to be **substitute commodities** if the demand Q_1 for the first increases as the price p_2 of the second increases and if the demand Q_2 for the second increases as the price p_1 of the first increases.
    (a) Give an example of a pair of substitute commodities.
    (b) If two commodities are substitutes, what must be true of the partial derivatives $\dfrac{\partial Q_1}{\partial p_2}$ and $\dfrac{\partial Q_2}{\partial p_1}$?
    (c) Suppose the demand functions for two commodities are $Q_1 = 3,000 + \dfrac{400}{p_1 + 3} + 50p_2$ and $Q_2 = 2,000 - 100p_1 + \dfrac{500}{p_2 + 4}$. Are the commodities substitutes?

Complementary Commodities

22. Two commodities are said to be **complementary commodities** if the demand Q_1 for the first decreases as the price p_2 of the second increases and if the demand Q_2 for the second decreases as the price p_1 of the first increases.
    (a) Give an example of a pair of complementary commodities.
    (b) If two commodities are complementary, what must be true of the partial derivatives $\dfrac{\partial Q_1}{\partial p_2}$ and $\dfrac{\partial Q_2}{\partial p_1}$?
    (c) Suppose the demand functions for two commodities are $Q_1 = 2,000 + \dfrac{400}{p_1 + 3} - 50p_2$ and $Q_2 = 2,000 - 100p_2 + \dfrac{500}{p_1 + 4}$. Are the commodities complementary?

In Problems 23 through 28, compute all the second-order partial derivatives of the given function.

23. $f(x, y) = 5x^4y^3 + 2xy$

24. $f(x, y) = \dfrac{x + 1}{y - 1}$

25. $f(x, y) = e^{x^2 y}$

26. $f(x, y) = \ln(x^2 + y^2)$

27. $f(x, y) = \sqrt{x^2 + y^2}$

28. $f(x, y) = x^2 y e^x$

Marginal Productivity 29. Suppose the output Q of a factory depends on the amount K of capital investment measured in units of $1,000 and on the size L of the labor force measured in worker-hours. Give an economic interpretation of the second-order partial derivative $\dfrac{\partial^2 Q}{\partial K^2}$.

Marginal Productivity 30. At a certain factory, the output is $Q = 120K^{1/2}L^{1/3}$ units, where K denotes the capital investment measured in units of $1,000 and L the size of the labor force measured in worker-hours.

(a) Determine the sign of the second-order partial derivative $\dfrac{\partial^2 Q}{\partial L^2}$ and give an economic interpretation.

(b) Determine the sign of the second-order partial derivative $\dfrac{\partial^2 Q}{\partial K^2}$ and give an economic interpretation.

Law of Diminishing Returns 31. Suppose the daily output Q of a factory depends on the amount K of capital investment and on the size L of the labor force. A **law of diminishing returns** states that in certain circumstances, there is a value L_0 such that the marginal product of labor will be increasing for $L < L_0$ and decreasing for $L > L_0$.

(a) Translate this law of diminishing returns into statements about the sign of a certain second-order partial derivative.

(b) Discuss the economic factors that might account for this phenomenon.

3 The Chain Rule: Approximation by the Total Differential

In many practical situations, a particular quantity is given as a function of two or more variables, each of which can be thought of as a function of yet another variable, and the goal is to find the rate of change of the quantity with respect to this other variable. For example, the demand for a certain commodity may depend on the price of the commodity itself and on the price of a competing commodity, both of which are increasing with time, and the goal may be to find the rate of change of the demand with respect to time. You can solve problems of this type by using a generalization of the chain rule that was introduced in Chapter 2, Section 5.

The Chain Rule Recall from Chapter 2 that if z is a function of x and x is a function of t, then z can be regarded as a function of t, and the rate of change of z with respect to t is given by the chain rule

$$\frac{dz}{dt} = \frac{dz}{dx}\frac{dx}{dt}$$

Here is the corresponding rule for functions of two variables.

Chain Rule for Partial Derivatives

Suppose z is a function of x and y, each of which is a function of t. Then z can be regarded as a function of t and

$$\frac{dz}{dt} = \frac{\partial z}{\partial x}\frac{dx}{dt} + \frac{\partial z}{\partial y}\frac{dy}{dt}$$

Observe that the expression for $\dfrac{dz}{dt}$ is the sum of two terms, each of which can be interpreted using the chain rule for a function of one variable. In particular,

$$\frac{\partial z}{\partial x}\frac{dx}{dt} = \text{rate of change of } z \text{ with respect to } t \text{ for fixed } y$$

and $\dfrac{\partial z}{\partial y}\dfrac{dy}{dt} = \text{rate of change of } z \text{ with respect to } t \text{ for fixed } x$

The chain rule for partial derivatives says that the total rate of change of z with respect to t is the sum of these two "partial" rates of change.

Here are two examples illustrating the use of the chain rule for partial derivatives.

EXAMPLE 3.1

Find $\dfrac{dz}{dt}$ if $z = x^2 + 3xy + 1$, $x = 2t + 1$, and $y = t^2$.

Solution

By the chain rule,

$$\frac{dz}{dt} = \frac{\partial z}{\partial x}\frac{dx}{dt} + \frac{\partial z}{\partial y}\frac{dy}{dt} = (2x + 3y)(2) + 3x(2t)$$

which you can rewrite in terms of t by substituting $x = 2t + 1$ and $y = t^2$ to get

$$\frac{dz}{dt} = 4(2t + 1) + 6t^2 + 3(2t + 1)(2t) = 18t^2 + 14t + 4$$

For practice, check this answer by first substituting $x = 2t + 1$ and $y = t^2$ into the formula for z and then differentiating directly with respect to t.

EXAMPLE 3.2

A health store carries two kinds of multiple vitamins, Brand A and Brand B. Sales figures indicate that if Brand A is sold for x dollars per bottle and Brand B for y dollars per bottle, the demand for Brand A will be

$$Q(x, y) = 300 - 20x^2 + 30y \qquad \text{bottles per month}$$

It is estimated that t months from now the price of Brand A will be

$$x = 2 + 0.05t \qquad \text{dollars per bottle}$$

and the price of Brand B will be

$$y = 2 + 0.1\sqrt{t} \qquad \text{dollars per bottle}$$

At what rate will the demand for Brand A be changing with respect to time 4 months from now?

Solution

Your goal is to find $\frac{dQ}{dt}$ when $t = 4$. Using the chain rule, you get

$$\frac{dQ}{dt} = \frac{\partial Q}{\partial x}\frac{dx}{dt} + \frac{\partial Q}{\partial y}\frac{dy}{dt} = -40x(0.05) + 30(0.05t^{-1/2})$$

When $t = 4$, $\qquad x = 2 + 0.05(4) = 2.2$

and hence,

$$\frac{dQ}{dt} = -40(2.2)(0.05) + 30(0.05)(0.5) = -3.65$$

That is, 4 months from now the monthly demand for Brand A will be decreasing at the rate of 3.65 bottles per month.

The Slope of a Level Curve

The slope of the line that is tangent to the level curve $f(x, y) = C$ at a particular point is given by the derivative $\frac{dy}{dx}$. One way to compute this derivative is to solve the equation $f(x, y) = C$ for y in terms of x and then differentiate the resulting expression with respect to x. Unfortunately, it is often difficult or even impossible to solve the equation $f(x, y) = C$ explicitly for y. In such cases, you can either differentiate the equation $f(x, y) = C$ implicitly (as in Chapter 2, Section 6) or use the following formula involving the partial derivatives of f.

Formula for the Slope of a Level Curve

If the level curve $f(x, y) = C$ is the graph of a differentiable function of x, the slope of its tangent is given by the formula

$$\frac{dy}{dx} = -\frac{f_x}{f_y}$$

We shall prove this formula using the chain rule for partial derivatives, but first, we shall illustrate the formula in two examples.

EXAMPLE 3.3

Find the slope $\dfrac{dy}{dx}$ of the level curve $f(x, y) = C$ if $f(x, y) = xy$ by

(a) Solving for y in terms of x and differentiating the resulting expression.
(b) Differentiating the equation $f(x, y) = C$ implicitly.
(c) Applying the formula.

Solution

(a) Solve the equation $xy = C$ for y to get

$$y = \frac{C}{x}$$

and differentiate: $\qquad \dfrac{dy}{dx} = -\dfrac{C}{x^2}$

(b) Use the product rule to differentiate both sides of the equation

$$xy = C$$

with respect to x, keeping in mind that y is really a function of x. You get

$$x\frac{dy}{dx} + y = 0 \qquad \text{or} \qquad \frac{dy}{dx} = -\frac{y}{x}$$

which is equivalent to the answer in part (a) since $y = \dfrac{C}{x}$.

(c) Since $f_x = y$ and $f_y = x$, the formula gives

$$\frac{dy}{dx} = -\frac{f_x}{f_y} = -\frac{y}{x}$$

which is the same answer obtained in part (b).

EXAMPLE 3.4

Find the slope of the level curve $f(x, y) = 6$ when $x = 2$ if $f(x, y) = x^2y^3 - 5y^3 - x$.

Solution

By the formula, $\dfrac{dy}{dx} = -\dfrac{f_x}{f_y} = -\dfrac{2xy^3 - 1}{3x^2y^2 - 15y^2}$

The desired slope is the value of this derivative when $x = 2$. Before you can compute this slope, you have to find the value of y that corresponds to $x = 2$ on the level curve $f(x, y) = 6$. To do this, substitute $x = 2$ and $f = 6$ into the equation for f to get

$$6 = 4y^3 - 5y^3 - 2 \qquad y^3 = -8 \qquad \text{or} \qquad y = -2$$

Now substitute $x = 2$ and $y = -2$ into the formula for $\dfrac{dy}{dx}$ to get

$$\text{Slope of level curve} = -\frac{2(2)(-2)^3 - 1}{3(2)^2(-2)^2 - 15(-2)^2} = -\frac{-33}{-12} = -\frac{11}{4}$$

Why the Formula Works   The proof that $\dfrac{dy}{dx} = -\dfrac{f_x}{f_y}$ is short but subtle. Here is the argument.

Suppose the equation $f(x, y) = C$ implicitly defines y as a differentiable function of x. Think of f as a function of the single variable x and differentiate both sides of the equation $f = C$ with respect to x to get

$$\frac{df}{dx} = \frac{dC}{dx} \qquad \text{or} \qquad \frac{df}{dx} = 0 \qquad \text{since } C \text{ is a constant}$$

But according to the chain rule for partial derivatives,

$$\frac{df}{dx} = \frac{\partial f}{\partial x}\frac{dx}{dx} + \frac{\partial f}{\partial y}\frac{dy}{dx} = \frac{\partial f}{\partial x} + \frac{\partial f}{\partial y}\frac{dy}{dx}$$

Substitute this expression for $\dfrac{df}{dx}$ into the equation $\dfrac{df}{dx} = 0$ and conclude, as required, that

$$\frac{\partial f}{\partial x} + \frac{\partial f}{\partial y}\frac{dy}{dx} = 0 \qquad \text{or} \qquad \frac{dy}{dx} = -\frac{\partial f/\partial x}{\partial f/\partial y} = -\frac{f_x}{f_y}$$

Applications   On the level curve $f(x, y) = C$, the derivative $\dfrac{dy}{dx}$ is the rate of change of y with respect to x and hence is approximately the amount by which the y coordinate of a point on the level curve changes when the x coordinate is increased by 1. (See Figure 9.10.) For example, if f represents output and x and y represent the levels of skilled and unskilled labor, respectively, the slope $\dfrac{dy}{dx}$ of the tangent to the level curve $f(x, y) = C$ is an approximation to the amount by which the manufacturer should change the level of unskilled labor y to compensate for a 1-unit increase in the level of skilled labor x so that output will remain unchanged. In each of the following two

Figure 9.10 The slope of the tangent is approximately the change in y.

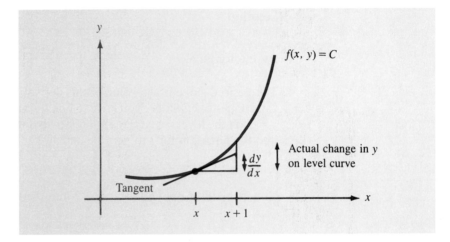

examples from economics, the slope of a level curve is used in the approximation of the change in one variable that is needed to offset a small change in the other variable so that the value of the function will remain unchanged. The procedure is based on the one-variable approximation formula from Chapter 2, Section 4, which is restated here for level curves using the partial derivative formula for $\dfrac{dy}{dx}$.

Approximation on Level Curves

The change in y needed to offset a small change Δx in x so that the value of the function $f(x, y)$ will remain unchanged is

$$\Delta y \simeq \frac{dy}{dx}\,\Delta x = -\frac{f_x}{f_y}\,\Delta x$$

Constant-Production Curves   The next example illustrates how the slope of a constant-production curve can be used to make decisions concerning the allocation of labor.

EXAMPLE 3.5

Using x worker-hours of skilled labor and y worker-hours of unskilled labor, a manufacturer can produce $f(x, y) = x^2y$ units. Currently 16 worker-hours of skilled labor and 32 worker-hours of unskilled labor are used, and the manufacturer is planning to increase skilled labor by $\frac{1}{4}$ worker-hour. Use calculus to estimate the corresponding change that the manufacturer should make in the level of unskilled labor so that the total output will remain the same.

Solution

The present output is $f(16, 32) = 8{,}192$ units. The combinations of x and y for which output will remain at this level are the coordinates of the points that lie on the constant-production curve $f(x, y) = 8{,}192$ (Figure 9.11).

Figure 9.11 The constant-production curve $f = 8{,}192$, where $f(x, y) = x^2y$.

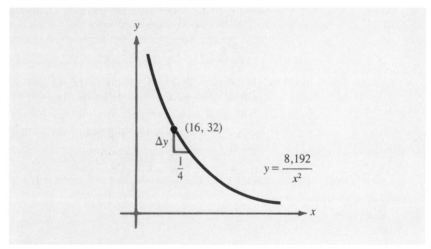

By the approximation formula, the change Δy in unskilled labor that should be made to offset a change of $\Delta x = \frac{1}{4}$ in skilled labor so that output will remain unchanged is

$$\Delta y \simeq \frac{dy}{dx}\,\Delta x = \left(-\frac{f_x}{f_y}\right)\!\left(\frac{1}{4}\right) = \left(-\frac{2xy}{x^2}\right)\!\left(\frac{1}{4}\right) = -\frac{y}{2x}$$

Evaluation at the current levels $x = 16$ and $y = 32$ gives

$$\Delta y \simeq -\frac{32}{2(16)} = -1$$

That is, to compensate for the proposed increase of $\frac{1}{4}$ worker-hour of skilled labor, the manufacturer should reduce the level of unskilled labor by approximately 1 worker-hour.

Indifference Curves   Recall from Section 1 that an indifference curve is a level curve $U(x, y) = C$ of a utility function $U(x, y)$. That is, an indifference curve gives all the combinations of x and y that lead to the same level of consumer satisfaction. Example 3.6 asks some additional questions about the indifference curve you first saw in Example 1.6.

EXAMPLE 3.6

Suppose the utility derived by a consumer from x units of one commodity and y units of a second commodity is given by the utility function $U(x, y) = x^{3/2}y$. The consumer currently owns $x = 16$ units of the first commodity and $y = 20$ units of the second. Use calculus to estimate how many units of the second commodity the consumer could substitute for 1 unit of the first commodity without changing the total utility.

Solution

The current level of utility is $U(16, 20) = 1,280$. The corresponding indifference curve is sketched in Figure 9.12.

The goal is to estimate the change Δy required to offset a change of $\Delta x = -1$ so that utility will remain at its current level. The approximation formula

$$\Delta y \simeq \frac{dy}{dx} \Delta x = -\frac{dy}{dx} = \frac{U_x}{U_y} = \frac{\frac{3}{2}x^{1/2}y}{x^{3/2}} = \frac{3y}{2x}$$

Figure 9.12 The indifference curve $U = 1,280$, where $U(x, y) = x^{3/2}y$.

with $x = 16$ and $y = 20$ gives

$$\Delta y \simeq \frac{3(20)}{2(16)} = 1.875 \text{ units}$$

That is, the consumer could substitute 1.875 units of the second commodity for 1 unit of the first commodity without affecting the level of utility.

The Total Differential   In Chapter 2, Section 4, you learned how to use the differential of a function to approximate the change in the function resulting from a small change in its independent variable. In particular, you saw that if y is a function of x,

$$\Delta y \simeq \frac{dy}{dx} \Delta x$$

where Δx is a small change in the variable x and Δy is the corresponding change in the function y. The expression $dy = \frac{dy}{dx} \Delta x$ that was used to approximate Δy was called the differential of y. Here is the analogous approximation formula for functions of two variables.

Approximation Formula

Suppose z is a function of x and y. If Δx denotes a small change in x and Δy a small change in y, the corresponding change in z is

$$\Delta z \simeq \frac{\partial z}{\partial x} \Delta x + \frac{\partial z}{\partial y} \Delta y$$

Observe that the expression used to approximate Δz is the sum of two terms, each of which is essentially a one-variable differential. In particular,

$$\frac{\partial z}{\partial x} \Delta x \simeq \text{change in } z \text{ due to the change in } x \text{ for fixed } y$$

and $$\frac{\partial z}{\partial y} \Delta y \simeq \text{change in } z \text{ due to the change in } y \text{ for fixed } x$$

The approximation formula says that the total change in z is approximately equal to the sum of these two partial changes.

The sum of the two one-variable differentials that appears in the approximation formula is called the **total differential** of z and is denoted by the symbol dz. Notice the similarity between the formula for the total differential dz and the chain rule.

The Total Differential

If z is a function of x and y, the total differential of z is

$$dz = \frac{\partial z}{\partial x} \Delta x + \frac{\partial z}{\partial y} \Delta y$$

The use of the total differential to approximate the change in a function is illustrated in the next example.

EXAMPLE 3.7

At a certain factory, the daily output is $Q = 60K^{1/2}L^{1/3}$ units, where K denotes the capital investment measured in units of $1,000 and L the size of the labor force measured in worker-hours. The current capital investment is $900,000, and 1,000 worker-hours of labor are used each day. Estimate the change in output that will result if capital investment is increased by $1,000 and labor is increased by 2 worker-hours.

Solution

Apply the approximation formula with $K = 900$, $L = 1,000$, $\Delta K = 1$, and $\Delta L = 2$ to get

$$\Delta Q \simeq \frac{\partial Q}{\partial K} \Delta K + \frac{\partial Q}{\partial L} \Delta L$$

$$= 30K^{-1/2}L^{1/3} \Delta K + 20K^{1/2}L^{-2/3} \Delta L$$

$$= 30\left(\frac{1}{30}\right)(10)\,(1) + 20(30)\left(\frac{1}{100}\right)(2)$$

$$= 22 \text{ units}$$

That is, output will increase by approximately 22 units.

Approximation of Percentage Change   The **percentage change** of a quantity expresses the change in the quantity as a percentage of its size prior to the change. In particular,

$$\text{Percentage change} = 100\,\frac{\text{change in quantity}}{\text{size of quantity}}$$

This formula can be combined with the approximation formula as follows.

Approximation of Percentage Change

Suppose z is a function of x and y. If Δx denotes a small change in x and Δy a small change in y, the corresponding percentage change in z is

$$\text{Percentage change in } z = 100\frac{\Delta z}{z} \simeq 100\frac{\dfrac{\partial z}{\partial x}\Delta x + \dfrac{\partial z}{\partial y}\Delta y}{z}$$

EXAMPLE 3.8

Use calculus to approximate the percentage by which the volume of a cylinder increases if the radius increases by 1 percent and the height increases by 2 percent.

Solution

The volume of a cylinder is given by the function $V(r, h) = \pi r^2 h$, where r is the radius and h the height. The fact that r increases by 1 percent means that $\Delta r = 0.01r$, and the fact that h increases by 2 percent means that $\Delta h = 0.02h$. By the approximation formula for percentage change,

$$\text{Percentage change in } V \simeq 100\frac{\dfrac{\partial V}{\partial r}\Delta r + \dfrac{\partial V}{\partial h}\Delta h}{V}$$

$$= 100\frac{2\pi rh(0.01r) + \pi r^2(0.02h)}{\pi r^2 h}$$

$$= 100\frac{0.02\ \pi r^2 h + 0.02\pi r^2 h}{\pi r^2 h}$$

$$= 100\frac{0.04\pi r^2 h}{\pi r^2 h} = 4 \text{ percent}$$

Problems 9.3

In Problems 1 through 8, use the chain rule to find $\dfrac{dz}{dt}$. Check your answer by writing z explicitly as a function of t and differentiating directly with respect to t.

1. $z = x + 2y; x = 3t, y = 2t + 1$

2. $z = 3x^2 + xy$; $x = t + 1$, $y = 1 - 2t$

3. $z = \dfrac{x}{y}$; $x = t^2$, $y = 3t$

4. $z = \dfrac{y}{x}$; $x = 2t$, $y = t^3$

5. $z = \dfrac{x + y}{x - y}$; $x = t^3 + 1$, $y = 1 - t^3$

6. $z = (2x + 3y)^2$; $x = 2t$, $y = 3t$

7. $z = (x - y^2)^3$; $x = t^2$, $y = 2t$

8. $z = xy$; $x = e^t$, $y = e^{-t}$

In Problems 9 through 13, use the chain rule to find $\dfrac{dz}{dt}$ for the specified value of t.

9. $z = 2x + 3y$; $x = t^2$, $y = 5t$; $t = 2$

10. $z = x^2y$; $x = 3t + 1$, $y = t^2 - 1$; $t = 1$

11. $z = \dfrac{3x}{y}$; $x = t$, $y = t^2$; $t = 3$

12. $z = x^{1/2}y^{1/3}$; $x = 2t$, $y = 2t^2$; $t = 2$

13. $z = xy$; $x = e^{2t}$, $y = e^{3t}$; $t = 0$

In Problems 14 through 19, use the formula $\dfrac{dy}{dx} = -\dfrac{f_x}{f_y}$ to find the slope of the level curve $f = C$, where C is a constant. Check your answer by differentiating implicitly.

14. $f(x, y) = x^2 - 4x - y$

15. $f(x, y) = \dfrac{x}{y}$

16. $f(x, y) = xe^y$

17. $f(x, y) = x^2y + 2y^3 - 3x - 2e^{-x}$

18. $f(x, y) = x + \ln\left(\dfrac{x^2}{y}\right)$

19. $f(x, y) = x \ln y$

In Problems 20 through 25, find the slope of the indicated level curve $f(x, y) = C$ for the specified value of x and the given function $f(x, y)$.

20. $f(x, y) = x^2 - y^3$; $C = 1$, $x = 3$

21. $f(x, y) = x^2 + xy + y^3$; $C = 8$, $x = 0$

22. $f(x, y) = x^2y - 3xy + 5$; $C = 9$, $x = 1$

23. $f(x, y) = (x^2 + y)^3$; $C = 8$, $x = -1$

24. $f(x, y) = \dfrac{e^y}{x}; C = 2, x = \dfrac{1}{2}$

25. $f(x, y) = e^{2x} \ln \dfrac{y}{x}; C = e^2, x = 1$

Allocation of Labor 26. Using x hours of skilled labor and y hours of unskilled labor, a manufacturer can produce $f(x, y) = 10xy^{1/2}$ units. Currently the manufacturer uses 30 hours of skilled labor and 36 hours of unskilled labor and is planning to use 1 additional hour of skilled labor. Use calculus to estimate the corresponding change that the manufacturer should make in the level of unskilled labor so that the total output will remain the same.

Allocation of Labor 27. Suppose the manufacturer in Problem 26 currently uses 30 hours of skilled labor and 36 hours of unskilled labor and is planning to use 1 additional hour of *unskilled* labor. Use calculus to estimate the corresponding change that should be made in the level of skilled labor so that the total output will remain the same. (*Hint:* Use the derivative $\dfrac{dx}{dy}$.)

Allocation of Resources 28. At a certain factory, the daily output is $Q = 200K^{1/2}L^{1/3}$ units, where K denotes the capital investment measured in units of $1,000 and L the size of the labor force measured in worker-hours. The current level of capital investment is $60,000, and the current size of the labor force is 10,000 worker-hours. The manufacturer is planning to increase the capital investment by $500. Use calculus to estimate the corresponding decrease that the manufacturer can make in the size of the labor force without affecting the daily output.

Allocation of Resources 29. At a certain factory, output Q is related to inputs x and y by the function $Q = 2x^3 + 3x^2y + y^3$. If the current levels of input are $x = 20$ and $y = 10$, use calculus to estimate the change in input x that should be made to offset an increase of 0.5 unit in input y so that output will be maintained at its current level.

Utility 30. Suppose the utility derived by a consumer from x units of one commodity and y units of a second commodity is given by the utility function $U(x, y) = 2x^3y^2$. The consumer currently owns $x = 5$ units of the first commodity and $y = 4$ units of the second. Use calculus to estimate how many units of the second commodity the consumer could substitute for 1 unit of the first commodity without affecting total utility.

Utility 31. Suppose the utility derived by a consumer from x units of one commodity and y units of a second commodity is given by the utility function $U(x, y) = (x + 1)(y + 2)$. The consumer currently owns

$x = 25$ units of the first commodity and $y = 8$ units of the second. Use calculus to estimate how many units of the first commodity the consumer could substitute for 1 unit of the second commodity without affecting total utility.

Consumer Demand   32. A paint store carries two brands of latex paint. Sales figures indicate that if the first brand is sold for x dollars per gallon and the second for y dollars per gallon, the demand for the first brand will be $Q(x, y) = 200 - 10x^2 + 20y$ gallons per month. It is estimated that t months from now the price of the first brand will be $x = 5 + 0.02t$ dollars per gallon and the price of the second brand will be $y = 6 + 0.4\sqrt{t}$ dollars per gallon. At what rate will the demand for the first brand of paint be changing with respect to time 9 months from now?

Consumer Demand   33. A bicycle dealer has found that if 10-speed bicycles are sold for x dollars apiece and the price of gasoline is y cents per gallon, approximately $f(x, y) = 200 - 24\sqrt{x} + 4(0.1y + 5)^{3/2}$ bicycles will be sold each month. It is estimated that t months from now the bicycles will be selling for $x = 129 + 5t$ dollars apiece and the price of gasoline will be $y = 80 + 10\sqrt{3t}$ cents per gallon. At what rate will the monthly demand for the bicycles be changing with respect to time 3 months from now?

Consumer Demand   34. The demand for a certain product is $Q(x, y) = 200 - 10x^2 + 20xy$ units per month, where x is the price of the product and y the price of a competing product. It is estimated that t months from now the price of the product will be $x = 10 + 0.5t$ dollars per unit while the price of the competing product will be $y = 12.8 + 0.2t^2$ dollars per unit.
(a) At what rate will the demand for the product be changing with respect to time 4 months from now?
(b) At what percentage rate will the demand for the product be changing with respect to time 4 months from now?

Manufacturing   35. The output at a certain plant is $Q(x, y) = 0.08x^2 + 0.12xy + 0.03y^2$ units per day, where x is the number of hours of skilled labor used and y is the number of hours of unskilled labor used. Currently 80 hours of skilled labor and 200 hours of unskilled labor are used each day. Use the total differential of Q to estimate the change in output that will result if an additional $\frac{1}{2}$ hour of skilled labor is used along with an additional 2 hours of unskilled labor.

Manufacturing   36. At a certain factory, the output is $Q = 120K^{1/2}L^{1/3}$ units, where K denotes the capital investment measured in units of $\$1,000$ and L the size of the labor force measured in worker-hours. The current capital investment is $\$400,000$ and 1,000 worker-hours of labor are currently used. Use the total differential of Q to estimate the change in output that will result if capital investment is increased by $500 and labor is increased by 4 worker-hours.

Retail Sales 37. A grocer's daily profit from the sale of two brands of orange juice is $P(x, y) = (x - 30)(70 - 5x + 4y) + (y - 40)(80 + 6x - 7y)$ cents, where x is the price per can of the first brand and y is the price per can of the second. Currently the first brand sells for 50 cents per can and the second for 52 cents per can. Use the total differential of P to estimate the change in the daily profit that will result if the grocer raises the price of the first brand by 1 cent per can and raises the price of the second brand by 2 cents per can.

Publishing 38. An editor estimates that if x thousand dollars is spent on development and y thousand on promotion, approximately $Q(x, y) = 20x^{3/2}y$ copies of a new book will be sold. Current plans call for the expenditure of $36,000 on development and $25,000 on promotion. Use the total differential of Q to estimate the change in sales that will result if the amount spent on development is increased by $500 and the amount spent on promotion is decreased by $500.

Landscaping 39. A rectangular garden that is 30 yards long and 40 yards wide is surrounded by a concrete path that is 0.8 yard wide. Use a total differential to estimate the area of the concrete path.

Packaging 40. A soft-drink can is 12 centimeters tall and has a radius of 3 centimeters. The manufacturer is planning to reduce the height of the can by 0.2 centimeter and the radius by 0.3 centimeter. Use a total differential to estimate how much less drink consumers will find in each can after the new cans are introduced. [*Hint:* Recall that the volume of a cylinder of radius r and height h is $V(r, h) = \pi r^2 h$.]

Volume 41. Use calculus to estimate the percentage by which the volume of a cylinder will change if the radius is increased by 1 percent and the height is decreased by 1.5 percent.

Manufacturing 42. At a certain factory, the daily output is $Q(K, L) = 60K^{1/2}L^{1/3}$ units, where K denotes the capital investment and L the size of the labor force. Use calculus to estimate the percentage by which the daily output will change if capital investment is increased by 1 percent and labor by 2 percent.

Manufacturing 43. At a certain factory, output is given by the Cobb-Douglas production function $Q(K, L) = AK^{\alpha}L^{1-\alpha}$, where A and α are positive constants with $0 < \alpha < 1$, and where K denotes the capital investment and L the size of the labor force. Use calculus to estimate the percentage by which output will change if both capital and labor are increased by 1 percent.

Manufacturing 44. Using x skilled workers and y unskilled workers, a manufacturer can produce $Q(x, y) = x^2y$ units per day. Currently the work force consists of 16 skilled workers and 32 unskilled workers, and the manufacturer is planning to hire 1 additional skilled worker. Use a total differential to estimate the corresponding change that the manufacturer should make in the level of unskilled labor so that the total output will remain the same. (*Hint:* The goal is to estimate Δy if $\Delta x = 1$ and $\Delta Q = 0$.)

45. Suppose $z = f(x, y)$, $x = at$, and $y = bt$, where a and b are constants. Think of z as a function of t and find an expression for the second derivative $\dfrac{d^2z}{dt^2}$ in terms of the constants a and b and the second-order partial derivatives f_{xx}, f_{yy}, and f_{xy}.

4 Relative Maxima and Minima

In this section you will learn how to use partial derivatives to find the relative maxima and minima of functions of two variables.

In geometric terms, a **relative maximum** of a function $f(x, y)$ is a peak, a point on the surface $z = f(x, y)$ that is higher than any nearby point on the surface. A **relative minimum** is the bottom of a valley, a point that is lower than any nearby point on the surface. For example, when $(x, y) = (a, b)$, the function sketched in Figure 9.13a has a relative minimum while the function sketched in Figure 9.13b has a relative maximum.

Figure 9.13 Relative extrema.

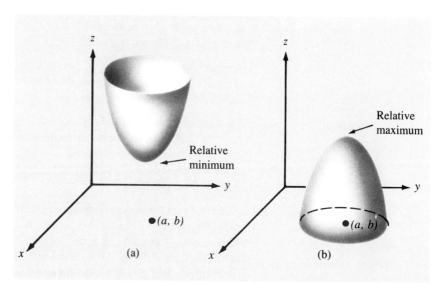

Critical Points   The points (a, b) in the domain of $f(x, y)$ for which both $f_x(a, b) = 0$ and $f_y(a, b) = 0$ are said to be **critical points** of f. Like the critical points for functions of one variable, these critical points play an important role in the study of relative maxima and minima.

To see the connection between critical points and relative extrema, suppose $f(x, y)$ has a relative maximum at (a, b). Then the curve formed by intersecting the surface $z = f(x, y)$ with the vertical plane $y = b$ has a relative maximum and hence a horizontal tangent when $x = a$. (See Figure 9.14a.) Since the partial derivative $f_x(a, b)$ is the slope of this tangent, it follows that $f_x(a, b) = 0$. Similarly, the curve formed by intersecting the surface $z = f(x, y)$ with the plane $x = a$ has a relative maximum when $y = b$ (see Figure 9.14b), and so $f_y(a, b) = 0$. This shows that a point at which a function of two variables has a relative maximum must be a critical point. A similar argument shows that a point at which a function of two variables has a relative minimum must also be a critical point.

Figure 9.14 The partial derivatives are zero at a relative extremum.

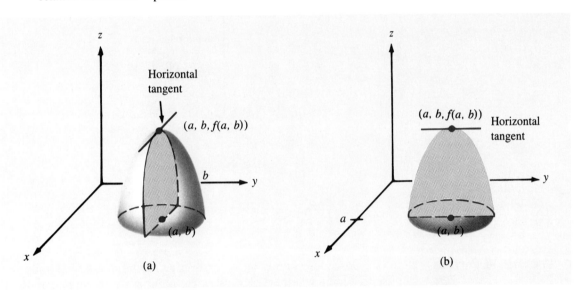

(a)   (b)

Here is a more precise statement of the situation.

Critical Points and Relative Extrema

A point (a, b) in the domain of $f(x, y)$ for which both

$$f_x(a, b) = 0 \quad \text{and} \quad f_y(a, b) = 0$$

is said to be a critical point of f.

If the first-order partial derivatives of f are defined at all points in some region in the xy plane, then the relative extrema of f in the region can occur only at critical points.

Although all the relative extrema of a function must occur at critical points, not every critical point of a function is necessarily a relative extremum. Consider, for example, the function $f(x, y) = y^2 - x^2$, whose graph, which resembles a saddle, is sketched in Figure 9.15. In this case, $f_x(0, 0) = 0$ because the surface has a relative *maximum* (and hence a horizontal tangent) "in the x direction," and $f_y(0, 0) = 0$ because the surface has a relative *minimum* (and hence a horizontal tangent) "in the y direction." Hence $(0, 0)$ is a critical point of f, but it is not a relative extremum. For a critical point to be a relative extremum, the nature of the extremum must be the same in *all directions*. A critical point that is neither a relative maximum nor a relative minimum is called a **saddle point.**

Figure 9.15 The surface $z = y^2 - x^2$.

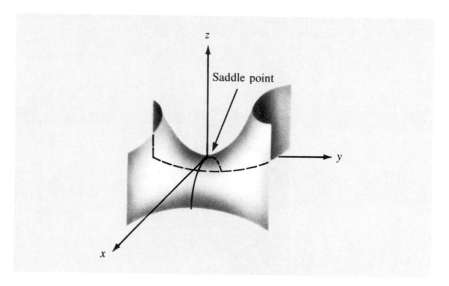

The Second Partials Test

Here is a procedure involving second-order partial derivatives that you can use to decide whether a given critical point is a relative maximum, a relative minimum, or a saddle point. This procedure is the two-variable version of the second derivative test for functions of a single variable that you saw in Chapter 3, Section 2.

The Second Partials Test

Suppose that (a, b) is a critical point of the function $f(x, y)$. Let

$$D = f_{xx}(a, b)f_{yy}(a, b) - [f_{xy}(a, b)]^2$$

If $D < 0$, then f has a saddle point at (a, b).

If $D > 0$ and $f_{xx}(a, b) < 0$, then f has a relative maximum at (a, b).

If $D > 0$ and $f_{xx}(a, b) > 0$, then f has a relative minimum at (a, b).

If $D = 0$, the test is inconclusive and f may have either a relative extremum or a saddle point at (a, b).

The sign of the quantity D that appears in the second partials test tells you whether the function has a relative extremum at (a, b): if D is positive, (a, b) is a relative extremum, and if D is negative, (a, b) is a saddle point.

Moreover, if D is positive [guaranteeing that (a, b) is a relative extremum], the function has either a relative maximum *in all directions* or a relative minimum *in all directions*. To decide which, you can restrict your attention to any *one* direction (say the x direction) and use the one-variable second derivative test to conclude that f has a relative maximum at (a, b) if $f_{xx}(a, b)$ is negative and a relative minimum at (a, b) if $f_{xx}(a, b)$ is positive.

The proof of the second partials test involves ideas that are beyond the scope of this text and will be omitted. Here are some examples illustrating the use of this test.

EXAMPLE 4.1

Classify the critical point of the function $f(x, y) = x^2 + y^2$.

Solution

Since
$$f_x = 2x \quad \text{and} \quad f_y = 2y$$
the only critical point of f is $(0, 0)$. To test this point, use the second-order partial derivatives
$$f_{xx} = 2 \quad f_{yy} = 2 \quad \text{and} \quad f_{xy} = 0$$
to get
$$D(x, y) = f_{xx}f_{yy} - (f_{xy})^2 = 2(2) - 0 = 4$$
That is, $D(x, y) = 4$ for *all* points (x, y) and, in particular,
$$D(0, 0) = 4 > 0$$
Hence, f has a relative extremum at $(0, 0)$. Moreover, since
$$f_{xx}(0, 0) = 2 > 0$$
it follows that the relative extremum at $(0, 0)$ is a relative minimum. For reference, the graph of f is sketched in Figure 9.16.

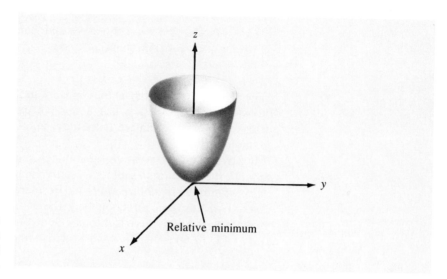

Figure 9.16 The surface
$z = x^2 + y^2$ with a
relative minimum at
$(0, 0)$.

Relative minimum

EXAMPLE 4.2

Classify the critical points of the function $f(x, y) = y^2 - x^2$.

Solution

Since
$$f_x = -2x \quad \text{and} \quad f_y = 2y$$
the only critical point of f is $(0, 0)$. To test this point, use the second-order
partial derivatives
$$f_{xx} = -2 \quad f_{yy} = 2 \quad \text{and} \quad f_{xy} = 0$$
to get $D(x, y) = f_{xx}f_{yy} - (f_{xy})^2 = -2(2) - 0 = -4$

That is, $D(x, y) = -4$ for all points (x, y) and, in particular,
$$D(0, 0) = -4 < 0$$
It follows that f must have a saddle point at $(0, 0)$.
    The graph of f is shown in Figure 9.17.

Solving the equations $f_x = 0$ and $f_y = 0$ simultaneously to find the
critical points of a function of two variables is rarely as simple as in
Examples 4.1 and 4.2. The algebra in the next example is more typical.
Before proceeding, you may wish to refer to the Algebra Review at the
back of the book, in which techniques for solving systems of two equations in two unknowns are discussed.

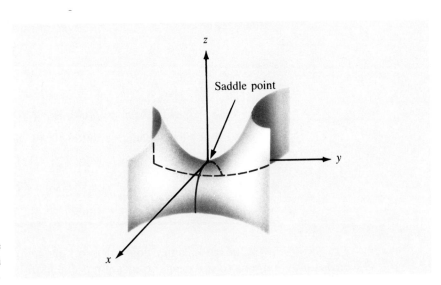

Figure 9.17 The surface
$z = y^2 - x^2$ with
a saddle point at $(0, 0)$.

EXAMPLE 4.3

Classify the critical points of the function $f(x, y) = x^3 - y^3 + 6xy$.

Solution

Since $\qquad f_x = 3x^2 + 6y \qquad$ and $\qquad f_y = -3y^2 + 6x$

you find the critical points of f by solving simultaneously the two equations

$$3x^2 + 6y = 0 \qquad \text{and} \qquad -3y^2 + 6x = 0$$

From the first equation, you get

$$y = -\frac{x^2}{2}$$

which you can substitute into the second equation to find

$$-\frac{3x^4}{4} + 6x = 0 \qquad \text{or} \qquad -x(x^3 - 8) = 0$$

The solutions of this equation are $x = 0$ and $x = 2$. These are the x coordinates of the critical points of f. To get the corresponding y coordinates, substitute these values of x into the equation $y = -\frac{x^2}{2}$ (or into either one of the two original equations). You will find that $y = 0$ when $x = 0$ and $y = -2$ when $x = 2$. It follows that the critical points of f are $(0, 0)$ and $(2, -2)$.

The second-order partial derivatives of f are

$$f_{xx} = 6x \qquad f_{yy} = -6y \qquad \text{and} \qquad f_{xy} = 6$$

Hence,

$$D(x, y) = f_{xx}f_{yy} - (f_{xy})^2 = -36xy - 36 = -36(xy + 1)$$

Since

$$D(0, 0) = -36[0(0) + 1] = -36 < 0$$

it follows that f has a saddle point at $(0, 0)$. Since

$$D(2, -2) = -36[2(-2) + 1] = 108 > 0$$

and

$$f_{xx}(2, -2) = 6(2) = 12 > 0$$

you see that f has a relative minimum at $(2, -2)$.

Practical Optimization Problems

In the next example, you will see how to apply the theory of relative extrema to solve an optimization problem from economics. Actually, you will be trying to find the *absolute* maximum of a certain function. It turns out, however, that the absolute and relative maxima of this function coincide. In fact, in the majority of two-variable optimization problems in the social sciences, the relative extrema and absolute extrema coincide. For this reason, the theory of absolute extrema for functions of two variables will not be developed in this text, and you may assume that the relative extremum you find as the solution to a practical optimization problem is actually the absolute extremum.

Maximization of Profit

EXAMPLE 4.4

The only grocery store in a small rural community carries two brands of frozen orange juice, a local brand that it obtains at the cost of 30 cents per can and a well-known national brand that it obtains at the cost of 40 cents per can. The grocer estimates that if the local brand is sold for x cents per can and the national brand for y cents per can, approximately $70 - 5x + 4y$ cans of the local brand and $80 + 6x - 7y$ cans of the national brand will be sold each day. How should the grocer price each brand to maximize the profit from the sale of the juice? (Assume that the absolute maximum and the relative maximum of the profit function are the same.)

Solution

Since

$$\text{Profit} = \frac{\text{profit from the sale}}{\text{of the local brand}} + \frac{\text{profit from the sale}}{\text{of the national brand}}$$

it follows that the total daily profit from the sale of the juice is given by the function

$$f(x, y) = (x - 30)(70 - 5x + 4y) + (y - 40)(80 + 6x - 7y)$$
$$= -5x^2 + 10xy - 20x - 7y^2 + 240y - 5{,}300$$

Compute the partial derivatives

$$f_x = -10x + 10y - 20 \quad \text{and} \quad f_y = 10x - 14y + 240$$

set them equal to zero to get

$$-10x + 10y - 20 = 0 \quad \text{and} \quad 10x - 14y + 240 = 0$$

or

$$-x + y = 2 \quad \text{and} \quad 5x - 7y = -120$$

and solve these equations simultaneously to get

$$x = 53 \quad \text{and} \quad y = 55$$

It follows that (53, 55) is the only critical point of f.

Now use the second-order partial derivatives

$$f_{xx} = -10 \qquad f_{yy} = -14 \quad \text{and} \quad f_{xy} = 10$$

to get $D(x, y) = f_{xx}f_{yy} - (f_{xy})^2 = -10(-14) - 100 = 40$

Since $D(53, 55) = 40 > 0 \quad \text{and} \quad f_{xx}(53, 55) = -10 < 0$

it follows that f has a (relative) maximum when $x = 53$ and $y = 55$. That is, the grocer can maximize profit by selling the local brand of juice for 53 cents per can and the national brand for 55 cents per can.

Problems 9.4

In Problems 1 through 12, find the critical points of the given function and classify them as relative maxima, relative minima, or saddle points.

1. $f(x, y) = 5 - x^2 - y^2$

2. $f(x, y) = 2x^2 - 3y^2$

3. $f(x, y) = xy$

4. $f(x, y) = xy + \dfrac{8}{x} + \dfrac{8}{y}$

5. $f(x, y) = 2x^3 + y^3 + 3x^2 - 3y - 12x - 4$

6. $f(x, y) = (x - 1)^2 + y^3 - 3y^2 - 9y + 5$

7. $f(x, y) = x^3 + y^2 - 6xy + 9x + 5y + 2$

8. $f(x, y) = -x^4 - 32x + y^3 - 12y + 7$

9. $f(x, y) = (x^2 + 2y^2)e^{1-x^2-y^2}$

10. $f(x, y) = (x - 4)\ln(xy)$

11. $f(x, y) = x^3 - 4xy + y^3$

12. $f(x, y) = \dfrac{x}{x^2 + y^2 + 4}$

In Problems 13 through 19, assume that the absolute maximum and the relative maximum of the profit function are the same.

Retail Sales   13. A T-shirt shop carries two competing shirts, one endorsed by Michael Jordan and the other by Magic Johnson. The owner of the store can obtain both types at a cost of $2 per shirt and estimates that if Jordan shirts are sold for x dollars apiece and Johnson shirts for y dollars apiece, consumers will buy approximately $40 - 50x + 40y$ Jordan shirts and $20 + 60x - 70y$ Johnson shirts each day. How should the owner price the shirts in order to generate the largest possible profit?

Pricing   14. The telephone company is planning to introduce two new types of executive communications systems that it hopes to sell to its largest commercial customers. It is estimated that if the first type of system is priced at x hundred dollars per system and the second type at y hundred dollars per system, approximately $40 - 8x + 5y$ consumers will buy the first type and $50 + 9x - 7y$ will buy the second type. If the cost of manufacturing the first type is $1,000 per system and the cost of manufacturing the second type is $3,000 per system, how should the telephone company price the systems to generate the largest possible profit?

Allocation of Funds   15. A manufacturer is planning to sell a new product at the price of $150 per unit and estimates that if x thousand dollars is spent on development and y thousand dollars is spent on promotion, consumers will buy approximately $\dfrac{320y}{y + 2} + \dfrac{160x}{x + 4}$ units of the product. If manufacturing costs for this product are $50 per unit, how much should the manufacturer spend on development and how much on promotion to generate the largest possible profit from the sale of this product? [*Hint:* Profit = (number of units)(price per unit − cost per unit) − total amount spent on development and promotion.]

Profit under Monopoly   16. A manufacturer with exclusive rights to a sophisticated new industrial machine is planning to sell a limited number of the machines to both foreign and domestic firms. The price the manufacturer can expect to receive for the machines will depend on the number of machines made available. (For example, if only a few of the machines are placed on the market, competitive bidding among prospective purchasers will tend to drive the price up.) It is estimated that if the manufacturer supplies x machines to the domestic market and y machines to the foreign market, the machines will sell for $60 - \dfrac{x}{5} + \dfrac{y}{20}$ thousand dollars apiece at home and for $50 - \dfrac{y}{10} + \dfrac{x}{20}$ thousand dollars apiece abroad. If the manufacturer can produce the machines at the cost of \$10,000 apiece, how many should be supplied to each market to generate the largest possible profit?

Profit under Monopoly   17. A manufacturer with exclusive rights to a new industrial machine is planning to sell a limited number of them and estimates that if x machines are supplied to the domestic market and y to the foreign market, the machines will sell for $150 - \dfrac{x}{6}$ thousand dollars apiece at home and for $100 - \dfrac{y}{10}$ thousand dollars apiece abroad.

     (a) How many machines should the manufacturer supply to the domestic market to generate the largest possible profit at home?

     (b) How many machines should the manufacturer supply to the foreign market to generate the largest possible profit abroad?

     (c) How many machines should the manufacturer supply to each market to generate the largest possible *total* profit?

     (d) Is the relationship between the answers in parts (a), (b), and (c) accidental? Explain. Does a similar relationship hold in Problem 16? What accounts for the difference between these two problems in this respect?

Retail Sales   18. A dairy produces whole milk and skim milk in quantities x and y gallons, respectively. Suppose that the price of whole milk is $p(x) = 100 - x$ and that of skim milk is $q(y) = 100 - y$ and assume that $C(x, y) = x^2 + xy + y^2$ is the joint-cost function of the commodities. What should x and y be in order to maximize profit?

Retail Sales   19. Repeat Problem 18 for the case where $p(x) = 20 - 5x$, $q(y) = 4 - 2y$, and $C(x, y) = 2xy + 4$.

Geometry   20. Find the shortest distance from the origin to the surface $y^2 - z^2 = 10$. [*Hint:* Express the distance $\sqrt{x^2 + y^2 + z^2}$ from the origin to a point (x, y, z) on the surface in terms of the two variables x and y, and, to eliminate the square root, minimize the *square* of the resulting distance function.]

Geometry   21. Find the shortest distance from the origin to the surface $z^2 = 3 + xy$. (See the hint to Problem 20.)

Level Curves   22. Sometimes you can classify the critical points of a function by inspecting its level curves. In each case shown in the figure below, determine the nature of the critical point of f at $(0, 0)$.

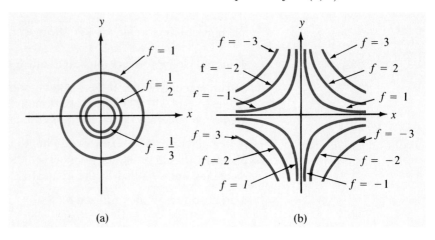

(a)   (b)

23. Find three positive numbers x, y, and z so that $x + y + z = 20$ and the product $P = xyz$ is a maximum. (*Hint:* Use the fact that $z = 20 - x - y$ to express P as a function of only two variables.)

24. Find three positive numbers x, y, and z so that $2x + 3y + z = 60$ and $S = x^2 + y^2 + z^2$ is minimized. (See the hint to Problem 23.)

25. Let $f(x, y) = x^2 + y^2 - 4xy$. Show that f does *not* have a relative minimum at its critical point $(0, 0)$, even though it does have a relative minimum at $(0, 0)$ in both the x and y directions. (*Hint:* Consider the direction defined by the line $y = x$. This is, substitute x for y in the formula for f and analyze the resulting function of x).

5 Lagrange Multipliers

Constrained Optimization Problems   In many applied problems, a function of two variables is to be optimized subject to a restriction or **constraint** on the variables. For example, an editor, constrained to stay within a fixed budget of $60,000, may wish to decide how to divide this money between development and promotion in order to maximize the future sales of a new book. If x denotes the amount of money allocated to development, y the amount allocated to promotion, and $f(x, y)$ the corresponding number of books that will be sold, the editor would like to maximize the sales function $f(x, y)$ subject to the budgetary constraint that $x + y = 60,000$.

For a geometric interpretation of the process of optimizing a function of two variables subject to a constraint, think of the function itself as a surface in three-dimensional space and of the constraint (which is an equation involving x and y) as a curve in the xy plane. When you find the maximum or minimum of the function subject to the given constraint, you are restricting your attention to the portion of the surface that lies directly above the constraint curve. The highest point on this portion of the surface is the constrained maximum, and the lowest point is the constrained minimum. The situation is illustrated in Figure 9.18.

Figure 9.18 Constrained and unconstrained extrema.

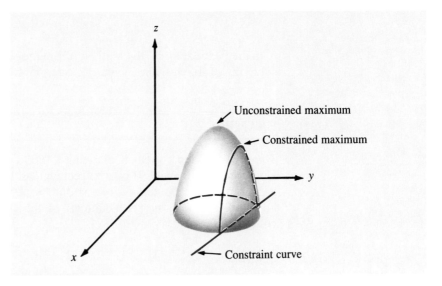

You have already seen some constrained optimization problems in Chapter 3 of this text. (See Chapter 3, Section 4, Example 4.2, for instance.) The technique you used in Chapter 3 to solve such a problem was to reduce it to a problem involving only one variable by solving the constraint equation for one of the variables and then substituting the resulting expression into the function that was to be optimized. The success of this technique depended on solving the constraint equation for one of the variables, which, in many problems, is difficult or even impossible to do. In this section you will see a more versatile technique called the **method of Lagrange multipliers,** in which the introduction of a *third* variable allows you to solve constrained optimization problems without first solving the constraint equation for one of the variables. Here is a statement of the method.

The Method of Lagrange Multipliers

Suppose $f(x, y)$ and $g(x, y)$ are functions whose first-order partial derivatives exist. To find the relative maximum and relative minimum of $f(x, y)$ subject to the constraint that $g(x, y) = k$ for some constant k, introduce a new variable λ (the Greek letter lambda) and solve the following three equations simultaneously:

$$f_x(x, y) = \lambda g_x(x, y) \qquad f_y(x, y) = \lambda g_y(x, y) \qquad g(x, y) = k$$

The desired relative extrema will be found among the resulting points (x, y).

In the next example, the method of Lagrange multipliers will be used to solve the problem from Chapter 3, Section 4, Example 4.2.

EXAMPLE 5.1

The highway department is planning to build a picnic area for motorists along a major highway. It is to be rectangular with an area of 5,000 square yards and is to be fenced off on the three sides not adjacent to the highway. What is the least amount of fencing that will be needed to complete the job?

Solution

Label the sides of the picnic area as indicated in Figure 9.19 and let f denote the amount of fencing required. Then,

$$f(x, y) = x + 2y$$

The goal is to minimize f subject to the constraint that the area must be 5,000 square yards, that is, subject to the constraint

$$xy = 5,000$$

Let $g(x, y) = xy$ and use the partial derivatives

$$f_x = 1 \qquad f_y = 2 \qquad g_x = y \qquad \text{and} \qquad g_y = x$$

to write the three Lagrange equations

$$1 = \lambda y \qquad 2 = \lambda x \qquad \text{and} \qquad xy = 5,000$$

Figure 9.19 Rectangular picnic area.

From the first and second equations you get

$$\lambda = \frac{1}{y} \quad \text{and} \quad \lambda = \frac{2}{x}$$

(since $y \neq 0$ and $x \neq 0$), which implies that

$$\frac{1}{y} = \frac{2}{x} \quad \text{or} \quad x = 2y$$

Now substitute $x = 2y$ into the third Lagrange equation to get

$$2y^2 = 5{,}000 \quad \text{or} \quad y = \pm 50$$

and finally use $y = 50$ in the equation $x = 2y$ to get $x = 100$. It follows that $x = 100$ and $y = 50$ are the values that minimize the function $f(x, y) = x + 2y$ subject to the constraint that $xy = 5{,}000$. That is, the optimal picnic area is 100 yards wide (along the highway), extends 50 yards back from the road, and requires $100 + 50 + 50 = 200$ yards of fencing.

Strictly speaking, the solution of the preceding example was incomplete since no attempt was made to verify that the optimal values $x = 100$ and $y = 50$ actually minimized the function f subject to the given constraint. There is a version of the second partials test that could have been used to show that the point $(100, 50)$ was indeed a constrained relative minimum of the function f. And, with a little more work, it could have been shown that this relative minimum was really the absolute constrained minimum. The techniques needed to carry out this analysis are discussed in more advanced texts. In this text, you may assume that the constrained maximum or minimum you get using the method of Lagrange multipliers is the extremum you are seeking.

Before we see why the method of Lagrange multipliers works, here is one more example illustrating its use.

EXAMPLE 5.2

Find the maximum and minimum values of the function $f(x, y) = xy$ subject to the constraint $x^2 + y^2 = 8$.

Solution

Let $g(x, y) = x^2 + y^2$ and use the partial derivatives

$$f_x = y \quad f_y = x \quad g_x = 2x \quad \text{and} \quad g_y = 2y$$

to get the three Lagrange equations

$$y = 2\lambda x \quad x = 2\lambda y \quad \text{and} \quad x^2 + y^2 = 8$$

Neither x nor y can be zero if all three of these equations are to hold (do you see why?), and so you can rewrite the first two equations as

$$2\lambda = \frac{y}{x} \quad \text{and} \quad 2\lambda = \frac{x}{y}$$

which implies that $\quad \dfrac{y}{x} = \dfrac{x}{y} \quad$ or $\quad x^2 = y^2$

Now substitute $x^2 = y^2$ into the third equation to get

$$2x^2 = 8 \quad \text{or} \quad x = \pm 2$$

If $x = 2$, it follows from the equation $x^2 = y^2$ that $y = 2$ or $y = -2$. Similarly, if $x = -2$, it follows that $y = 2$ or $y = -2$. Hence, the four points at which the constrained extrema can occur are $(2, 2)$, $(2, -2)$, $(-2, 2)$, and $(-2, -2)$. Since

$$f(2, 2) = f(-2, -2) = 4 \quad \text{and} \quad f(2, -2) = f(-2, 2) = -4$$

it follows that when $x^2 + y^2 = 8$, the maximum value of $f(x, y)$ is 4, which occurs at the points $(2, 2)$ and $(-2, -2)$, and the minimum value is -4, which occurs at $(2, -2)$ and $(-2, 2)$.

For practice, check these answers by solving the optimization problem using the methods of Chapter 3.

A Word of Advice

Notice that in each of the preceding examples, the first two Lagrange equations were used to eliminate the new variable λ, and then the resulting expression relating x and y was substituted into the third Lagrange equation. For most constrained optimization problems you will encounter, this particular sequence of steps will lead you quickly to the desired solution.

Why the Method of Lagrange Multipliers Works

Although a rigorous explanation of why the method of Lagrange multipliers works involves advanced ideas beyond the scope of this text, there is a rather simple geometric argument that you should find convincing. Suppose the constraint curve $g(x, y) = k$ and the level curves $f(x, y) = C$ are drawn in the xy plane as shown in Figure 9.20.

To maximize $f(x, y)$ subject to the constraint $g(x, y) = k$, you must find the highest level curve of f that intersects the constraint curve. As the sketch in Figure 9.20 suggests, this critical intersection will occur at a point at which the constraint curve is tangent to a level curve, that is, at which the slope of the constraint curve $g(x, y) = k$ is equal to the slope of a level curve $f(x, y) = C$. According to the formula you learned in Section 3 of this chapter,

$$\text{Slope of constraint curve} = -\frac{g_x}{g_y}$$

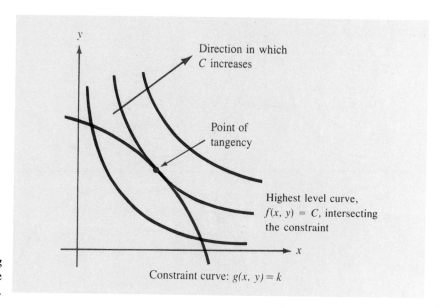

Figure 9.20 Increasing level curves and the constraint curve.

and $\qquad$ Slope of level curve $= -\dfrac{f_x}{f_y}$

Hence, the condition that the slopes be equal can be expressed by the equation

$$-\frac{f_x}{f_y} = -\frac{g_x}{g_y} \qquad \text{or, equivalently,} \qquad \frac{f_x}{g_x} = \frac{f_y}{g_y}$$

If you let λ denote this common ratio, you have

$$\lambda = \frac{f_x}{g_x} \qquad \text{and} \qquad \lambda = \frac{f_y}{g_y}$$

from which you get the first two Lagrange equations

$$f_x = \lambda g_x \qquad \text{and} \qquad f_y = \lambda g_y$$

The third Lagrange equation

$$g(x, y) = k$$

is simply a statement of the fact that the point in question actually lies on the constraint curve.

Suppose all three of the Lagrange equations are satisfied at a certain point (a, b). Then f will reach its constrained *maximum* at (a, b) if the *highest* level curve that intersects the constraint curve does so at this point. On the other hand, if the *lowest* level curve that intersects the constraint curve does so at (a, b), then f will achieve its constrained *minimum* at this point. The situation is illustrated further in Figure 9.21,

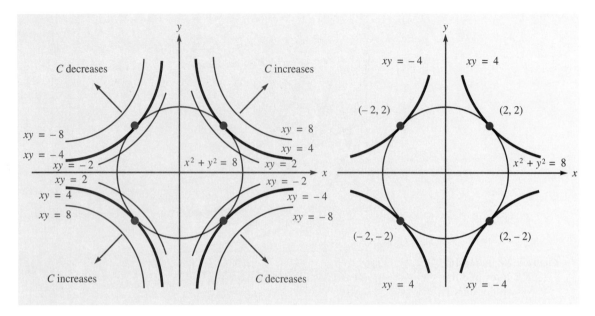

Figure 9.21 (a) Constraint curve $x^2 + y^2 = 8$ and family of level curves $xy = C$. (b) Constraint curve and optimal level curves.

which shows the constraint curve $x^2 + y^2 = 8$ and the optimal level curves $xy = 4$ and $xy = -4$ from Example 5.2. Notice that in this case there are four points at which the constraint curve is tangent to a level curve. Two of these points, $(2, 2)$ and $(-2, -2)$, maximize f subject to the given constraint, while the other two, $(2, -2)$ and $(-2, 2)$, minimize f subject to the constraint.

Applications to Economics

In the next two examples, the method of Lagrange multipliers is used to solve constrained optimization problems from economics.

Allocation of Funds

EXAMPLE 5.3

An editor has been allotted \$60,000 to spend on the development and promotion of a new book. It is estimated that if x thousand dollars is spent on development and y thousand on promotion, approximately $f(x, y) = 20x^{3/2}y$ copies of the book will be sold. How much money should the editor allocate to development and how much to promotion in order to maximize sales?

Solution

The goal is to maximize the function $f(x, y) = 20x^{3/2}y$ subject to the constraint $g(x, y) = 60$, where $g(x, y) = x + y$. The corresponding Lagrange equations are

$$30x^{1/2}y = \lambda \qquad 20x^{3/2} = \lambda \qquad \text{and} \qquad x + y = 60$$

From the first two equations you get

$$30x^{1/2}y = 20x^{3/2}$$

Since the maximum value of f clearly does not occur when $x = 0$, you may assume that $x \neq 0$ and divide both sides of this equation by $30x^{1/2}$ to get

$$y = \frac{2}{3}x$$

Substituting this expression into the third equation, you get

$$x + \frac{2}{3}x = 60 \quad \text{or} \quad \frac{5}{3}x = 60$$

from which it follows that

$$x = 36 \quad \text{and} \quad y = 24$$

That is, to maximize sales, the editor should spend \$36,000 on development and \$24,000 on promotion. If this is done, approximately $f(36, 24) = 103,680$ copies of the book will be sold.

Figure 9.22 Budgetary constraint and optimal sales level.

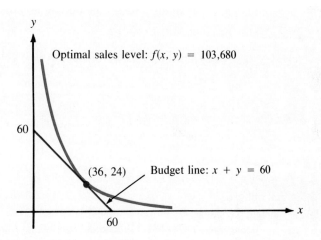

A graph showing the relationship between the budgetary constraint and the level curve for optimal sales is sketched in Figure 9.22.

Maximization of Utility

In Section 3 you were introduced to a classic problem in economics in which a consumer is considering the purchase of a number of units of each of two commodities. Associated with the consumer is a **utility function** $U(x, y)$, which measures the total satisfaction or **utility** the consumer receives from having x units of the first commodity and y units of the second. The problem is to determine how many units of each commodity the consumer should buy to maximize utility while staying within a fixed budget. The application of the method of Lagrange multipliers to the utility problem is illustrated in the next example.

EXAMPLE 5.4

A consumer has $600 to spend on two commodities, the first of which costs $20 per unit and the second $30 per unit. Suppose that the utility derived by the consumer from x units of the first commodity and y units of the second commodity is given by the **Cobb-Douglas utility function** $U(x, y) = 10x^{0.6}y^{0.4}$. How many units of each commodity should the consumer buy to maximize utility?

Solution

The total cost of buying x units of the first commodity at $20 per unit and y units of the second at $30 per unit is $20x + 30y$. Since the consumer has only $600 to spend, the goal is to maximize utility $U(x, y)$ subject to the budgetary constraint that $20x + 30y = 600$.

The three Lagrange equations are

$$6x^{-0.4}y^{0.4} = 20\lambda \qquad 4x^{0.6}y^{-0.6} = 30\lambda \qquad \text{and} \qquad 20x + 30y = 600$$

From the first two equations you get

$$\frac{6x^{-0.4}y^{0.4}}{20} = \frac{4x^{0.6}y^{-0.6}}{30}$$

$$9x^{-0.4}y^{0.4} = 4x^{0.6}y^{-0.6}$$

$$9y = 4x \qquad \text{or} \qquad y = \frac{4}{9}x$$

Figure 9.23 Budgetary constraint and optimal indifference curve.

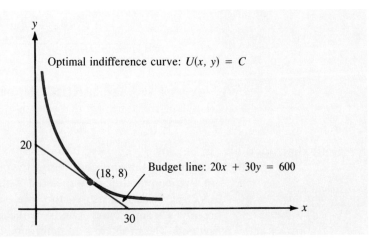

Substituting this into the third equation, you get

$$20x + 30\left(\frac{4}{9}x\right) = 600$$

from which it follows that

$$x = 18 \quad \text{and} \quad y = 8$$

That is, to maximize utility, the consumer should buy 18 units of the first commodity and 8 units of the second.

Recall from Section 1 that the level curves of a utility function are known as **indifference curves**. A graph showing the relationship between the optimal indifference curve $U(x, y) = C$, where $C = U(18, 8)$ and the budgetary constraint $20x + 30y = 600$, is sketched in Figure 9.23.

The Significance of the Lagrange Multiplier λ

You can solve most constrained optimization problems by the method of Lagrange multipliers without actually obtaining a numerical value for the **Lagrange multiplier** λ. In some problems, however, you may want to compute λ. This is because λ has the following useful interpretation.

The Lagrange Multiplier

Suppose M is the maximum (or minimum) value of $f(x, y)$ subject to the constraint $g(x, y) = k$. The Lagrange multiplier λ is the rate of change of M with respect to k. That is,

$$\lambda = \frac{dM}{dk}$$

Hence,

$$\lambda \simeq \text{change in } M \text{ resulting from a 1-unit increase in } k$$

The proof of this fact will be given after the following examples illustrating its use.

EXAMPLE 5.5

Suppose the editor in Example 5.3 is allotted \$60,200 instead of \$60,000 to spend on the development and promotion of the new book. Estimate how the additional \$200 will affect the maximum sales level.

Solution

In Example 5.3, you solved the three Lagrange equations

$$30x^{1/2}y = \lambda \qquad 20x^{3/2} = \lambda \qquad \text{and} \qquad x + y = 60$$

to conclude that the maximum value M of $f(x, y)$ subject to the constraint $x + y = 60$ occurred when $x = 36$ and $y = 24$. To find λ, substitute these values of x and y into the first or second Lagrange equation. Using the second equation, you get

$$\lambda = 20(36)^{3/2} = 4{,}320$$

The goal is to estimate the change ΔM in the maximal sales that will result from an increase of $\Delta k = 0.2$ (thousand dollars) in the available funds. Since $\lambda = \dfrac{dM}{dk}$, the one-variable approximation formula from Chapter 2, Section 4, gives

$$\Delta M \simeq \frac{dM}{dk} \Delta k = \lambda \, \Delta k = 4{,}320(0.2) = 864$$

That is, maximal sales of the book will increase by approximately 864 copies if the budget is increased from \$60,000 to \$60,200 and the money is allocated optimally.

In the context of the maximization of utility subject to a budgetary constraint, the Lagrange multiplier λ is the approximate change in maximum utility resulting from a 1-unit increase in the budget and is known as the **marginal utility of money**. Here is an example.

EXAMPLE 5.6

Suppose the consumer in Example 5.4 has \$601 instead of \$600 to spend on the two commodities. Estimate how the additional \$1 will affect the maximum utility.

Solution

From the first Lagrange equation

$$6x^{-0.4}y^{0.4} = 20\lambda$$

in Example 5.4 it follows that

$$\lambda = \frac{6x^{-0.4}y^{0.4}}{20} = \frac{3y^{0.4}}{10x^{0.4}} = 0.3\left(\frac{y}{x}\right)^{0.4}$$

Since the maximum value M of utility when \$600 was available occurred when $x = 18$ and $y = 8$, substitute these values into the formula for λ to get

$$\lambda = 0.3\left(\frac{8}{18}\right)^{0.4} \approx 0.22$$

which is approximately the increase ΔM in maximum utility resulting from the $1 increase in available funds.

Proof That $\lambda = \dfrac{dM}{dk}$ The proof that $\lambda = \dfrac{dM}{dk}$ is based on the chain rule for partial derivatives from Section 3. Here is the argument.

Suppose M is the optimal value of f subject to the constraint $g = k$. Then $M = f(x, y)$ for some ordered pair (x, y) that satisfies the three Lagrange equations:

$$f_x = \lambda g_x \qquad f_y = \lambda g_y \qquad \text{and} \qquad g = k$$

Moreover, the coordinates of the optimal ordered pair (x, y) depend on k (since different constraint levels will generally lead to different optimal combinations of x and y). Thus,

$$M = f(x, y) \qquad \text{where } x \text{ and } y \text{ are functions of } k$$

By the chain rule for partial derivatives,

$$\frac{dM}{dk} = \frac{\partial M}{\partial x}\frac{dx}{dk} + \frac{\partial M}{\partial y}\frac{dy}{dk}$$

$$= f_x\frac{dx}{dk} + f_y\frac{dy}{dk} \qquad \text{since } M = f(x, y)$$

$$= \lambda g_x\frac{dx}{dk} + \lambda g_y\frac{dy}{dk} \qquad \text{by the first two Lagrange equations}$$

$$= \lambda\left(g_x\frac{dx}{dk} + g_y\frac{dy}{dk}\right)$$

$$= \lambda\frac{dg}{dk} \qquad \text{by the chain rule applied to } g(x, y)$$

$$= \lambda(1) = \lambda \qquad \text{since } \frac{dg}{dk} = 1 \text{ (remember, } g = k\text{)}$$

and the proof is complete.

Problems 9.5

Use the method of Lagrange multipliers to solve the following problems.

1. Find the maximum value of the function $f(x, y) = xy$ subject to the constraint $x + y = 1$.

2. Find the maximum and minimum values of the function $f(x, y) = xy$ subject to the constraint $x^2 + y^2 = 1$.

3. Find the minimum value of the function $f(x, y) = x^2 + y^2$ subject to the constraint $xy = 1$.

4. Find the minimum value of the function $f(x, y) = x^2 + 2y^2 - xy$ subject to the constraint $2x + y = 22$.

5. Find the minimum value of the function $f(x, y) = x^2 - y^2$ subject to the constraint $x^2 + y^2 = 4$.

6. Find the maximum and minimum values of the function $f(x, y) = 8x^2 - 24xy + y^2$ subject to the constraint $x^2 + y^2 = 1$.

7. Find the maximum and minimum values of the function $f(x, y) = x^2 - y^2 - 2y$ subject to the constraint $x^2 + y^2 = 1$.

8. Find the maximum value of the function $f(x, y) = xy$ subject to the constraint $x + y = 1$.

Construction
9. A farmer wishes to fence off a rectangular pasture along the bank of a river. The area of the pasture is to be 3,200 square meters, and no fencing is needed along the river bank. Find the dimensions of the pasture that will require the least amount of fencing.

Construction
10. There are 320 meters of fencing available to enclose a rectangular field. How should the fencing be used so that the enclosed area is as large as possible?

Postal Service
11. According to postal regulations, the girth plus length of parcels sent by fourth-class mail may not exceed 108 inches. What is the largest possible volume of a rectangular parcel with two square sides that can be sent by fourth-class mail?

Packaging
12. Use the fact that 12 fluid ounces is (approximately) 6.89π cubic inches to find the dimensions of the 12-ounce soda can that can be constructed using the least amount of metal. (Recall that the volume of a cylinder of radius r and height h is $\pi r^2 h$, that the circumference of

a circle of radius r is $2\pi r$, and that the area of a circle of radius r is πr^2.)

Packaging 13. A cylindrical can is to hold 4π cubic inches of frozen orange juice. The cost per square inch of constructing the metal top and bottom is twice the cost per square inch of constructing the cardboard side. What are the dimensions of the least expensive can?

Allocation of Funds 14. A manufacturer has \$8,000 to spend on the development and promotion of a new product. It is estimated that if x thousand dollars is spent on development and y thousand is spent on promotion, sales will be approximately $f(x, y) = 50x^{1/2}y^{3/2}$ units. How much money should the manufacturer allocate to development and how much to promotion to maximize sales?

Allocation of Funds 15. If x thousand dollars is spent on labor and y thousand dollars is spent on equipment, the output at a certain factory will be $Q(x, y) = 60x^{1/3}y^{2/3}$ units. If \$120,000 is available, how should this be allocated between labor and equipment to generate the largest possible output?

Allocation of Funds 16. A manufacturer is planning to sell a new product at the price of \$150 per unit and estimates that if x thousand dollars is spent on development and y thousand dollars is spent on promotion, approximately $\dfrac{320y}{y + 2} + \dfrac{160x}{x + 4}$ units of the product will be sold. The cost of manufacturing the product is \$50 per unit. If the manufacturer has a total of \$8,000 to spend on development and promotion, how should this money be allocated to generate the largest possible profit? [*Hint:* Profit = (number of units)(price per unit − cost per unit) − total amount spent on development and promotion.]

Marginal Analysis 17. Use the Lagrange multiplier λ to estimate the change in the maximum output of the factory in Problem 15 that will result if the money available for labor and equipment is increased by \$1,000.

Marginal Analysis 18. Suppose the manufacturer in Problem 16 decides to spend \$8,100 instead of \$8,000 on the development and promotion of the new product. Use the Lagrange multiplier λ to estimate how this change will affect the maximum possible profit.

Allocation of Unrestricted Funds 19. (a) If unlimited funds are available, how much should the manufacturer in Problem 16 spend on development and how much on promotion in order to generate the largest possible profit? (*Hint:* Use the methods of Section 4.)

(b) What is the value of the Lagrange multiplier λ that corresponds to the optimal budget in part (a)? Explain your answer in light of the interpretation of λ as $\dfrac{dM}{dk}$.

(c) Your answer to part (b) should suggest another method for solving the problem in part (a). Solve the problem using this new method.

Utility   20. A consumer has $280 to spend on two commodities, the first of which costs $2 per unit and the second $5 per unit. Suppose that the utility derived by the consumer from x units of the first commodity and y units of the second commodity is $U(x, y) = 100x^{0.25}y^{0.75}$.
(a) How many units of each commodity should the consumer buy to maximize utility?
(b) Compute the marginal utility of money and interpret the result in economic terms.

Utility   21. A consumer has k dollars to spend on two commodities, the first of which costs a dollars per unit and the second b dollars per unit. Suppose that the utility derived by the consumer from x units of the first commodity and y units of the second commodity is given by the Cobb-Douglas utility function $U(x, y) = x^{\alpha}y^{\beta}$, where $0 < \alpha < 1$ and $\alpha + \beta = 1$. Show that utility will be maximized when $x = \dfrac{k\alpha}{a}$ and $y = \dfrac{k\beta}{b}$.

22. In Problem 21, how does the maximum output change if k is increased by 1 dollar?

In Problems 23 through 25, let $Q(x, y)$ be a production function, where x and y represent units of labor and capital, respectively. If unit costs of labor and capital are given by p and q, respectively, then $px + qy$ represents the total cost of production.

Minimum Cost   23. Use Lagrange multipliers to show that subject to a fixed production level c, the total cost is minimized when

$$\frac{Q_x}{p} = \frac{Q_y}{q} \quad \text{and} \quad Q(x, y) = c$$

provided Q_x and Q_y are not both 0 and $p \neq 0$, $q \neq 0$. (This is often referred to as the **minimum cost problem,** and its solution is called the **least cost combination of inputs.**)

Fixed Budget   24. Show that the inputs x, y that maximize the production level $Q(x, y)$ subject to a fixed cost k satisfy

$$\frac{Q_x}{p} = \frac{Q_y}{q} \quad \text{with } px + qy = k$$

(Assume that neither p nor q is 0.) This is called a **fixed budget problem.**

Minimum Cost

25. Show that subject to the fixed production level $Ax^\alpha y^\beta = k$, with $\alpha + \beta = 1$, the cost function $C(x, y) = px + qy$ is minimized when

$$x = \frac{k}{A}\left(\frac{\alpha q}{\beta p}\right)^\beta, \, y = \frac{k}{A}\left(\frac{\beta p}{\alpha q}\right)^\alpha$$

26. Use Lagrange multipliers to find the possible maximum or minimum points on that part of the surface $z = x - y$ for which $y = x^5 + x - 2$. Then sketch the curve $y = x^5 + x - 2$ and the level curves to the surface $f(x, y) = x - y$ and show that the points you have just found do not represent local maxima or minima. What do you conclude from this observation?

6 The Method of Least Squares

Throughout this text, you have seen examples in which functions relating two or more variables were differentiated or integrated to obtain useful information about practical situations. In this section, you will see one of the techniques that researchers use to determine such functions from observed data.

Suppose data consisting of n points $(x_1, y_1), (x_2, y_2), \ldots, (x_n, y_n)$ are known and the goal is to find a function $y = f(x)$ that fits the data reasonably well. The first step is to decide what type of function to try. Sometimes this can be done by a theoretical analysis of the underlying practical situation and sometimes by inspection of the graph of the n points. Two sets of data are plotted in Figure 9.24. In Figure 9.24a, the points lie roughly along a straight line, so a linear function $y = mx + b$ would be an appropriate choice in this case. In Figure 9.24b, the points

Figure 9.24 Data that are (a) approximately linear and (b) approximately exponential.

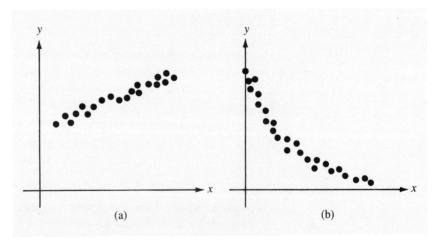

appear to follow an exponential curve, and a function of the form $y = Ae^{-kx}$ would be reasonable.

The Least-Squares Criterion

Once the type of function has been chosen, the next step is to determine the particular function of this type whose graph is "closest" to the given set of points. A convenient way to measure how close a curve is to a set of points is to compute the sum of the squares of the vertical distances from the points to the curve. In Figure 9.25, for example, this is the sum $d_1^2 + d_2^2 + d_3^2$. The closer the curve is to the points, the smaller this sum will be, and the curve for which this sum is smallest is said to be **closest** to the set of points **according to the least-squares criterion.**

Figure 9.25 Sum of the squares of the vertical distances: $d_1^2 + d_2^2 + d_3^2$.

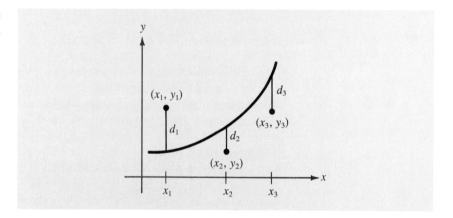

The use of the least-squares criterion to fit a linear function to a set of points is illustrated in the following example. The computation involves the technique from Section 4 for minimizing a function of two variables.

EXAMPLE 6.1

Use the least-squares criterion to find the equation of the line that is closest to the three points (1, 1), (2, 3), and (4, 3).

Solution

As indicated in Figure 9.26, the sum of the squares of the vertical distances from the three given points to the line $y = mx + b$ is

$$d_1^2 + d_2^2 + d_3^2 = (m + b - 1)^2 + (2m + b - 3)^2 + (4m + b - 3)^2$$

This sum depends on the coefficients m and b that define the line, and so the sum can be thought of as a function $S(m, b)$ of the two variables m and b. The goal, therefore, is to find the values of m and b that minimize the function

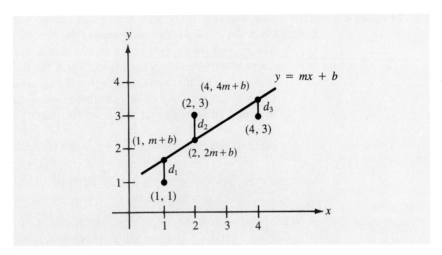

Figure 9.26 Minimize the sum $d_1^2 + d_2^2 + d_3^2$

$$S(m, b) = (m + b - 1)^2 + (2m + b - 3)^2 + (4m + b - 3)^2$$

You do this by setting the partial derivatives $\dfrac{\partial S}{\partial m}$ and $\dfrac{\partial S}{\partial b}$ equal to zero to get

$$\frac{\partial S}{\partial m} = 2(m + b - 1) + 4(2m + b - 3) + 8(4m + b - 3)$$

$$= 42m + 14b - 38 = 0$$

and

$$\frac{\partial S}{\partial b} = 2(m + b - 1) + 2(2m + b - 3) + 2(4m + b - 3)$$

$$= 14m + 6b - 14 = 0$$

and solving the resulting simplified equations

$$21m + 7b = 19$$

$$7m + 3b = 7$$

simultaneously for m and b to conclude that

$$m = \frac{4}{7} \quad \text{and} \quad b = 1$$

It can be shown that the critical point $(m, b) = \left(\dfrac{4}{7}, 1\right)$ does indeed minimize the function $S(m, b)$, and so it follows that

$$y = \frac{4}{7}x + 1$$

is the equation of the line that is closest to the three given points.

The Least-Squares Line   The line that is closest to a set of points according to the least-squares criterion is called the **least-squares line** for the points. (The term **regression line** is also used, especially in statistical work.) The procedure used in Example 6.1 can be generalized to give formulas for the slope m and the y intercept b of the least-squares line for an arbitrary set of n points (x_1, y_1), $(x_2, y_2), \ldots, (x_n, y_n)$. The formulas involve sums of the x and y values. All the sums run from $j = 1$ to $j = n$. To simplify the notation, the indices are omitted and, for example, Σx is used instead of $\sum\limits_{j=1}^{n} x_j$.

The Least-Squares Line

The equation of the least-squares line for the n points (x_1, y_1), $(x_2, y_2), \ldots, (x_n, y_n)$ is $y = mx + b$, where

$$m = \frac{n\Sigma xy - \Sigma x \Sigma y}{n\Sigma x^2 - (\Sigma x)^2} \quad \text{and} \quad b = \frac{\Sigma x^2 \Sigma y - \Sigma x \Sigma xy}{n\Sigma x^2 - (\Sigma x)^2}$$

EXAMPLE 6.2

Use the formulas to find the least-squares line for the points $(1, 1)$, $(2, 3)$, and $(4, 3)$ from Example 6.1.

Solution

Arrange your calculations as follows:

x	y	xy	x^2
1	1	1	1
2	3	6	4
4	3	12	16
$\Sigma x = 7$	$\Sigma y = 7$	$\Sigma xy = 19$	$\Sigma x^2 = 21$

Then use the formulas with $n = 3$ to get

$$m = \frac{3(19) - 7(7)}{3(21) - (7)^2} = \frac{4}{7} \quad \text{and} \quad b = \frac{21(7) - 7(19)}{3(21) - (7)^2} = 1$$

from which it follows that the equation of the least-squares line is

$$y = \frac{4}{7}x + 1$$

Least-Squares Prediction   The least-squares line (or curve) that fits data collected in the past can be used to make rough predictions about the future. This is illustrated in Figure 9.27, which shows the least-squares line for a company's annual sales for its first 4 years of operation. A reasonable estimate of fifth-year sales is the value of y obtained from the equation of the line when $x = 5$.

Least-squares prediction is illustrated further in the next example. To keep the calculations relatively simple, an unrealistically small set of data is used.

Figure 9.27 Projected fifth-year sales: $y = 5m + b$.

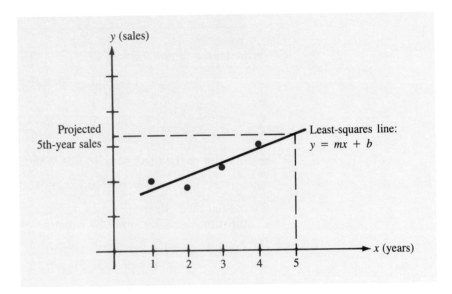

EXAMPLE 6.3

A college admissions officer has compiled the following data relating students' high-school and college grade-point averages:

High-school GPA	2.0	2.5	3.0	3.0	3.5	3.5	4.0	4.0
College GPA	1.5	2.0	2.5	3.5	2.5	3.0	3.0	3.5

Find the equation of the least-squares line for these data and use it to predict the college GPA of a student whose high-school GPA is 3.75.

Solution

Let x denote the high-school GPA and y the college GPA and arrange the calculations as follows:

x	y	xy	x^2
2.0	1.5	3.0	4.0
2.5	2.0	5.0	6.25
3.0	2.5	7.5	9.0
3.0	3.5	10.5	9.0
3.5	2.5	8.75	12.25
3.5	3.0	10.5	12.25
4.0	3.0	12.0	16.0
4.0	3.5	14.0	16.0
$\Sigma x = 25.5$	$\Sigma y = 21.5$	$\Sigma xy = 71.25$	$\Sigma x^2 = 84.75$

Use the least-squares formula with $n = 8$ to get

$$m = \frac{8(71.25) - 25.5(21.5)}{8(84.75) - (25.5)^2} \approx 0.78$$

and

$$b = \frac{84.75(21.5) - 25.5(71.25)}{8(84.75) - (25.5)^2} \approx 0.19$$

The equation of the least-squares line is therefore

$$y = 0.78x + 0.19$$

To predict the college GPA y of a student whose high-school GPA x is 3.75, substitute $x = 3.75$ into the equation of the least-squares line. This gives

$$y = 0.78(3.75) + 0.19 = 3.12$$

which suggests that the student's college GPA might be about 3.1.

A graph of the original data and of the corresponding least-squares line is shown in Figure 9.28. Actually, in practice, it is a good idea to plot the

Figure 9.28 The least-squares line for high-school and college GPAs.

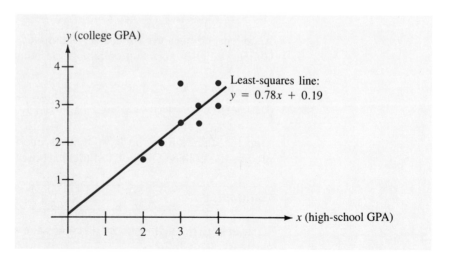

Least-squares line: $y = 0.78x + 0.19$

y (college GPA)

x (high-school GPA)

data *before* proceeding with the calculations. By looking at the graph you will usually be able to tell whether approximation by a straight line is appropriate or whether a curve of some other shape should be used instead.

Nonlinear Curve Fitting

In each of the preceding examples, the least-squares criterion was used to fit a linear function to a set of data. With appropriate modifications, the procedure can also be used to fit nonlinear functions to data. For example, to find the quadratic function $y = Ax^2 + Bx + C$ whose graph is closest to a set of points, you proceed as in Example 6.1 and form the sum of the squares of the vertical distances from the given points to the graph. In this case, the sum is a function $S(A, B, C)$ of the coefficients A, B, and C. To minimize this function of three variables, you set all three of its first-order partial derivatives equal to zero and solve the resulting system of three equations in three unknowns. A computer is often used to perform the necessary calculations.

Problems 9.6

In Problems 1 through 4, plot the given points and use the method of Example 6.1 to find the corresponding least-squares line.

1. $(0, 1)$, $(2, 3)$, $(4, 2)$

2. $(1, 1)$, $(2, 2)$, $(6, 0)$

3. $(1, 2)$, $(2, 4)$, $(4, 4)$, $(5, 2)$

4. $(1, 5)$, $(2, 4)$, $(3, 2)$, $(6, 0)$

In Problems 5 through 10, plot the given points and use the formula to find the corresponding least-squares line.

5. $(1, 2)$, $(2, 2)$, $(2, 3)$, $(5, 5)$

6. $(-4, -1)$, $(-3, 0)$, $(-1, 0)$, $(0, 1)$, $(1, 2)$

7. $(-2, 5)$, $(0, 4)$, $(2, 3)$, $(4, 2)$, $(6, 1)$

8. $(-6, 2)$, $(-3, 1)$, $(0, 0)$, $(0, -3)$, $(1, -1)$, $(3, -2)$

9. $(0, 1)$, $(1, 1.6)$, $(2.2, 3)$, $(3.1, 3.9)$, $(4, 5)$

10. $(3, 5.72)$, $(4, 5.31)$, $(6.2, 5.12)$, $(7.52, 5.32)$, $(8.03, 5.67)$

College Admissions

11. Over the past 4 years, a college admissions officer has compiled the following data (measured in units of 1,000) relating the number of college catalogs requested by high-school students by December 1 to the number of completed applications received by March 1:

Catalogs requested	4.5	3.5	4.0	5.0
Applications received	1.0	0.8	1.0	1.5

(a) Plot these data on a graph.

(b) Find the equation of the least-squares line.

(c) Use the least-squares line to predict the number of completed applications that will be received by March 1 if 4,800 catalogs are requested by December 1.

Sales 12. A company's annual sales (in units of 1 billion dollars) for its first 5 years of operation are shown in the following table:

Year	1	2	3	4	5
Sales	0.9	1.5	1.9	2.4	3.0

(a) Plot these data on a graph.

(b) Find the equation of the least-squares line.

(c) Use the least-squares line to predict the company's sixth-year sales.

Voter Turnout 13. On election day, the polls in a certain state open at 8:00 A.M. Every 2 hours after that, an election official determines what percentage of the registered voters have already cast their ballots. The data through 6:00 P.M. are shown below:

Time	10:00	12:00	2:00	4:00	6:00
Percentage turnout	12	19	24	30	37

(a) Plot these data on a graph.

(b) Find the equation of the least-squares line. (Let x denote the number of hours after 8:00 A.M.)

(c) Use the least-squares line to predict what percentage of the registered voters will have cast their ballots by the time the polls close at 8:00 P.M.

Public Health 14. In a study of five industrial areas, a researcher obtained the following data relating the average number of units of a certain pollutant in the air and the incidence (per 100,000 people) of a certain disease:

Units of pollutant	3.4	4.6	5.2	8.0	10.7
Incidence of disease	48	52	58	76	96

(a) Plot these data on a graph.

(b) Find the equation of the least-squares line.

(c) Use the least-squares line to estimate the incidence of the disease in an area with an average pollution level of 7.3 units.

Population Prediction 15. The following table gives the approximate U.S. census figures (in millions) for the 48 contiguous states for the first half of the twentieth century:

Year	1900	1910	1920	1930	1940	1950
Population	75	91.97	105.7	122.78	131.7	178.5

(a) Find the least-squares regression line for the given population data and use this line to "predict" the population for 1970. (The actual population for the 48 contiguous states in 1970 was approximately 200 million.)

(b) Use the least-squares linear approximation to estimate the population of the 48 contiguous states in 1990. Look up the actual U.S. population in an atlas and adjust the value by subtracting the populations of Alaska and Hawaii.

(c) Use the least-squares linear approximation to predict the population of the 48 contiguous states in the year 2000.

Chapter Summary and Review Problems

Important Terms, Symbols, and Formulas

Function of two variables: $z = f(x, y)$

Surface

Level curve: $f(x, y) = C$

Constant-production curve (isoquant)

Indifference curve

Partial derivatives:

$$f_x = \frac{\partial z}{\partial x} \qquad f_y = \frac{\partial z}{\partial y}$$

Second-order partial derivatives:

$$f_{xx} = \frac{\partial^2 z}{\partial x^2} \qquad f_{xy} = \frac{\partial^2 z}{\partial y \partial x} \qquad f_{yx} = \frac{\partial^2 z}{\partial x \partial y} \qquad f_{yy} = \frac{\partial^2 z}{\partial y^2}$$

Mixed second-order partial derivatives: $f_{xy} = f_{yx}$

Chain rule: $\dfrac{dz}{dt} = \dfrac{\partial z}{\partial x} \dfrac{dx}{dt} + \dfrac{\partial z}{\partial y} \dfrac{dy}{dt}$

Approximation formula; total differential:

$$\Delta z \approx \frac{\partial z}{\partial x} \Delta x + \frac{\partial z}{\partial y} \Delta y = dz$$

Percentage change in $z = 100\dfrac{\Delta z}{z} \approx 100 \dfrac{\dfrac{\partial z}{\partial x} \Delta x + \dfrac{\partial z}{\partial y} \Delta y}{z}$

Slope of a level curve: $\dfrac{dy}{dx} = -\dfrac{f_x}{f_y}$

Approximation on level curves: $\Delta y \approx \dfrac{dy}{dx}\,\Delta x = -\dfrac{f_x}{f_y}\,\Delta x$

Relative maximum; relative minimum; saddle point

Critical point: $f_x = f_y = 0$

Second partials test at a critical point: Let $D = f_{xx}f_{yy} - (f_{xy})^2$

      If $D < 0$, f has a saddle point.
      If $D > 0$ and $f_{xx} < 0$, f has a relative maximum.
      If $D > 0$ and $f_{xx} > 0$, f has a relative minimum.
      If $D = 0$, the test is inconclusive.

Method of Lagrange multipliers: To find the relative extremum of $f(x, y)$ subject to $g(x, y) = k$, solve the equations

$$f_x = \lambda g_x \qquad f_y = \lambda g_y \qquad \text{and} \qquad g = k$$

The Lagrange multiplier: $\lambda = \dfrac{dM}{dk} \approx$ change in M resulting from a 1-unit increase in k, where M is the optimal value of $f(x, y)$ subject to $g(x, y) = k$.

Least-squares criterion

Least-squares line: $y = mx + b$, where

$$m = \frac{n\Sigma xy - \Sigma x \Sigma y}{n\Sigma x^2 - (\Sigma x)^2} \qquad \text{and} \qquad b = \frac{\Sigma x^2 \Sigma y - \Sigma x \Sigma xy}{n\Sigma x^2 - (\Sigma x)^2}$$

Review Problems

1. For each of the following functions, compute the first-order partial derivatives f_x and f_y.

    (a) $f(x, y) = 2x^3y + 3xy^2 + \dfrac{y}{x}$

    (b) $f(x, y) = (xy^2 + 1)^5$

    (c) $f(x, y) = xye^{xy}$

    (d) $f(x, y) = \dfrac{x^2 - y^2}{2x + y}$

    (e) $f(x, y) = \ln\!\left(\dfrac{xy}{x + 3y}\right)$

2. For each of the following functions, compute the second-order partial derivatives f_{xx}, f_{yy}, f_{xy}, and f_{yx}.

    (a) $f(x, y) = x^2 + y^3 - 2xy^2$

    (b) $f(x, y) = e^{x^2+y^2}$

    (c) $f(x, y) = x \ln y$

3. At a certain factory, the daily output is approximately $40K^{1/3}L^{1/2}$ units, where K denotes the capital investment measured in units of $1,000 and L denotes the size of the labor force measured in worker-hours. Suppose that the current capital investment is $125,000 and that 900 worker-hours of labor are used each day. Use marginal analysis to estimate the effect that an additional capital investment of $1,000 will have on the daily output if the size of the labor force is not changed.

4. In economics, the marginal product of labor is the rate at which output Q changes with respect to labor L for a fixed level of capital investment K. An economic law states that, under certain circumstances, the marginal product of labor increases as the level of capital investment increases. Translate this law into a mathematical statement involving a second-order partial derivative.

5. Use the chain rule to find $\dfrac{dz}{dt}$.

   (a) $z = x^3 - 3xy^2$; $x = 2t$, $y = t^2$

   (b) $z = x \ln y$; $x = 2t$, $y = e^t$

6. A grocery store carries two brands of diet cola. Sales figures indicate that if the first brand is sold for x cents per can and the second brand for y cents per can, consumers will buy $Q(x, y) = 240 + 0.1y^2 - 0.2x^2$ cans of the first brand per week. Currently, the first brand sells for 45 cents per can and the second brand sells for 48 cents per can. Use the total differential to estimate how the demand for the first brand of diet cola will change if the price of the first brand is increased by 2 cents per can while the price of the second brand is decreased by 1 cent per can.

7. At a certain factory the daily output is $Q(K, L) = 120K^{1/3}L^{2/3}$ units, where K denotes the capital investment and L the size of the labor force. Use calculus to estimate the percentage by which the daily output will change if capital investment is increased by 2 percent and labor by 1 percent.

8. Suppose that when apples sell for x cents per pound and bakers earn y dollars per hour, the price of apple pies at the local supermarket is $p(x, y) = \dfrac{1}{2}x^{1/3}y^{1/2}$ dollars per pie. Suppose also that t months from now, the price of apples will be $x = 23 + \sqrt{8t}$ cents per pound and bakers' wages will be $y = 3.96 + 0.02t$ dollars per hour. If the supermarket can sell $Q(p) = \dfrac{3,600}{p}$ pies per week when the price is p dollars per pie, at what rate will the weekly demand Q be changing with respect to time 2 months from now?

9. For each of the following functions, sketch the indicated level curves.

    (a) $f(x, y) = x^2 - y; f = 2, f = -2$
    (b) $f(x, y) = 6x + 2y; f = 0, f = 1, f = 2$

10. For each of the following functions, find the slope of the indicated level curve at the specified value of x.

    (a) $f(x, y) = x^2 - y^3; f = 2; x = 1$
    (b) $f(x, y) = xe^y; f = 2; x = 2$

11. Using x skilled workers and y unskilled workers, a manufacturer can produce $Q(x, y) = 60x^{1/3}y^{2/3}$ units per day. Currently the manufacturer employs 10 skilled workers and 40 unskilled workers and is planning to hire 1 additional skilled worker. Use calculus to estimate the corresponding change that the manufacturer should make in the level of unskilled labor so that the total output will remain the same.

12. Find the critical points of each of the following functions and classify them as relative maxima, relative minima, or saddle points.

    (a) $f(x, y) = x^3 + y^3 + 3x^2 - 18y^2 + 81y + 5$

    (b) $f(x, y) = x^2 + y^3 + 6xy - 7x - 6y$

    (c) $f(x, y) = x^3 + y^3 + 3x^2 - 18y^2 + 81y + 5$

13. Use the method of Lagrange multipliers to find the maximum and minimum values of the function $f(x, y) = x^2 + 2y^2 + 2x + 3$ subject to the constraint $x^2 + y^2 = 4$.

14. Use the method of Lagrange multipliers to prove that of all rectangles with a given perimeter, the square has the largest area.

15. A manufacturer is planning to sell a new product at the price of $350 per unit and estimates that if x thousand dollars is spent on development and y thousand dollars is spent on promotion, consumers will buy approximately $\dfrac{250y}{y + 2} + \dfrac{100x}{x + 5}$ units of the product. If manufacturing costs for this product are $150 per unit, how much should the manufacturer spend on development and how much on promotion to generate the largest possible profit if unlimited funds are available?

16. Suppose the manufacturer in Problem 15 has only $11,000 to spend on the development and promotion of the new product. How should this money be allocated to generate the largest possible profit?

17. Suppose the manufacturer in Problem 16 decides to spend $12,000 instead of $11,000 on the development and promotion of the new product. Use the Lagrange multiplier λ to estimate how this change will affect the maximum possible profit.

18. Let $f(x, y) = \dfrac{12}{x} + \dfrac{18}{y} + xy$, where $x > 0$, $y > 0$. How do you know that f has to have a minimum in the region $x > 0$, $y > 0$? Find the minimum.

19. The difference between an animal's surface temperature and that of the surrounding air causes a transfer of energy by convection. The coefficient of convection h is given by

$$h = \frac{kV^{1/3}}{D^{2/3}}$$

where V (cm/sec) is wind velocity, D (cm) is the diameter of the animal's body, and k is a constant. Find the ratio of the rate of change of the convection coefficient with respect to wind velocity to the rate of change of the convection coefficient with respect to the diameter of the animal's body.

20. For the production function given by $Q = x^a y^b$, where $a > 0$ and $b > 0$, show that

$$x \frac{\partial Q}{\partial x} + y \frac{\partial Q}{\partial y} = (a + b)Q$$

In particular, if $b = 1 - a$ with $0 < a < 1$,

$$x \frac{\partial Q}{\partial x} + y \frac{\partial Q}{\partial y} = Q$$

21. An elastic cylindrical container (squeeze tube) is filled with air so that the radius of the base is 2.02 centimeters and the height is 6.04 centimeters. If the container is deflated so that the radius of the base is reduced to 2 centimeters and the height to 6 centimeters, approximately how much air has been removed? (Ignore the thickness of the container.)

22. Plot the points $(1, 1)$, $(1, 2)$, $(3, 2)$, and $(4, 3)$ and use partial derivatives to find the corresponding least-squares line.

23. The marketing manager for a certain company has compiled the following data relating monthly advertising expenditure and monthly sales (both measured in units of $1,000):

Advertising	3	4	7	9	10
Sales	78	86	138	145	156

(a) Plot these data on a graph.
(b) Find the least-squares line.
(c) Use the least-squares line to predict monthly sales if the monthly advertising expenditure is $5,000.

CHAPTER 10
DOUBLE INTEGRALS

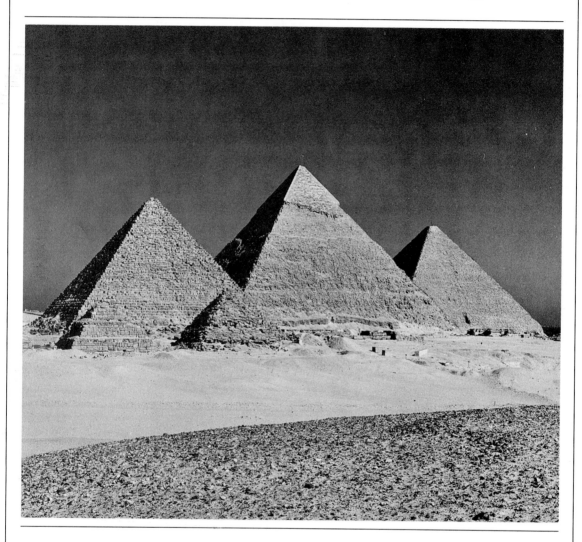

1   Double Integrals

This chapter is about definite integrals of functions of two variables. They are known as **double integrals** and are evaluated by a process involving repeated partial antidifferentiation. In this first section you will see how to evaluate double integrals once their limits of integration are known. In Section 2 you will learn how to determine the limits of integration. And in Section 3 you will see a sampling of applications, all of which are generalizations of applications of definite integrals of functions of one variable with which you are already familiar.

The discussion in this chapter will be kept at an introductory level. Additional detail and theory can be found in more advanced calculus texts.

Evaluation of Double Integrals   The symbol

$$\int_a^b \int_c^d f(x, y)\, dy\, dx$$

is called a double integral (or **iterated integral**) and is an abbreviation for

$$\int_a^b \left[\int_c^d f(x, y)\, dy \right] dx$$

To evaluate a double integral you first compute the inner definite integral

$$\int_c^d f(x, y)\, dy$$

taking the antiderivative of f with respect to y while keeping x fixed. The result will be a function of the single variable x, which you then integrate with respect to x between $x = a$ and $x = b$. Here is an example.

EXAMPLE 1.1

Evaluate $\int_0^1 \int_{-1}^2 xy^2\, dy\, dx$.

Solution

First perform the inner integration with respect to y, treating x as a constant to get

$$\int_{-1}^2 xy^2\, dy = \frac{1}{3}xy^3 \Big|_{y=-1}^{y=2} = \frac{8}{3}x + \frac{1}{3}x = 3x$$

(Note that the limits of integration -1 and 2 refer to the variable y.) Now integrate the result of this calculation with respect to x from $x = 0$ to $x = 1$ to conclude that

$$\int_0^1 \int_{-1}^2 xy^2 \, dy \, dx = \int_0^1 3x \, dx = \frac{3}{2}x^2 \Big|_0^1 = \frac{3}{2}$$

To give your solution a more professional appearance, arrange your work compactly as follows:

$$\int_0^1 \int_{-1}^2 xy^2 \, dy \, dx = \int_0^1 \left(\frac{1}{3}xy^3 \Big|_{y=-1}^{y=2}\right) dx = \int_0^1 3x \, dx = \frac{3}{2}x^2 \Big|_0^1 = \frac{3}{2}$$

Double Integrals with Variable Limits of Integration

In the preceding example, all four of the limits of integration were constants. In the next example, the limits of integration on the inner integral are not constants but functions of x. The technique for evaluating the integral remains the same.

EXAMPLE 1.2

Evaluate $\int_0^1 \int_{x^2}^{\sqrt{x}} 160xy^3 \, dy \, dx$.

Solution

$$\int_0^1 \int_{x^2}^{\sqrt{x}} 160xy^3 \, dy \, dx = \int_0^1 \left(40xy^4 \Big|_{y=x^2}^{y=\sqrt{x}}\right) dx \qquad \text{since } x \text{ is treated as a constant}$$

$$= \int_0^1 [40x(\sqrt{x})^4 - 40x(x^2)^4] \, dx$$

$$= \int_0^1 (40x^3 - 40x^9) \, dx$$

$$= (10x^4 - 4x^{10}) \Big|_0^1$$

$$= 6$$

In Example 1.2, the variable limits of integration on the inner integral were functions of x. These functions, x^2 and $\sqrt{x}$, were substituted for y during the first integration, resulting in an expression containing only the variable x, which was then integrated between the constant limits $x = 0$ and $x = 1$. In general, the outer limits of integration must be constants (so that the final answer will be a constant), while the inner limits may be functions of the variable with respect to which the second integration is to be performed.

In the next example, x is the first variable of integration and y is the second. Notice that in this case the variable limit of integration on the inner integral is a function of y.

EXAMPLE 1.3

Evaluate $\displaystyle\int_0^1 \int_0^y y^2 e^{xy} \, dx \, dy$.

Solution

$$\int_0^1 \int_0^y y^2 e^{xy} \, dx \, dy = \int_0^1 \left(y e^{xy} \Big|_{x=0}^{x=y} \right) dy \qquad \text{since } \int e^{xy} \, dx = \frac{1}{y} e^{xy}$$

$$= \int_0^1 (y e^{y^2} - y) \, dy$$

$$= \left(\frac{1}{2} e^{y^2} - \frac{1}{2} y^2 \right) \Big|_0^1$$

$$= \left(\frac{1}{2} e - \frac{1}{2} \right) - \left(\frac{1}{2} - 0 \right) = \frac{1}{2} e - 1$$

Techniques of Integration In performing the repeated antidifferentiation required to evaluate a double integral, you may have to use one or more of the special techniques of integration you learned in Chapter 5. The next example involves integration by parts.

EXAMPLE 1.4

Evaluate $\displaystyle\int_0^1 \int_{1-y}^1 e^y \, dx \, dy$.

Solution

$$\int_0^1 \int_{1-y}^1 e^y \, dx \, dy = \int_0^1 \left(x e^y \Big|_{x=1-y}^{x=1} \right) dy$$

$$= \int_0^1 [e^y - (1 - y)e^y] \, dy = \int_0^1 ye^y \, dy$$

$$= ye^y \Big|_0^1 - \int_0^1 e^y \, dy \qquad \text{integration by parts}$$

$$= (ye^y - e^y) \Big|_0^1 = 1$$

Problems 10.1

Evaluate the following double integrals.

1. $\int_0^1 \int_1^2 x^2 y \, dx \, dy$

2. $\int_1^2 \int_0^1 x^2 y \, dy \, dx$

3. $\int_0^{\ln 2} \int_{-1}^0 2xe^y \, dx \, dy$

4. $\int_2^3 \int_{-1}^1 (x + 2y) \, dy \, dx$

5. $\int_1^3 \int_0^1 \frac{2xy}{x^2 + 1} \, dx \, dy$

6. $\int_0^1 \int_0^1 x^2 e^{xy} \, dy \, dx$

7. $\int_0^4 \int_0^{\sqrt{x}} x^2 y \, dy \, dx$

8. $\int_0^1 \int_1^5 y\sqrt{1 - y^2} \, dx \, dy$

9. $\int_0^1 \int_{y-1}^{1-y} (2x + y) \, dx \, dy$

10. $\int_0^1 \int_{x^2}^x 2xy \, dy \, dx$

11. $\int_0^2 \int_{x^2}^4 xe^y \, dy \, dx$

12. $\int_0^1 \int_{e^y}^e \frac{1}{x} \, dx \, dy$

13. $\int_0^1 \int_0^x \frac{2y}{x^3 + 1} \, dy \, dx$

14. $\int_0^2 \int_0^{\sqrt{4-y^2}} y \, dx \, dy$

15. $\int_0^1 \int_0^x xe^y \, dy \, dx$

16. $\int_1^2 \int_x^{x^2} e^{y/x} \, dy \, dx$

17. $\int_0^1 \int_y^1 ye^{x+y} \, dx \, dy$

18. $\int_1^e \int_0^{\ln x} 2 \, dy \, dx$

19. $\int_0^1 \int_0^4 \sqrt{xy} \, dy \, dx$

20. $\int_0^1 \int_x^{2x} e^{y-x} \, dy \, dx$

21. $\int_1^e \int_0^{\ln x} xy \, dy \, dx$

22. $\int_0^3 \int_{y^2/4}^{\sqrt{10-y^2}} dx \, dy$

2   Finding Limits of Integration

The limits of integration for the definite integral of a function of one variable come from the inequality defining a corresponding interval on the x axis. In particular, the integral of $f(x)$ over the interval $a \leq x \leq b$ is $\int_a^b f(x)\, dx$. A double integral of a function $f(x, y)$ of two variables is an integral associated with a region R in the xy plane. The notation

$$\iint\limits_R f(x, y)\, dA$$

is sometimes used to denote such an integral. It can be shown that the corresponding limits of integration come from inequalities involving x and y that define the region R.

In this section, the procedure for finding limits of integration for double integrals will be discussed for regions of two fundamental types. More complicated regions of integration can usually be divided into two or more of these elementary regions.

Regions Described in Terms of Vertical Cross Sections The region R in Figure 10.1 is bounded below by the graph of the function $y = g_1(x)$ and above by the graph of the function $y = g_2(x)$, and it extends from $x = a$ on the left to $x = b$ on the right. The inequalities

$$a \leq x \leq b \qquad \text{and} \qquad g_1(x) \leq y \leq g_2(x)$$

can be used to describe such a region. The first inequality specifies the interval in which x must lie, and the second indicates the lower and upper bounds of the vertical cross section of R for each x in this interval. Roughly speaking, the inequalities state that "y goes from $g_1(x)$ to $g_2(x)$ for each x between a and b."

Figure 10.1 Region described by $a \leq x \leq b$ and $g_1(x) \leq y \leq g_2(x)$.

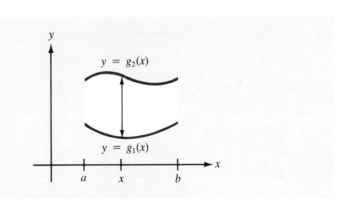

EXAMPLE 2.1

Let R be the region bounded by the curve $y = x^2$ and the line $y = 2x$. Use inequalities to describe R in terms of its vertical cross sections.

Solution

Begin with a sketch of the curve and line as shown in Figure 10.2, identify the region R, and, for reference, draw a vertical cross section. Solve the equations $y = x^2$ and $y = 2x$ simultaneously to find the points of intersection, $(0, 0)$ and $(2, 4)$. Observe that in the region R, the variable x takes on all values from $x = 0$ to $x = 2$ and that for each such value of x, the vertical cross section is bounded below by $y = x^2$ and above by $y = 2x$. Hence, R can be described by the inequalities

$$0 \le x \le 2 \quad \text{and} \quad x^2 \le y \le 2x$$

Figure 10.2 Region bounded by $y = x^2$ and $y = 2x$.

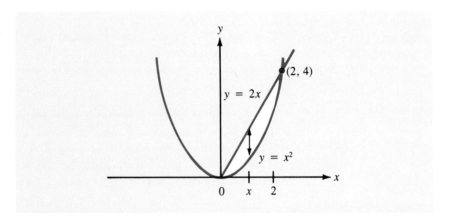

Regions Described in Terms of Horizontal Cross Sections

The region R in Figure 10.3 is bounded on the left by the graph of $x = h_1(y)$ and on the right by the graph of $x = h_2(y)$, and it extends from $y = c$ on the bottom to $y = d$ on top. It can be described by the inequalities

$$c \le y \le d \quad \text{and} \quad h_1(y) \le x \le h_2(y)$$

where the first inequality specifies the interval in which y must lie and the second indicates the left-hand and right-hand bounds of a horizontal cross section. Roughly speaking, the inequalities state that "x goes from $h_1(y)$ to $h_2(y)$ for each y between c and d."

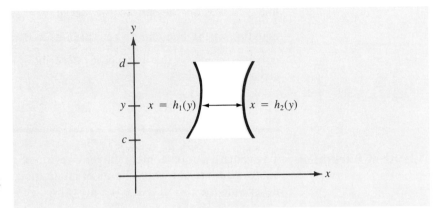

Figure 10.3 Region bounded by $c \le y \le d$ and $h_1(y) \le x \le h_2(y)$.

EXAMPLE 2.2

Let R be the region from Example 2.1 bounded by the curve $y = x^2$ and the line $y = 2x$. Use inequalities to describe R in terms of its horizontal cross sections.

Solution

As in Example 2.1, sketch the region and find the points of intersection of the line and curve, but this time draw a horizontal cross section (Figure 10.4).

In the region R, the variable y takes on all values from $y = 0$ to $y = 4$, and for each such value of y, the horizontal cross section extends from the

Figure 10.4 Region bounded by $y = x^2$ and $y = 2x$.

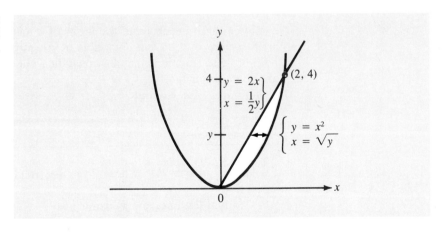

line $y = 2x$ on the left to the curve $y = x^2$ on the right. Since the equation of the line can be rewritten as $x = \frac{1}{2}y$ and the equation of the curve as $x = \sqrt{y}$, the inequalities describing R in terms of its horizontal cross sections are

$$0 \le y \le 4 \quad \text{and} \quad \frac{1}{2}y \le x \le \sqrt{y}$$

Limits of Integration   To evaluate a double integral over a region of one of these types, you use an iterated integral whose limits of integration come from the inequalities describing the region. Here is a more precise statement of the procedure.

Limits of Integration for Double Integrals

If R can be described by the inequalities

$$a \le x \le b \quad \text{and} \quad g_1(x) \le y \le g_2(x)$$

then

$$\iint_R f(x, y) \, dA = \int_a^b \int_{g_1(x)}^{g_2(x)} f(x, y) \, dy \, dx$$

If R can be described by the inequalities

$$c \le y \le d \quad \text{and} \quad h_1(y) \le x \le h_2(y)$$

then

$$\iint_R f(x, y) \, dA = \int_c^d \int_{h_1(y)}^{h_2(y)} f(x, y) \, dx \, dy$$

Notice that in each case, the limits of integration on the inner integral are functions of the *second* variable of integration. (In some cases, one or both of these functions may be constant.) The limits on the outer integral are always constants.

Here are some examples.

EXAMPLE 2.3

Evaluate $\iint_R 40x^2y \, dA$, where R is the region bounded by the curve $y = \sqrt{x}$ and the line $y = x$, using

(a) Vertical cross sections

(b) Horizontal cross sections

Solution

(a) From the sketch in Figure 10.5, observe that R can be described in terms of vertical cross sections by the inequalities

$$0 \le x \le 1 \qquad \text{and} \qquad x \le y \le \sqrt{x}$$

Hence,

$$\iint_R 40x^2y \, dA = \int_0^1 \int_x^{\sqrt{x}} 40x^2y \, dy \, dx$$

$$= \int_0^1 \left(20x^2y^2 \Big|_{y=x}^{y=\sqrt{x}} \right) dx$$

$$= \int_0^1 (20x^3 - 20x^4) \, dx$$

$$= (5x^4 - 4x^5) \Big|_0^1$$

$$= 1$$

Figure 10.5 Region bounded by $y = \sqrt{x}$ and $y = x$.

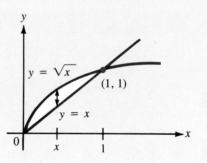

(b) Observe that R can be described in terms of horizontal cross sections by the inequalities

$$0 \le y \le 1 \qquad \text{and} \qquad y^2 \le x \le y$$

The corresponding double integral is

$$\int_0^1 \int_{y^2}^y 40x^2y \, dx \, dy$$

For practice (and to check your answer), evaluate this integral. The answer, of course, should also be 1.

EXAMPLE 2.4

Evaluate $\iint\limits_R (x + y)\, dA$, where R is the triangle with vertices $(0, 0)$, $(0, 1)$, and $(1, 1)$.

Solution

The triangle is sketched in Figure 10.6. It is bounded above by the horizontal line $y = 1$ and below by the line $y = x$ and can be described in terms of vertical cross sections by the inequalities

$$0 \leq x \leq 1 \qquad \text{and} \qquad x \leq y \leq 1$$

Hence,

$$\iint\limits_R (x + y)\, dA = \int_0^1 \int_x^1 (x + y)\, dy\, dx$$

$$= \int_0^1 \left[\left(xy + \frac{1}{2}y^2 \right) \Big|_{y=x}^{y=1} \right] dx$$

$$= \int_0^1 \left[\left(x + \frac{1}{2} \right) - \left(x^2 + \frac{1}{2}x^2 \right) \right] dx = \int_0^1 \left(\frac{1}{2} + x - \frac{3}{2}x^2 \right) dx$$

$$= \left(\frac{1}{2}x + \frac{1}{2}x^2 - \frac{1}{2}x^3 \right) \Big|_0^1 = \frac{1}{2}$$

For practice, check this answer by evaluating the corresponding double integral with the order of integration reversed.

Figure 10.6 Triangle with vertices $(0, 0)$, $(0, 1)$, and $(1, 1)$.

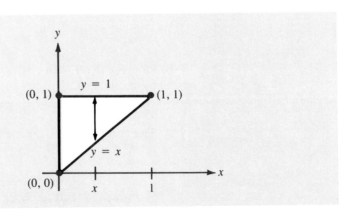

Integration over More Complex Regions Many regions that are not of one of the fundamental types can be divided into two or more subregions that are. The next example illustrates how to integrate over such a region.

EXAMPLE 2.5

Evaluate $\iint_R 1 \, dA$, where R is the region in the first quadrant that lies under the curve $y = \dfrac{1}{x}$ and is bounded by this curve and the lines $y = x$, $y = 0$, and $x = 2$.

Solution

The region is sketched in Figure 10.7. Observe that to the left of $x = 1$, vertical cross sections of R are bounded above by the line $y = x$, while to the right of $x = 1$, they are bounded above by the curve $y = \dfrac{1}{x}$. This prevents a simple description of R in terms of vertical cross sections and suggests that R be broken into two subregions, R_1 and R_2, as shown in Figure 10.7.

Figure 10.7 Region in the first quadrant bounded by $y = \dfrac{1}{x}$, $y = x$, $y = 0$, and $x = 2$.

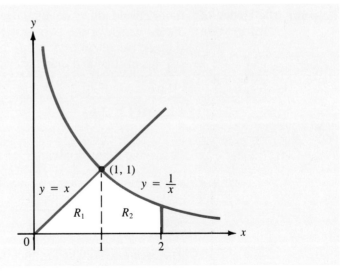

R_1 can be described by the inequalities

$$0 \le x \le 1 \quad \text{and} \quad 0 \le y \le x$$

while R_2 can be described by the inequalities

$$1 \le x \le 2 \quad \text{and} \quad 0 \le y \le \frac{1}{x}$$

Hence,
$$\iint_R 1 \, dA = \iint_{R_1} 1 \, dA + \iint_{R_2} 1 \, dA$$

$$= \int_0^1 \int_0^x 1 \, dy \, dx + \int_1^2 \int_0^{1/x} 1 \, dy \, dx$$

$$= \int_0^1 x \, dx + \int_1^2 \frac{1}{x} \, dx$$

$$= \frac{1}{2}x^2 \Big|_0^1 + \ln x \Big|_1^2$$

$$= \frac{1}{2} + \ln 2$$

For practice, do the problem again, this time using horizontal cross sections. Notice that this solution also involves the use of two iterated integrals.

Selecting the Order of Integration

Before evaluating a double integral, it is a good idea to take a moment to decide which of the two possible orders of integration will lead to the simpler calculation. A wise choice of order can result in substantial savings of time and effort.

EXAMPLE 2.6

Set up the iterated integral you would prefer to use to evaluate $\iint_R y \, dA$, where R is the region bounded by $y = \ln x$, $y = 1$, $x = 0$, and $y = 0$.

Solution

The region is sketched in Figure 10.8. To the left of $x = 1$, vertical cross sections are bounded below by the x axis ($y = 0$), while to the right of $x = 1$, they are bounded below by the curve $y = \ln x$. As a result, the use of vertical cross sections will involve the two iterated integrals

$$\iint_R y \, dA = \int_0^1 \int_0^1 y \, dy \, dx + \int_1^e \int_{\ln x}^1 y \, dy \, dx$$

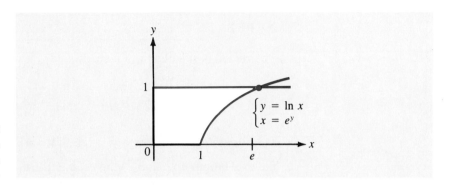

Figure 10.8 Region bounded by $y = \ln x$, $y = 1$, $x = 0$, and $y = 0$.

the second of which will eventually require the integration of $(\ln x)^2$, which is not easy.

On the other hand, the horizontal cross sections are always bounded on the left by the y axis ($x = 0$) and on the right by the curve $y = \ln x$, whose equation can be rewritten as $x = e^y$. The inequalities that describe R in terms of its horizontal cross sections are

$$0 \le y \le 1 \quad \text{and} \quad 0 \le x \le e^y$$

and the corresponding integral is

$$\iint\limits_R y \, dA = \int_0^1 \int_0^{e^y} y \, dx \, dy$$

Clearly, this second integral, in which x is the first variable of integration, is preferable to the pair of integrals that is needed if y is to be the first variable of integration.

For practice, evaluate $\displaystyle\iint\limits_R y \, dA$ using both methods. Each calculation will involve integration by parts. The answer is 1.

In the next example, the choice of the order of integration is critical. The wrong choice leads to an integral that is impossible to evaluate by elementary methods.

EXAMPLE 2.7

Evaluate $\displaystyle\iint\limits_R e^{y^2} \, dA$, where R is the triangle bounded by the lines $y = \frac{1}{2}x$, $y = 1$, and $x = 0$.

Solution

The region is sketched in Figure 10.9. The two possible iterated integrals are

$$\int_0^2 \int_{x/2}^1 e^{y^2} \, dy \, dx \qquad \text{and} \qquad \int_0^1 \int_0^{2y} e^{y^2} \, dx \, dy$$

The first integral, in which y is the first variable of integration, requires the evaluation of $\int e^{y^2} \, dy$, which cannot be done by elementary methods. The second, on the other hand, is easy:

$$\int_0^1 \int_0^{2y} e^{y^2} \, dx \, dy = \int_0^1 \left(x e^{y^2} \Big|_{x=0}^{x=2y} \right) dy$$

$$= \int_0^1 2y e^{y^2} \, dy = e^{y^2} \Big|_0^1 = e - 1$$

Figure 10.9 Region bounded by $y = \dfrac{1}{2}x$, $y = 1$, and $x = 0$.

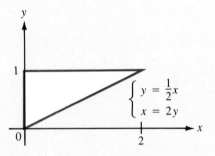

Problems 10.2

In Problems 1 through 6, use inequalities to describe R in terms of its vertical and horizontal cross sections.

1. R is the region bounded by $y = x^2$ and $y = 3x$.

2. R is the region bounded by $y = \sqrt{x}$ and $y = x^2$.

3. R is the rectangle with vertices $(-1, 1)$, $(2, 1)$, $(2, 2)$, and $(-1, 2)$.

4. R is the triangle with vertices $(1, 0)$, $(1, 1)$, and $(2, 0)$.

5. R is the region bounded by $y = \ln x$, $y = 0$, and $x = e$.

6. R is the region bounded by $y = e^x$, $y = 2$, and $x = 0$.

In Problems 7 through 22, evaluate the given double integral for the specified region R.

7. $\displaystyle\iint_R 3xy^2\, dA$, where R is the rectangle bounded by the lines $x = -1$, $x = 2$, $y = -1$, and $y = 0$.

8. $\displaystyle\iint_R (x + 2y)\, dA$, where R is the triangle with vertices $(0, 0)$, $(1, 0)$, and $(0, 2)$.

9. $\displaystyle\iint_R xe^y\, dA$, where R is the triangle with vertices $(0, 0)$, $(1, 0)$, and $(1, 1)$.

10. $\displaystyle\iint_R 48xy\, dA$, where R is the region bounded by $y = x^3$ and $y = \sqrt{x}$.

11. $\displaystyle\iint_R (2y - x)\, dA$, where R is the region bounded by $y = x^2$ and $y = 2x$.

12. $\displaystyle\iint_R 12x\, dA$, where R is the region bounded by $y = x^2$ and $y = 6 - x$.

13. $\displaystyle\iint_R 2\, dA$, where R is the region bounded by $y = \dfrac{16}{x}$, $y = x$, and $x = 2$.

14. $\displaystyle\iint_R y\, dA$, where R is the region bounded by $y = \sqrt{x}$, $y = 2 - x$, and $y = 0$.

15. $\displaystyle\iint_R 4x\, dA$, where R is the region in the first quadrant bounded by $y = 4 - x^2$, $y = 3x$, and $x = 0$.

16. $\displaystyle\iint_R 4x\, dA$, where R is the region in the first quadrant bounded by $y = 4 - x^2$, $y = 3x$, and $y = 0$.

17. $\displaystyle\iint_R (2x + 1)\, dA$, where R is the triangle with vertices $(-1, 0)$, $(1, 0)$, and $(0, 1)$.

18. $\displaystyle\iint_R 2x\, dA$, where R is the region bounded by $y = \dfrac{1}{x^2}$, $y = x$, $x = 2$, and $y = 0$.

19. $\iint\limits_{R} \dfrac{1}{y^2 + 1} \, dA$, where R is the triangle bounded by the lines $y = \dfrac{1}{2}x$, $y = -x$, and $y = 2$.

20. $\iint\limits_{R} e^{y^3} \, dA$, where R is the region bounded by $y = \sqrt{x}$, $y = 1$, and $x = 0$.

21. $\iint\limits_{R} 12x^2 e^{y^2} \, dA$, where R is the region in the first quadrant bounded by $y = x^3$ and $y = x$.

22. $\iint\limits_{R} y \, dA$, where R is the region bounded by $y = \ln x$, $y = 0$, and $x = e$.

In Problems 23 through 34, sketch the region of integration for the given integral and set up an equivalent integral with the order of integration reversed. (In some cases, two or more integrals may be needed.)

23. $\displaystyle\int_{0}^{2} \int_{0}^{4 - x^2} f(x, y) \, dy \, dx$

24. $\displaystyle\int_{0}^{1} \int_{0}^{2y} f(x, y) \, dx \, dy$

25. $\displaystyle\int_{0}^{1} \int_{x^3}^{\sqrt{x}} f(x, y) \, dy \, dx$

26. $\displaystyle\int_{0}^{4} \int_{y/2}^{\sqrt{y}} f(x, y) \, dx \, dy$

27. $\displaystyle\int_{1}^{e^2} \int_{\ln x}^{2} f(x, y) \, dy \, dx$

28. $\displaystyle\int_{0}^{\ln 3} \int_{e^x}^{3} f(x, y) \, dy \, dx$

29. $\displaystyle\int_{-1}^{1} \int_{x^2 + 1}^{2} f(x, y) \, dy \, dx$

30. $\displaystyle\int_{-1}^{1} \int_{-\sqrt{y+1}}^{\sqrt{y+1}} f(x, y) \, dx \, dy$

31. $\displaystyle\int_{0}^{1} \int_{x}^{2 - x} f(x, y) \, dy \, dx$

32. $\displaystyle\int_{0}^{3} \int_{y/3}^{\sqrt{4 - y}} f(x, y) \, dx \, dy$

33. $\displaystyle\int_{-3}^{2} \int_{x^2}^{6 - x} f(x, y) \, dy \, dx$

34. $\displaystyle\int_{2}^{3} \int_{x}^{16/x} f(x, y) \, dy \, dx$

35. Sketch the region of integration for

$$\int_{1}^{2} \left[\int_{x}^{x^3} f(x, y) \, dy \right] dx + \int_{2}^{8} \left[\int_{x}^{8} f(x, y) \, dy \right] dx$$

and reverse the order of integration.

36. Sketch the region of integration for

$$\int_{0}^{1} \left[\int_{0}^{y} (x^2 + y^2) \, dx \right] dy + \int_{1}^{2} \left[\int_{0}^{2 - y} (x^2 + y^2) \, dx \right] dy$$

and evaluate by reversing the order of integration.

3 Applications of Double Integrals

In this section, you will see a sampling of applications of double integrals, most of which are generalizations of applications of definite integrals of functions of one variable with which you are already familiar.

The Area of a Plane Region The area of a region R in the xy plane can be computed as the double integral over R of the constant function $f(x, y) = 1$.

Area Formula

The area of a region R in the xy plane is given by the formula

$$\text{Area of } R = \iint\limits_{R} 1 \, dA$$

To get a feel for why the area formula holds, consider the elementary region R shown in Figure 10.10, which is bounded above by the curve $y = g_2(x)$ and below by the curve $y = g_1(x)$, and which extends from $x = a$ to $x = b$.

Figure 10.10 Area of $R = \iint\limits_{R} 1 \, dA$.

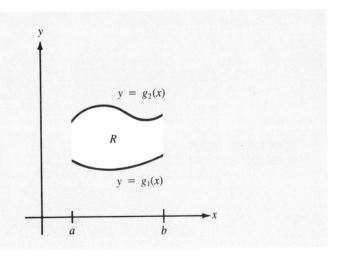

According to the double-integral formula for area,

$$\text{Area of } R = \iint\limits_{R} 1 \, dA$$

$$= \int_{a}^{b} \int_{g_1(x)}^{g_2(x)} 1 \, dy \, dx$$

$$= \int_a^b \left[y \Big|_{y=g_1(x)}^{y=g_2(x)} \right] dx$$

$$= \int_a^b \left[g_2(x) - g_1(x) \right] dx$$

which is precisely the formula for the area between two curves that you saw in Chapter 6, Section 2.

Here are two examples illustrating the use of the area formula.

EXAMPLE 3.1

Find the area of the region R bounded by the curves $y = x^3$ and $y = x^2$.

Solution

The region is shown in Figure 10.11. Using the area formula, you get

$$\text{Area of } R = \iint_R 1 \, dA = \int_0^1 \int_{x^3}^{x^2} 1 \, dy \, dx$$

$$= \int_0^1 \left(y \Big|_{y=x^3}^{y=x^2} \right) dx$$

$$= \int_0^1 (x^2 - x^3) \, dx$$

$$= \left(\frac{1}{3}x^3 - \frac{1}{4}x^4 \right) \Big|_0^1$$

$$= \frac{1}{12}$$

Figure 10.11 Region bounded by $y = x^3$ and $y = x^2$.

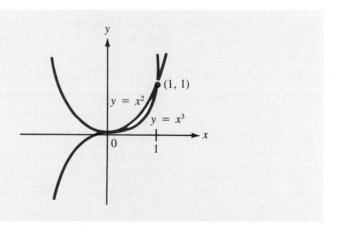

The area calculated in Example 3.1 using a double integral is the same as that calculated in Example 2.6 of Chapter 6 using a single integral. Compare the two solutions. Observe, in particular, that after the integration with respect to y, the double integral in Example 3.1 reduces to the single integral that was used in Chapter 6.

EXAMPLE 3.2

Find the area of the region R in the first quadrant bounded by $y = x^3$, $y = x - 2$, $y = 0$, and $y = 1$.

Solution

From the sketch in Figure 10.12 you see that the use of vertical cross sections to describe R will lead to three iterated integrals. To avoid this, use horizontal cross sections.

$$\text{Area of } R = \int_0^1 \int_{y^{1/3}}^{y+2} 1 \; dx \; dy$$

$$= \int_0^1 (y + 2 - y^{1/3}) \; dy = \left(\frac{1}{2} y^2 + 2y - \frac{3}{4} y^{4/3} \right) \Big|_0^1$$

$$= \frac{1}{2} + 2 - \frac{3}{4} = \frac{7}{4}$$

Figure 10.12 Region bounded by $y = x^3$, $y = x - 2$, $y = 0$, and $y = 1$.

The Volume under a Surface   Recall from Chapter 6, Section 2, that the definite integral $\int_a^b f(x) \; dx$ of a nonnegative function $f(x)$ of one variable gives the area of the region under the graph of f and above the interval $a \le x \le b$. It can be shown that the double integral $\iint_R f(x, y) \; dA$ of a nonnegative function $f(x, y)$ gives the **volume of the solid** under the graph of f and above the region R.

Volume Formula

If $f(x, y) \geq 0$ for all points (x, y) in R, and if S is the solid bounded above by the surface $z = f(x, y)$ and below by R, then

$$\text{Volume of } S = \iint\limits_{R} f(x, y) \, dA$$

Here is an example illustrating the use of this formula.

EXAMPLE 3.3

Find the volume under the surface $z = e^{-x}e^{-y}$ and above the triangle with vertices $(0, 0)$, $(1, 0)$, and $(0, 1)$.

Solution

The triangle is shown in Figure 10.13a. The surface under which you are to find the volume is the graph of the function $f(x, y) = e^{-x}e^{-y}$, which is nonnegative for all values of x and y. Hence,

$$
\begin{aligned}
\text{Volume} &= \int_0^1 \int_0^{1-x} e^{-x}e^{-y} \, dy \, dx \\
&= \int_0^1 \left(-e^{-x}e^{-y} \Big|_{y=0}^{y=1-x} \right) dx \\
&= \int_0^1 (-e^{-x}e^{x-1} + e^{-x}) \, dx = \int_0^1 (-e^{-1} + e^{-x}) \, dx
\end{aligned}
$$

Figure 10.13 Triangle and solid for Example 3.3.

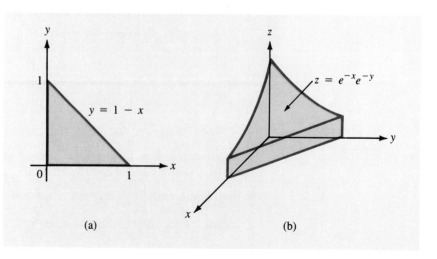

(a)

(b)

$$= (-xe^{-1} - e^{-x})\Big|_0^1 = 1 - 2e^{-1}$$

Notice that you do not have to draw the three-dimensional solid to compute its volume (although you do have to check that the function is nonnegative for all points in the region of integration). The picture of the solid in Figure 10.13b is shown here simply to help you visualize the situation.

The Average Value of a Function of Two Variables

In Chapter 6, Section 4, you saw that the average value of a continuous function $f(x)$ over an interval $a \le x \le b$ is given by the integral formula

$$\text{Average value} = \frac{1}{b-a} \int_a^b f(x) \, dx$$

That is, to find the average value of a function of one variable over an interval, you integrate the function over the interval and divide by the length of the interval. The two-variable procedure is similar. In particular, to find the average value of a function of two variables over a region, you integrate the function over the region and divide by the area of the region.

Average-Value Formula

The average value of a continuous function $f(x, y)$ over a region R is given by the formula

$$\text{Average value} = \frac{1}{\text{area of } R} \iint_R f(x, y) \, dA$$

EXAMPLE 3.4

Suppose R represents the surface of a lake and $f(x, y)$ the depth of the lake at the point (x, y). Find an expression for the average depth of the lake.

Solution

According to the average-value formula,

$$\text{Average depth} = \frac{1}{\text{area of } R} \iint_R f(x, y) \, dA$$

Moreover, the area of R is itself a double integral:

$$\text{Area of } R = \iint_R 1 \, dA$$

Hence, $$\text{Average depth} = \dfrac{\displaystyle\iint_R f(x,\ y)\ dA}{\displaystyle\iint_R 1\ dA}$$

EXAMPLE 3.5

Find the average value of the function $f(x,\ y) = xe^y$ on the triangle with vertices $(0,\ 0)$, $(1,\ 0)$, and $(1,\ 1)$.

Solution

The triangle is shown in Figure 10.14. Its area can be calculated without integration using the familiar formula from plane geometry:

$$\text{Area} = \frac{1}{2}(\text{base})(\text{height}) = \frac{1}{2}(1)(1) = \frac{1}{2}$$

Hence, $$\text{Average value} = \frac{1}{1/2}\int_0^1\int_0^x xe^y\ dy\ dx$$

$$= 2\int_0^1 \left(xe^y\Big|_{y=0}^{y=x}\right) dx$$

$$= 2\int_0^1 (xe^x - x)\ dx$$

$$= 2\left(xe^x - e^x - \frac{1}{2}x^2\right)\Big|_0^1$$

$$= 2\left[\left(e - e - \frac{1}{2}\right) - (-1)\right]$$

$$= 1$$

Joint Probability Density Functions   One of the most important applications of integration in the social, managerial, and life sciences is the computation of probabilities. The technique of integrating probability density functions to find probabilities was introduced in Chapter 6, Section 2, and discussed in more detail in Chapter 8, Section 3. Recall that a probability density function for a single variable x is a nonnegative function $f(x)$ such that the probability that x is between a and b is given by the formula

$$P(a \le x \le b) = \int_a^b f(x)\ dx$$

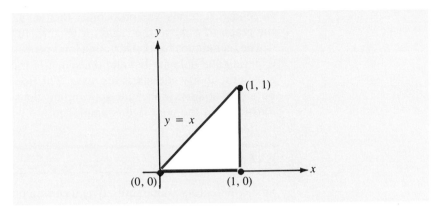

Figure 10.14 Triangle with vertices $(0, 0)$, $(1, 0)$, and $(1, 1)$.

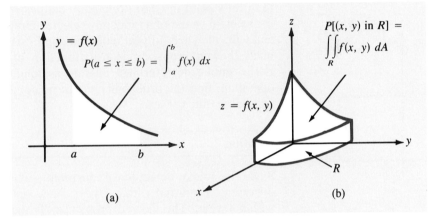

Figure 10.15 Probability as area and volume.

In geometric terms, the probability $P(a \le x \le b)$ is the area under the graph of f from $x = a$ to $x = b$ (Figure 10.15a).

In situations involving two variables, you compute probabilities by evaluating double integrals of a two-variable density function. In particular, you integrate a **joint probability density function,** which is a nonnegative function $f(x, y)$ such that the probability that x is between a and b and y is between c and d is given by the formula

$$P(a \le x \le b \text{ and } c \le y \le d) = \int_c^d \int_a^b f(x, y) \, dx \, dy$$

$$= \int_a^b \int_c^d f(x, y) \, dy \, dx$$

More generally, the probability that the ordered pair (x, y) lies in a region R is given by the formula

$$P[(x, y) \text{ in } R] = \iint_R f(x, y) \, dA$$

In geometric terms, the probability that (x, y) is in R is the volume under the graph of f above the region R (Figure 10.15b).

The techniques for constructing joint probability density functions from experimental data are beyond the scope of this book and are discussed in most probability and statistics texts. The use of double integrals to compute probabilities once the appropriate density functions are known is illustrated in the next two examples.

EXAMPLE 3.6

Smoke detectors manufactured by a certain firm contain two independent circuits, one manufactured at the firm's California plant and the other at the firm's plant in Ohio. Reliability studies suggest that if x denotes the life span (in years) of a randomly selected circuit from the California plant and y the life span (in years) of a randomly selected circuit from the Ohio plant, the joint probability density function for x and y is $f(x, y) = e^{-x}e^{-y}$. If the smoke detector will operate as long as either of its circuits is operating, find the probability that a randomly selected smoke detector will fail within 1 year.

Solution

Since the smoke detector will operate as long as either of its circuits is operating, it will fail within 1 year if and only if *both* of its circuits fail within 1 year. The desired probability is therefore the probability that both $0 \leq x \leq 1$ and $0 \leq y \leq 1$. The points (x, y) for which both these inequalities hold form the square R shown in Figure 10.16. The corresponding probability is the double integral of the density function f over this region R. That is,

$$P(0 \leq x \leq 1 \text{ and } 0 \leq y \leq 1) = \int_0^1 \int_0^1 e^{-x}e^{-y} \, dx \, dy$$

$$= \int_0^1 \left(-e^{-x}e^{-y} \Big|_{x=0}^{x=1} \right) dy = \int_0^1 -(e^{-1} - 1)e^{-y} \, dy$$

$$= (e^{-1} - 1)e^{-y} \Big|_0^1 = (e^{-1} - 1)^2$$

$$\approx 0.3996$$

Figure 10.16 Square consisting of all points (x, y) for which $0 \leq x \leq 1$ and $0 \leq y \leq 1$.

EXAMPLE 3.7

Suppose x denotes the time (in minutes) that a person sits in the waiting room of a certain dentist and y the time (in minutes) that a person stands in line at a certain bank. You have an appointment with the dentist, after which you are planning to cash a check at the bank. If $f(x, y)$ is the joint probability density function for x and y, write down a double integral that gives the probability that your *total* waiting time will be no more than 20 minutes.

Solution

The goal is to find the probability that $x + y \le 20$. The points (x, y) for which $x + y \le 20$ lie on or below the line $x + y = 20$. Moreover, since x and y stand for nonnegative quantities, only those points in the first quadrant are meaningful in this particular context. The problem, then, is to find the probability that a randomly selected point (x, y) lies in R, where R is the region in the first quadrant bounded by the line $x + y = 20$ and the coordinate axes (Figure 10.17). This probability is given by the double integral of the density function f over the region R. That is,

$$P[(x, y) \text{ in } R] = \iint_R f(x, y) \, dA = \int_0^{20} \int_0^{20-x} f(x, y) \, dy \, dx$$

Figure 10.17 Triangle consisting of all points (x, y) for which $x + y \le 20$.

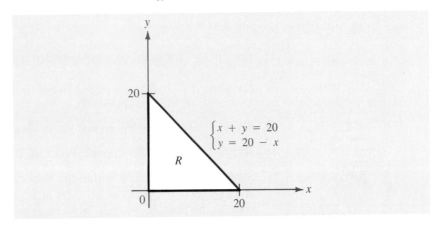

Problems 10.3

In Problems 1 through 10, use a double integral to find the area of R.

1. R is the triangle with vertices $(-4, 0)$, $(2, 0)$, and $(2, 6)$.

2. R is the triangle with vertices $(0, -1)$, $(-2, 1)$, and $(2, 1)$.

3. R is the region bounded by $y = \frac{1}{2}x^2$ and $y = 2x$.

4. R is the region bounded by $y = \sqrt{x}$ and $y = x^2$.

5. R is the region bounded by $y = x^2 - 4x + 3$ and the x axis.

6. R is the region bounded by $y = x^2 + 6x + 5$ and the x axis.

7. R is the region bounded by $y = \ln x$, $y = 0$, and $x = e$.

8. R is the region bounded by $y = x$, $y = \ln x$, $y = 0$, and $y = 1$.

9. R is the region in the first quadrant bounded by $y = 4 - x^2$, $y = 3x$, and $y = 0$.

10. R is the region in the first quadrant that lies under the curve $y = \dfrac{16}{x}$ and is bounded by this curve and the lines $y = x$, $y = 0$, and $x = 8$.

In Problems 11 through 16, find the volume of the solid bounded above by the surface $z = f(x, y)$ and below by the region R.

11. $f(x, y) = 6 - 2x - y$; R is the rectangle with vertices $(0, 0)$, $(1, 0)$, $(0, 2)$, and $(1, 2)$.

12. $f(x, y) = 2x + y$; R is the triangle bounded by $y = x$, $y = 2 - x$, and the x axis.

13. $f(x, y) = e^{y^2}$; R is the triangle bounded by $y = \frac{1}{2}x$, $x = 0$, and $y = 1$.

14. $f(x, y) = x + 1$; R is the region in the first quadrant bounded by $y = 8 - x^2$, $y = x^2$, and the y axis.

15. $f(x, y) = e^{2x+y}$; R is the triangle with vertices $(0, 0)$, $(1, 0)$, and $(0, 1)$.

16. $f(x, y) = 4xe^y$; R is the triangle bounded by $y = 2x$, $y = 2$, and $x = 0$.

Air Pollution   17. Suppose R is the region within the boundary of Los Angeles County and $f(x, y)$ is the number of units of carbon monoxide in the air at noon at the point (x, y) in R. Find an expression for the average level of carbon monoxide in the county at noon.

Elevation   18. Suppose the rectangle R in the first quadrant bounded by the coordinate axes and the lines $x = 3$ and $y = 1$ represents the region inside the boundary of a certain national park. Suppose the elevation above sea level at any point (x, y) in the park is $f(x, y) = 900(2x + y^2)$ feet. Find the average elevation in the park.

Property Value   19. Suppose the triangle R with vertices $(0, 0)$, $(1, 0)$, and $(1, 1)$ represents the region inside the boundary of a certain rural congressional district. Suppose the property value at any point (x, y) in the district is $f(x, y) = 400xe^{-y}$ dollars per acre. Find the average property value in the district.

Average Value

In Problems 20 through 25, find the average value of the given function over the specified region R.

20. $f(x, y) = 6xy$; R is the triangle with vertices $(0, 0)$, $(0, 1)$, and $(3, 1)$.

21. $f(x, y) = 3y$; R is the triangle with vertices $(0, 0)$, $(4, 0)$, and $(2, 2)$.

22. $f(x, y) = e^{x^2}$; R is the triangle with vertices $(0, 0)$, $(1, 0)$, and $(1, 1)$.

23. $f(x, y) = x$; R is the region bounded by $y = 4 - x^2$ and the x axis.

24. $f(x, y) = e^{x^3}$; R is the region in the first quadrant bounded by $y = x^2$, $y = 0$, and $x = 1$.

25. $f(x, y) = e^x y^{-1/2}$; R is the region in the first quadrant bounded by $y = x^2$, $x = 0$, and $y = 1$.

Probability

26. Suppose the joint probability density function for the nonnegative variables x and y is $f(x, y) = 2e^{-2x}e^{-y}$. Find the probability that $0 \leq x \leq 1$ and $1 \leq y \leq 2$.

Probability

27. Suppose the joint probability density function for the nonnegative variables x and y is $f(x, y) = xe^{-x}e^{-y}$. Find the probability that $0 \leq x \leq 1$ and $0 \leq y \leq 2$.

Probability

28. Suppose the joint probability density function for the nonnegative variables x and y is $f(x, y) = 2e^{-2x}e^{-y}$. Find the probability that $x + y \leq 1$.

Probability

29. Suppose the joint probability density function for the nonnegative variables x and y is $f(x, y) = xe^{-x}e^{-y}$. Find the probability that $x + y \leq 1$.

Health Care

30. Suppose x is the length of time (in days) that a person stays in the hospital after abdominal surgery and y is the length of time (in days) that a person stays in the hospital after orthopedic surgery. On Monday, the patient in bed 107A undergoes an emergency appendectomy (abdominal surgery), while the patient's roommate in bed 107B undergoes (orthopedic) surgery for the repair of torn knee cartilage. If the joint probability density function for x and y is $f(x, y) = \frac{1}{12}e^{-x/4}e^{-y/3}$, find the probability that both patients will be discharged from the hospital within 3 days.

Warranty Protection

31. A certain appliance consisting of two independent electronic components will be usable as long as either one of its components is still operating. The appliance carries a warranty from the manufacturer guaranteeing replacement if the appliance becomes unusable within 1 year of the date of purchase. Let x denote the life span (in years) of the first component and y the life span (in years) of the second, and suppose that the joint probability density function for x and y is

$f(x, y) = \frac{1}{4}e^{-x/2}e^{-y/2}$. You purchase one of these appliances, selected at random from the manufacturer's stock. Find the probability
that the warranty will expire before your appliance becomes unusable. (*Hint:* You want the probability that the appliance does *not* fail
during the first year, which you can compute as 1 minus the probability that it *does* fail during this period.)

Customer Service   32. Suppose *x* denotes the time (in minutes) that a person stands in line at
a certain bank and *y* the duration (in minutes) of a routine transaction
at the teller's window. You arrive at the bank to deposit a check. If
the joint probability density function for *x* and *y* is $f(x, y) = \frac{1}{8}e^{-x/4}e^{-y/2}$, find the probability that you will complete your business
at the bank within 8 minutes.

Insurance Sales   33. Suppose *x* denotes the time (in minutes) that a person spends with an
agent choosing a life insurance policy and *y* the time (in minutes) that
the agent spends doing the paperwork once the client has decided.
You arrange to meet with an insurance agent to buy a life insurance
policy. If the joint probability density function for *x* and *y* is $f(x, y) = \frac{1}{300}e^{-x/30}e^{-y/10}$, find the probability that the entire transaction will
take more than half an hour. (*Hint:* The probability you want is 1
minus the probability that the transaction will be completed within 30
minutes.)

In the America's Cup races of 1987, the winning boat, Stars and Stripes, benefited from
sophisticated, computer-enhanced construction techniques (see *Scientific American*,
August, 1987). Here is a problem that illustrates a principle of scientific sail design.

Sailboat Design   34. We can define the *center of pressure* on a boat's sail as the point $(\bar{x}, \bar{y})$
where all aerodynamic forces appear to act. Suppose the sail occupies
a region *R* in the plane. Then it can be shown that $\bar{x}$ and $\bar{y}$ are given by
the formulas

$$\bar{x} = \frac{\displaystyle\iint_R xy \, dA}{\displaystyle\iint_R y \, dA} \qquad \bar{y} = \frac{\displaystyle\iint_R y^2 \, dA}{\displaystyle\iint_R y \, dA}$$

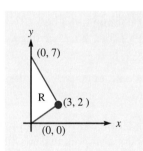

Calculate the center of pressure on the triangular sail represented in
the accompanying figure.

Chapter Summary and Review Problems

Important Terms,
Symbols, and
Formulas

Double integrals: If R is described by vertical cross sections,

$$\iint_R f(x, y)\, dA = \int_a^b \int_{g_1(x)}^{g_2(x)} f(x, y)\, dy\, dx$$

If R is described by horizontal cross sections,

$$\iint_R f(x, y)\, dA = \int_c^d \int_{h_1(y)}^{h_2(y)} f(x, y)\, dx\, dy$$

$$\text{Area of } R = \iint_R 1\, dA$$

$$\text{Volume of } S = \iint_R f(x, y)\, dA$$

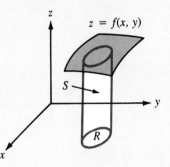

$$\text{Average value of } f \text{ over } R = \frac{1}{\text{area of } R} \iint_R f(x, y)\, dA$$

Joint probability:

$$P[(x, y) \text{ in } R] = \iint_R f(x, y)\, dA$$

where f is the joint probability density function

Review Problems   In Problems 1 through 5, evaluate the given double integral.

1. $\int_0^1 \int_{-2}^0 (2x + 3y) \, dy \, dx$

2. $\int_0^1 \int_0^y x\sqrt{1-y^3} \, dx \, dy$

3. $\int_0^1 \int_{-x}^x \frac{6xy^2}{x^5 + 1} \, dy \, dx$

4. $\int_0^1 \int_0^2 e^{-x-y} \, dy \, dx$

5. $\int_0^1 \int_0^x xe^{2y} \, dy \, dx$

In Problems 6 through 10, evaluate the given double integral for the specified region R.

6. $\iint_R 6x^2y \, dA$, where R is the rectangle with vertices $(-1, 0)$, $(2, 0)$, $(2, 3)$, and $(-1, 3)$.

7. $\iint_R (x + 2y) \, dA$, where R is the triangle with vertices $(0, 0)$, $(0, 3)$, and $(1, 1)$.

8. $\iint_R 40x^2y \, dA$, where R is the region bounded by $y = \sqrt{x}$ and $y = \frac{1}{2}x$.

9. $\iint_R 6y^2e^{x^2} \, dA$, where R is the region in the first quadrant bounded by $y = \sqrt[3]{x}$ and $y = x$.

10. $\iint_R 32xy^3 \, dA$, where R is the region bounded by $y = \frac{8}{x}$, $y = \sqrt{x}$, and $y = 1$.

In Problems 11 through 14, sketch the region of integration for the given iterated integral and set up an equivalent iterated integral (or integrals) with the order of integration reversed.

11. $\int_0^4 \int_{x^2/8}^{\sqrt{x}} f(x, y) \, dy \, dx$

12. $\int_0^2 \int_1^{e^y} f(x, y) \, dx \, dy$

13. $\int_0^2 \int_{x^2}^{8-2x} f(x, y) \, dy \, dx$

14. $\int_{-3}^1 \int_{x^2}^{6-x} f(x, y) \, dy \, dx$

15. Find the area of the region bounded by $y = x^2 - 4$ and $y = x - 2$.

16. Find the area of the region bounded by $y = \ln x$, $y = 0$, $y = 1$, and $x = 0$.

17. Find the area of the region in the first quadrant bounded by $y = x^2$, $y = 2 - x$, and $y = 0$.

18. Find the volume under the surface $z = 2xy$ and above the triangle with vertices $(0, 0)$, $(2, 0)$, and $(0, 1)$.

19. Find the volume under the surface $z = xe^y$ and above the region bounded by $y = x$ and $y = x^2$.

20. Find the average value of $f(x, y) = 2xy$ over the region bounded by $y = x^2$ and $y = x$.

21. Find the volume under the surface $z = x + 2y + 4$ and above the region bounded by the lines $y = 2x$, $y = 3 - x$, and $y = 0$.

22. Suppose R is the region inside the boundary of a certain county in the Midwest. After a winter storm, the depth of snow at the point (x, y) in R was $f(x, y)$ feet.
    (a) Assuming x and y are measured in feet, find an expression for the number of cubic feet of snow that fell on the county.
    (b) Find an expression for the average depth of snow in the county.

23. Suppose the joint probability density function for the nonnegative variables x and y is $f(x, y) = 6e^{-2x}e^{-3y}$.
    (a) Find the probability that $0 \le x \le 1$ and $0 \le y \le 2$.
    (b) Find the probability that $x + y \le 2$.

24. Suppose x denotes the time (in minutes) that a person spends in a certain doctor's waiting room and y the duration (in minutes) of a complete physical examination. You arrive at the doctor's office for a physical 50 minutes before you are due to leave for a meeting. If the joint probability density function for x and y is $f(x, y) = \frac{1}{500}e^{-x/10}e^{-y/50}$, find the probability that you will be late leaving for your meeting.

CHAPTER 11
INFINITE SERIES AND TAYLOR APPROXIMATION

1   Infinite Series

The sum of infinitely many numbers may be finite. This statement, which may seem paradoxical at first, plays a central role in mathematics and has a variety of important applications. The purpose of this chapter is to explore its meaning and some of its consequences.

You are already familiar with the phenomenon of a finite-valued infinite sum. You know, for example, that the repeating decimal $0.333\ldots$ stands for the infinite sum $\dfrac{3}{10} + \dfrac{3}{100} + \dfrac{3}{1,000} + \cdots$ and that its value is the finite number $\dfrac{1}{3}$. What may not be familiar, however, is exactly what it means to say that this infinite sum "adds up to" $\dfrac{1}{3}$. The situation will be clarified in this introductory section.

Infinite Series   An expression of the form

$$a_1 + a_2 + \cdots + a_n + \cdots$$

is called an **infinite series.** It is customary to use summation notation to write series compactly as follows:

$$a_1 + a_2 + \cdots + a_n + \cdots = \sum_{n=1}^{\infty} a_n$$

The use of this notation is illustrated in the following example.

EXAMPLE 1.1

(a) Write out some representative terms of the series $\displaystyle\sum_{n=1}^{\infty} \dfrac{1}{2^n}$.

(b) Use summation notation to write the series $1 - \dfrac{1}{4} + \dfrac{1}{9} - \dfrac{1}{16} + \cdots$ in compact form.

Solution

(a) $\displaystyle\sum_{n=1}^{\infty} \dfrac{1}{2^n} = \dfrac{1}{2} + \dfrac{1}{4} + \dfrac{1}{8} + \cdots + \dfrac{1}{2^n} + \cdots$

(b) The nth term of this series is

$$\frac{(-1)^{n+1}}{n^2}$$

where the factor $(-1)^{n+1}$ generates the alternating signs starting with a plus sign when $n = 1$. Thus,

$$1 - \frac{1}{4} + \frac{1}{9} - \frac{1}{16} + \cdots = \sum_{n=1}^{\infty} \frac{(-1)^{n+1}}{n^2}$$

Applications of Infinite Series

Infinite series arise in a variety of practical situations. In Section 2 you will see applications of series to such fields as economics, medicine, and probability. In Section 3 you will see how infinite series are used in the approximation of functions that may be difficult or impossible to evaluate directly. As a preview of these applications and to give you a more concrete feel for how series arise in practice, here is an investment problem that leads to an infinite series.

EXAMPLE 1.2

Find an expression for the amount of money you should invest today at an annual interest rate of 10 percent compounded continuously so that, starting next year, you can make annual withdrawals of $400 in perpetuity.

Solution

The amount you should invest today to generate the desired sequence of withdrawals is the sum of the present values of the individual withdrawals. You compute the present value of each withdrawal using the formula $P = Be^{-rt}$ from Chapter 4, Section 5, with $B = 400$, $r = 0.1$, and t being the time (in years) at which the withdrawal is made. Thus,

P_1 = present value of 1st withdrawal = $400e^{-0.1(1)}$ = $400e^{-0.1}$

P_2 = present value of 2nd withdrawal = $400e^{-0.1(2)}$ = $400e^{-0.2}$

.

.

.

P_n = present value of nth withdrawal = $400e^{-0.1(n)}$

and so on. The situation is illustrated in Figure 11.1. The amount that you

Figure 11.1 The present value of the nth withdrawal.

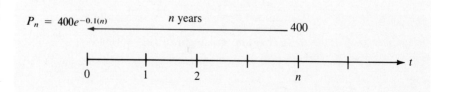

$P_n = 400e^{-0.1(n)}$ n years   400

should invest today is the sum of these (infinitely many) present values. That is,

$$\text{Amount to be invested} = \sum_{n=1}^{\infty} P_n = \sum_{n=1}^{\infty} 400e^{-0.1(n)}$$

In Section 2 you will learn how to show that this infinite series "adds up to" $3,803.33. As an experiment, use your calculator to add the first several terms of this series and see what happens as you include more and more terms.

Convergence and Divergence of Infinite Series

Roughly speaking, an infinite series is said to **converge** if it "adds up to" a finite number and to **diverge** if it does not. A more precise statement of the criterion for convergence involves the sum

$$S_n = a_1 + a_2 + \cdots + a_n$$

of the first n terms of the series. This (finite) sum is called the **nth partial sum** of the series, and its behavior as n approaches infinity determines the convergence or divergence of the series. Here is a summary of the criterion.

Convergence and Divergence

An infinite series $\sum_{n=1}^{\infty} a_n$ with nth partial sum

$$S_n = a_1 + a_2 + \cdots + a_n$$

is said to converge if there is a (finite) number S such that

$$\lim_{n \to \infty} S_n = S$$

In this case, S is said to be the sum of the series, and one writes

$$\sum_{n=1}^{\infty} a_n = S$$

If S_n does not have a finite limit as n approaches infinity, the series is said to diverge.

According to this criterion, to determine the convergence or divergence of an infinite series, you start by finding an expression for the sum of the first n terms of the series and then take the limit of this finite sum as n approaches infinity. This procedure should be reminiscent of the one that was used in Chapter 8, Section 2, to determine the convergence or divergence of an improper integral.

The notions of convergence and divergence of infinite series are illustrated in the following examples.

EXAMPLE 1.3

Investigate the possible convergence of the series $\displaystyle\sum_{n=1}^{\infty} \frac{1}{n(n+1)}$.

Solution

To get a feel for the series, you might begin by computing a few of its partial sums.

$$S_1 = \frac{1}{1 \cdot 2} = \frac{1}{2}$$

$$S_2 = \frac{1}{1 \cdot 2} + \frac{1}{2 \cdot 3} = \frac{1}{2} + \frac{1}{6} = \frac{2}{3}$$

$$S_3 = \left(\frac{1}{1 \cdot 2} + \frac{1}{2 \cdot 3}\right) + \frac{1}{3 \cdot 4} = \frac{2}{3} + \frac{1}{12} = \frac{3}{4}$$

$$S_4 = \left(\frac{1}{1 \cdot 2} + \frac{1}{2 \cdot 3} + \frac{1}{3 \cdot 4}\right) + \frac{1}{4 \cdot 5} = \frac{3}{4} + \frac{1}{20} = \frac{4}{5}$$

$$S_5 = \left(\frac{1}{1 \cdot 2} + \frac{1}{2 \cdot 3} + \frac{1}{3 \cdot 4} + \frac{1}{4 \cdot 5}\right) + \frac{1}{5 \cdot 6} = \frac{4}{5} + \frac{1}{30} = \frac{5}{6}$$

These calculations *suggest* that $S_n = \dfrac{n}{n+1}$ and that the sum of the series is 1. However, since the reason for the observed pattern among the partial sums is probably not clear, it is advisable to seek an algebraic solution to the problem.

The trick is to use the fact that

$$\frac{1}{n(n+1)} = \frac{1}{n} - \frac{1}{n+1}$$

to write the series as

$$\sum_{n=1}^{\infty} \frac{1}{n(n+1)} = \sum_{n=1}^{\infty} \left(\frac{1}{n} - \frac{1}{n+1}\right)$$

and the nth partial sum as

$$S_n = \left(\frac{1}{1} - \frac{1}{2}\right) + \left(\frac{1}{2} - \frac{1}{3}\right) + \left(\frac{1}{3} - \frac{1}{4}\right) + \cdots + \left(\frac{1}{n} - \frac{1}{n+1}\right)$$

Since all but the first and last terms in this sum cancel, S_n can be rewritten as $$S_n = 1 - \frac{1}{n + 1} = \frac{n}{n + 1}$$

Finally, since $$\lim_{n \to \infty} S_n = \lim_{n \to \infty} \frac{n}{n + 1} = 1$$

it follows that the series converges to 1. That is,

$$\sum_{n=1}^{\infty} \frac{1}{n(n + 1)} = 1$$

The series in the preceding example is known as a **telescoping series** because of the cancellation that takes place within the nth partial sum.

The series in the next example is the one that corresponds to the decimal representation of the fraction $\frac{1}{3}$. It is an example of an important type of series that will be examined in more detail in Section 2.

EXAMPLE 1.4

Investigate the possible convergence of the series $\sum_{n=1}^{\infty} \frac{3}{10^n}$.

Solution

Notice that each term of the series

$$\sum_{n=1}^{\infty} \frac{3}{10^n} = \frac{3}{10} + \frac{3}{10^2} + \frac{3}{10^3} + \cdots$$

is $\frac{1}{10}$ times the preceding term. This leads to the following trick for finding a compact formula for S_n.

Begin with the nth partial sum in the form

$$S_n = \frac{3}{10} + \frac{3}{10^2} + \frac{3}{10^3} + \cdots + \frac{3}{10^n}$$

and multiply both sides of this equation by $\frac{1}{10}$ to get

$$\frac{1}{10} S_n = \frac{3}{10^2} + \frac{3}{10^3} + \cdots + \frac{3}{10^n} + \frac{3}{10^{n+1}}$$

Now subtract the expression for $\frac{1}{10} S_n$ from the expression for S_n. Most of the terms cancel, and you are left with

$$\frac{9}{10} S_n = \frac{3}{10} - \frac{3}{10^{n+1}} = \frac{3}{10}\left(1 - \frac{1}{10^n}\right)$$

or

$$S_n = \frac{1}{3}\left(1 - \frac{1}{10^n}\right)$$

Then,

$$\lim_{n \to \infty} S_n = \lim_{n \to \infty} \frac{1}{3}\left(1 - \frac{1}{10^n}\right) = \frac{1}{3}$$

and so

$$\sum_{n=1}^{\infty} \frac{3}{10^n} = \frac{1}{3}$$

as expected.

A Necessary Condition for Convergence

The individual terms a_n of a convergent series $\sum_{n=1}^{\infty} a_n$ must approach zero as n increases without bound. To see this, write a_n as the difference of two partial sums as follows:

$$a_n = (a_1 + a_2 + \cdots + a_n) - (a_1 + a_2 + \cdots + a_{n-1})$$
$$= S_n - S_{n-1}$$

Then, if $\sum_{n=1}^{\infty} a_n$ converges to some number S, both partial sums S_n and S_{n-1} approach S as n approaches infinity, and so

$$\lim_{n \to \infty} a_n = \lim_{n \to \infty} S_n - \lim_{n \to \infty} S_{n-1} = S - S = 0$$

This observation can be rephrased as a test for *divergence* as follows.

Test for Divergence

If $\lim_{n \to \infty} a_n \neq 0$, then $\sum_{n=1}^{\infty} a_n$ diverges.

The use of this test is illustrated in the next example.

EXAMPLE 1.5

Investigate the possible convergence of the series $\sum_{n=1}^{\infty} \frac{n+3}{2n+1}$.

Solution

The limit of the nth term is

$$\lim_{n \to \infty} a_n = \lim_{n \to \infty} \frac{n + 3}{2n + 1} = \frac{1}{2}$$

which is not equal to zero. Hence, the series cannot converge and must diverge.

Look again at the series in Example 1.5. Its individual terms a_n decrease as n increases. In particular,

$$a_1 = \frac{4}{3} \simeq 1.3333$$

$$a_2 = \frac{5}{5} = 1$$

$$a_3 = \frac{6}{7} \simeq 0.8571$$

$$a_4 = \frac{7}{9} \simeq 0.7778$$

$$a_5 = \frac{8}{11} \simeq 0.7273$$

.
.
.

$$a_{50} = \frac{53}{101} \simeq 0.5248$$

.
.
.

$$a_{100} = \frac{103}{201} \simeq 0.5124$$

and so on. However, it is because the terms do not decrease *to zero* (but approach $\frac{1}{2}$ instead) that the series $\sum\limits_{n=1}^{\infty} a_n$ cannot converge.

Warning against Misuse of the Divergence Test   The test for divergence states that a series whose individual terms do not approach zero must diverge. The converse of this test is *not* true. That is, the fact that the individual terms of a series approach zero does not guarantee that the series converges. Here is an example.

EXAMPLE 1.6

Investigate the possible convergence of the series $\sum\limits_{n=1}^{\infty} \frac{1}{n}$.

Solution

Since

$$\lim_{n\to\infty} a_n = \lim_{n\to\infty} \frac{1}{n} = 0$$

it is possible, but not guaranteed, that the series converges.

It is natural to proceed by computing a few of the partial sums to get a better feel for the series.

$$S_1 = 1$$

$$S_2 = 1 + \frac{1}{2} = \frac{3}{2} = 1.5$$

$$S_3 = \left(1 + \frac{1}{2}\right) + \frac{1}{3} = \frac{3}{2} + \frac{1}{3} = \frac{11}{6} \simeq 1.8333$$

$$S_4 = \left(1 + \frac{1}{2} + \frac{1}{3}\right) + \frac{1}{4} = \frac{11}{6} + \frac{1}{4} = \frac{25}{12} \simeq 2.0833$$

$$S_5 = \left(1 + \frac{1}{2} + \frac{1}{3} + \frac{1}{4}\right) + \frac{1}{5} = \frac{25}{12} + \frac{1}{5} = \frac{137}{60} \simeq 2.2833$$

$$S_6 = \left(1 + \frac{1}{2} + \frac{1}{3} + \frac{1}{4} + \frac{1}{5}\right) + \frac{1}{6} = \frac{137}{60} + \frac{1}{6} = \frac{147}{60} = 2.45$$

Unfortunately, there is no obvious pattern among these partial sums, nor is it even clear whether they are approaching a finite limit.

Actually, this series diverges. To see this, group the terms as follows:

$$1 + \underbrace{\frac{1}{2}}_{} + \underbrace{\frac{1}{3} + \frac{1}{4}}_{2 \text{ terms}} + \underbrace{\frac{1}{5} + \frac{1}{6} + \frac{1}{7} + \frac{1}{8}}_{2^2 \text{ terms}} + \underbrace{\frac{1}{9} + \cdots + \frac{1}{16}}_{2^3 \text{ terms}} + \frac{1}{17} + \cdots$$

The sum of the terms in each group is greater than or equal to $\frac{1}{2}$. For example,

$$\frac{1}{3} + \frac{1}{4} > \frac{1}{4} + \frac{1}{4} = \frac{1}{2}$$

and

$$\frac{1}{5} + \frac{1}{6} + \frac{1}{7} + \frac{1}{8} > \frac{1}{8} + \frac{1}{8} + \frac{1}{8} + \frac{1}{8} = \frac{1}{2}$$

so that

$$\underbrace{1}_{\substack{1\text{st}\\ \text{group}}} + \underbrace{\frac{1}{2}}_{\substack{2\text{nd}\\ \text{group}}} + \underbrace{\frac{1}{3} + \frac{1}{4}}_{\substack{3\text{rd}\\ \text{group}}} + \underbrace{\frac{1}{5} + \frac{1}{6} + \frac{1}{7} + \frac{1}{8}}_{4\text{th group}} + \cdots >$$

$$\underbrace{1}_{\substack{1\text{st}\\ \text{group}}} + \underbrace{\frac{1}{2}}_{\substack{2\text{nd}\\ \text{group}}} + \underbrace{\frac{1}{2}}_{\substack{3\text{rd}\\ \text{group}}} + \underbrace{\frac{1}{2}}_{\substack{4\text{th}\\ \text{group}}} + \cdots$$

By including sufficiently many of these groups, you can make the corresponding partial sum as large as you like, from which it follows that the infinite series must diverge.

The series $\sum\limits_{n=1}^{\infty} \dfrac{1}{n}$ in Example 1.6 is known as the **harmonic series** and is one of the most well-known examples of a divergent series whose individual terms approach zero.

It is instructive to compare the convergent series $\sum\limits_{n=1}^{\infty} \dfrac{1}{n(n+1)}$ from Example 1.3 with the divergent harmonic series $\sum\limits_{n=1}^{\infty} \dfrac{1}{n}$ from Example 1.6. Each is a series of positive terms that approach zero as n increases. Yet one converges and has a finite sum, while the other diverges and has an "infinite sum." The reason for the difference is that, as n increases, the terms $\dfrac{1}{n(n+1)}$ approach zero more quickly than the terms $\dfrac{1}{n}$ do. The situation is illustrated in the following table and should be reminiscent of the difference between convergent and divergent improper integrals.

n	1	2	3	4	5	10	50
$\dfrac{1}{n}$	1	0.5	0.3333	0.25	0.2	0.1	0.02
$\dfrac{1}{n(n+1)}$	0.5	0.1667	0.0833	0.05	0.0333	0.0091	0.0004

A Useful Property of Convergent Series

Like finite sums, convergent infinite series obey a distributive law with respect to multiplication. According to this law, if you multiply each term of a convergent series by a constant c, you get a new series whose sum is c times the sum of the original series.

The Distributive Law for Series

For any constant c,

$$\sum_{n=1}^{\infty} ca_n = c \sum_{n=1}^{\infty} a_n$$

This law follows from the fact that the sum of an infinite series is determined by its nth partial sum, which, being a finite sum, obeys the familiar distributive law from elementary algebra. In particular,

$$\sum_{n=1}^{\infty} ca_n = \lim_{n \to \infty} (ca_1 + ca_2 + \cdots + ca_n)$$

$$= \lim_{n \to \infty} c(a_1 + a_2 + \cdots + a_n)$$

$$= c \lim_{n \to \infty} (a_1 + a_2 + \cdots + a_n)$$

$$= c \sum_{n=1}^{\infty} a_n$$

The distributive law also holds for divergent series in the sense that if $\sum_{n=1}^{\infty} a_n$ diverges, so does $\sum_{n=1}^{\infty} ca_n$ (provided, of course, that $c \neq 0$).

The distributive law allows you to factor constants out of infinite series. Such factorization will be used several times in the next section to simplify calculations.

Problems 11.1

In Problems 1 through 6, use summation notation to write the given series in compact form.

1. $\dfrac{1}{3} + \dfrac{1}{9} + \dfrac{1}{27} + \dfrac{1}{81} + \cdots$

2. $1 + \dfrac{1}{8} + \dfrac{1}{27} + \dfrac{1}{64} + \cdots$

3. $\dfrac{1}{2} + \dfrac{2}{3} + \dfrac{3}{4} + \dfrac{4}{5} + \cdots$

4. $\dfrac{1}{3} + \dfrac{2}{5} + \dfrac{3}{7} + \dfrac{4}{9} + \cdots$

5. $\dfrac{1}{2} - \dfrac{4}{3} + \dfrac{9}{4} - \dfrac{16}{5} + \cdots$

6. $-3 + \dfrac{9}{4} - \dfrac{27}{9} + \dfrac{81}{16} - \cdots$

In Problems 7 through 10, find the fourth partial sum S_4 of the given series.

7. $\displaystyle\sum_{n=1}^{\infty} \dfrac{1}{2^n}$

8. $\displaystyle\sum_{n=1}^{\infty} \dfrac{n}{n+1}$

9. $\displaystyle\sum_{n=1}^{\infty} \dfrac{(-1)^n}{n}$

10. $\displaystyle\sum_{n=1}^{\infty} \dfrac{(-1)^{n+1}}{n(n+1)}$

In Problems 11 through 18, find the sum of the given convergent series by taking the limit of a compact expression for the nth partial sum.

11. $\displaystyle\sum_{n=1}^{\infty} \left(\dfrac{1}{n+3} - \dfrac{1}{n+4} \right)$

12. $\displaystyle\sum_{n=1}^{\infty} \left(\dfrac{1}{n} - \dfrac{1}{n+2} \right)$

13. $\displaystyle\sum_{n=1}^{\infty} \dfrac{1}{(n+1)(n+2)}$

14. $\displaystyle\sum_{n=1}^{\infty} \dfrac{1}{(n+2)(n+3)}$

15. $\displaystyle\sum_{n=1}^{\infty} \dfrac{6}{10^n}$

16. $\displaystyle\sum_{n=1}^{\infty} \dfrac{1}{2^n}$

17. $\displaystyle\sum_{n=1}^{\infty} \frac{4}{3^n}$

18. $\displaystyle\sum_{n=1}^{\infty} \left(-\frac{1}{5}\right)^n$

In Problems 19 through 25, determine whether the given series converges, and if it does, find its sum.

19. $\displaystyle\sum_{n=1}^{\infty} \frac{n}{2n+1}$

20. $\displaystyle\sum_{n=1}^{\infty} \frac{1}{n+1}$

21. $\displaystyle\sum_{n=1}^{\infty} \left(\frac{1}{n+1} - \frac{1}{n}\right)$

22. $\displaystyle\sum_{n=1}^{\infty} \frac{n^2}{50n^2+1}$

23. $\displaystyle\sum_{n=1}^{\infty} \left(-\frac{2}{3}\right)^n$

24. $\displaystyle\sum_{n=1}^{\infty} \left(-\frac{3}{2}\right)^n$

25. $\displaystyle\sum_{n=1}^{\infty} \frac{1}{\sqrt{n}}$ (*Hint:* Show that $S_n \geq \dfrac{n}{\sqrt{n}} = \sqrt{n}$.)

2   The Geometric Series

A **geometric series** is an infinite series in which the ratio of successive terms is constant. For example, the series

$$3 + \frac{3}{2} + \frac{3}{4} + \frac{3}{8} + \cdots$$

is geometric because each term is one-half of the preceding term. In general, a geometric series is a series of the form

$$\sum_{n=m}^{\infty} ar^n = ar^m + ar^{m+1} + ar^{m+2} + \cdots$$

The constant r, which is the ratio of the successive terms of a geometric series, is known as the **ratio** of the series. The ratio of a geometric series may be positive or negative. For example,

$$\sum_{n=0}^{\infty} \frac{2}{(-3)^n} = 2 - \frac{2}{3} + \frac{2}{9} - \frac{2}{27} + \cdots$$

is a geometric series with ratio $r = -\dfrac{1}{3}$.

Elementary Geometric Series   The most elementary geometric series are those for which $m = 0$ and $a = 1$, that is, series of the form

$$\sum_{n=0}^{\infty} r^n = 1 + r + r^2 + \cdots$$

Because of the distributive law, every geometric series can be written as a multiple of one of these elementary geometric series. In particular,

$$\sum_{n=m}^{\infty} ar^n = ar^m + ar^{m+1} + ar^{m+2} + \cdots$$

$$= ar^m(1 + r + r^2 + \cdots) \qquad \text{by the distributive law}$$

$$= ar^m \sum_{n=0}^{\infty} r^n$$

For this reason, it suffices to study the elementary geometric series.

Convergence of Geometric Series

A geometric series $\sum_{n=0}^{\infty} r^n$ converges if the absolute value of its ratio r is less than 1 and diverges otherwise. The divergence when $|r| \geq 1$ follows from the fact that, in this case, the individual terms r^n do not approach zero as n increases. When $|r| < 1$, you can establish the convergence of the series and derive a formula for its sum by means of a calculation similar to the one you saw in Example 1.4 of the preceding section. In particular, begin with the nth partial sum in the form

$$S_n = 1 + r + r^2 + \cdots + r^{n-1} \qquad (r \neq 0)$$

(Note that the power of r in the nth term is only $n - 1$ since the first term in the sum is $1 = r^0$.) Now multiply both sides of this equation by r to get

$$rS_n = r + r^2 + \cdots + r^{n-1} + r^n$$

and subtract the expression for rS_n from the expression for S_n to get

$$(1 - r)S_n = 1 - r^n$$

or
$$S_n = \frac{1 - r^n}{1 - r}$$

Since $|r| < 1$,
$$\lim_{n \to \infty} r^n = 0$$

and so
$$\lim_{n \to \infty} S_n = \lim_{n \to \infty} \frac{1 - r^n}{1 - r} = \frac{1}{1 - r}$$

Hence, the series converges, and its sum is $\dfrac{1}{1 - r}$.

The Sum of a Geometric Series

If $|r| < 1$, with $r \neq 0$, then

$$\sum_{n=0}^{\infty} r^n = \frac{1}{1 - r}$$

Notice that this formula for the sum of a geometric series is valid only for elementary series of the form

$$\sum_{n=0}^{\infty} r^n = 1 + r + r^2 + \cdots$$

that start with the term $r^0 = 1$. To apply the formula to any other geometric series, you will first have to rewrite the series as a constant multiple of an elementary one. The use of the formula is illustrated in the following examples.

EXAMPLE 2.1

Find the sum of the series $\displaystyle\sum_{n=0}^{\infty} \left(-\frac{2}{3}\right)^n$.

Solution

This is an elementary geometric series with $r = -\frac{2}{3}$. Hence,

$$\sum_{n=0}^{\infty} \left(-\frac{2}{3}\right)^n = \frac{1}{1 - \left(-\frac{2}{3}\right)} = \frac{1}{\frac{5}{3}} = \frac{3}{5}$$

EXAMPLE 2.2

Find the sum of the series $\displaystyle\sum_{n=0}^{\infty} \frac{3}{2^n}$.

Solution

$$\sum_{n=0}^{\infty} \frac{3}{2^n} = 3 \sum_{n=0}^{\infty} \left(\frac{1}{2}\right)^n = 3\left(\frac{1}{1 - \frac{1}{2}}\right) = 3\left(\frac{1}{\frac{1}{2}}\right) = 6$$

EXAMPLE 2.3

Find the sum of the series $\displaystyle\sum_{n=2}^{\infty} \frac{2}{5^n}$.

Solution

$$\sum_{n=2}^{\infty} \frac{2}{5^n} = \frac{2}{5^2} + \frac{2}{5^3} + \frac{2}{5^4} + \cdots$$

$$= \frac{2}{5^2}\left(1 + \frac{1}{5} + \frac{1}{5^2} + \cdots\right) \qquad \text{by the distributive law}$$

$$= \frac{2}{25}\sum_{n=0}^{\infty}\left(\frac{1}{5}\right)^n$$

$$= \frac{2}{25}\left(\frac{1}{1 - \frac{1}{5}}\right) = \frac{2}{25}\left(\frac{5}{4}\right) = \frac{1}{10}$$

EXAMPLE 2.4

Find the sum of the series $\displaystyle\sum_{n=1}^{\infty}\frac{(-2)^n}{3^{n+1}}$.

Solution

$$\sum_{n=1}^{\infty}\frac{(-2)^n}{3^{n+1}} = \frac{-2}{3^2} + \frac{(-2)^2}{3^3} + \frac{(-2)^3}{3^4} + \cdots$$

$$= \frac{-2}{3^2}\left[1 + \left(-\frac{2}{3}\right) + \left(-\frac{2}{3}\right)^2 + \cdots\right]$$

$$= -\frac{2}{9}\sum_{n=0}^{\infty}\left(-\frac{2}{3}\right)^n$$

$$= -\frac{2}{9}\left(\frac{1}{1 + \frac{2}{3}}\right) = -\frac{2}{9}\left(\frac{3}{5}\right) = -\frac{2}{15}$$

Applications of Geometric Series   Geometric series arise in many branches of mathematics and in a variety of applications. Here are four illustrations taken from number theory, economics, medicine, and probability.

Repeating Decimals   A repeating decimal is a geometric series, and its value is the sum of the series. Here is an example.

EXAMPLE 2.5

Express the repeating decimal 0.232323 . . . as a fraction.

Solution

Write the decimal as a geometric series as follows:

$$0.232323 \ldots = \frac{23}{100} + \frac{23}{10,000} + \frac{23}{1,000,000} + \cdots$$

$$= \frac{23}{100}\left[1 + \frac{1}{100} + \left(\frac{1}{100}\right)^2 + \cdots\right]$$

$$= \frac{23}{100}\sum_{n=0}^{\infty}\left(\frac{1}{100}\right)^n$$

$$= \frac{23}{100}\left(\frac{1}{1 - \frac{1}{100}}\right) = \frac{23}{100}\left(\frac{100}{99}\right) = \frac{23}{99}$$

The Multiplier Effect in Economics   A tax rebate that returns a certain amount of money to taxpayers can result in spending that is many times this amount. This phenomenon is known in economics as the **multiplier effect.** It occurs because the portion of the rebate that is spent by one individual becomes income for one or more others who, in turn, spend some of it again, creating income for yet other individuals to spend. If the fraction of income that is saved remains constant as this process continues indefinitely, the total amount spent as a result of the rebate is the sum of a geometric series. Here is an example.

EXAMPLE 2.6

Suppose that nationwide, approximately 90 percent of all income is spent and 10 percent saved. How much additional spending will be generated by a 40 billion dollar tax rebate if savings habits do not change?

Solution

The amount (in billions) spent by original recipients of the rebate is

$$0.9(40)$$

This becomes new income, of which 90 percent or

$$0.9[0.9(40)] = (0.9)^2(40)$$

is spent. This, in turn, generates additional spending of

$$0.9[(0.9)^2(40)] = (0.9)^3(40)$$

and so on. The total amount spent if this process continues indefinitely is

$$0.9(40) + (0.9)^2(40) + (0.9)^3(40) + \cdots$$

$$= 0.9(40)[1 + 0.9 + (0.9)^2 + \cdots]$$

$$= 36\sum_{n=0}^{\infty}(0.9)^n = 36\left(\frac{1}{1 - 0.9}\right) = 360 \text{ billion}$$

Accumulation of Medication in the Body

In Example 2.5 of Chapter 8, a nuclear power plant generated radioactive waste, which decayed exponentially. The problem was to determine the long-run accumulation of waste from the plant. Because the waste was being generated *continuously*, the solution involved an (improper) integral. The situation in the next example is similar. A patient receives medication which is eliminated from the body exponentially, and the goal is to determine the long-run accumulation of medication in the patient's body. In this case, however, the medication is being administered in *discrete* doses, and the solution to the problem involves an infinite series rather than an integral.

EXAMPLE 2.7

A patient is given an injection of 10 units of a certain drug every 24 hours. The drug is eliminated exponentially so that the fraction that remains in the patient's body after t days is $f(t) = e^{-t/5}$. If the treatment is continued indefinitely, approximately how many units of the drug will eventually be in the patient's body just prior to an injection?

Solution

Of the original dose of 10 units, only $10e^{-1/5}$ units are left in the patient's body after 1 day (just prior to the second injection). That is,

$$\text{Amount in body after 1 day} = S_1 = 10e^{-1/5}$$

The medication in the patient's body after 2 days consists of what remains from the first *two* doses. Of the original dose, only $10e^{-2/5}$ units are left (since 2 days have elapsed), and of the second dose, $10e^{-1/5}$ units remain. (See Figure 11.2.) Hence,

$$\text{Amount in body after 2 days} = S_2 = 10e^{-1/5} + 10e^{-2/5}$$

Similarly,

$$\begin{matrix}\text{Amount in body} \\ \text{after } n \text{ days}\end{matrix} = S_n = 10e^{-1/5} + 10e^{-2/5} + \cdots + 10e^{-n/5}$$

Figure 11.2 Amount of medication in the body after 2 days.

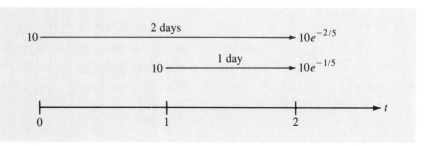

The amount S of medication in the patient's body in the long run is the limit of S_n as n approaches infinity. That is,

$$S = \lim_{n \to \infty} S_n = \sum_{n=1}^{\infty} 10e^{-n/5}$$

$$= \sum_{n=1}^{\infty} 10(e^{-1/5})^n = 10e^{-1/5} \sum_{n=0}^{\infty} (e^{-1/5})^n$$

$$= 10e^{-1/5}\left(\frac{1}{1 - e^{-1/5}}\right) \approx 45.17 \text{ units}$$

Geometric Random Variables

Suppose a random experiment with two possible outcomes (traditionally called "success" and "failure") is performed repeatedly until the first success occurs. The number x of repetitions required to produce the first success is known as a **geometric random variable.**

A geometric random variable is a discrete random variable that can take on any positive integer value. For any positive integer n, the probability that $x = n$ is the probability that the first $n - 1$ trials of the experiment result in failure and the nth in success:

$$\underbrace{F \quad F \quad F \cdots F}_{n - 1 \text{ failures}} \quad \underbrace{S}_{\text{1st success}}$$

If, on each trial, the probability of success is p, then the probability of failure must be $1 - p$, and the probability of $n - 1$ failures followed by 1 success is the product

$$(1 - p)(1 - p)(1 - p) \cdots (1 - p)p$$

That is,

$$P(x = n) = (1 - p)^{n-1}p$$

If $p \neq 0$, success will eventually occur if the experiment is repeated indefinitely. Hence, $P(x \geq 1)$ ought to be 1. To compute this probability, you find the sum of a certain geometric series as follows:

$$P(x \geq 1) = \sum_{n=1}^{\infty} P(x = n)$$

$$= \sum_{n=1}^{\infty} (1 - p)^{n-1}p$$

$$= p + (1 - p)p + (1 - p)^2p + \cdots$$

$$= p[1 + (1 - p) + (1 - p)^2 + \cdots]$$

$$= p \sum_{n=0}^{\infty} (1 - p)^n$$

$$= \frac{p}{1 - (1 - p)} = \frac{p}{p} = 1$$

A simple application of geometric random variables is given in the next example.

EXAMPLE 2.8

In a certain two-person board game played with 1 die, each player needs a six to start play. You and your opponent take turns rolling the die. Find the probability that you will get the first six if you roll first.

Solution

In this context, "success" is getting a six and "failure" is getting any of the other five possible numbers. Hence, the probability of success is $p = \frac{1}{6}$ and the probability of failure is $1 - p = \frac{5}{6}$.

Let x denote the number of the roll on which the first six comes up. If x is odd, the first six will occur on one of your rolls, while if x is even, the first six will occur on one of your opponent's rolls. Hence, the probability that you get the first six is

$$P(x \text{ is odd}) = P(x = 1) + P(x = 3) + P(x = 5) + \cdots$$

$$= \frac{1}{6} + \left(\frac{5}{6}\right)^2 \left(\frac{1}{6}\right) + \left(\frac{5}{6}\right)^4 \left(\frac{1}{6}\right) + \cdots$$

$$= \frac{1}{6} + \left(\frac{25}{36}\right)\left(\frac{1}{6}\right) + \left(\frac{25}{36}\right)^2\left(\frac{1}{6}\right) + \cdots$$

$$= \frac{1}{6} \sum_{n=0}^{\infty} \left(\frac{25}{36}\right)^n$$

$$= \frac{1}{6}\left(\frac{1}{1 - \frac{25}{36}}\right) = \frac{1}{6}\left(\frac{36}{11}\right) = \frac{6}{11} \simeq 0.5455$$

Problems 11.2

In Problems 1 through 14, determine whether the given geometric series converges, and if so, find its sum.

1. $\displaystyle\sum_{n=0}^{\infty} \left(\frac{4}{5}\right)^n$

2. $\displaystyle\sum_{n=0}^{\infty} \left(-\frac{4}{5}\right)^n$

3. $\sum_{n=0}^{\infty} \frac{2}{3^n}$

4. $\sum_{n=0}^{\infty} \frac{2}{(-3)^n}$

5. $\sum_{n=1}^{\infty} \left(\frac{3}{2}\right)^n$

6. $\sum_{n=1}^{\infty} \frac{3}{2^n}$

7. $\sum_{n=2}^{\infty} \frac{3}{(-4)^n}$

8. $\sum_{n=2}^{\infty} \left(-\frac{4}{3}\right)^n$

9. $\sum_{n=1}^{\infty} 5(0.9)^n$

10. $\sum_{n=1}^{\infty} e^{-0.2n}$

11. $\sum_{n=1}^{\infty} \frac{3^n}{4^{n+2}}$

12. $\sum_{n=2}^{\infty} \frac{(-2)^{n-1}}{3^{n+1}}$

13. $\sum_{n=0}^{\infty} \frac{4^{n+1}}{5^{n-1}}$

14. $\sum_{n=2}^{\infty} (-1)^n \frac{2^{n+1}}{3^{n-3}}$

In Problems 15 through 18, express the given decimal as a fraction.

15. 0.3333 . . .

16. 0.5555 . . .

17. 0.252525 . . .

18. 1.405405405 . . .

Multiplier Effect 19. Suppose that nationwide, approximately 92 percent of all income is spent and 8 percent saved. How much additional spending will be generated by a 50 billion dollar tax cut if savings habits do not change?

Present Value 20. An investment guarantees annual payments of $1,000 in perpetuity, with the payments beginning immediately. Find the present value of this investment if the prevailing annual interest rate remains fixed at 12 percent compounded continuously. (*Hint:* The present value of the investment is the sum of the present values of the individual payments.)

Present Value 21. How much should you invest today at an annual interest rate of 15 percent compounded continuously so that, starting next year, you can make annual withdrawals of $2,000 in perpetuity?

Accumulation of Medication 22. A patient is given an injection of 20 units of a certain drug every 24 hours. The drug is eliminated exponentially so that the fraction that remains in the patient's body after t days is $f(t) = e^{-t/2}$. If the treatment is continued indefinitely, approximately how many units of the drug will eventually be in the patient's body just prior to an injection?

Public Health 23. A health problem, especially in emerging countries, is that the food available to people often contains toxins or poisons. Suppose that a person consumes a dose d of a certain poison every day and that the person's body excretes q percent of the accumulated poison every day. How much poison is eventually stored in the person's body? (Assume $q < 1$.)

Group Membership

24. Each January first, the administration of a certain private college adds 6 new members to its board of trustees. If the fraction of trustees who remain active for at least t years is $f(t) = e^{-0.2t}$, approximately how many active trustees will the college have on December thirty-first in the long run?

Zeno's Paradox

25. The Greek philosopher Zeno of Elea (495–435 B.C.) presented a number of paradoxes that caused a great deal of concern to the mathematicians of his day. Perhaps the most famous is the **racecourse paradox,** which can be stated as follows:

"A runner can never reach the end of a race. As he runs the course, he must first run $\frac{1}{2}$ its length, then $\frac{1}{2}$ the remaining distance, then $\frac{1}{2}$ the distance that remains after that, and so on."

Suppose that the runner runs at a constant pace and takes T minutes to run the first half of the course. Set up an infinite series that gives the *total* time required to run the course and verify that the total time is $2T$, as intuition would lead us to expect.

Games of Chance

26. You and a friend take turns rolling a die until one of you wins by getting a three or a four. If your friend rolls first, find the probability that you will win.

Numismatics

27. In 1959, the design on the reverse side of U.S. pennies was changed from one depicting a pair of wheat stalks to one depicting the Lincoln Memorial. Today, very few of the old pennies are still in circulation. If only 0.2 percent of the pennies now in circulation were minted before 1959, find the probability that you will have to examine at least 100 pennies to find one with the old design. (Incidentally, there is an easy way to do this problem that does not involve infinite series. For practice, do it both ways.)

Quality Control

28. Three inspectors take turns checking electronic components as they come off an assembly line. If 10 percent of all the components produced on the assembly line are defective, find the probability that the inspector who checks the first component will be the one who finds the first defective component.

Depletion of Reserves

29. A certain rare gas used in industrial processes had known reserves of 3×10^{11} m^3 in 1990. In 1991, 1.7×10^9 m^3 of the gas were consumed with a yearly increase of 7.3 percent. When will all known reserves of the gas be consumed?

Difference Equation   30. Suppose $f(n)$ gives the value of a certain quantity after n periods, for $n = 0, 1, 2, \ldots$. (For example, a census of the U.S. population is taken every 10 years.)

(a) If it is known that $f(n + 1) = cf(n) + r$, where c and r are constants, show that

$$f(n) = c^n f(0) + \left(\sum_{k=0}^{n-1} c^k \right) r \quad \text{for } n = 1, 2, \ldots$$

(b) If $-1 < c < 1$, what quantity does $f(n)$ approach as n increases without bound?

Population   31. Suppose the number of females in the nth generation of the population of a certain species is given by $P(n)$, for $n = 1, 2, \ldots$. Further assume that on the average, each female member of the $(n - 1)$st generation gives birth to b female offspring and that in each generation, d females can be expected to die.

(a) If the species starts out with 100 females [that is, $P(0) = 100$], what is $P(n)$? (See Problem 30.)

(b) Suppose it is known that $b = 0.2$ and $d = 15$. What will the female population of the 10th generation be? What happens to the population in "the long run" (as n increases without bound)?

National Income   32. In the Harrod-Hicks model of an expanding national economy, it is found that if $I(n)$ denotes the national income after n years, then the change in national income during the $(n + 1)$st year is given by

$$I(n + 1) - I(n) = kI(n) - h$$

where k and h are positive constants. Find $I(n)$ in terms of k, h, and the initial national income $I(0)$. (See Problem 30.)

3   Taylor Approximation

The purpose of this section is to develop a general procedure for finding infinite series that converge to given functions and to illustrate how the partial sums of these series can be used to approximate the functions. Even if the original functions are difficult or impossible to evaluate directly, the partial sums of the corresponding infinite series turn out to be polynomials and can be evaluated with ease. The approximation of functions by the partial sums of their infinite series is how electronic computers and calculators are programmed to generate values of functions such as e^x, $\ln x$, and the trigonometric functions.

Power Series   The infinite series that are used in the approximation of functions are series of the form

$$\sum_{n=0}^{\infty} a_n x^n = a_0 + a_1 x + a_2 x^2 + \cdots$$

in which each term is a constant times a power of x. Series of this type are called **power series**. Notice that the nth partial sum of such a power series is a polynomial

$$S_n = a_0 + a_1 x + a_2 x^2 + \cdots + a_{n-1} x^{n-1}$$

Convergence of Power Series

If the variable x in a power series is given a specific numerical value, the series becomes a series of constant terms which either converges or diverges. A power series may converge for some values of x and diverge for other values of x. It can be shown that to each power series there corresponds a symmetric interval of the form $-R \le x \le R$ in the interior of which the series converges and outside of which it diverges (Figure 11.3). (At the endpoints $x = -R$ and $x = R$, convergence or divergence is possible.) The number R is called the **radius of convergence** of the series, and the set of all points for which the series converges is called its **interval of convergence**. Thus the interval of convergence of a power series consists of the interval $-R < x < R$, where R is the radius of convergence, and possibly one or both of the endpoints $x = -R$ and $x = R$. (If the series happens to converge for all values of x, the radius of convergence is said to be ∞.)

Figure 11.3 The interval of convergence for a power series.

Factorial Notation

The construction of power series that converge to given functions involves products of the form

$$n(n - 1)(n - 2) \cdots 3 \cdot 2 \cdot 1$$

It is customary to abbreviate such a product using the symbol $n!$, which is read "n **factorial**." For example, $3! = 3 \cdot 2 \cdot 1 = 6$. Moreover, $0!$ is *defined* to be 1 (because, as it turns out, this definition leads to compact representations of certain formulas involving factorials).

Factorial Notation

If n is a positive integer,

$$n! = n(n - 1)(n - 2) \cdots 3 \cdot 2 \cdot 1$$

Moreover,

$$0! = 1$$

Power Series That Converge to Given Functions

Imagine that a function $f(x)$ is given and that you would like to find the corresponding coefficients a_n such that the power series $\sum\limits_{n=0}^{\infty} a_n x^n$ converges to $f(x)$ on some interval. To discover what these coefficients might be, suppose that

$$f(x) = \sum_{n=0}^{\infty} a_n x^n = a_0 + a_1 x + a_2 x^2 + a_3 x^3 + \cdots + a_n x^n + \cdots$$

If $x = 0$, only the first term in the sum is nonzero and so

$$a_0 = f(0)$$

Now differentiate the series term by term. It can be shown that if the original power series converges to $f(x)$ on the interval $-R < x < R$, then the differentiated series converges to $f'(x)$ on this interval. Hence,

$$f'(x) = a_1 + 2a_2 x + 3a_3 x^2 + \cdots + na_n x^{n-1} + \cdots$$

and, if $x = 0$, it follows that $\quad a_1 = f'(0)$

Differentiate again to get

$$f''(x) = 2a_2 + 3 \cdot 2a_3 x + \cdots + n(n-1)a_n x^{n-2} + \cdots$$

and let $x = 0$ to conclude that

$$f''(0) = 2a_2 \quad \text{or} \quad a_2 = \frac{f''(0)}{2}$$

Similarly, the third derivative of f is

$$f^{(3)}(x) = 3 \cdot 2a_3 + \cdots + n(n-1)(n-2)a_n x^{n-3} + \cdots$$

and

$$f^{(3)}(0) = 3 \cdot 2a_3 \quad \text{or} \quad a_3 = \frac{f^{(3)}(0)}{3!}$$

and so on. In general,

$$f^{(n)}(0) = n!a_n \quad \text{or} \quad a_n = \frac{f^{(n)}(0)}{n!}$$

where $f^{(n)}(0)$ denotes the nth derivative of f evaluated at $x = 0$, and $f^{(0)}(0) = f(0)$.

The preceding argument shows that *if* there is any power series that converges to $f(x)$, it must be the one whose coefficients are obtained from the derivatives of f by the formula

$$a_n = \frac{f^{(n)}(0)}{n!}$$

This series is known as the **Taylor series of f** (about $x = 0$), and the corresponding coefficients a_n are called the **Taylor coefficients of f**.

Taylor Series

The Taylor series of $f(x)$ about $x = 0$ is the power series $\displaystyle\sum_{n=0}^{\infty} a_n x^n$, where

$$a_n = \frac{f^{(n)}(0)}{n!}$$

Convergence of Taylor Series   Although the Taylor series of a function is the only power series that can possibly converge to the function, it need not actually do so. There are, in fact, some functions whose Taylor series converge, but not to the values of the function. Fortunately, such functions rarely arise in practice. Most of the functions you are likely to encounter are equal to their Taylor series wherever the series converge.

Calculation of Taylor Series   The use of the formula for the Taylor coefficients is illustrated in the following example.

The Taylor Series for e^x

EXAMPLE 3.1

Find the Taylor series of the function $f(x) = e^x$ about $x = 0$.

Solution

Compute the Taylor coefficients as follows:

$$f(x) = e^x \qquad f(0) = 1 \qquad a_0 = \frac{f(0)}{0!} = \frac{1}{0!}$$

$$f'(x) = e^x \qquad f'(0) = 1 \qquad a_1 = \frac{f'(0)}{1!} = \frac{1}{1!}$$

$$f''(x) = e^x \qquad f''(0) = 1 \qquad a_2 = \frac{f''(0)}{2!} = \frac{1}{2!}$$

$$f^{(3)}(x) = e^x \qquad f^{(3)}(0) = 1 \qquad a_3 = \frac{f^{(3)}(0)}{3!} = \frac{1}{3!}$$

$$\cdot \qquad\qquad\qquad \cdot \qquad\qquad\qquad \cdot$$
$$\cdot \qquad\qquad\qquad \cdot \qquad\qquad\qquad \cdot$$
$$\cdot \qquad\qquad\qquad \cdot \qquad\qquad\qquad \cdot$$

$$f^{(n)}(x) = e^x \qquad f^{(n)}(0) = 1 \qquad a_n = \frac{f^{(n)}(0)}{n!} = \frac{1}{n!}$$

The corresponding Taylor series is

$$\sum_{n=0}^{\infty} a_n x^n = \sum_{n=0}^{\infty} \frac{x^n}{n!}$$

which can be shown to converge to e^x for all x. That is,

$$e^x = \sum_{n=0}^{\infty} \frac{x^n}{n!} = 1 + x + \frac{x^2}{2!} + \frac{x^3}{3!} + \cdots \qquad \text{for all } x$$

The Series for
$$f(x) = \frac{1}{1 - x}$$

Recall from the discussion of the geometric series in Section 2 that if $|r| < 1$, $\sum_{n=0}^{\infty} r^n = \frac{1}{1 - r}$. It should come as no surprise, therefore, that the Taylor series for the function $f(x) = \frac{1}{1 - x}$ is $\sum_{n=0}^{\infty} x^n$. Here is the calculation.

EXAMPLE 3.2

Find the Taylor series of the function $f(x) = \frac{1}{1 - x}$ about $x = 0$.

Solution

Compute the Taylor coefficients as follows:

$$f(x) = \frac{1}{1 - x} \qquad f(0) = 1 \qquad a_0 = \frac{f(0)}{0!} = \frac{1}{0!} = 1$$

$$f'(x) = \frac{1}{(1 - x)^2} \qquad f'(0) = 1 \qquad a_1 = \frac{f'(0)}{1!} = \frac{1}{1!} = 1$$

$$f''(x) = \frac{2 \cdot 1}{(1 - x)^3} \qquad f''(0) = 2! \qquad a_2 = \frac{f''(0)}{2!} = \frac{2!}{2!} = 1$$

$$f^{(3)}(x) = \frac{3 \cdot 2 \cdot 1}{(1 - x)^4} \qquad f^{(3)}(0) = 3! \qquad a_3 = \frac{f^{(3)}(0)}{3!} = \frac{3!}{3!} = 1$$

$$\qquad \cdot \qquad\qquad\qquad \cdot \qquad\qquad\qquad \cdot$$

$$\qquad \cdot \qquad\qquad\qquad \cdot \qquad\qquad\qquad \cdot$$

$$\qquad \cdot \qquad\qquad\qquad \cdot \qquad\qquad\qquad \cdot$$

$$f^{(n)}(x) = \frac{n!}{(1 - x)^{n+1}} \qquad f^{(n)}(0) = n! \qquad a_n = \frac{f^{(n)}(0)}{n!} = \frac{n!}{n!} = 1$$

The corresponding Taylor series is $\sum_{n=0}^{\infty} x^n$, which, as you already know, converges to $f(x) = \dfrac{1}{1 - x}$ for $|x| < 1$. Thus,

$$\frac{1}{1 - x} = \sum_{n=0}^{\infty} x^n = 1 + x + x^2 + \cdots \qquad \text{for } |x| < 1$$

Modification of Taylor Series

As you have seen in the preceding two examples, the use of the formula for the Taylor coefficients to find the Taylor series of a function can be somewhat tedious. Sometimes you will be able to avoid this procedure and find the Taylor series of a function by modifying another Taylor series that you already know. One technique that is particularly useful is to replace x in the series that is known by a more complicated expression, say $g(x)$. In particular, if $f(x) = \sum_{n=0}^{\infty} a_n x^n$ for $|x| < R$ and $g(x)$ is any function of x, you get the Taylor series of the composite function $f[g(x)]$ by replacing x by $g(x)$ in the series for $f(x)$. The new series will converge for any value of x for which $|g(x)| < R$.

Substitution in Taylor Series

If $f(x) = \sum_{n=0}^{\infty} a_n x^n$ for $|x| < R$, then

$$f[g(x)] = \sum_{n=0}^{\infty} a_n [g(x)]^n \qquad \text{for } |g(x)| < R$$

Here is an example.

EXAMPLE 3.3

Starting with an appropriate geometric series, find the Taylor series for the function $\dfrac{x}{1 + 3x^2}$ and specify the interval of convergence.

Solution

Start with the fact that

$$\frac{1}{1 - x} = \sum_{n=0}^{\infty} x^n \qquad \text{for } |x| < 1$$

and replace x by $-3x^2$ to get

$$\frac{1}{1 + 3x^2} = \sum_{n=0}^{\infty} (-3x^2)^n = \sum_{n=0}^{\infty} (-3)^n x^{2n}$$

The new interval of convergence is $|-3x^2| < 1$, which can be rewritten as $|x^2| < \frac{1}{3}$ or $|x| < \frac{1}{\sqrt{3}}$.

Now multiply by x (which does not affect the interval of convergence) to get

$$\frac{x}{1 + 3x^2} = \sum_{n=0}^{\infty} (-3)^n x^{2n+1} \qquad \text{for } |x| < \frac{1}{\sqrt{3}}$$

Taylor Series about $x = a$

In certain situations it will be inappropriate (or impossible) to represent a given function by a power series of the form $\sum_{n=0}^{\infty} a_n x^n$, and you will have to use a series of the form $\sum_{n=0}^{\infty} a_n (x - a)^n$ instead, where a is a constant. This is the case, for example, for the function $f(x) = \ln x$, which is undefined for $x \leq 0$ and so cannot possibly be the sum of a series of the form $\sum_{n=0}^{\infty} a_n x^n$ on a symmetric interval about $x = 0$.

Using an argument similar to the one at the beginning of this section, you can show that if $f(x) = \sum_{n=0}^{\infty} a_n (x - a)^n$, the coefficients of the series are related to the derivatives of f by the formula

$$a_n = \frac{f^{(n)}(a)}{n!}$$

The power series in $(x - a)$ with these coefficients is called the **Taylor series of $f(x)$ about $x = a$**. It can be shown that series of this type converge on symmetric intervals of the form $a - R < x < a + R$, including possibly one or both of the endpoints $x = a - R$ and $x = a + R$. Notice that the Taylor series about $x = 0$ is simply a special case of the more general Taylor series about $x = a$.

Taylor Series about $x = a$

The Taylor series of $f(x)$ about $x = a$ is the power series

$$\sum_{n=0}^{\infty} a_n (x - a)^n$$

where

$$a_n = \frac{f^{(n)}(a)}{n!}$$

Here is an example.

The Taylor Series for ln x

EXAMPLE 3.4

Find the Taylor series of the function $f(x) = \ln x$ about $x = 1$.

Solution

Compute the Taylor coefficients as follows:

$f(x) = \ln x$ $f(1) = 0$ $a_0 = \dfrac{f(1)}{0!} = 0$

$f'(x) = \dfrac{1}{x}$ $f'(1) = 1$ $a_1 = \dfrac{f'(1)}{1!} = 1$

$f''(x) = \dfrac{-1}{x^2}$ $f''(1) = -1$ $a_2 = \dfrac{f''(1)}{2!} = -\dfrac{1}{2}$

$f^{(3)}(x) = \dfrac{2!}{x^3}$ $f^{(3)}(1) = 2!$ $a_3 = \dfrac{f^{(3)}(1)}{3!} = \dfrac{1}{3}$

$f^{(4)}(x) = \dfrac{-3!}{x^4}$ $f^{(4)}(1) = -3!$ $a_4 = \dfrac{f^{(4)}(1)}{4!} = -\dfrac{1}{4}$

$\qquad \cdot \qquad\qquad\qquad\qquad \cdot \qquad\qquad\qquad\qquad \cdot$

$\qquad \cdot \qquad\qquad\qquad\qquad \cdot \qquad\qquad\qquad\qquad \cdot$

$\qquad \cdot \qquad\qquad\qquad\qquad \cdot \qquad\qquad\qquad\qquad \cdot$

$f^{(n)}(x) = (-1)^{n+1}\dfrac{(n-1)!}{x^n}$ $f^{(n)}(1) = (-1)^{n+1}(n-1)!$ $a_n = \dfrac{(-1)^{n+1}}{n}$

The corresponding Taylor series is

$$\sum_{n=1}^{\infty} \frac{(-1)^{n+1}}{n}(x-1)^n$$

which can be shown to converge to $\ln x$ for $0 < x \le 2$. That is,

$$\ln x = \sum_{n=1}^{\infty} \frac{(-1)^{n+1}}{n}(x-1)^n$$

$$= (x-1) - \frac{1}{2}(x-1)^2 + \frac{1}{3}(x-1)^3 - \cdots \qquad \text{for } 0 < x \le 2$$

Taylor Polynomials   The partial sums of a Taylor series of a function are polynomials that can be used to approximate the function. In general, the more terms there are in the partial sum, the better the approximation will be. It is customary to let $P_n(x)$ denote the $(n + 1)$st partial sum, which is a polynomial of degree (at most) n. In particular,

$$P_n(x) = f(a) + f'(a)(x - a) + \frac{f''(a)}{2!}(x - a)^2 + \cdots + \frac{f^{(n)}(a)}{n!}(x - a)^n$$

The polynomial $P_n(x)$ is known as the **nth Taylor polynomial** of $f(x)$ about $x = a$.

Approximation by Taylor Polynomials

The approximation of $f(x)$ by its Taylor polynomials $P_n(x)$ is most accurate near $x = a$ and for large values of n. Here is a geometric argument that should give you some additional insight into the situation.

Observe first that at $x = a$, f and P_n are equal, as are, respectively, their first n derivatives. For example,

$$P_2(x) = f(a) + f'(a)(x - a) + \frac{f''(a)}{2!}(x - a)^2 \quad \text{and} \quad P_2(a) = f(a)$$

$$P_2'(x) = f'(a) + f''(a)(x - a) \quad \text{and} \quad P_2'(a) = f'(a)$$

$$P_2''(x) = f''(a) \quad \text{and} \quad P_2''(a) = f''(a)$$

The fact that $P_n(a) = f(a)$ implies that the graphs of P_n and f intersect at $x = a$. The fact that $P_n'(a) = f'(a)$ implies further that the graphs have the same slope at $x = a$. The fact that $P_n''(a) = f''(a)$ imposes the further restriction that the graphs have the same concavity at $x = a$. In general, as n increases, the number of matching derivatives increases, and the graph of P_n approximates more closely that of f near $x = a$. The situation is illustrated in Figure 11.4, which shows the graphs of e^x and its first three Taylor polynomials.

Figure 11.4 The graphs of e^x and its first three Taylor polynomials.

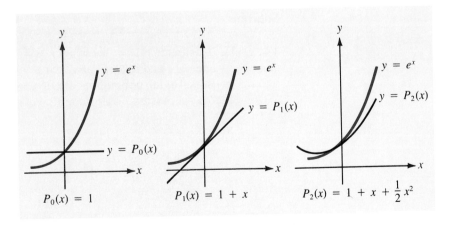

$P_0(x) = 1$ $P_1(x) = 1 + x$ $P_2(x) = 1 + x + \frac{1}{2}x^2$

The use of Taylor polynomials to approximate functions is illustrated in the following examples.

EXAMPLE 3.5

Use a Taylor polynomial of degree 6 to approximate e.

Solution

From Example 3.1

$$e^x = \sum_{n=0}^{\infty} \frac{x^n}{n!} \qquad \text{for all } x$$

Hence, $P_6(x) = 1 + x + \frac{x^2}{2!} + \frac{x^3}{3!} + \frac{x^4}{4!} + \frac{x^5}{5!} + \frac{x^6}{6!}$

and $e = e^1 \approx P_6(1) = 1 + 1 + \frac{1}{2!} + \frac{1}{3!} + \frac{1}{4!} + \frac{1}{5!} + \frac{1}{6!}$

$$= 1 + 1 + \frac{1}{2} + \frac{1}{6} + \frac{1}{24} + \frac{1}{120} + \frac{1}{720} \approx 2.71806$$

By analysis of the error that results from this sort of approximation (based on techniques beyond the scope of this text), it can be shown that the actual value of e, rounded off to five decimal places, is 2.71828.

EXAMPLE 3.6

Use an appropriate Taylor polynomial of degree 3 to approximate $\sqrt{4.1}$.

Solution

The goal is to estimate $f(x) = \sqrt{x}$ when $x = 4.1$. Since 4.1 is close to 4 and since the values of f and its derivatives at $x = 4$ are easy to compute, it is natural to use a Taylor polynomial about $x = 4$.

Compute the necessary Taylor coefficients about $x = 4$ as follows:

$$f(x) = \sqrt{x} \qquad f(4) = 2 \qquad a_0 = \frac{2}{0!} = 2$$

$$f'(x) = \frac{1}{2}x^{-1/2} \qquad f'(4) = \frac{1}{4} \qquad a_1 = \frac{\frac{1}{4}}{1!} = \frac{1}{4}$$

$$f''(x) = -\frac{1}{4}x^{-3/2} \qquad f''(4) = -\frac{1}{32} \qquad a_2 = \frac{-\frac{1}{32}}{2!} = -\frac{1}{64}$$

$$f^{(3)}(x) = \frac{3}{8}x^{-5/2} \qquad f^{(3)}(4) = \frac{3}{256} \qquad a_3 = \frac{\frac{3}{256}}{3!} = \frac{1}{512}$$

The corresponding Taylor polynomial of degree 3 is

$$P_3(x) = 2 + \frac{1}{4}(x - 4) - \frac{1}{64}(x - 4)^2 + \frac{1}{512}(x - 4)^3$$

and so

$$\sqrt{4.1} \simeq P_3(4.1) = 2 + \frac{1}{4}(0.1) - \frac{1}{64}(0.1)^2 + \frac{1}{512}(0.1)^3 \simeq 2.02485$$

Incidentally, all five decimal places of this estimate are correct. Rounded off to seven decimal places, the true value of $\sqrt{4.1}$ is 2.0248457.

Approximation of Definite Integrals   In the next example, a Taylor polynomial is used to estimate a definite integral whose exact value cannot be calculated by elementary methods.

EXAMPLE 3.7

Use a Taylor polynomial of degree 8 to approximate $\int_0^1 e^{-x^2} dx$.

Solution

The easiest way to get the Taylor series of e^{-x^2} is to start with the series

$$e^x = \sum_{n=0}^{\infty} \frac{x^n}{n!}$$

for e^x and replace x by $-x^2$ to get

$$e^{-x^2} = \sum_{n=0}^{\infty} \frac{(-1)^n x^{2n}}{n!}$$

Thus,

$$e^{-x^2} \simeq 1 - x^2 + \frac{1}{2}x^4 - \frac{1}{6}x^6 + \frac{1}{24}x^8$$

and so

$$\int_0^1 e^{-x^2} dx \simeq \int_0^1 \left(1 - x^2 + \frac{1}{2}x^4 - \frac{1}{6}x^6 + \frac{1}{24}x^8\right) dx$$

$$= \left(x - \frac{1}{3}x^3 + \frac{1}{10}x^5 - \frac{1}{42}x^7 + \frac{1}{216}x^9\right)\Big|_0^1$$

$$= 1 - \frac{1}{3} + \frac{1}{10} - \frac{1}{42} + \frac{1}{216} \simeq 0.7475$$

Problems 11.3

In Problems 1 through 8, use the formula for the Taylor coefficients to find the Taylor series of f about $x = 0$.

1. $f(x) = e^{3x}$

2. $f(x) = e^{-2x}$

3. $f(x) = \ln(1 + x)$

4. $f(x) = \dfrac{1}{2 - x}$

5. $f(x) = \dfrac{e^x + e^{-x}}{2}$

6. $f(x) = \dfrac{e^x - e^{-x}}{2}$

7. $f(x) = (1 + x)e^x$

8. $f(x) = (3x + 2)e^x$

In Problems 9 through 14, use the formula for the Taylor coefficients to find the Taylor series of f about $x = a$.

9. $f(x) = e^{2x};\ a = 1$

10. $f(x) = e^{-3x};\ a = -1$

11. $f(x) = \dfrac{1}{x};\ a = 1$

12. $f(x) = \ln 2x;\ a = \dfrac{1}{2}$

13. $f(x) = \dfrac{1}{2 - x};\ a = 1$

14. $f(x) = \dfrac{1}{1 + x};\ a = 2$

In Problems 15 through 20, modify one of the Taylor series you already know to find the Taylor series of the given function and specify an interval on which the series converges.

15. $f(x) = \dfrac{x}{1 - 5x}$

16. $f(x) = \dfrac{1}{1 + 8x^3}$

17. $f(x) = e^{-x/2}$

18. $f(x) = 2xe^{x^2/2}$

19. $f(x) = \dfrac{1}{x}$ about $x = 2$

20. $f(x) = \dfrac{e^{x^2} + e^{-x^2}}{2}$

In Problems 21 through 26, use a Taylor polynomial of degree n to approximate the given number.

21. $\sqrt{3.8};\ n = 3$

22. $\sqrt{1.2};\ n = 3$

23. $\ln 1.1;\ n = 5$

24. $\ln 0.7;\ n = 5$

25. $e^{0.3};\ n = 4$

26. $\dfrac{1}{\sqrt{e}};\ n = 4$

In Problems 27 through 30, use a Taylor polynomial of degree n to approximate the given integral.

27. $\displaystyle\int_0^{1/2} e^{-x^2}\,dx;\ n = 6$

28. $\displaystyle\int_{-0.2}^{0.1} e^{-x^2}\,dx;\ n = 4$

29. $\int_0^{0.1} \dfrac{1}{1 + x^2} \, dx; \; n = 4$ 30. $\int_{-1/2}^0 \dfrac{1}{1 - x^3} \, dx; \; n = 9$

Chapter Summary and Review Problems

Important Terms, Symbols, and Formulas

Infinite series: $\displaystyle\sum_{n=1}^{\infty} a_n = a_1 + a_2 + a_3 + \cdots$

nth partial sum: $S_n = a_1 + a_2 + \cdots + a_n$

Convergence and divergence:

$$\sum_{n=1}^{\infty} a_n \text{ converges to } S \text{ if and only if } \lim_{n \to \infty} S_n = S$$

Test for divergence: If $\displaystyle\lim_{n \to \infty} a_n \neq 0$, then $\displaystyle\sum_{n=1}^{\infty} a_n$ diverges.

Distributive law: $\displaystyle\sum_{n=1}^{\infty} ca_n = c \sum_{n=1}^{\infty} a_n$

Harmonic series: $\displaystyle\sum_{n=1}^{\infty} \frac{1}{n}$ diverges (even though $\displaystyle\lim_{n \to \infty} a_n = 0$).

Geometric series: If $|r| < 1$, $\displaystyle\sum_{n=0}^{\infty} r^n = \frac{1}{1 - r}$ $(r \neq 0)$

Power series: $\displaystyle\sum_{n=0}^{\infty} a_n x^n; \; \sum_{n=0}^{\infty} a_n(x - a)^n$

Radius of convergence; interval of convergence

Factorial notation: $n! = n(n - 1)(n - 2) \cdots 3 \cdot 2 \cdot 1; \; 0! = 1$

Taylor series of $f(x)$ about $x = a$:

$$\sum_{n=0}^{\infty} a_n(x - a)^n \quad \text{where } a_n = \frac{f^{(n)}(a)}{n!}$$

Important Taylor series:

$$e^x = \sum_{n=0}^{\infty} \frac{x^n}{n!} \quad \text{for all } x$$

$$\frac{1}{1 - x} = \sum_{n=0}^{\infty} x^n \quad \text{for } |x| < 1$$

$$\ln x = \sum_{n=1}^{\infty} \frac{(-1)^{n+1}}{n}(x - 1)^n \quad \text{for } 0 < x \leq 2$$

Substitution in power series

nth Taylor polynomial: $P_n(x)$

Review Problems

In Problems 1 through 3, find the sum of the given convergent series by taking the limit of a compact expression for the nth partial sum.

1. $\displaystyle\sum_{n=1}^{\infty} \left(\frac{1}{n+1} - \frac{1}{n+3} \right)$

2. $\displaystyle\sum_{n=2}^{\infty} \frac{1}{n(n-1)}$

3. $\displaystyle\sum_{n=1}^{\infty} \frac{2}{(-3)^n}$

In Problems 4 through 7, determine whether the given geometric series converges, and if so, find its sum.

4. $\displaystyle\sum_{n=1}^{\infty} \frac{3}{(-5)^n}$

5. $\displaystyle\sum_{n=0}^{\infty} \left(-\frac{3}{2} \right)^n$

6. $\displaystyle\sum_{n=1}^{\infty} e^{-0.5n}$

7. $\displaystyle\sum_{n=2}^{\infty} \frac{2^{n+1}}{3^{n-3}}$

8. Express the repeating decimal $1.545454\ldots$ as a fraction.

9. A ball is dropped from a height of 6 feet and allowed to bounce indefinitely. How far will the ball travel if it always rebounds to 80 percent of its previous height?

10. How much should you invest today at an annual interest rate of 12 percent compounded continuously so that, starting next year, you can make annual withdrawals of $500 in perpetuity?

11. A patient is given an injection of 10 units of a certain drug every 24 hours. The drug is eliminated exponentially so that the fraction that remains in the patient's body after t days is $f(t) = e^{-0.8t}$. If the treatment is continued indefinitely, approximately how many units of the drug will eventually be in the patient's body immediately following an injection?

12. You and two friends take turns rolling a die until one of you wins by getting a six. Find the probability that you will win if you roll second.

In Problems 13 through 16, use the formula for the Taylor coefficients to find the Taylor series of f about $x = a$.

13. $f(x) = \dfrac{1}{x+3}$; $a = 0$

14. $f(x) = e^{-3x}$; $a = 0$

15. $f(x) = (1 + 2x)e^x; a = 0$ 16. $f(x) = \dfrac{1}{(1 + x)^2}; a = -2$

17. Modify a geometric series to find the Taylor series about $x = 0$ for the function $f(x) = \dfrac{4x}{1 + 4x^2}$.

18. Use a Taylor polynomial of degree 3 to approximate $\sqrt{0.9}$.

19. Use a Taylor polynomial of degree 10 to approximate $\displaystyle\int_0^{1/2} \dfrac{x}{1 + x^3}\, dx$.

CHAPTER 12
TRIGONOMETRIC FUNCTIONS

1   The Trigonometric Functions

In this chapter you will be introduced to some functions that are widely used in the natural sciences to study periodic or rhythmic phenomena such as oscillations, the periodic motion of planets, and the respiratory cycle and heartbeat of animals. These functions are also related to the measurement of angles and hence play an important role in such fields as architecture, navigation, and surveying.

Angles   An **angle** is formed when one line segment in the plane is rotated into another about their common endpoint. The resulting angle is said to be a **positive angle** if the rotation is in a counterclockwise direction and a **negative angle** if the rotation is in a clockwise direction. The situation is illustrated in Figure 12.1.

Figure 12.1 Positive and negative angles.

Measurement of Angles   You are probably already familiar with the use of degrees to measure angles. A **degree** is the amount by which a line segment must be rotated so

Figure 12.2 Angles measured in degrees.

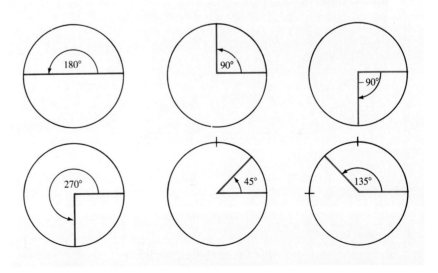

that its free endpoint traces out $\frac{1}{360}$ of a circle. Thus, for example, a complete counterclockwise rotation generating an entire circle contains 360°, one-half of a complete counterclockwise rotation contains 180°, and one-sixth of a complete counterclockwise rotation contains 60°. Some important angles are shown in Figure 12.2.

Radian Measure   Although measurement of angles in degrees is convenient for many geometric applications, there is another unit of angle measurement called a **radian** that leads to simpler rules for the differentiation and integration of trigonometric functions. One radian is defined to be the amount by which a line segment of length 1 must be rotated so that its free endpoint traces out a circular arc of length 1. The situation is illustrated in Figure 12.3.

The total circumference of a circle of radius 1 is 2π. Hence, 2π radians is equal to 360°, π radians is equal to 180°, and, in general, the relationship between radians and degrees is given by the following proportion.

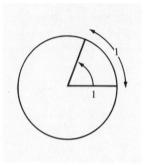

Figure 12.3 An angle of 1 radian.

Conversion Formula

$$\frac{\text{Degrees}}{180} = \frac{\text{radians}}{\pi}$$

Here is an example.

EXAMPLE 1.1

(a) Convert 45° to radians.

(b) Convert $\frac{\pi}{6}$ radian to degrees.

Solution

(a) From the proportion $\dfrac{45}{180} = \dfrac{\text{radians}}{\pi}$

it follows that $\text{Radians} = \dfrac{\pi}{4}$

That is, 45° equals $\frac{\pi}{4}$ radian.

(b) From the proportion $\dfrac{\text{Degrees}}{180} = \dfrac{\pi/6}{\pi}$

it follows that   Degrees = 30

That is, $\frac{\pi}{6}$ radian equals 30°.

For reference, six of the most important angles are shown in Figure 12.4 along with their measurements in degrees and in radians.

Figure 12.4 Six important angles in degrees and radians.

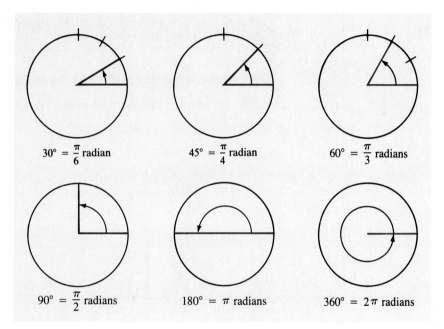

$30° = \frac{\pi}{6}$ radian $45° = \frac{\pi}{4}$ radian $60° = \frac{\pi}{3}$ radians

$90° = \frac{\pi}{2}$ radians $180° = \pi$ radians $360° = 2\pi$ radians

The Sine and the Cosine

Suppose the line segment joining the points $(0, 0)$ and $(1, 0)$ on the x axis is rotated through an angle of θ radians so that the free endpoint of the segment moves from $(1, 0)$ to a point (x, y) as in Figure 12.5. The x and y coordinates of the point (x, y) are known, respectively, as the **cosine** and

Figure 12.5 The sine and cosine of an angle.

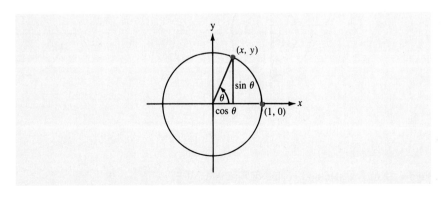

sine of the angle θ. The symbol cos θ is used to denote the cosine of θ, and sin θ is used to denote its sine.

> ## The Sine and Cosine
>
> For any angle θ,
>
> $$\cos \theta = x \quad \text{and} \quad \sin \theta = y$$
>
> where (x, y) is the point to which $(1, 0)$ is carried by a rotation of θ radians about the origin.

The values of the sine and cosine for multiples of $\frac{\pi}{2}$ can be read from Figure 12.6 and are summarized in the table on page 649.

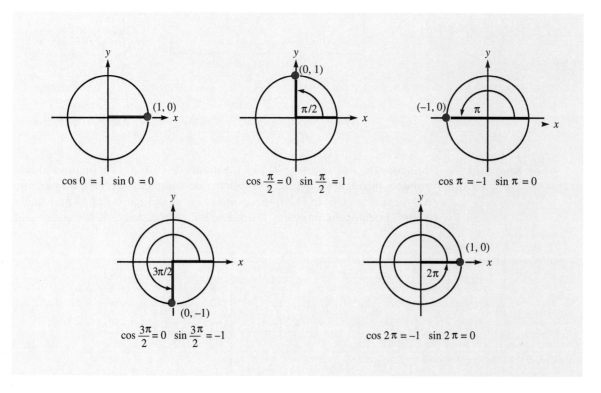

Figure 12.6 The sine and cosine of multiples of $\frac{\pi}{2}$.

θ	0	$\dfrac{\pi}{2}$	π	$\dfrac{3\pi}{2}$	2π
$\cos \theta$	1	0	-1	0	1
$\sin \theta$	0	1	0	-1	0

The sine and cosine of a few other important angles are also easy to obtain geometrically, as you will see shortly. For other angles, use Table III at the back of the book or your calculator to find the sine and cosine.

Elementary Properties of the Sine and Cosine

Since there are 2π radians in a complete rotation, it follows that

$$\sin (\theta + 2\pi) = \sin \theta \quad \text{and} \quad \cos (\theta + 2\pi) = \cos \theta$$

That is, the sine and cosine functions are **periodic** with period 2π. The situation is illustrated in Figure 12.7.

Figure 12.7 The periodicity of $\sin \theta$ and $\cos \theta$.

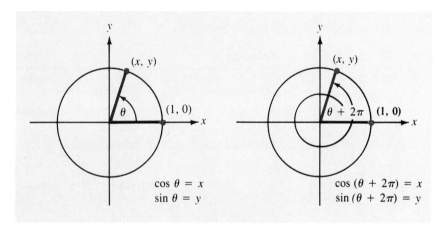

Since negative angles correspond to clockwise rotations, it follows that

$$\sin (-\theta) = -\sin \theta \quad \text{and} \quad \cos (-\theta) = \cos \theta$$

This is illustrated in Figure 12.8.

Figure 12.8 The sine and cosine of negative angles.

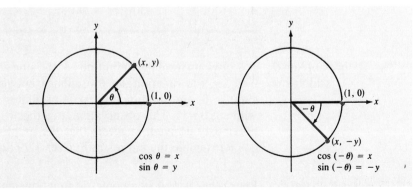

> **Properties of the Sine and Cosine**
>
> $$\sin (\theta + 2\pi) = \sin \theta \quad \text{and} \quad \cos (\theta + 2\pi) = \cos \theta$$
> $$\sin (-\theta) = -\sin \theta \quad \text{and} \quad \cos (-\theta) = \cos \theta$$

The use of these properties is illustrated in the following example.

EXAMPLE 1.2

Evaluate the following expressions.

(a) $\cos (-\pi)$ (b) $\sin \left(-\dfrac{\pi}{2}\right)$ (c) $\cos 3\pi$ (d) $\sin \dfrac{5\pi}{2}$

Solution

(a) Since $\cos \pi = -1$, it follows that

$$\cos (-\pi) = \cos \pi = -1$$

(b) Since $\sin \dfrac{\pi}{2} = 1$, it follows that

$$\sin \left(-\frac{\pi}{2}\right) = -\sin \frac{\pi}{2} = -1$$

(c) Since $3\pi = \pi + 2\pi$ and $\cos \pi = -1$, it follows that

$$\cos 3\pi = \cos (\pi + 2\pi) = \cos \pi = -1$$

(d) Since $\dfrac{5\pi}{2} = \dfrac{\pi}{2} + 2\pi$ and $\sin \dfrac{\pi}{2} = 1$, it follows that

$$\sin \frac{5\pi}{2} = \sin \left(\frac{\pi}{2} + 2\pi\right) = \sin \frac{\pi}{2} = 1$$

The Graphs of $\sin \theta$ and $\cos \theta$ It is obvious from the definitions of the sine and cosine that as θ goes from 0 to 2π, the function $\sin \theta$ oscillates between 1 and -1, starting with $\sin 0 = 0$, and the function $\cos \theta$ oscillates between 1 and -1, starting with $\cos 0 = 1$. This observation, together with the elementary properties previously derived, suggests that the graphs of the functions $\sin \theta$ and $\cos \theta$ resemble the curves in Figures 12.9 and 12.10, respectively.

Other Trigonometric Functions   Four other useful trigonometric functions can be defined in terms of the sine and cosine as follows.

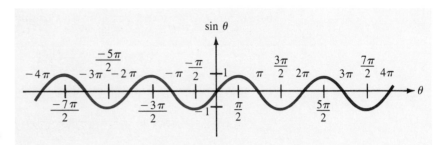

Figure 12.9 The graph of sin θ.

Figure 12.10 The graph of cos θ.

The Tangent, Cotangent, Secant, and Cosecant

For any angle θ,

$$\tan \theta = \frac{\sin \theta}{\cos \theta} \qquad \cot \theta = \frac{1}{\tan \theta} = \frac{\cos \theta}{\sin \theta}$$

$$\sec \theta = \frac{1}{\cos \theta} \qquad \csc \theta = \frac{1}{\sin \theta}$$

provided that the denominators are not zero.

EXAMPLE 1.3

Evaluate the following expressions.

(a) $\tan \pi$ (b) $\cot \dfrac{\pi}{2}$ (c) $\sec(-\pi)$ (d) $\csc\left(-\dfrac{5\pi}{2}\right)$

Solution

(a) Since $\sin \pi = 0$ and $\cos \pi = -1$, it follows that

$$\tan \pi = \frac{\sin \pi}{\cos \pi} = \frac{0}{-1} = 0$$

(b) Since $\sin \dfrac{\pi}{2} = 1$ and $\cos \dfrac{\pi}{2} = 0$, it follows that

$$\cot \frac{\pi}{2} = \frac{\cos (\pi/2)}{\sin (\pi/2)} = \frac{0}{1} = 0$$

(c) Since $\cos \pi = -1$, it follows that

$$\sec (-\pi) = \frac{1}{\cos (-\pi)} = \frac{1}{\cos \pi} = -1$$

(d) Since $\dfrac{5\pi}{2} = \dfrac{\pi}{2} + 2\pi$ and $\sin \dfrac{\pi}{2} = 1$, it follows that

$$\csc \left(-\frac{5\pi}{2}\right) = \frac{1}{\sin (-5\pi/2)} = \frac{1}{-\sin (5\pi/2)} = \frac{1}{-\sin (\pi/2)} = -1$$

Right Triangles

If you had a high-school course in trigonometry, you may remember the following definitions of the sine, cosine, and tangent involving the sides of a right triangle like the one in Figure 12.11.

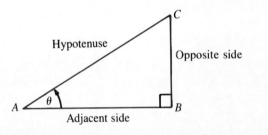

Figure 12.11 Triangle used to define trigonometric functions.

The Trigonometry of Right Triangles

$$\sin \theta = \frac{\text{opposite}}{\text{hypotenuse}} \qquad \cos \theta = \frac{\text{adjacent}}{\text{hypotenuse}} \qquad \tan \theta = \frac{\text{opposite}}{\text{adjacent}}$$

The definitions of trigonometric functions that you have seen in this section involving the coordinates of points on a circle of radius 1 are equivalent to the definitions from high-school trigonometry. To see this, superimpose an xy coordinate system over the triangle ABC as shown in Figure 12.12 and draw the circle of radius 1 that is centered at the origin. Since ABC and ADE are similar triangles, it follows that

$$\frac{AD}{AE} = \frac{AB}{AC}$$

But the length AD is the x coordinate of point E on the circle, and so, by

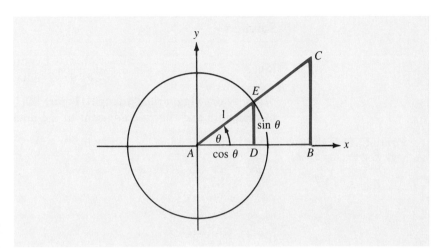

Figure 12.12 Similar triangles.

the definition of the cosine,

$$AD = \cos\theta$$

Moreover,

$$AE = 1 \quad AB = \text{adjacent side} \quad \text{and} \quad AC = \text{hypotenuse}$$

Hence,

$$\frac{\cos\theta}{1} = \frac{\text{adjacent}}{\text{hypotenuse}} \quad \text{or} \quad \cos\theta = \frac{\text{adjacent}}{\text{hypotenuse}}$$

For practice, convince yourself that similar arguments lead to the formulas for the sine and tangent.

Calculations with Right Triangles

Many calculations involving trigonometric functions can be performed easily and quickly with the aid of appropriate right triangles. For example, from the well-known 30-60-90° triangle in Figure 12.13, you see immediately that

$$\sin 30° = \frac{1}{2} \quad \text{and} \quad \cos 30° = \frac{\sqrt{3}}{2}$$

or, using radian measure,

$$\sin\frac{\pi}{6} = \frac{1}{2} \quad \text{and} \quad \cos\frac{\pi}{6} = \frac{\sqrt{3}}{2}$$

Here is an example that further illustrates the use of right triangles.

Figure 12.13 A 30-60-90° triangle.

EXAMPLE 1.4

Find $\tan\theta$ if $\sec\theta = \frac{3}{2}$.

Solution

Since
$$\sec \theta = \frac{1}{\cos \theta} = \frac{\text{hypotenuse}}{\text{adjacent}}$$

begin by drawing a right triangle (Figure 12.14) in which the hypotenuse has length 3 and the side adjacent to the angle θ has length 2.

Figure 12.14 Right triangle for Example 1.4.

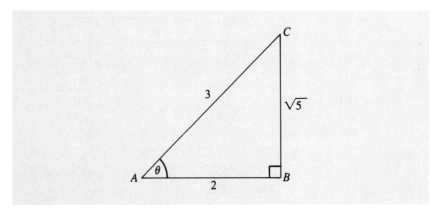

By the pythagorean theorem,
$$\text{Length of } BC = \sqrt{3^2 - 2^2} = \sqrt{5}$$

and so
$$\tan \theta = \frac{\text{opposite}}{\text{adjacent}} = \frac{\sqrt{5}}{2}$$

Important Angles   Using appropriate right triangles and the definitions and elementary properties of the sine and cosine, you should now be able to find the sine and cosine of the important angles listed in the following table. Before proceeding, take a few minutes to fill in the table so that you will have this information summarized for easy reference in the future (see page 680).

A Table of Important Values

θ	0	$\dfrac{\pi}{6}$	$\dfrac{\pi}{4}$	$\dfrac{\pi}{3}$	$\dfrac{\pi}{2}$	$\dfrac{2\pi}{3}$	$\dfrac{3\pi}{4}$	$\dfrac{5\pi}{6}$	π
$\sin \theta$									
$\cos \theta$									

Trigonometric   An **identity** is an equation that holds for all values of its variable or
Identities   variables. Identities involving trigonometric functions are called **trig-**

onometric identities and can be used to simplify trigonometric expressions. You have already seen some elementary trigonometric identities in this section. For example, the formulas

$$\cos(-\theta) = \cos\theta \quad \text{and} \quad \sin(-\theta) = -\sin\theta$$

are identities because they hold for all values of θ. Most trigonometry texts list dozens of trigonometric identities. Fortunately, only a few of them will be needed in this book.

One of the most important and well-known trigonometric identities is a simple consequence of the pythagorean theorem. It states that

$$\sin^2\theta + \cos^2\theta = 1$$

where $\sin^2\theta$ stands for $(\sin\theta)^2$ and $\cos^2\theta$ stands for $(\cos\theta)^2$. To see why this identity holds, look at Figure 12.15, which shows a point (x, y) on a circle of radius 1. By definition,

$$y = \sin\theta \quad \text{and} \quad x = \cos\theta$$

and, by the pythagorean theorem,

$$y^2 + x^2 = 1$$

from which the identity follows immediately.

Figure 12.15
$\sin^2\theta + \cos^2\theta = 1.$

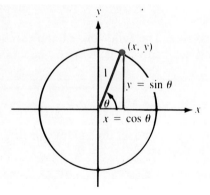

The Pythagorean Identity

For any angle θ,

$$\sin^2\theta + \cos^2\theta = 1$$

The following formulas for the cosine and sine of the sum of two angles are particularly useful. Proofs of these identities can be found in most trigonometry texts and will be omitted here.

> **The Addition Formulas**
>
> For any angles a and b,
>
> $$\cos(a + b) = \cos a \cos b - \sin a \sin b$$
>
> and $$\sin(a + b) = \sin a \cos b + \cos a \sin b$$

If the angles a and b in the addition formulas are equal, the identities can be rewritten as follows.

> **The Double-Angle Formulas**
>
> For any angle θ,
>
> $$\cos 2\theta = \cos^2 \theta - \sin^2 \theta$$
>
> and $$\sin 2\theta = 2 \sin \theta \cos \theta$$

Some of these trigonometric identities will be used in Section 2 to derive the formulas for the derivatives of the sine and cosine functions. They can also be used in the solution of certain trigonometric equations.

Trigonometric Equations   The following examples illustrate the use of trigonometric identities to solve trigonometric equations.

EXAMPLE 1.5

Find all the values of θ in the interval $0 \leq \theta \leq \pi$ that satisfy the equation $\sin 2\theta = \sin \theta$.

Solution

Use the double-angle formula

$$\sin 2\theta = 2 \sin \theta \cos \theta$$

to rewrite the given equation as

$$2 \sin \theta \cos \theta = \sin \theta$$

or $$2 \sin \theta \cos \theta - \sin \theta = 0$$

and factor to get $$\sin \theta (2 \cos \theta - 1) = 0$$

which is satisfied if

$$\sin \theta = 0 \quad \text{or} \quad 2 \cos \theta - 1 = 0$$

From the graphs of the sine and cosine functions and from the table of important values, which you completed earlier in this section, it should be clear that the only solutions of

$$\sin \theta = 0$$

in the interval $0 \le \theta \le \pi$ are

$$\theta = 0 \quad \text{and} \quad \theta = \pi$$

and that the only solution of

$$2 \cos \theta - 1 = 0 \quad \text{or} \quad \cos \theta = \frac{1}{2}$$

in the interval $0 \le \theta \le \pi$ is $\qquad \theta = \frac{\pi}{3}$

Hence, the only solutions of the original equation in the specified interval are $\theta = 0$, $\theta = \frac{\pi}{3}$, and $\theta = \pi$.

EXAMPLE 1.6

Find all the values of θ in the interval $0 \le \theta \le 2\pi$ that satisfy the equation $\sin^2 \theta - \cos^2 \theta + \sin \theta = 0$.

Solution

Use the pythagorean identity

$$\sin^2 \theta + \cos^2 \theta = 1 \quad \text{or} \quad \cos^2 \theta = 1 - \sin^2 \theta$$

to rewrite the original equation without any cosine terms as

$$\sin^2 \theta - (1 - \sin^2 \theta) + \sin \theta = 0$$

or $\qquad\qquad 2 \sin^2 \theta + \sin \theta - 1 = 0$

and factor to get

$$(2 \sin \theta - 1)(\sin \theta + 1) = 0$$

The only solutions of

$$2 \sin \theta - 1 = 0 \quad \text{or} \quad \sin \theta = \frac{1}{2}$$

in the interval $0 \le \theta \le 2\pi$ are

$$\theta = \frac{\pi}{6} \quad \text{and} \quad \theta = \frac{5\pi}{6}$$

and the only solution of

$$\sin \theta + 1 = 0 \quad \text{or} \quad \sin \theta = -1$$

in the interval is

$$\theta = \frac{3\pi}{2}$$

Hence, the solutions of the original equation in the specified interval are $\theta = \frac{\pi}{6}$, $\theta = \frac{5\pi}{6}$, and $\theta = \frac{3\pi}{2}$.

Problems 12.1

In Problems 1 through 6, specify the number of degrees in the given angle.

1.

2.

3.

4.

5.

6.
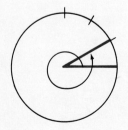

In Problems 7 through 12, show the given angle graphically.

7. 60°   8. −45°   9. 240°

10. 120°   11. −150°   12. 405°

In Problems 13 through 18, convert from degrees to radians.

13. 15°   14. 270°   15. −150°

16. −240°   17. 540°   18. 1°

In Problems 19 through 24, convert from radians to degrees.

19. $\dfrac{5\pi}{6}$ 20. $\dfrac{2\pi}{3}$ 21. $\dfrac{3\pi}{2}$

22. $-\dfrac{\pi}{12}$ 23. $-\dfrac{3\pi}{4}$ 24. 1

In Problems 25 through 30, specify the number of radians in the given angle.

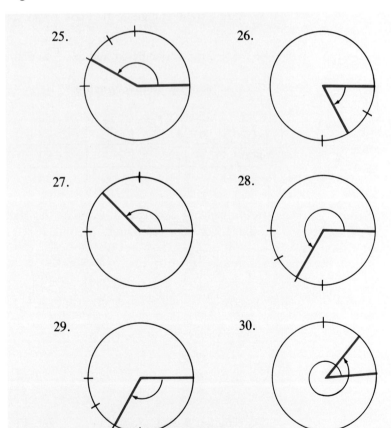

25.   26.

27.   28.

29.   30.

In Problems 31 through 36, show graphically the angles with the given radian measure.

31. $\dfrac{3\pi}{4}$

32. $-\dfrac{\pi}{6}$

33. $\dfrac{2\pi}{3}$

34. $\dfrac{5\pi}{6}$

35. $-\dfrac{5\pi}{6}$

36. $\dfrac{3\pi}{2}$

In Problems 37 through 44, evaluate the given expression without using Table III or your calculator.

37. $\cos \dfrac{7\pi}{2}$

38. $\sin 5\pi$

39. $\cos \left(-\dfrac{5\pi}{2}\right)$

40. $\sin \left(-\dfrac{7\pi}{2}\right)$

41. $\cot \dfrac{5\pi}{2}$

42. $\sec 3\pi$

43. $\tan (-\pi)$

44. $\csc \left(-\dfrac{7\pi}{2}\right)$

45. Use a right triangle to find $\cos \dfrac{\pi}{3}$ and $\sin \dfrac{\pi}{3}$.

46. Use a right triangle to find $\cos \dfrac{\pi}{4}$ and $\sin \dfrac{\pi}{4}$.

47. Complete the following table:

θ	$\dfrac{7\pi}{6}$	$\dfrac{5\pi}{4}$	$\dfrac{4\pi}{3}$	$\dfrac{3\pi}{2}$	$\dfrac{5\pi}{3}$	$\dfrac{7\pi}{4}$	$\dfrac{11\pi}{6}$	2π
$\sin \theta$								
$\cos \theta$								

In Problems 48 through 58, evaluate the given expression without using Table III or your calculator.

48. $\cos \left(-\dfrac{2\pi}{3}\right)$

49. $\sin \left(-\dfrac{\pi}{6}\right)$

50. $\sin \left(-\dfrac{3\pi}{4}\right)$

51. $\cos \dfrac{7\pi}{3}$

52. $\sin \left(-\dfrac{7\pi}{6}\right)$

53. $\tan \dfrac{\pi}{6}$

54. $\cot \dfrac{\pi}{3}$

55. $\sec \dfrac{\pi}{3}$

56. $\csc \dfrac{2\pi}{3}$

57. $\tan \left(-\dfrac{\pi}{4}\right)$

58. $\cot \left(-\dfrac{5\pi}{3}\right)$

59. Find $\tan \theta$ if $\sec \theta = \dfrac{5}{4}$.

60. Find $\tan \theta$ if $\cos \theta = \dfrac{3}{5}$.

61. Find $\tan \theta$ if $\csc \theta = \dfrac{5}{3}$.

62. Find $\sin \theta$ if $\tan \theta = \dfrac{2}{3}$.

63. Find $\cos \theta$ if $\tan \theta = \dfrac{3}{4}$.

64. Find $\sec \theta$ if $\cot \theta = \dfrac{4}{3}$.

In Problems 65 through 73, find all the values of θ in the specified interval that satisfy the given equation.

65. $\cos \theta - \sin 2\theta = 0; 0 \le \theta \le 2\pi$

66. $3 \sin^2 \theta + \cos 2\theta = 2; 0 \le \theta \le 2\pi$

67. $\sin 2\theta = \sqrt{3} \cos \theta; 0 \le \theta \le \pi$

68. $\cos 2\theta = \cos \theta; 0 \le \theta \le \pi$

69. $2 \cos^2 \theta = \sin 2\theta; 0 \le \theta \le \pi$

70. $3 \cos^2 \theta - \sin^2 \theta = 2; 0 \le \theta \le \pi$

71. $2 \cos^2 \theta + \sin \theta = 1; 0 \le \theta \le \pi$

72. $\cos^2 \theta - \sin^2 \theta + \cos \theta = 0; 0 \le \theta \le \pi$

73. $\sin^2 \theta - \cos^2 \theta + 3 \sin \theta = 1; 0 \le \theta \le \pi$

74. Starting with the identity $\sin^2 \theta + \cos^2 \theta = 1$, derive the identity $1 + \tan^2 \theta = \sec^2 \theta$.

75. Starting with the addition formula

$$\sin (a + b) = \sin a \cos b + \cos a \sin b$$

derive the subtraction formula

$$\sin (a - b) = \sin a \cos b - \cos a \sin b$$

76. Starting with the addition formula

$$\cos (a + b) = \cos a \cos b - \sin a \sin b$$

derive the subtraction formula

$$\cos (a - b) = \cos a \cos b + \sin a \sin b$$

77. Starting with the addition formulas for the sine and cosine, derive the addition formula

$$\tan (a + b) = \frac{\tan a + \tan b}{1 - \tan a \tan b}$$

for the tangent.

78. Use the addition formula for the sine to derive the identity

$$\sin \left(\frac{\pi}{2} - \theta \right) = \cos \theta$$

79. Use the addition formula for the cosine to derive the identity

$$\cos\left(\frac{\pi}{2} - \theta\right) = \sin \theta$$

80. Give geometric arguments involving the coordinates of points on a circle of radius 1 to show why the identities in Problems 78 and 79 are valid.

2 Differentiation of Trigonometric Functions

The trigonometric functions are relatively easy to differentiate. In this section you will learn how. In the following section you will see a variety of rate-of-change and optimization problems that can be solved using derivatives of trigonometric functions.

The Derivatives of the Sine and Cosine   Here are the formulas for the derivatives of the sine and cosine functions. The proofs will be discussed later in this section, after you have had a chance to practice using the formulas.

> **The Derivatives of sin θ and cos θ**
>
> If θ is measured in radians,
>
> $$\frac{d}{d\theta} \sin \theta = \cos \theta \quad \text{and} \quad \frac{d}{d\theta} \cos \theta = -\sin \theta$$

These formulas are usually used in conjunction with the chain rule in the following forms:

> **Chain Rule for Sine and Cosine**
>
> If h is a differentiable function of θ and θ is measured in radians,
>
> $$\frac{d}{d\theta} \sin h(\theta) = h'(\theta) \cos h(\theta)$$
>
> and $$\frac{d}{d\theta} \cos h(\theta) = -h'(\theta) \sin h(\theta)$$

Here are some examples.

EXAMPLE 2.1

Differentiate the function $f(\theta) = \sin(3\theta + 1)$.

Solution

Using the chain rule for the sine function with $h(\theta) = 3\theta + 1$, you get

$$f'(\theta) = 3 \cos (3\theta + 1)$$

EXAMPLE 2.2

Differentiate the function $f(\theta) = \cos^2 \theta$.

Solution

Since
$$\cos^2 \theta = (\cos \theta)^2$$

you use the chain rule for powers and the formula for the derivative of the cosine to get

$$f'(\theta) = 2 \cos \theta \, (-\sin \theta) = -2 \cos \theta \sin \theta$$

The Derivatives of Other Trigonometric Functions The differentiation formulas for the sine and cosine can be used to obtain differentiation formulas for the other trigonometric functions. The procedure is illustrated in the next example.

EXAMPLE 2.3

Show that $\dfrac{d}{d\theta} \tan \theta = \sec^2 \theta$.

Solution

Write
$$\tan \theta = \frac{\sin \theta}{\cos \theta}$$

and use the quotient rule to get

$$\frac{d}{d\theta} \tan \theta = \frac{\cos \theta \, (\cos \theta) - \sin \theta \, (-\sin \theta)}{\cos^2 \theta}$$

$$= \frac{\cos^2 \theta + \sin^2 \theta}{\cos^2 \theta}$$

$$= \frac{1}{\cos^2 \theta} \qquad \text{since } \sin^2 \theta + \cos^2 \theta = 1$$

$$= \sec^2 \theta \qquad \text{since } \sec \theta = \frac{1}{\cos \theta}$$

The Derivative of tan θ

If θ is measured in radians,

$$\frac{d}{d\theta} \tan \theta = \sec^2 \theta$$

Chain Rule for the Tangent

If h is a differentiable function of θ and θ is measured in radians,

$$\frac{d}{d\theta} \tan h(\theta) = h'(\theta) \sec^2 h(\theta)$$

Here is an example illustrating the use of the chain rule for the tangent.

EXAMPLE 2.4

Differentiate the function $f(\theta) = \tan \theta^2$.

Solution

Using the chain rule for tangents, you get

$$f'(\theta) = 2\theta \sec^2 \theta^2$$

The derivations of the differentiation formulas for the other trigonometric functions are left as exercises for you to do. (See Problems 21, 22, and 23 at the end of this section.) Only the formulas for the derivatives of the sine, cosine, and tangent will be needed for the applications in Section 3.

The Proof That $\frac{d}{d\theta} \sin \theta = \cos \theta$ Since the function $f(\theta) = \sin \theta$ is unrelated to any of the functions whose derivatives were obtained in previous chapters, it will be necessary to return to the *definition* of the derivative (from Chapter 2, Section 1) to find the derivative in this case. In terms of θ, the definition states that

$$f'(\theta) = \lim_{\Delta\theta \to 0} \frac{f(\theta + \Delta\theta) - f(\theta)}{\Delta\theta}$$

In performing the required calculations when $f(\theta) = \sin \theta$, you will need to use the addition formula

$$\sin (a + b) = \sin a \cos b + \cos a \sin b$$

from the preceding section as well as the following two facts about the behavior of $\sin \theta$ and $\cos \theta$ as θ approaches zero.

Two Important Limits

If θ is measured in radians,

$$(1) \quad \lim_{\theta \to 0} \frac{\sin \theta}{\theta} = 1$$

and

$$(2) \quad \lim_{\theta \to 0} \frac{\cos \theta - 1}{\theta} = 0$$

Proofs of the Limits   To see why the first of these limits holds, look at Figure 12.16. From the definition of the sine of an angle, it follows that

$$\sin \theta = y = \text{length of segment } PQ$$

and since θ is measured in radians, it follows that

$$\theta = \text{length of arc } PR$$

(You will recall that 1 radian is the amount by which a line segment of length 1 must be rotated so that its free endpoint traces out a circular arc of length 1. Hence, θ radians is the amount by which a line segment of length 1 must be rotated so that its free endpoint traces out a circular arc of length θ.) Thus,

$$\frac{\sin \theta}{\theta} = \frac{\text{length of segment } PQ}{\text{length of arc } PR}$$

The picture suggests that as θ approaches zero, the length of the

Figure 12.16 $\dfrac{\sin \theta}{\theta} =$

$\dfrac{\text{length of } PQ}{\text{length of } PR} \to 1.$

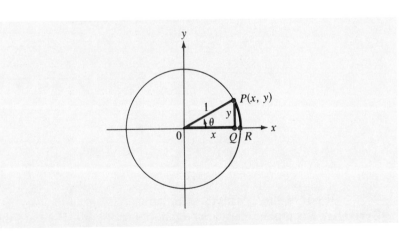

segment PQ and that of the arc PR will get closer and closer to one another so that their ratio will approach 1. (A formal proof of this fact can be found in more advanced texts.) Hence,

$$\lim_{\theta \to 0} \frac{\sin \theta}{\theta} = 1$$

and the first limit is established.

The second limit,

$$\lim_{\theta \to 0} \frac{\cos \theta - 1}{\theta} = 0$$

can be derived algebraically from the first one. (See Problem 24 at the end of this section.) It also has the following simple geometric interpretation, which should give you additional insight into why it holds.

Figure 12.17 shows a portion of the graph of the function $\cos \theta$ near $\theta = 0$. The graph has a relative maximum when $\theta = 0$, and so, assuming that $\cos \theta$ is differentiable at $\theta = 0$ (which we have not yet proved), it follows that the derivative of $\cos \theta$ at $\theta = 0$ is zero. But the difference quotient you would use to find this derivative is

$$\frac{\cos (0 + \Delta\theta) - \cos 0}{\Delta\theta} = \frac{\cos \Delta\theta - 1}{\Delta\theta}$$

Hence,

$$\lim_{\Delta\theta \to 0} \frac{\cos \Delta\theta - 1}{\Delta\theta} = 0$$

which, except for the symbol used to denote the variable ($\Delta\theta$ instead of θ), is precisely the limit under consideration.

Figure 12.17
$$\lim_{\theta \to 0} \frac{\cos \theta - 1}{\theta} =$$
(slope of tangent at 0) = 0.

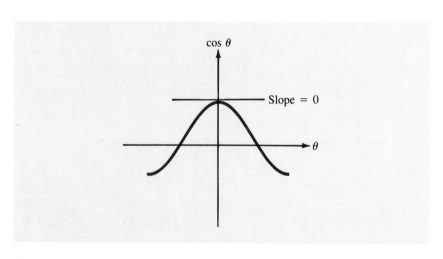

Proof of the Derivative Formula With these preliminaries out of the way, the proof of the formula for the derivative of $\sin \theta$ is not hard. Here are the details.

Let $f(\theta) = \sin\theta$. Then,

$$\frac{f(\theta + \Delta\theta) - f(\theta)}{\Delta\theta} = \frac{\sin(\theta + \Delta\theta) - \sin\theta}{\Delta\theta}$$

$$= \frac{\sin\theta\cos\Delta\theta + \cos\theta\sin\Delta\theta - \sin\theta}{\Delta\theta}$$

by the addition formula for the sine

$$= \frac{\sin\theta\cos\Delta\theta - \sin\theta}{\Delta\theta} + \frac{\cos\theta\sin\Delta\theta}{\Delta\theta}$$

$$= \sin\theta\frac{\cos\Delta\theta - 1}{\Delta\theta} + \cos\theta\frac{\sin\Delta\theta}{\Delta\theta}$$

Now let $\Delta\theta$ approach zero to get the derivative. Since

$$\lim_{\Delta\theta\to 0}\frac{\cos\Delta\theta - 1}{\Delta\theta} = 0 \qquad \text{by limit (2)}$$

and

$$\lim_{\Delta\theta\to 0}\frac{\sin\Delta\theta}{\Delta\theta} = 1 \qquad \text{by limit (1)}$$

it follows that

$$f'(\theta) = \lim_{\Delta\theta\to 0}\sin\theta\frac{\cos\Delta\theta - 1}{\Delta\theta} + \lim_{\Delta\theta\to 0}\cos\theta\frac{\sin\Delta\theta}{\Delta\theta}$$

$$= (\sin\theta)(0) + (\cos\theta)(1) = \cos\theta$$

That is,

$$\frac{d}{d\theta}\sin\theta = \cos\theta$$

The Proof That $\dfrac{d}{d\theta}\cos\theta = -\sin\theta$ The formula for the derivative of $\cos\theta$ can be obtained from the definition of the derivative by a calculation similar to the one just given for $\sin\theta$. (Problem 25 at the end of this section will suggest that you try it.) Here is an attractive alternative derivation that takes advantage of the formula for the derivative of the sine function and uses the identities

$$\cos\theta = \sin\left(\frac{\pi}{2} - \theta\right) \qquad \text{and} \qquad \sin\theta = \cos\left(\frac{\pi}{2} - \theta\right)$$

which relate the sine and cosine functions. (These identities are simple consequences of the addition formulas for the sine and cosine. They can also be obtained geometrically, directly from the definitions of the sine and cosine. See Problems 78, 79, and 80 of the preceding section.)

To obtain the derivative of $\cos\theta$, start with the identity

$$\cos\theta = \sin\left(\frac{\pi}{2} - \theta\right)$$

and differentiate using the chain rule for the sine to get

$$\frac{d}{d\theta} \cos \theta = (-1) \cos \left(\frac{\pi}{2} - \theta\right)$$

Then apply the identity $\sin \theta = \cos \left(\frac{\pi}{2} - \theta\right)$

to the right-hand side to conclude that

$$\frac{d}{d\theta} \cos \theta = -\sin \theta$$

Problems 12.2

In Problems 1 through 20, differentiate the given function.

1. $f(\theta) = \sin 3\theta$
2. $f(\theta) = \cos 2\theta$
3. $f(\theta) = \sin (1 - 2\theta)$
4. $f(\theta) = \sin \theta^2$
5. $f(\theta) = \cos (\theta^3 + 1)$
6. $f(\theta) = \sin^2 \theta$
7. $f(\theta) = \cos^2 \left(\frac{\pi}{2} - \theta\right)$
8. $f(\theta) = \sin (2\theta + 1)^2$
9. $f(\theta) = \cos (1 + 3\theta)^2$
10. $f(\theta) = e^{-\theta} \sin \theta$
11. $f(\theta) = e^{-\theta/2} \cos 2\pi\theta$
12. $f(\theta) = \dfrac{\cos \theta}{1 - \cos \theta}$
13. $f(\theta) = \dfrac{\sin \theta}{1 + \sin \theta}$
14. $f(\theta) = \tan (5\theta + 2)$
15. $f(\theta) = \tan (1 - \theta^5)$
16. $f(\theta) = \tan^2 \theta$
17. $f(\theta) = \tan^2 \left(\frac{\pi}{2} - 2\pi\theta\right)$
18. $f(\theta) = \tan (\pi - 4\theta)^2$
19. $f(\theta) = \ln \sin^2 \theta$
20. $f(\theta) = \ln \tan^2 \theta$

21. Show that $\dfrac{d}{d\theta} \sec \theta = \sec \theta \tan \theta$.

22. Show that $\dfrac{d}{d\theta} \csc \theta = -\csc \theta \cot \theta$.

23. Show that $\dfrac{d}{d\theta} \cot \theta = -\csc^2 \theta$.

24. Use the fact that $\qquad \lim_{\theta \to 0} \dfrac{\sin \theta}{\theta} = 1$

to prove that $\qquad \lim_{\theta \to 0} \dfrac{\cos \theta - 1}{\theta} = 0$

Hint: Write

$$\frac{\cos\theta - 1}{\theta} = \frac{\cos\theta - 1}{\theta} \cdot \frac{\cos\theta + 1}{\cos\theta + 1}$$

and multiply out the terms on the right-hand side. Simplify the resulting expression using an appropriate trigonometric identity and then let θ approach zero, keeping in mind that

$$\lim_{\theta \to 0} \sin\theta = 0 \quad \text{and} \quad \lim_{\theta \to 0} \cos\theta = 1$$

25. Use the definition of the derivative to prove that $\dfrac{d}{d\theta}\cos\theta = -\sin\theta$.

3   Applications of Trigonometric Functions

In this section, you will see a sampling of applications involving trigonometric functions and their derivatives. The first example is a rate-of-change problem, and the subsequent examples are optimization problems.

Related Rates   In the first example, the goal is to find the rate of change of an angle θ with respect to time t. An explicit formula relating θ and t is not given. However, the relationship between θ and another variable x is known, as is the rate of change of x with respect to t. Using the chain rule and implicit differentiation, you will be able to put the information together to find the rate of change of θ with respect to t. Problems of this type are known as **related rates problems**.

EXAMPLE 3.1

An observer watches a plane approach at a speed of 400 miles per hour and at an altitude of 4 miles. At what rate is the angle of elevation of the observer's line of sight changing with respect to time when the horizontal distance between the plane and the observer is 3 miles?

Solution

Let x denote the horizontal distance between the plane and the observer, let t denote time (in hours), and draw a diagram representing the situation as in Figure 12.18.

You know that $\dfrac{dx}{dt} = -400$ (the minus sign indicating that the distance x is decreasing), and your goal is to find $\dfrac{d\theta}{dt}$ when $x = 3$. From the right

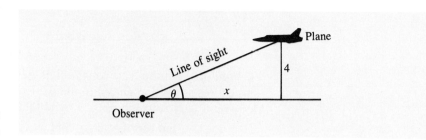

Figure 12.18 Observation of an approaching plane.

triangle in Figure 12.18 you see that

$$\tan \theta = \frac{4}{x}$$

Differentiate both sides of this equation with respect to t (remembering to use the chain rule) to get

$$\sec^2 \theta \, \frac{d\theta}{dt} = -\frac{4}{x^2} \frac{dx}{dt}$$

or

$$\frac{d\theta}{dt} = -\frac{4}{x^2} \cos^2 \theta \, \frac{dx}{dt}$$

The values of x and $\frac{dx}{dt}$ are given. You get the value of $\cos^2 \theta$ when $x = 3$ from the right triangle in Figure 12.19. In particular, when $x = 3$,

$$\cos^2 \theta = \left(\frac{3}{5}\right)^2 = \frac{9}{25}$$

Substituting $x = 3$, $\frac{dx}{dt} = -400$, and $\cos^2 \theta = \frac{9}{25}$ into the formula for $\frac{d\theta}{dt}$, you get

$$\frac{d\theta}{dt} = -\frac{4}{9}\left(\frac{9}{25}\right)(-400) = 64 \text{ radians per hour}$$

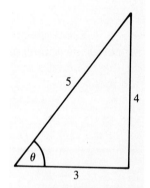

Figure 12.19 Triangle for the computation of $\cos \theta$ when $x = 3$.

Optimization Problems

The next two examples are optimization problems involving trigonometric functions. To solve problems of this type, proceed as in Chapter 3, Section 4. In particular, (1) construct a function representing the quantity to be optimized in terms of a convenient variable, (2) identify an interval on which the function has a practical interpretation, (3) differentiate the function and set the derivative equal to zero to find the critical points, and (4) compare the values of the function at the critical points in the interval and at the endpoints (or use the second derivative test) to verify that the desired absolute extremum has been found.

EXAMPLE 3.2

Two sides of a triangle are 4 inches long. What should the angle between these sides be to make the area of the triangle as large as possible?

Solution

The triangle is shown in Figure 12.20. In general, the area of a triangle is given by the formula

$$\text{Area} = \frac{1}{2}(\text{base})(\text{height})$$

In this case, $\qquad$ Base $= 4$

and, since $\sin \theta = \dfrac{h}{4}$, $\quad$ Height $= h = 4 \sin \theta$

Hence, the area of the triangle is given by the function

$$A(\theta) = \frac{1}{2}(4)(4 \sin \theta) = 8 \sin \theta$$

Since only values of θ between $\theta = 0$ and $\theta = \pi$ radians are meaningful in the context of this problem, the goal is to find the absolute maximum of the function $A(\theta)$ on the interval $0 \le \theta \le \pi$.

The derivative of $A(\theta)$ is

$$A'(\theta) = 8 \cos \theta$$

which is zero on the interval $0 \le \theta \le \pi$ only when $\theta = \dfrac{\pi}{2}$. Comparing

$$A(0) = 8 \sin 0 = 0$$

$$A\left(\frac{\pi}{2}\right) = 8 \sin \frac{\pi}{2} = 8(1) = 8$$

Figure 12.20 Triangle for Example 3.2.

and $$A(\pi) = 8 \sin \pi = 0$$

you conclude that the area is maximized when the angle θ measures $\dfrac{\pi}{2}$ radians (or 90°), that is, when the triangle is a right triangle.

Illumination from a Light Source

EXAMPLE 3.3

A lamp with adjustable height hangs directly above the center of a circular kitchen table that is 8 feet in diameter. The illumination at the edge of the table is directly proportional to the cosine of the angle θ and inversely proportional to the square of the distance d, where θ and d are as shown in Figure 12.21. How close to the table should the lamp be pulled to maximize the illumination at the edge of the table?

Figure 12.21 Hanging lamp and table for Example 3.3.

Solution

Let L denote the illumination at the edge of the table. Then,

$$L = \frac{k \cos \theta}{d^2}$$

where k is a (positive) constant of proportionality. Moreover, from the right triangle in Figure 12.21,

$$\sin \theta = \frac{4}{d} \quad \text{or} \quad d = \frac{4}{\sin \theta}$$

Hence, $$L(\theta) = \frac{k \cos \theta}{(4/\sin \theta)^2} = \frac{k}{16} \cos \theta \sin^2 \theta$$

Only values of θ between 0 and $\dfrac{\pi}{2}$ radians are meaningful in the context of this problem. Hence, the goal is to find the absolute maximum of the function $L(\theta)$ on the interval $0 \le \theta \le \frac{\pi}{2}$.

Using the product rule and chain rule to differentiate $L(\theta)$, you get

$$L'(\theta) = \frac{k}{16}[\cos \theta \, (2 \sin \theta \cos \theta) + \sin^2 \theta \, (-\sin \theta)]$$

$$= \frac{k}{16}(2 \cos^2 \theta \sin \theta - \sin^3 \theta)$$

$$= \frac{k}{16} \sin \theta \, (2 \cos^2 \theta - \sin^2 \theta)$$

which is zero when

$$\sin \theta = 0 \qquad \text{or} \qquad 2 \cos^2 \theta - \sin^2 \theta = 0$$

The only value of θ in the interval $0 \le \theta \le \frac{\pi}{2}$ for which $\sin \theta = 0$ is the endpoint $\theta = 0$. To solve the equation

$$2 \cos^2 \theta - \sin^2 \theta = 0$$

rewrite it as

$$2 \cos^2 \theta = \sin^2 \theta$$

divide both sides by $\cos^2 \theta$

$$2 = \frac{\sin^2 \theta}{\cos^2 \theta} = \left(\frac{\sin \theta}{\cos \theta}\right)^2$$

and take the square root of each side to get

$$\pm\sqrt{2} = \frac{\sin \theta}{\cos \theta} = \tan \theta$$

Since both $\sin \theta$ and $\cos \theta$ are nonnegative on the interval $0 \le \theta \le \frac{\pi}{2}$, you may discard the negative square root and conclude that the critical point occurs when

$$\tan \theta = \sqrt{2}$$

Although it would not be hard to use your calculator (or Table III at the back of the book) to find (or, at least, to estimate) the angle θ for which $\tan \theta = \sqrt{2}$, it is not necessary to do so. Instead, look at the right triangle in Figure 12.22, in which

$$\tan \theta = \frac{\text{opposite}}{\text{adjacent}} = \frac{\sqrt{2}}{1} = \sqrt{2}$$

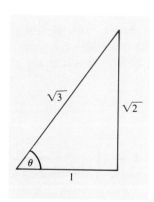

Figure 12.22 Right triangle with $\tan \theta = \sqrt{2}$.

and observe that

$$\sin \theta = \frac{\text{opposite}}{\text{hypotenuse}} = \frac{\sqrt{2}}{\sqrt{3}} \quad \text{and} \quad \cos \theta = \frac{\text{adjacent}}{\text{hypotenuse}} = \frac{1}{\sqrt{3}}$$

Substitute these values into the equation for $L(\theta)$ to conclude that, at the critical point,

$$L(\theta) = \frac{k}{16} \cos \theta \sin^2 \theta = \frac{k}{16}\left(\frac{1}{\sqrt{3}}\right)\left(\frac{\sqrt{2}}{\sqrt{3}}\right)^2 = \frac{k}{24\sqrt{3}}$$

Compare this value with the values of $L(\theta)$ at the endpoints $\theta = 0$ and $\theta = \frac{\pi}{2}$, which are

$$L(0) = \frac{k}{16} (\cos 0)(\sin^2 0) = 0 \quad \text{since } \sin 0 = 0$$

and

$$L\left(\frac{\pi}{2}\right) = \frac{k}{16}\left(\cos \frac{\pi}{2}\right)\left(\sin^2 \frac{\pi}{2}\right) = 0 \quad \text{since } \cos \frac{\pi}{2} = 0$$

The largest of these possible maximum values is $\dfrac{k}{24\sqrt{3}}$, which occurs when $\tan \theta = \sqrt{2}$.

Finally, to find the height h that maximizes the illumination L, observe from Figure 12.21 that

$$\tan \theta = \frac{4}{h}$$

Since $\tan \theta = \sqrt{2}$, it follows that

$$h = \frac{4}{\tan \theta} = \frac{4}{\sqrt{2}} \approx 2.83 \text{ feet}$$

Minimization of Travel Time

In the next example, calculus is used to establish a general principle about the minimization of travel time. As you will see subsequently, the principle can be rephrased to give a well-known law about the reflection of light.

EXAMPLE 3.4

Two off-shore oil wells are, respectively, a and b miles out to sea. A motorboat that travels at a constant speed s transports workers from the first well to the shore and then proceeds to the second well. Show that the total travel time is minimized if the angle α between the motorboat's path of arrival and the shoreline is equal to the angle β between the shoreline and the motorboat's path of departure.

Solution

Even though the goal is to prove that two angles are equal, it turns out that the easiest way to solve this problem is to let the variable x represent a convenient distance and to introduce trigonometry only at the end.

Begin with a sketch of the situation as in Figure 12.23 and define the variable x and the constant distance d as indicated. Then, by the pythagorean theorem,

$$\text{Distance from first well to } P = \sqrt{a^2 + x^2}$$

and

$$\text{Distance from second well to } P = \sqrt{b^2 + (d - x)^2}$$

Since (for constant speeds),

$$\text{Time} = \frac{\text{distance}}{\text{speed}}$$

the total travel time is given by the function

$$T(x) = \frac{\sqrt{a^2 + x^2}}{s} + \frac{\sqrt{b^2 + (d - x)^2}}{s}$$

where $0 \leq x \leq d$.

Then,
$$T'(x) = \frac{x}{s\sqrt{a^2 + x^2}} - \frac{d - x}{s\sqrt{b^2 + (d - x)^2}}$$

If you set $T'(x)$ equal to zero and do a little algebra, you get

$$\frac{x}{\sqrt{a^2 + x^2}} = \frac{d - x}{\sqrt{b^2 + (d - x)^2}}$$

Now look again at the right triangles in Figure 12.23 and observe that

Figure 12.23 Path of motorboat from oil well to shore to oil well.

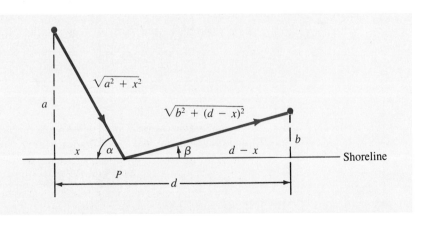

$$\frac{x}{\sqrt{a^2 + x^2}} = \cos \alpha \qquad \text{and} \qquad \frac{d - x}{\sqrt{b^2 + (d - x)^2}} = \cos \beta$$

Hence, $T'(x) = 0$ implies that

$$\cos \alpha = \cos \beta$$

In the context of this problem, both α and β must measure between 0 and $\frac{\pi}{2}$ radians, and for angles in this range, equality of the cosines implies equality of the angles themselves. (See, for example, the graph of the cosine function in Figure 12.10.) Hence, $T'(x) = 0$ implies that

$$\alpha = \beta$$

Moreover, it should be clear from the geometry of Figure 12.23 that no matter what the relative sizes of a, b, and d may be, there is a (unique) point P for which $\alpha = \beta$. It follows that the function $T(x)$ has a critical point in the interval $0 \le x \le d$ and that at this critical point, $\alpha = \beta$.

To verify that this critical point is really the absolute minimum, you can use the second derivative test. A routine calculation gives

$$T''(x) = \frac{a^2}{s(a^2 + x^2)^{3/2}} + \frac{b^2}{s[b^2 + (d - x)^2]^{3/2}}$$

(For practice, check this calculation.) Since $T''(x)$ is positive for all values of x, and since there is only one critical point in the interval $0 \le x \le d$, it follows that this critical point does indeed correspond to the absolute minimum of the total time T on this interval.

Reflection of Light

According to **Fermat's principle** in optics, light traveling from one point to another takes the path that requires the least amount of time. Suppose (as

Figure 12.24
The reflection of light: $\theta_1 = \theta_2$.

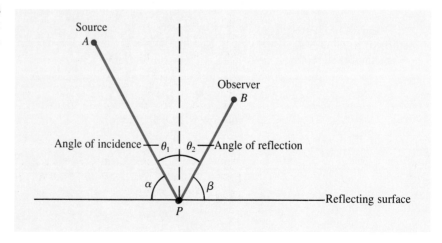

illustrated in Figure 12.24) that a ray of light is transmitted from a source at a point A, strikes a reflecting surface (such as a mirror) at point P, and is subsequently received by an observer at point B.

Since, by Fermat's principle, the path from A to P to B minimizes time, it follows from the calculation in Example 3.4 that the angles α and β are equal. This, in turn, implies the **law of reflection,** which states that the **angle of incidence** (θ_1 in Figure 12.24) must be equal to the **angle of reflection** (θ_2 in Figure 12.24).

Problems 12.3

Related Rates

1. An observer watches a plane approach at a speed of 500 miles per hour and at an altitude of 3 miles. At what rate is the angle of elevation of the observer's line of sight changing with respect to time when the horizontal distance between the plane and the observer is 4 miles?

Related Rates

2. A person 6 feet tall is watching a streetlight 18 feet high while walking toward it at a speed of 5 feet per second. At what rate is the angle of elevation of the person's line of sight changing with respect to time when the person is 9 feet from the base of the light?

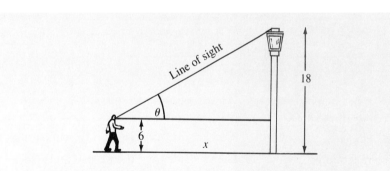

Related Rates

3. An attendant is standing at the end of a pier 12 feet above the water and is pulling a rope attached to a rowboat at the rate of 4 feet of rope per minute. At what rate is the angle that the rope makes with the surface of the water changing with respect to time when the boat is 16 feet from the pier?

Related Rates

4. A revolving searchlight in a lighthouse 2 miles off shore is following a jogger running along the shore. When the jogger is 1 mile from the point on the shore that is closest to the lighthouse, the searchlight is turning at the rate of 0.25 revolution per hour. How fast is the jogger running at that moment? (*Hint:* Since 0.25 revolution per hour is $\dfrac{\pi}{2}$

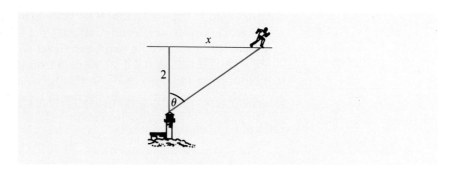

radians per hour, the problem is to find $\dfrac{dx}{dt}$ for the accompanying

picture when $x = 1$ and $\dfrac{d\theta}{dt} = \dfrac{\pi}{2}$.)

Maximization of
Volume

5. You have a piece of metal that is 20 meters long and 6 meters wide, which you are going to bend to form a trough as indicated in the following diagram. At what angle should the sides meet so that the volume of the trough is as large as possible? (*Hint:* The volume is the length of the trough times the area of its triangular cross section.)

Maximization of Area

6. Prove that of all isosceles triangles whose equal sides are of a specified length, the triangle of greatest area is the right triangle.

Maximization of Area

7. The two sides and the base of an isosceles trapezoid are each 5 inches long. At what angle should the sides meet the horizontal top to

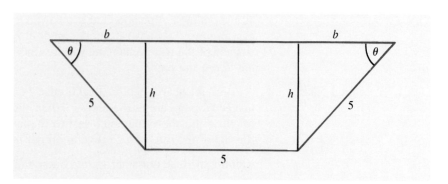

maximize the area of the trapezoid? [*Hints:* (1) As indicated in the accompanying picture, the area of the trapezoid is the sum of the areas of two right triangles and the area of a rectangle. (2) You should be able to factor the derivative of the area function if you first simplify it using an appropriate trigonometric identity.]

Illumination from a Light Source

8. In Example 3.3, an alternative characterization of the illumination is that it is directly proportional to the sine of the angle ϕ and inversely proportional to the square of the distance d, where ϕ is the angle at which the ray of light meets the table. (That is, ϕ is the angle in the left-hand corner of the triangle in Figure 12.21 between the hypotenuse labeled d and the base labeled 4.) For practice, do Example 3.3 again, this time using this alternative characterization of the illumination. (Your final answer, of course, should be the same as before.)

Minimization of Horizontal Clearance

9. Find the length of the longest pipe that can be carried horizontally around a corner joining two corridors that are $2\sqrt{2}$ feet wide. [*Hint:* Show that the horizontal clearance $C = x + y$ can be written as

$$C(\theta) = \frac{2\sqrt{2}}{\sin \theta} + \frac{2\sqrt{2}}{\cos \theta}$$

and find the absolute *minimum* of $C(\theta)$ on an appropriate interval.]

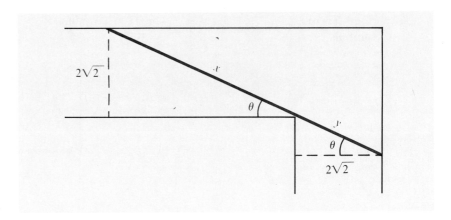

Refraction of Light

10. The accompanying figure shows a ray of light emitted from a source at point A under water and subsequently received by an observer at point B above the surface of the water. If v_1 is the speed of light in water and v_2 is the speed of light in air, show that

$$\frac{\sin \theta_1}{\sin \theta_2} = \frac{v_1}{v_2}$$

where θ_1 is the **angle of incidence** and θ_2 is the **angle of refraction.** (*Hint:* Apply Fermat's principle, which states that light takes the path

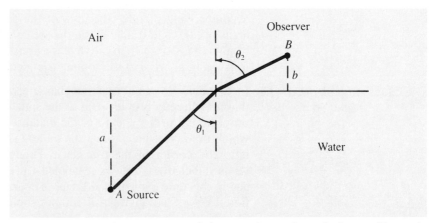

requiring the least amount of time, and use the solution to Example 3.4 as a guide. You may assume without proof that the total time is minimized when its derivative is equal to zero.)

Chapter Summary and Review Problems

Important Terms, Symbols, and Formulas

Angle measurement: $\dfrac{\text{degrees}}{180} = \dfrac{\text{radians}}{\pi}$

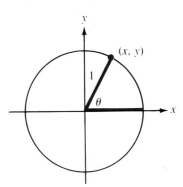

Sine and cosine: $\sin \theta = y$ $\cos \theta = x$

Important values:

θ	0	$\dfrac{\pi}{6}$	$\dfrac{\pi}{4}$	$\dfrac{\pi}{3}$	$\dfrac{\pi}{2}$	$\dfrac{2\pi}{3}$	$\dfrac{3\pi}{4}$	$\dfrac{5\pi}{6}$	π
$\sin \theta$	0	$\dfrac{1}{2}$	$\dfrac{\sqrt{2}}{2}$	$\dfrac{\sqrt{3}}{2}$	1	$\dfrac{\sqrt{3}}{2}$	$\dfrac{\sqrt{2}}{2}$	$\dfrac{1}{2}$	0
$\cos \theta$	1	$\dfrac{\sqrt{3}}{2}$	$\dfrac{\sqrt{2}}{2}$	$\dfrac{1}{2}$	0	$-\dfrac{1}{2}$	$-\dfrac{\sqrt{2}}{2}$	$-\dfrac{\sqrt{3}}{2}$	-1

Other trigonometric functions:

$$\tan \theta = \frac{\sin \theta}{\cos \theta} \qquad \cot \theta = \frac{\cos \theta}{\sin \theta}$$

$$\sec \theta = \frac{1}{\cos \theta} \qquad \csc \theta = \frac{1}{\sin \theta}$$

Trigonometric identities:

Periodicity: $\sin (\theta + 2\pi) = \sin \theta \qquad \cos (\theta + 2\pi) = \cos \theta$

Negative angles: $\sin (-\theta) = -\sin \theta \qquad \cos (-\theta) = \cos \theta$

Pythagorean identity: $\sin^2 \theta + \cos^2 \theta = 1$

Addition formulas:

$$\cos (a + b) = \cos a \cos b - \sin a \sin b$$

$$\sin (a + b) = \sin a \cos b + \cos a \sin b$$

Double-angle formulas:

$$\cos 2\theta = \cos^2 \theta - \sin^2 \theta$$

$$\sin 2\theta = 2 \sin \theta \cos \theta$$

Right triangles:

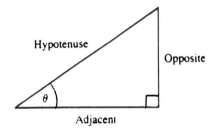

$$\sin \theta = \frac{\text{opposite}}{\text{hypotenuse}}$$

$$\cos \theta = \frac{\text{adjacent}}{\text{hypotenuse}}$$

$$\tan \theta = \frac{\text{opposite}}{\text{adjacent}}$$

Differentiation formulas:

$$\frac{d}{d\theta} \sin \theta = \cos \theta \qquad \frac{d}{d\theta} \cos \theta = -\sin \theta \qquad \frac{d}{d\theta} \tan \theta = \sec^2 \theta$$

Chain rules: $\qquad \dfrac{d}{d\theta} \sin h(\theta) = h'(\theta) \cos h(\theta)$

$$\frac{d}{d\theta} \cos h(\theta) = -h'(\theta) \sin h(\theta)$$

$$\frac{d}{d\theta} \tan h(\theta) = h'(\theta) \sec^2 h(\theta)$$

Review Problems

1. Specify the radian measurement and degree measurement for each of the following angles.

 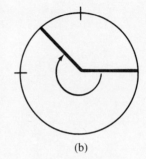

(a)   (b)

2. Convert $50°$ to radians.

3. Convert 0.25 radian to degrees.

4. Evaluate the given expressions without using Table III or your calculator.

   (a) $\sin\left(-\dfrac{5\pi}{3}\right)$ (b) $\cos\dfrac{15\pi}{4}$

   (c) $\sec\dfrac{7\pi}{3}$ (d) $\cot\dfrac{2\pi}{3}$

5. Find $\tan \theta$ if $\sin \theta = \dfrac{4}{5}$.

6. Find $\csc \theta$ if $\cot \theta = \dfrac{\sqrt{5}}{2}$.

In Problems 7 through 10, find all the values of θ in the specified interval that satisfy the given equation.

7. $2 \cos \theta + \sin 2\theta = 0; \; 0 \le \theta \le 2\pi$

8. $3 \sin^2 \theta - \cos^2 \theta = 2; \; 0 \le \theta \le \pi$

9. $2 \sin^2 \theta = \cos 2\theta; \; 0 \le \theta \le \pi$

10. $5 \sin \theta - 2 \cos^2 \theta = 1; \; 0 \le \theta \le 2\pi$

11. Starting with the identity $\sin^2 \theta + \cos^2 \theta = 1$, derive the identity $1 + \cot^2 \theta = \csc^2 \theta$.

12. Starting with the double-angle formulas for the sine and cosine, derive the double-angle formula

$$\tan 2\theta = \frac{2 \tan \theta}{1 - \tan^2 \theta}$$

for the tangent.

13. (a) Starting with the addition formulas for the sine and cosine, derive the identities

$$\cos\left(\frac{\pi}{2} + \theta\right) = -\sin\theta \quad \text{and} \quad \sin\left(\frac{\pi}{2} + \theta\right) = \cos\theta$$

(b) Give geometric arguments to establish the identities in part (a).

In Problems 14 through 19, differentiate the given function.

14. $f(\theta) = \sin(3\theta + 1)^2$

15. $f(\theta) = \cos^2(3\theta + 1)$

16. $f(\theta) = \tan(3\theta^2 + 1)$

17. $f(\theta) = \tan^2(3\theta^2 + 1)$

18. $f(\theta) = \dfrac{\sin\theta}{1 - \cos\theta}$

19. $f(\theta) = \ln\cos^2\theta$

20. Show that $\dfrac{d}{d\theta}\sin\theta\cos\theta = \cos 2\theta$.

21. On New Year's Eve, a holiday reveler is watching the descent of a lighted ball from atop a tall building that is 600 feet away. The ball is falling at the rate of 20 feet per minute. At what rate is the angle of elevation of the reveler's line of sight changing with respect to time when the ball is 800 feet from the ground?

22. A trough 9 meters long is to have a cross section consisting of an isosceles trapezoid in which the base and two sides are all 4 meters long. At what angle should the sides of the trapezoid meet the horizontal top to maximize the capacity of the trough?

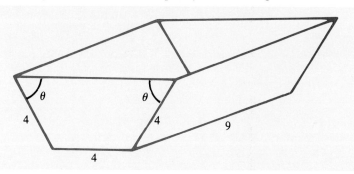

23. Find the length of the longest pipe that can be carried horizontally around a corner joining a corridor that is 8 feet wide and one that is $5\sqrt{5}$ feet wide.

24. A cable is run in a straight line from a power plant on one side of a

river 900 meters wide to a point P on the other side and then along the river bank to a factory, 3,000 meters downstream from the power plant. The cost of running the cable under the water is $5 per meter, while the cost over land is $4 per meter. If θ is the (smaller) angle between the segment of cable under the river and the opposite bank, show that $\cos \theta = \dfrac{4}{5}$ (the ratio of the per-meter costs) for the route that minimizes the total installation cost. (You may assume without proof that the absolute minimum occurs when the derivative of the cost function is zero.)

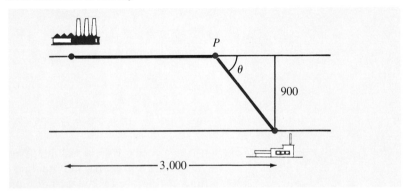

Appendix: Algebra Review

This appendix contains a review of certain algebraic topics that are used throughout the book. A set of practice problems appears at the end, and the answers to the odd-numbered problems are given at the back of the book.

The Real Numbers
An **integer** is a "whole number," either positive or negative. For example, 1, 2, 875, -15, -83, and 0 are integers, while $\frac{2}{3}$, 8.71, and $\sqrt{2}$ are not.

A **rational number** is a number that can be expressed as the quotient $\frac{a}{b}$ of two integers, where $b \neq 0$. For example,

$$\frac{2}{3} \qquad \frac{8}{3} \qquad -6\frac{1}{2} = \frac{-13}{2} \qquad \text{and} \qquad 0.25 = \frac{25}{100} = \frac{1}{4}$$

are rational numbers. Every integer is a rational number since it can be expressed as itself divided by 1. When expressed in decimal form, rational numbers are either terminating or infinitely repeating decimals. For example,

$$\frac{5}{8} = 0.625 \qquad \frac{1}{3} = 0.333 \ldots \qquad \text{and} \qquad \frac{13}{11} = 1.181818 \ldots$$

A number that cannot be expressed as the quotient of two integers is called an **irrational number.** For example,

$$\sqrt{2} \simeq 1.41421356 \qquad \text{and} \qquad \pi \simeq 3.14159265$$

are irrational numbers.

The rational numbers and irrational numbers form the **real numbers** and can be visualized geometrically as points on a **number line** as illustrated in Figure A.1.

Inequalities If a and b are real numbers and a is to the right of b on the number line, we say that a **is greater than** b and write $a > b$. If a is to the left of b, we say that a **is less than** b and write $a < b$ (Figure A.2). For example,

$$5 > 2 \qquad -12 < 0 \qquad \text{and} \qquad -8.2 < -2.4$$

Figure A.2 Inequalities.

Moreover,

$$\frac{6}{7} < \frac{7}{8}$$

as you can see by writing

$$\frac{6}{7} = \frac{48}{56} \qquad \text{and} \qquad \frac{7}{8} = \frac{49}{56}$$

The symbol $\geq$ stands for **greater than or equal to,** and the symbol $\leq$ stands for **less than or equal to.** Thus, for example,

$$-3 \geq -4 \qquad -3 \geq -3 \qquad -4 \leq -3 \qquad \text{and} \qquad -4 \leq -4$$

Intervals A set of real numbers that can be represented on the number line by a line segment is called an **interval.** Inequalities can be used to describe intervals. For example, the interval $a \leq x < b$ consists of all real numbers x that are between a and b, including a but excluding b. This interval is shown in Figure A.3. The numbers a and b are known as the **endpoints** of the interval. The square bracket at a indicates that a is included in the interval, while the rounded bracket at b indicates that b is excluded.

Figure A.3 The interval $a \leq x < b$.

Intervals may be finite or infinite in extent and may or may not contain any endpoints. The possibilities (including customary notation and terminology) are illustrated in Figure A.4.

Figure A.4 Intervals of real numbers.

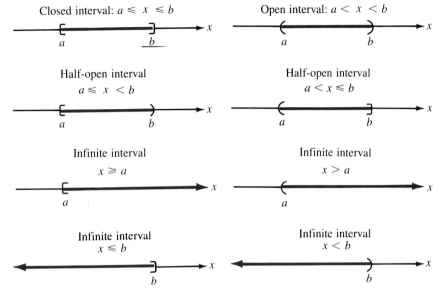

EXAMPLE A.1

Use inequalities to describe the following intervals.

(a)

(b)

(c)

Solution

(a) $x \leq 3$ (b) $x > -2$ (c) $-2 < x \leq 3$

EXAMPLE A.2

Represent each of the following intervals as a line segment on a number line.

(a) $x < -1$ (b) $-1 \le x \le 2$ (c) $x > 2$

Solution

(a)

(b)

(c)

Absolute Value

The **absolute value** of a real number is its nonnegative value or magnitude. Thus, the absolute value of a nonnegative number is the number itself, while the absolute value of a negative number is the number without the minus sign. The absolute value of a number x is denoted by $|x|$. Here is the formal definition.

> **Absolute Value**
>
> For any real number x, the absolute value of x is
>
> $$|x| = \begin{cases} x & \text{if } x \ge 0 \\ -x & \text{if } x < 0 \end{cases}$$

For example,

$$|-3| = 3 \qquad |3| = 3 \qquad |0| = 0 \qquad \text{and} \qquad |5 - 9| = |-4| = 4$$

Distance

The absolute value of a number can be interpreted geometrically as its distance from zero on the number line. More generally, the distance between any two numbers is the absolute value of their difference (taken in either order). The situation is illustrated in Figure A.5.

Figure A.5 Distance between a and $b = |a - b|$ $= |b - a|$.

EXAMPLE A.3

Find the distance on the number line between -2 and 3.

Solution

The distance between two numbers is the absolute value of their difference. Hence,

$$\text{Distance} = |-2 - 3| = |-5| = 5$$

The situation is illustrated in Figure A.6.

Figure A.6 Distance between -2 and 3.

Absolute Values and Intervals   The geometric interpretation of absolute value as distance can be used to simplify certain algebraic inequalities involving absolute values. Here is an example.

EXAMPLE A.4

Find the interval consisting of all real numbers x such that $|x - 1| \leq 3$.

Solution

In geometric terms, the numbers x for which $|x - 1| \leq 3$ are those whose distance from 1 is less than or equal to 3. As illustrated in Figure A.7, these are the numbers in the interval $-2 \leq x \leq 4$.

Figure A.7 The interval on which $|x - 1| \leq 3$ is $-2 \leq x \leq 4$.

To find this interval algebraically, without relying on the geometry, rewrite the inequality

$$|x - 1| \leq 3$$

as

$$-3 \leq x - 1 \leq 3$$

and add 1 to each term to get

$$-3 + 1 \leq x - 1 + 1 \leq 3 + 1$$

or

$$-2 \leq x \leq 4$$

Exponential Notation   The following rules define the expression a^x for $a > 0$ and all rational values of x.

> **Definition of a^x for $x > 0$**
>
> Integer powers: If n is a positive integer,
>
> $$a^n = a \cdot a \cdots a$$
>
> where the product $a \cdot a \cdots a$ contains n factors.
>
> Fractional powers: If n and m are positive integers,
>
> $$a^{n/m} = (\sqrt[m]{a})^n$$
>
> where $\sqrt[m]{}$ denotes the positive mth root.
>
> Negative powers: $a^{-x} = \dfrac{1}{a^x}$
>
> Zero power: $a^0 = 1$

EXAMPLE A.5

Evaluate the following expressions (without using your calculator).

(a) $9^{1/2}$ (b) $27^{2/3}$ (c) $8^{-1/3}$ (d) $\left(\dfrac{1}{100}\right)^{-3/2}$ (e) 5^0

Solution

(a) $9^{1/2} = \sqrt{9} = 3$

(b) $27^{2/3} = (\sqrt[3]{27})^2 = 3^2 = 9$

(c) $8^{-1/3} = \dfrac{1}{8^{1/3}} = \dfrac{1}{\sqrt[3]{8}} = \dfrac{1}{2}$

(d) $\left(\dfrac{1}{100}\right)^{-3/2} = 100^{3/2} = (\sqrt{100})^3 = 10^3 = 1,000$

(e) $5^0 = 1$

Laws of Exponents   Exponents obey the following useful laws.

> **Laws of Exponents**
>
> The product law: $a^r a^s = a^{r+s}$
>
> The quotient law: $\dfrac{a^r}{a^s} = a^{r-s}$
>
> The power law: $(a^r)^s = a^{rs}$

The use of these laws is illustrated in the next two examples.

EXAMPLE A.6

Evaluate the following expressions (without using a calculator).

(a) $(2^{-2})^3$ (b) $\dfrac{3^3}{3^{1/3}(3^{2/3})}$ (c) $2^{7/4}(8^{-1/4})$

Solution

(a) $(2^{-2})^3 = 2^{-6} = \dfrac{1}{2^6} = \dfrac{1}{64}$

(b) $\dfrac{3^3}{3^{1/3}(3^{2/3})} = \dfrac{3^3}{3^{1/3+2/3}} = \dfrac{3^3}{3^1} = 3^2 = 9$

(c) $2^{7/4}(8^{-1/4}) = 2^{7/4}(2^3)^{-1/4} = 2^{7/4}(2^{-3/4}) = 2^{7/4-3/4} = 2^1 = 2$

EXAMPLE A.7

Solve each of the following equations for n.

(a) $\dfrac{a^5}{a^2} = a^n$ (b) $(a^n)^5 = a^{20}$

Solution

(a) Since $\dfrac{a^5}{a^2} = a^{5-2} = a^3$, it follows that $n = 3$.

(b) Since $(a^n)^5 = a^{5n}$, it follows that $5n = 20$ or $n = 4$.

Factoring   To factor an expression is to write it as a product of two or more terms, called **factors.** Factoring is used to simplify complicated expressions and to solve equations, and is based on the distributive law for addition and multiplication.

> **The Distributive Law**
>
> For any real numbers a, b, and c,
> $$ab + ac = a(b + c)$$

The factoring techniques you will need in this book are illustrated in the following examples.

Factoring Out Common Terms

EXAMPLE A.8

Factor the expression $3x^4 - 6x^3$.

Solution

Since $3x^3$ is a factor of each of the terms in this expression, you can use the distributive law to "factor out" $3x^3$ and write
$$3x^4 - 6x^3 = 3x^3(x - 2)$$

EXAMPLE A.9

Factor the expression $1,000 - 200(x - 5)$.

Solution

The greatest factor common to the terms $1,000$ and $200(x - 5)$ is 200. Factor this out to get

$$1,000 - 200(x - 5) = 200[5 - (x - 5)]$$
$$= 200(5 - x + 5)$$
$$= 200(10 - x)$$

EXAMPLE A.10

Simplify the expression $10(1 - x)^4(x + 1)^4 + 8(x + 1)^5(1 - x)^3$.

Solution

The greatest common factor is $2(1 - x)^3(x + 1)^4$. Factor this out to get

$10(1 - x)^4(x + 1)^4 + 8(x + 1)^5(1 - x)^3$
$$= 2(1 - x)^3(x + 1)^4[5(1 - x) + 4(x + 1)]$$

Since no further factorization is possible, do the multiplication in the square brackets and combine the resulting terms to conclude that

$10(1 - x)^4(x + 1)^4 + 8(x + 1)^5(1 - x)^3$
$$= 2(1 - x)^3(x + 1)^4(5 - 5x + 4x + 4)$$
$$= 2(1 - x)^3(x + 1)^4(9 - x)$$

Simplification of Quotients by Factoring and Canceling

The next example illustrates how you can combine factoring and canceling to simplify certain types of quotients that arise frequently in calculus.

EXAMPLE A.11

Simplify the quotient

$$\frac{4(x + 3)^4(x - 2)^2 - 6(x + 3)^3(x - 2)^3}{(x + 3)(x - 2)^3}$$

Solution

First simplify the numerator to get

$$\frac{4(x + 3)^4(x - 2)^2 - 6(x + 3)^3(x - 2)^3}{(x + 3)(x - 2)^3}$$

$$= \frac{2(x + 3)^3(x - 2)^2[2(x + 3) - 3(x - 2)]}{(x + 3)(x - 2)^3}$$

$$= \frac{2(x + 3)^3(x - 2)^2(2x + 6 - 3x + 6)}{(x + 3)(x - 2)^3}$$

$$= \frac{2(x + 3)^3(x - 2)^2(12 - x)}{(x + 3)(x - 2)^3}$$

and then "cancel" the common factor of $(x + 3)(x - 2)^2$ from the numerator and denominator to conclude that

$$\frac{4(x + 3)^4(x - 2)^2 - 6(x + 3)^3(x - 2)^3}{(x + 3)(x - 2)^3} = \frac{2(x + 3)^2(12 - x)}{x - 2}$$

Polynomials   A **polynomial** is an expression of the form

$$a_0 + a_1x + a_2x^2 + \cdots + a_nx^n$$

where n is a nonnegative integer and $a_0, a_1, a_2, \ldots, a_n$ are real numbers known as the **coefficients** of the polynomial. If $a_n \neq 0$, n is said to be the **degree** of the polynomial. For example, $3x^5 - 7x^2 + 12$ is a polynomial of degree 5.

Factoring Polynomials with Integer Coefficients   Many of the polynomials that arise in practice have integer coefficients (or are closely related to polynomials that do). Techniques for factoring polynomials with integer coefficients are illustrated in the following examples. In each, the goal will be to rewrite the given polynomial as a product of polynomials of lower degree that also have integer coefficients.

EXAMPLE A.12

Factor the polynomial $x^2 - 2x - 3$ using integer coefficients.

Solution

The goal is to write the polynomial as a product of the form

$$x^2 - 2x - 3 = (x + a)(x + b)$$

where a and b are integers. The distributive law implies that

$$(x + a)(x + b) = x^2 + (a + b)x + ab$$

Hence, the goal is to find integers a and b such that

$$x^2 - 2x - 3 = x^2 + (a + b)x + ab$$

or, equivalently, such that

$$a + b = -2 \quad \text{and} \quad ab = -3$$

From the list

$$1, -3 \quad \text{and} \quad -1, 3$$

of pairs of integers whose product is -3, choose $a = -3$ and $b = 1$ as the only pair whose sum is -2. It follows that

$$x^2 - 2x - 3 = (x - 3)(x + 1)$$

which you should check by multiplying out the right-hand side.

EXAMPLE A.13

Factor the polynomial $x^3 - 8$ using integer coefficients.

Solution

The fact that $2^3 = 8$ tells you that $x - 2$ must be a factor of this expression. That is, there are integers a and b for which

$$x^3 - 8 = (x - 2)(x^2 + ax + b)$$

Since

$$(x - 2)(x^2 + ax + b) = x^3 + (a - 2)x^2 + (b - 2a)x - 2b$$

(which you should check for yourself), the goal is to find integers a and b for which

$$a - 2 = 0 \quad b - 2a = 0 \quad \text{and} \quad 2b = 8$$

Clearly, the only such integers are $a = 2$ and $b = 4$. Hence,

$$x^3 - 8 = (x - 2)(x^2 + 2x + 4)$$

Convince yourself by examining pairs of integers whose product is 4 that the polynomial $x^2 + 2x + 4$ cannot be factored further with integer coefficients.

The Difference of Two Squares

Here is the formula (which you can verify by multiplication) for the factorization of the difference between two perfect squares. It is particularly useful, and you should memorize it.

> **The Difference of Two Squares**
>
> For any real numbers a and b,
> $$a^2 - b^2 = (a + b)(a - b)$$

The use of this formula is illustrated in the next example.

EXAMPLE A.14

Factor the polynomial $x^5 - 4x^3$ using integer coefficients.

Solution

First factor out x^3 to get
$$x^5 - 4x^3 = x^3(x^2 - 4)$$
and then factor $x^2 - 4$ (which is the difference of two squares) to conclude that
$$x^5 - 4x^3 = x^3(x + 2)(x - 2)$$

The Solution of Equations by Factoring

The **solutions** of an equation are the values of the variable that make the equation true. For example, $x = 2$ is a solution of the equation
$$x^3 - 6x^2 + 12x - 8 = 0$$
because substitution of 2 for x gives
$$2^3 - 6(2^2) + 12(2) - 8 = 8 - 24 + 24 - 8 = 0$$

In the following two examples, you will see how factoring can be used to solve certain equations. The technique is based on the fact that if the product of two (or more) terms is equal to zero, then at least one of the terms must be equal to zero. For example, if $ab = 0$, then either $a = 0$ or $b = 0$ (or both).

EXAMPLE A.15

Solve the equation $x^2 - 3x = 10$.

Solution

First subtract 10 from both sides to get

$$x^2 - 3x - 10 = 0$$

and then factor the resulting polynomial on the left-hand side to get

$$(x - 5)(x + 2) = 0$$

Since the product $(x - 5)(x + 2)$ can be zero only if one (or both) of its factors is zero, it follows that the solutions are $x = 5$ (which makes the first factor zero) and $x = -2$ (which makes the second factor zero).

EXAMPLE A.16

Solve the equation $1 - \dfrac{1}{x} - \dfrac{2}{x^2} = 0$.

Solution

Put the fractions on the left-hand side over a common denominator and add to get

$$\frac{x^2}{x^2} - \frac{x}{x^2} - \frac{2}{x^2} = 0$$

or

$$\frac{x^2 - x - 2}{x^2} = 0$$

Now factor the polynomial in the numerator to get

$$\frac{(x + 1)(x - 2)}{x^2} = 0$$

Since a quotient is zero only if its numerator is zero, it follows that the solutions are $x = -1$ and $x = 2$.

The Solution of Quadratic Equations

An equation of the form

$$ax^2 + bx + c = 0 \qquad \text{(for } a \neq 0\text{)}$$

is said to be a **quadratic equation.** A quadratic equation can have at most two solutions. As you have seen, one way to find the solutions is to factor the equation. When the factors are not obvious or when the equation cannot be factored at all, you can use the following special formula to solve quadratic equations.

The Quadratic Formula

The solutions of the quadratic equation

$$ax^2 + bx + c = 0 \quad \text{(for } a \neq 0\text{)}$$

are given by the formula

$$x = \frac{-b \pm \sqrt{b^2 - 4ac}}{2a}$$

The term $b^2 - 4ac$ in the quadratic formula is called the **discriminant** of the quadratic equation. If the discriminant is positive, the equation has two solutions, one coming from the formula with the sign $\pm$ replaced by $+$ and the other with $\pm$ replaced by $-$. If the discriminant is zero, the equation has only one solution since the formula reduces to $x = \frac{-b}{2a}$. If the discriminant is negative, the equation has no real solutions since negative numbers do not have real square roots.

The use of the quadratic formula is illustrated in the following examples.

EXAMPLE A.17

Solve the equation $x^2 + 3x + 1 = 0$.

Solution

This is a quadratic equation with $a = 1$, $b = 3$, and $c = 1$. Using the quadratic formula, you get

$$x = \frac{-3 + \sqrt{5}}{2} \simeq -0.38 \quad \text{and} \quad x = \frac{-3 - \sqrt{5}}{2} \simeq -2.62$$

EXAMPLE A.18

Solve the equation $x^2 + 18x + 81 = 0$.

Solution

This is a quadratic equation with $a = 1$, $b = 18$, and $c = 81$. Using the quadratic formula, you find that the discriminant is zero and that the formula for x gives

$$x = \frac{-18 \pm \sqrt{0}}{2} = -\frac{18}{2} = -9.$$

EXAMPLE A.19

Solve the equation $x^2 + x + 1 = 0$.

Solution

This is a quadratic equation with $a = 1$, $b = 1$, and $c = 1$. Using the quadratic formula, you get

$$x = \frac{-1 \pm \sqrt{-3}}{2}$$

Since there is no real square root of -3, it follows that the equation has no real solution.

Systems of Equations   A collection of equations that are to be solved simultaneously is called a **system of equations.** Some of the calculus problems in Chapter 9 involve the solution of systems of two (or more) equations in two (or more) unknowns. A typical example is to find the real numbers x and y that satisfy the system

$$2x + 3y = 5$$
$$x + 2y = 4$$

The procedure for solving a system of two equations in two unknowns is to (temporarily) eliminate one of the variables, thereby reducing the problem to a single equation in one variable, which you then solve for its variable. Once you have found the value of one of the variables, you can substitute it into either of the original equations and solve to get the value of the other variable.

The most common techniques for the elimination of variables are illustrated in the next two examples.

Elimination by Multiplication and Addition   ## EXAMPLE A.20

Solve the system

$$4x + 3y = 13$$
$$3x + 2y = 7$$

Solution

To eliminate y, multiply both sides of the first equation by 2 and both sides of the second equation by -3 so that the system becomes

$$8x + 6y = 26$$
$$-9x - 6y = -21$$

and add the equations to get

$$-x + 0 = 5 \quad \text{or} \quad x = -5$$

To find y, you can substitute $x = -5$ into either of the original equations. If you choose the second equation, you get

$$3(-5) + 2y = 7 \quad 2y = 22 \quad \text{or} \quad y = 11$$

That is, the solution of the system is $x = -5$ and $y = 11$.

To check this answer, substitute $x = -5$ and $y = 11$ into each of the original equations. From the first equation you get

$$4(-5) + 3(11) = -20 + 33 = 13$$

and from the second equation you get

$$3(-5) + 2(11) = -15 + 22 = 7$$

as required.

Elimination by Substitution

EXAMPLE A.21

Solve the system

$$2y^2 - x^2 = 14$$
$$x - y = 1$$

Solution

Solve the second equation for x to get

$$x = y + 1$$

and substitute this into the first equation to eliminate x. This gives

$$2y^2 - (y + 1)^2 = 14$$
$$2y^2 - (y^2 + 2y + 1) = 14$$
$$2y^2 - y^2 - 2y - 1 = 14$$
$$y^2 - 2y - 15 = 0$$

or

$$(y + 3)(y - 5) = 0$$

from which it follows that

$$y = -3 \quad \text{or} \quad y = 5$$

If $y = -3$, the second equation gives

$$x - (-3) = 1 \quad \text{or} \quad x = -2$$

and if $y = 5$, the second equation gives

$$x - 5 = 1 \quad \text{or} \quad x = 6$$

Hence the system has two solutions,

$$x = 6, y = 5 \quad \text{and} \quad x = -2, y = -3$$

To check these answers, substitute each pair x, y into the first equation. If $x = 6$ and $y = 5$, you get

$$2(5^2) - 6^2 = 50 - 36 = 14$$

and if $x = -2$ and $y = -3$, you get

$$2(-3)^2 - (-2)^2 = 18 - 4 = 14$$

as required.

Summation Notation   Sums of the form

$$a_1 + a_2 + \cdots + a_n$$

appear in Chapters 6 and 11. To describe such a sum, it suffices to specify the general term a_j and to indicate that n terms of this form are to be added, starting with the term in which $j = 1$ and ending with the term in which $j = n$. It is customary to use the Greek letter Σ (sigma) to denote summation and to express the sum compactly as follows.

Summation Notation

$$a_1 + a_2 + \cdots + a_n = \sum_{j=1}^{n} a_j$$

The use of summation notation is illustrated in the following examples.

EXAMPLE A.22

Use summation notation to express the following sums.

(a) $1 + 4 + 9 + 16 + 25 + 36 + 49 + 64$

(b) $(1 - x_1)^2 \, \Delta x + (1 - x_2)^2 \, \Delta x + \cdots + (1 - x_{15})^2 \, \Delta x$

Solution

(a) This is a sum of 8 terms of the form j^2, starting with $j = 1$ and ending with $j = 8$. Hence,

$$1 + 4 + 9 + 16 + 25 + 36 + 49 + 64 = \sum_{j=1}^{8} j^2$$

(b) The jth term of this sum is $(1 - x_j)^2 \, \Delta x$. Hence,

$$(1 - x_1)^2 \, \Delta x + (1 - x_2)^2 \, \Delta x + \cdots + (1 - x_{15})^2 \, \Delta x$$

$$= \sum_{j=1}^{15} (1 - x_j)^2 \, \Delta x$$

EXAMPLE A.23

Evaluate the following sums.

(a) $\displaystyle\sum_{j=1}^{4} (j^2 + 1)$

(b) $\displaystyle\sum_{j=1}^{3} (-2)^j$

Solution

(a) $\displaystyle\sum_{j=1}^{4} (j^2 + 1) = (1^2 + 1) + (2^2 + 1) + (3^2 + 1) + (4^2 + 1)$

$$= 2 + 5 + 10 + 17 = 34$$

(b) $\displaystyle\sum_{j=1}^{3} (-2)^j = (-2)^1 + (-2)^2 + (-2)^3 = -2 + 4 - 8 = -6$

Problems

Intervals   In Problems 1 through 4, use inequalities to describe the given interval.

1.

2.

3.

4.

In Problems 5 through 8, represent the given interval as a line segment on a number line.

5. $x \geq 2$ 6. $-6 \leq x < 4$

7. $-2 < x \leq 0$ 8. $x > 3$

Distance   In Problems 9 through 12, find the distance on the number line between the given pair of real numbers.

9. 0 and -4 10. 2 and 5

11. -2 and 3   12. -3 and -1

Absolute Value and   In Problems 13 through 18, find the interval or intervals consisting of all
Intervals   real numbers x that satisfy the given inequality.

13. $|x| \leq 3$ 14. $|x - 2| \leq 5$

15. $|x + 4| \leq 2$ 16. $|1 - x| < 3$

17. $|x + 2| \geq 5$ 18. $|x - 1| > 3$

Exponential Notation   In Problems 19 through 26, evaluate the given expression without using a calculator.

19. 5^3 20. 2^{-3}

21. $16^{1/2}$ 22. $36^{-1/2}$

23. $8^{2/3}$ 24. $27^{-4/3}$

25. $\left(\dfrac{1}{4}\right)^{1/2}$ 26. $\left(\dfrac{1}{4}\right)^{-3/2}$

In Problems 27 through 34, evaluate the given expression without using a calculator.

27. $\dfrac{2^5(2^2)}{2^8}$ 28. $\dfrac{3^4(3^3)}{(3^2)^3}$

29. $\dfrac{2^{4/3}(2^{5/3})}{2^5}$ 30. $\dfrac{5^{-3}(5^2)}{(5^{-2})^3}$

31. $\dfrac{2(16^{3/4})}{2^3}$ 32. $\dfrac{\sqrt{27}\,(\sqrt{3})^3}{9}$

33. $[\sqrt{8}\,(2^{5/2})]^{-1/2}$ 34. $[\sqrt{27}\,(3^{5/2})]^{1/2}$

In Problems 35 through 42, solve the given equation for n. (Assume $a > 0$ and $a \neq 1$.)

35. $a^3 a^7 = a^n$

36. $\dfrac{a^5}{a^2} = a^n$

37. $a^4 a^{-3} = a^n$

38. $a^2 a^n = \dfrac{1}{a}$

39. $(a^3)^n = a^{12}$

40. $(a^n)^5 = \dfrac{1}{a^{10}}$

41. $a^{3/5} a^{-n} = \dfrac{1}{a^2}$

42. $(a^n)^3 = \dfrac{1}{\sqrt{a}}$

Factoring   In Problems 43 through 48, factor the given expression.

43. $x^5 - 4x^4$

44. $3x^3 - 12x^4$

45. $100 - 25(x - 3)$

46. $60 - 20(4 - x)$

47. $8(x + 1)^3(x - 2)^2 + 6(x + 1)^2(x - 2)^3$

48. $12(x + 3)^5(x - 1)^3 - 8(x + 3)^6(x - 1)^2$

Simplification of Quotients   In Problems 49 through 52, simplify the given quotient by factoring the numerator and then canceling.

49. $\dfrac{(x + 3)^3(x + 1) - (x + 3)^2(x + 1)^2}{(x + 3)(x + 1)}$

50. $\dfrac{3(x - 2)^2(x + 1)^2 - 2(x - 2)(x + 1)^3}{(x - 2)^4}$

51. $\dfrac{4(1 - x)^2(x + 3)^3 + 2(1 - x)(x + 3)^4}{(1 - x)^4}$

52. $\dfrac{6(x + 2)^5(1 - x)^4 - 4(x + 2)^6(1 - x)^3}{(x + 2)^8(1 - x)^2}$

Factoring Polynomials with Integer Coefficients   In Problems 53 through 66, factor the given polynomial using integer coefficients.

53. $x^2 + x - 2$

54. $x^2 + 3x - 10$

55. $x^2 - 7x + 12$

56. $x^2 + 8x + 12$

57. $x^2 - 2x + 1$

58. $x^2 + 6x + 9$

59. $x^2 - 4$

60. $64 - x^2$

61. $x^3 - 1$

62. $x^3 - 27$

63. $x^7 - x^5$

64. $x^3 + 2x^2 + x$

65. $2x^3 - 8x^2 - 10x$

66. $x^4 + 5x^3 - 14x^2$

Solution of Equations
by Factoring

In Problems 67 through 80, solve the given equation by factoring.

67. $x^2 - 2x - 8 = 0$ 68. $x^2 - 4x + 3 = 0$

69. $x^2 + 10x + 25 = 0$ 70. $x^2 + 8x + 16 = 0$

71. $x^2 - 16 = 0$ 72. $x^2 - 25 = 0$

73. $2x^2 + 3x + 1 = 0$ 74. $x^2 - 2x + 1 = 0$

75. $4x^2 + 12x + 9 = 0$ 76. $6x^2 + 7x - 3 = 0$

77. $1 + \dfrac{4}{x} - \dfrac{5}{x^2} = 0$ 78. $\dfrac{9}{x^2} - \dfrac{6}{x} + 1 = 0$

79. $2 + \dfrac{2}{x} - \dfrac{4}{x^2} = 0$ 80. $\dfrac{3}{x^2} - \dfrac{5}{x} - 2 = 0$

Quadratic Formula

In Problems 81 through 86, use the quadratic formula to solve the given equation.

81. $2x^2 + 3x + 1 = 0$ 82. $-x^2 + 3x - 1 = 0$

83. $x^2 - 2x + 3 = 0$ 84. $x^2 - 2x + 1 = 0$

85. $4x^2 + 12x + 9 = 0$ 86. $x^2 + 12 = 0$

Systems of Equations

In Problems 87 through 92, solve the given system of equations.

87. $x + 5y = 13$ 88. $2x - 3y = 4$
    $3x - 10y = -11$ $3x - 5y = 2$

89. $5x - 4y = 12$ 90. $3x^2 - 9y = 0$
    $2x - 3y = 2$ $3y^2 - 9x = 0$

91. $2y^2 - x^2 = 1$ 92. $2x^2 - y^2 = -7$
    $x - 2y = 3$ $2x + y = 1$

Summation Notation

In Problems 93 through 96, evaluate the given sum.

93. $\displaystyle\sum_{j=1}^{4} (3j + 1)$ 94. $\displaystyle\sum_{j=1}^{5} j^2$

95. $\displaystyle\sum_{j=1}^{10} (-1)^j$ 96. $\displaystyle\sum_{j=1}^{5} 2^j$

In Problems 97 through 102, use summation notation to express the given sum.

97. $1 + \dfrac{1}{2} + \dfrac{1}{3} + \dfrac{1}{4} + \dfrac{1}{5} + \dfrac{1}{6}$

98. $3 + 6 + 9 + 12 + 15 + 18 + 21 + 24 + 27 + 30$

99. $2x_1 + 2x_2 + 2x_3 + 2x_4 + 2x_5 + 2x_6$

100. $1 - 1 + 1 - 1 + 1 - 1$

101. $1 - 2 + 3 - 4 + 5 - 6 + 7 - 8$

102. $x - x^2 + x^3 - x^4 + x^5$

Tables

Table I. Powers of e

x	e^x	e^{-x}	x	e^x	e^{-x}	x	e^x	e^{-x}
0.00	1.0000	1.00000	0.50	1.6487	.60653	1.00	2.7183	.36788
0.01	1.0101	0.99005	0.51	1.6653	.60050	1.10	3.0042	.33287
0.02	1.0202	.98020	0.52	1.6820	.59452	1.20	3.3201	.30119
0.03	1.0305	.97045	0.53	1.6989	.58860	1.30	3.6693	.27253
0.04	1.0408	.96079	0.54	1.7160	.58275	1.40	4.0552	.24660
0.05	1.0513	.95123	0.55	1.7333	.57695	1.50	4.4817	.22313
0.06	1.0618	.94176	0.56	1.7507	.57121	1.60	4.9530	.20190
0.07	1.0725	.93239	0.57	1.7683	.56553	1.70	5.4739	.18268
0.08	1.0833	.92312	0.58	1.7860	.55990	1.80	6.0496	.16530
0.09	1.0942	.91393	0.59	1.8040	.55433	1.90	6.6859	.14957
0.10	1.1052	.90484	0.60	1.8221	.54881	2.00	7.3891	.13534
0.11	1.1163	.89583	0.61	1.8404	.54335	3.00	20.086	.04979
0.12	1.1275	.88692	0.62	1.8589	.53794	4.00	54.598	.01832
0.13	1.1388	.87809	0.63	1.8776	.53259	5.00	148.41	.00674
0.14	1.1503	.86936	0.64	1.8965	.52729	6.00	403.43	.00248
0.15	1.1618	.86071	0.65	1.9155	.52205	7.00	1096.6	.00091
0.16	1.1735	.85214	0.66	1.9348	.51685	8.00	2981.0	.00034
0.17	1.1853	.84366	0.67	1.9542	.51171	9.00	8103.1	.00012
0.18	1.1972	.83527	0.68	1.9739	.50662	10.00	22026.5	.00005
0.19	1.2092	.82696	0.69	1.9937	.50158			
0.20	1.2214	.81873	0.70	2.0138	.49659			
0.21	1.2337	.81058	0.71	2.0340	.49164			
0.22	1.2461	.80252	0.72	2.0544	.48675			
0.23	1.2586	.79453	0.73	2.0751	.48191			
0.24	1.2712	.78663	0.74	2.0959	.47711			
0.25	1.2840	.77880	0.75	2.1170	.47237			
0.26	1.2969	.77105	0.76	2.1383	.46767			
0.27	1.3100	.76338	0.77	2.1598	.46301			
0.28	1.3231	.75578	0.78	2.1815	.45841			
0.29	1.3364	.74826	0.79	2.2034	.45384			
0.30	1.3499	.74082	0.80	2.2255	.44933			
0.31	1.3634	.73345	0.81	2.2479	.44486			
0.32	1.3771	.72615	0.82	2.2705	.44043			
0.33	1.3910	.71892	0.83	2.2933	.43605			
0.34	1.4049	.71177	0.84	2.3164	.43171			
0.35	1.4191	.70469	0.85	2.3396	.42741			
0.36	1.4333	.69768	0.86	2.3632	.42316			
0.37	1.4477	.69073	0.87	2.3869	.41895			
0.38	1.4623	.68386	0.88	2.4109	.41478			
0.39	1.4770	.67706	0.89	2.4351	.41066			
0.40	1.4918	.67032	0.90	2.4596	.40657			
0.41	1.5068	.66365	0.91	2.4843	.40252			
0.42	1.5220	.65705	0.92	2.5093	.39852			
0.43	1.5373	.65051	0.93	2.5345	.39455			
0.44	1.5527	.64404	0.94	2.5600	.39063			
0.45	1.5683	.63763	0.95	2.5857	.38674			
0.46	1.5841	.63128	0.96	2.6117	.38289			
0.47	1.6000	.62500	0.97	2.6379	.37908			
0.48	1.6161	.61878	0.98	2.6645	.37531			
0.49	1.6323	.61263	0.99	2.6912	.37158			

Excerpted from *Handbook of Mathematical Tables and Formulas*, 5th ed., by R. S. Burington. Copyright © 1973 by McGraw-Hill, Inc. Used with permission of McGraw-Hill Book Company.

Table II. The Natural Logarithm (Base e)

x	ln x	x	ln x	x	ln x	x	ln x
.01	− 4.60517	0.50	− 0.69315	1.00	0.00000	1.5	0.40547
.02	− 3.91202	.51	.67334	1.01	.00995	1.6	7000
.03	.50656	.52	.65393	1.02	.01980	1.7	0.53063
.04	.21888	.53	.63488	1.03	.02956	1.8	8779
		.54	.61619	1.04	.03922	1.9	0.64185
.05	− 2.99573	.55	.59784	1.05	.04879	2.0	9315
.06	.81341	.56	.57982	1.06	.05827	2.1	0.74194
.07	.65926	.57	.56212	1.07	.06766	2.2	8846
.08	.52573	.58	.54473	1.08	.07696	2.3	0.83291
.09	.40795	.59	.52763	1.09	.08618	2.4	7547
0.10	− 2.30259	0.60	− 0.51083	1.10	.09531	2.5	0.91629
.11	.20727	.61	.49430	1.11	.10436	2.6	5551
.12	.12026	.62	.47804	1.12	.11333	2.7	9325
.13	.04022	.63	.46204	1.13	.12222	2.8	1.02962
.14	− 1.96611	.64	.44629	1.14	.13103	2.9	6471
.15	.89712	.65	.43078	1.15	.13976	3.0	9861
.16	.83258	.66	.41552	1.16	.14842	4.0	1.38629
.17	.77196	.67	.40048	1.17	.15700	5.0	1.60944
.18	.71480	.68	.38566	1.18	.16551	10.0	2.30258
.19	.66073	.69	.37106	1.19	.17395		
0.20	− 1.60944	0.70	− 0.35667	1.20	.18232		
.21	.56065	.71	.34249	1.21	.19062		
.22	.51413	.72	.32850	1.22	.19885		
.23	.46968	.73	.31471	1.23	.20701		
.24	.42712	.74	.30111	1.24	.21511		
.25	.38629	.75	.28768	1.25	.22314		
.26	.34707	.76	.27444	1.26	.23111		
.27	.30933	.77	.26136	1.27	.23902		
.28	.27297	.78	.24846	1.28	.24686		
.29	.23787	.79	.23572	1.29	.25464		
0.30	− 1.20397	0.80	− 0.22314	1.30	.26236		
.31	.17118	.81	.21072	1.31	.27003		
.32	.13943	.82	.19845	1.32	.27763		
.33	.10866	.83	.18633	1.33	.28518		
.34	.07881	.84	.17435	1.34	.29267		
.35	− 1.04982	.85	− 0.16252	1.35	.30010		
.36	.02165	.86	.15032	1.36	.30748		
.37	− 0.99425	.87	.13926	1.37	.31481		
.38	.96758	.88	.12783	1.38	.32208		
.39	.94161	.89	.11653	1.39	.32930		
0.40	− 0.91629	0.90	− 0.10536	1.40	.33647		
.41	.89160	.91	.09431	1.41	.34359		
.42	.86750	.92	.08338	1.42	.35066		
.43	.84397	.93	.07257	1.43	.35767		
.44	.82098	.94	.06188	1.44	.36464		
.45	.79851	.95	.05129	1.45	.37156		
.46	.77653	.96	.04082	1.46	.37844		
.47	.75502	.97	.03046	1.47	.38526		
.48	.73397	.98	.02020	1.48	.39204		
.49	.71335	.99	.01005	1.49	.39878		

Table III. Trigonometric Functions

Degrees	Radians	sin	cos	tan	Degrees	Radians	sin	cos	tan
0	0.0000	0.0000	1.000	0.0000	45	0.7854	0.7071	0.7071	1.000
1	0.01745	0.01745	0.9998	0.01746	46	0.8028	0.7193	0.6947	1.036
2	0.03491	0.03490	0.9994	0.03492	47	0.8203	0.7314	0.6820	1.072
3	0.05236	0.05234	0.9986	0.05241	48	0.8378	0.7431	0.6691	1.111
4	0.06981	0.06976	0.9976	0.06993	49	0.8552	0.7547	0.6561	1.150
5	0.08727	0.08716	0.9962	0.08749	50	0.8727	0.7660	0.6428	1.192
6	0.1047	0.1045	0.9945	0.1051	51	0.8901	0.7772	0.6293	1.235
7	0.1222	0.1219	0.9926	0.1228	52	0.9076	0.7880	0.6157	1.280
8	0.1396	0.1392	0.9903	0.1405	53	0.9250	0.7986	0.6018	1.327
9	0.1571	0.1564	0.9877	0.1584	54	0.9425	0.8090	0.5878	1.376
10	0.1745	0.1736	0.9848	0.1763	55	0.9599	0.8192	0.5736	1.428
11	0.1920	0.1908	0.9816	0.1944	56	0.9774	0.8290	0.5592	1.483
12	0.2094	0.2079	0.9782	0.2126	57	0.9948	0.8387	0.5446	1.540
13	0.2269	0.2250	0.9744	0.2309	58	1.012	0.8480	0.5299	1.600
14	0.2444	0.2419	0.9703	0.2493	59	1.030	0.8572	0.5150	1.664
15	0.2618	0.2588	0.9659	0.2680	60	1.047	0.8660	0.5000	1.732
16	0.2792	0.2756	0.9613	0.2868	61	1.065	0.8746	0.4848	1.804
17	0.2967	0.2924	0.9563	0.3057	62	1.082	0.8830	0.4695	1.881
18	0.3142	0.3090	0.9511	0.3249	63	1.100	0.8910	0.4540	1.963
19	0.3316	0.3256	0.9455	0.3443	64	1.117	0.8988	0.4384	2.050
20	0.3491	0.3420	0.9397	0.3640	65	1.134	0.9063	0.4226	2.144
21	0.3665	0.3584	0.9336	0.3839	66	1.152	0.9136	0.4067	2.246
22	0.3840	0.3746	0.9272	0.4040	67	1.169	0.9205	0.3907	2.356
23	0.4014	0.3907	0.9205	0.4245	68	1.187	0.9272	0.3746	2.475
24	0.4189	0.4067	0.9136	0.4452	69	1.204	0.9336	0.3584	2.605
25	0.4363	0.4226	0.9063	0.4663	70	1.222	0.9397	0.3420	2.748
26	0.4538	0.4384	0.8988	0.4877	71	1.239	0.9455	0.3256	2.904
27	0.4712	0.4540	0.8910	0.5095	72	1.257	0.9511	0.3090	3.078
28	0.4887	0.4695	0.8830	0.5317	73	1.274	0.9563	0.2924	3.271
29	0.5062	0.4848	0.8746	0.5543	74	1.292	0.9613	0.2756	3.487
30	0.5236	0.5000	0.8660	0.5774	75	1.309	0.9659	0.2588	3.732
31	0.5410	0.5150	0.8572	0.6009	76	1.326	0.9703	0.2419	4.011
32	0.5585	0.5299	0.8480	0.6249	77	1.344	0.9744	0.2250	4.332
33	0.5760	0.5446	0.8387	0.6494	78	1.361	0.9782	0.2079	4.705
34	0.5934	0.5592	0.8290	0.6745	79	1.379	0.9816	0.1908	5.145
35	0.6109	0.5736	0.8192	0.7002	80	1.396	0.9848	0.1736	5.671
36	0.6283	0.5878	0.8090	0.7265	81	1.414	0.9877	0.1564	6.314
37	0.6458	0.6018	0.7986	0.7536	82	1.431	0.9903	0.1392	7.115
38	0.6632	0.6157	0.7880	0.7813	83	1.449	0.9926	0.1219	8.144
39	0.6807	0.6293	0.7772	0.8098	84	1.466	0.9945	0.1045	9.514
40	0.6981	0.6428	0.7660	0.8391	85	1.484	0.9962	0.08716	11.43
41	0.7156	0.6561	0.7547	0.8693	86	1.501	0.9976	0.06976	14.30
42	0.7330	0.6691	0.7431	0.9004	87	1.518	0.9986	0.05234	19.08
43	0.7505	0.6820	0.7314	0.9325	88	1.536	0.9994	0.03490	28.64
44	0.7679	0.6947	0.7193	0.9657	89	1.553	0.9998	0.01745	57.29
45	0.7854	0.7071	0.7071	1.000	90	1.571	1.000	0.0000	———

ANSWERS TO
ODD-NUMBERED PROBLEMS AND
REVIEW PROBLEMS

Answers to

Odd-Numbered Problems

and Review Problems

CHAPTER 1, SECTION 1 (PAGE 9)

1. $f(1) = 6, f(0) = -2, f(-2) = 0$

3. $g(-1) = -2, g(1) = 2, g(2) = \dfrac{5}{2}$

5. $h(2) = 2\sqrt{3}, h(0) = 2, h(-4) = 2\sqrt{3}$

7. $f(1) = 1, f(5) = \dfrac{1}{27}, f(13) = \dfrac{1}{125}$

9. $f(1) = 0, f(2) = 2, f(3) = 2$

11. $f(-6) = 3, f(-5) = -4, f(16) = 4$

13. All real numbers x except $x = -2$.

15. All real numbers x for which $x \geq 5$.

17. All real numbers t.

19. All real numbers t for which $t \geq 2$.

21. All real numbers x for which $|x| > 3$.

23. All real numbers t except $t = 1$.

25. (a) \$4,500   (b) \$371

27. (a) $33\dfrac{1}{3}°$ Celsius   (b) Decreased by 7.5° Celsius

29. (a) All real numbers n except $n = 0$.
    (b) All positive integers n.
    (c) 7 minutes   (d) 12th trial
    (e) The time required will approach but never exceed 3 minutes.

31. (a) All real numbers x except $x = 300$.
    (b) All real numbers x for which $0 \leq x \leq 100$.
    (c) 120   (d) 300   (e) 60

33. (a) 192 feet   (b) 80 feet   (c) 256 feet
    (d) After 4 seconds.

35. $g[h(x)] = 3x^2 + 14x + 10$

37. $g[h(x)] = x^3 + 2x^2 + 4x + 2$

39. $g[h(x)] = \dfrac{1}{(x-1)^2}$

41. $g[h(x)] = |x|$

43. $f(x - 2) = 2x^2 - 11x + 15$

45. $f(x - 1) = x^5 - 3x^2 + 6x - 3$

47. $f(x^2 + 3x - 1) = \sqrt{x^2 + 3x - 1}$

49. $f(x + 1) = \dfrac{x}{x + 1}$

51. $h(x) = 3x - 5, g(u) = \sqrt{u}$

53. $h(x) = x^2 + 1, g(u) = \dfrac{1}{u}$

55. $h(x) = x + 3, g(u) = \sqrt{u} - \dfrac{1}{(u + 1)^3}$

57. (a) $C[q(t)] = 625t^2 + 25t + 900$ (b) \$6,600
    (c) After 4 hours.

CHAPTER 1, SECTION 2 (PAGE 26)

1.

3.

5.

7.

9.

11.

13.

15.

17.

19.

21.

23.

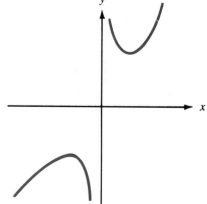

25. $P(x) = (x - 20)(120 - x)$

Optimal price = $70 per recorder

27. (a)

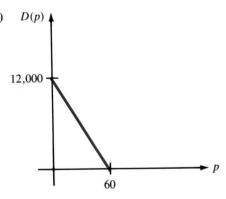

(b) $E(p) = -200p^2 + 12,000p = -200p(p - 60)$

(c)

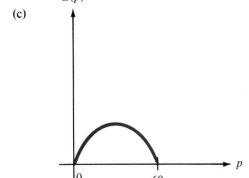

(d) $E(0) = 0$ because price is zero.
$E(60) = 0$ because demand is zero.

(e) Optimal price = $30 per unit.

29. The graph has a practical interpretation for $0 \le x \le 100$.

31. (a)

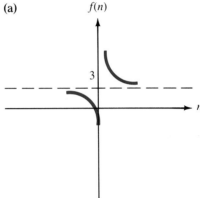

(b) The graph has a practical interpretation for $n = 1, 2, 3, \ldots$

(c) As n increases without bound, the height of the graph decreases and approaches 3. That is, as the number of trials increases, the time required for the rat to traverse the maze decreases, approaching a lower bound of 3 minutes.

33. Optimal number of machines = 10.

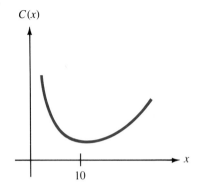

35. $A(x) = x + 6 + \dfrac{19}{x}$

37. (3, 1)

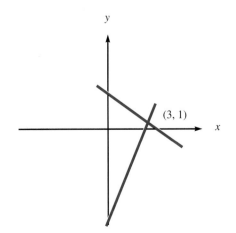

39. $(-3, 9)$, $(2, 4)$

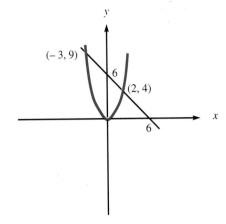

41. $(0, 0)$, $(5, -25)$

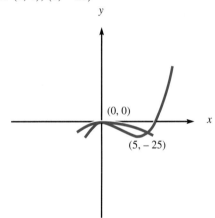

43. $(1 + \sqrt{3}, 4 + 2\sqrt{3})$, $(1 - \sqrt{3}, 4 - 2\sqrt{3})$

45. (2, 3)

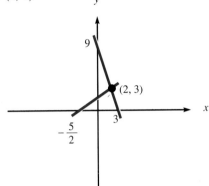

51. $\left(\dfrac{3 + \sqrt{5}}{2}, \dfrac{1 + \sqrt{5}}{2}\right)$, $\left(\dfrac{3 - \sqrt{5}}{2}, \dfrac{1 - \sqrt{5}}{2}\right)$,

47. (1, 1)

53. (a)

49. (1, 1)

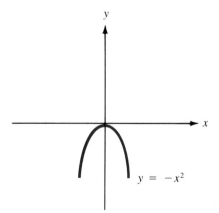

(b) The graph of g is obtained by reflecting the graph of f across the x axis.

55. (a)

(b)

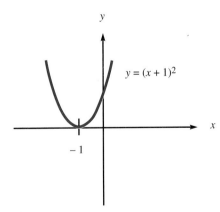

(c) The graph of g is obtained by translating the graph of f horizontally by c units.

57. Hint: Where is (x_2, y_1)?

59. $(x_1, y_1) = (a, b)$ and $(x_2, y_2) = (x, y)$

61. $(0, 15)$, $(-5, 0)$, $(3, 0)$, $V(-1, 16)$

63. $(0, -5)$, $\left(-\frac{5}{2}, 0\right)$, $\left(\frac{1}{3}, 0\right)$, $V\left(-\frac{13}{12}, -\frac{289}{24}\right)$

CHAPTER 1, SECTION 3 (PAGE 41)

1. $-\dfrac{7}{2}$

3. -1

5. Undefined

7. $m = 3$, $b = 0$

9. $m = 3$, $b = 6$

11. $m = -\dfrac{3}{2}$, $b = 3$

13. $m = \dfrac{3}{5}$, $b = \dfrac{4}{5}$

15. $m = -\dfrac{5}{2}$, $b = 5$

17. m is undefined; no y intercept.

19. $y = x - 2$

21. $y = -\dfrac{1}{2}x + \dfrac{1}{2}$

23. $y = 5$

25. $y = -x + 1$

27. $45x + 52y = 43$

29. $y = 5$

31. $y = 60x + 5{,}000$

33. (a) $y = 35x + 220$, (b) 325, (c) 220

35. $y = -150x + 1{,}500$

37. (a) $y = -4x + 248$ (b) 216 million gallons

39. (a) $y = \dfrac{9}{5}x + 32$ (b) 59°F   (c) 20°C

41. $400(10 - x)$, $0 \le x \le 10$

43. (a) $y = -2x + 5$ (b) $y = \dfrac{3}{2}x + 2$

(c) $y = -\dfrac{3}{5}x + \dfrac{19}{5}$

CHAPTER 1, SECTION 4 (PAGE 57)

1. $A = 2y(500 - y)$ if the stream is along the length. $A = (1,000 - y) \, y/2$ if the stream is along the width.

3. $P = x(18 - x)$

5. $R = x(35x + 15)$

7. $6.50

9. 80 trees

11. 10 days

13. $A = x(160 - x)$; 80 m

15. $V = 2x^3$

17. $V = x(12 - x^2)/3$ cubic meters

19. $S = 13.78 \, \pi/r + 2\pi r^2$

21. (a) $V = \dfrac{r}{2}(S - 2\pi r^2)$ or

$V = [h/(4\pi)](-\pi h + \sqrt{\pi^2 h^2 + 2\pi S})^2$
(b) $S = 2(\pi r^2 + V/r)$ or
$S = 2V/h + 2\sqrt{\pi h V}$

23. $V = \pi r(27 - r^2)/2$

25. $f(x) = \begin{cases} 10x & x \le 5 \\ 5x + 25 & x = 6, 7, 8, 9 \end{cases}$

27. $R(x) = \begin{cases} 2,400 & 0 < x \le 40 \\ 80x - 0.5x^2 & 40 < x < 80 \\ 40x & 80 \le x \end{cases}$

29. $R = kq$

31. $R = kq(n - q)$

33. $C(x) = k_1 x + k_2/x$

35. $D = 100t$

37. $C(x) = 4(3,000 - x) + 5\sqrt{810,000 + x^2}$

39. $C(x) = 20x + 1,280/x$

41. (a) 4   (b) 7

43. Join the second club if fewer than 200 hours of tennis will be played and the first if more than 200 hours will be played.

45. $p = \$40, q = 360$ units

47. (a) $p = \$80, q = 70$ units

(b)

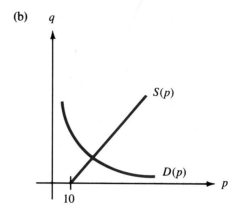

(c) $S(10) = 0$. Manufacturers will not supply any units unless the market price exceeds \$10.

49. 2 hours 45 minutes after second plane leaves.

CHAPTER 1, SECTION 5 (PAGE 77)

1. $\lim\limits_{x \to a} f(x) = b$ 3. $\lim\limits_{x \to a} f(x) = b$

5. Limit does not exist.   7. 4

9. 7   11. 16

13. $\dfrac{3}{4}$ 15. Limit does not exist.

17. 2   19. 7

21. $\dfrac{5}{3}$ 23. 5

25. $\dfrac{1}{4}$ 27. 2

29. Yes   31. Yes   33. No   35. No

37. No   39. No   41. Yes   43. None

45. $x = 3$

47. $x = -1$

49. $x = -5$ and $x = 1$

51. $x = -1$ and $x = 2$

53. $x = 2$

55. $A = 6$

57. $A = 0, B = -4$

59. Continuous on $0 < x < 1$; not continuous on $0 \le x \le 1$.

61. Continuous on both, assuming $f(3) = 6$.

63. Use intermediate value property.

65. $f(x) = x^2 + 2x - 1 - \sqrt[3]{x}. \; f(0) < 0, f(1) > 0.$

CHAPTER 1, REVIEW PROBLEMS (PAGE 82)

1. (a) All real numbers x.
   (b) All real numbers x except $x = 1$ and $x = -2$.
   (c) All real numbers x for which $|x| \ge 3$.

2. (a) \$45   (b) \$1   (c) 9 months from now.
   (d) The price will approach \$40.

3. (a) $g[h(x)] = x^2 - 4x + 4$
   (b) $g[h(x)] = \dfrac{1}{2x + 5}$
   (c) $g[h(x)] = \sqrt{-2x - 3}$

4. (a) $f(x - 2) = x^2 - 5x + 10$
   (b) $f(x^2 + 1) = \sqrt{x^2 + 1} + \dfrac{2}{x^2}$
   (c) $f(x + 1) - f(x) = 2x + 1$

5. (a) $g(u) = u^5$ and $h(x) = x^2 + 3x + 4$
   (b) $g(u) = u^2 + \dfrac{5}{2(u + 1)^3}$ and $h(x) = 3x + 1$

6. (a) $Q[p(t)] = \sqrt{23.4 + 0.1t^2}$
   (b) 4.93 units   (c) 4 years from now.

7. $c = -4$

8. (a)

(b)

(c)

(d)

(e)

10. (a)

9. (a)

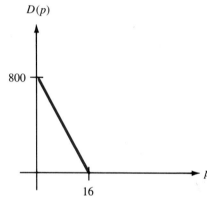

(b) $E(p) = -50p(p - 16)$

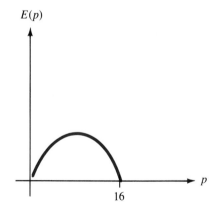

(c) Optimal price = $8 per unit.

(b) 5 weeks
(c) 20 weeks

11. (a) $m = 3, b = 2$

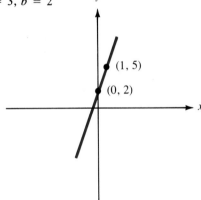

(b) $m = \dfrac{5}{4}, b = -5$

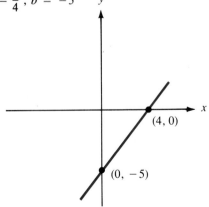

(c) $m = -\dfrac{3}{2}, b = 0$

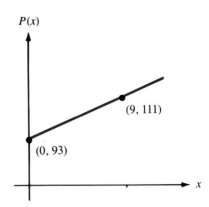

16. (a) $C(x) = 400x + 3,200$ where x is the number of months since the paper first appeared.
    (b) 5,200

(d) $m = -\dfrac{2}{3}, b = 8$

17. (a)

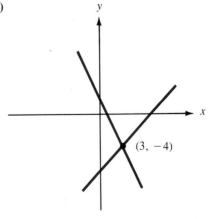

12. $y = 5x - 4$

13. $y = -2x + 5$

14. $y = 7x - 10$

15. (a) $P(x) = 2x + 93$, where x is the number of months since the beginning of the year.
    (b) 93 cents per gallon
    (c) \$1.11 per gallon

(b)

(e)

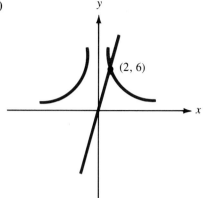

(2, 6)

18. $P(x) = (50 - x)(x - 10)$
    Optimal price = \$30

(c)

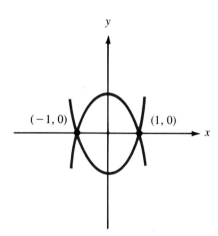

$(-1, 0)$ $(1, 0)$

19. $P(x) = 2(100 - x)(x - 50)$
    Optimal price = \$75

(d)

$(-5, 25)$

$(3, 9)$

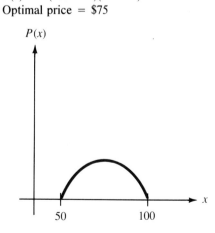

20. $V(r) = 20r - \frac{3}{4}\pi r^3$

21. $C(x) = 80x + \frac{11,520}{x}$

Optimal number of machines = 12

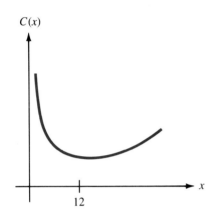

22. Call the first plumber if the work will take less than $1\frac{1}{2}$ hours and the second if the work will take more than $1\frac{1}{2}$ hours.

23. (a) 150   (b) $1,500 profit   (c) 180

24. $D(p) = -4p + 32$
$S(p) = p^2 - 2p - 3$
$p_E = 5; S(6) - D(6) = 13$

25. First publisher if $n \le 12,000$ copies.

26. $R(x) = k(n - x)$, where n is the total number of relevant facts in the subject's memory.

27. (a) $-\frac{2}{3}$ (b) $\frac{3}{2}$ (c) 0   (d) -12

28. (a) $x < 0$ (b) $x = -3$ (c) $2, -\frac{3}{2}$

(d) Continuous for all x, assuming $h(3) = 0$.

29. (a) $A = 6$ (b) $A = 2$

30. (a) 9 minutes   (b) $f(2) = 5,000$ and $f(5^-) = 26,000^-$

CHAPTER 2, SECTION 1 (PAGE 98)

1. $f'(x) = 5, m = 5$

3. $f'(x) = 4x - 3, m = -3$

5. $f'(x) = -\frac{2}{x^2}, m = -8$

7. $f'(x) = \frac{1}{2\sqrt{x}}, m = \frac{1}{6}$

9. $f'(2) = 5, y = 5x - 3$

11. $f'\left(\frac{1}{2}\right) = -48, y = -48x + 36$

13.

$\left(\frac{3}{2}, -\frac{9}{4}\right)$

15.

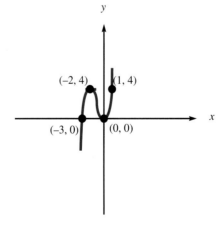

$(-2, 4)$ $(1, 4)$

$(-3, 0)$ $(0, 0)$

$f'(0) = f'(-2) = 0$
horizontal tangent line at $(-2, 4), (0, 0)$

17. $70

19. (a) $m_s = -3.9$ (b) $f'(-2) = -4$

21. (a) $f'(x) = 3$ (b) $y = 3x - 2$ (c) line = tangent line

23. (a) $f'(x) = 2x + 3$ (b) $2x, 3$ (c) $f'(x) = (2x) + (3)$ (d) $(f + g)' = f' + g'$

25. Positive angle of inclination of the tangent line if $f'(x) > 0$.

27.

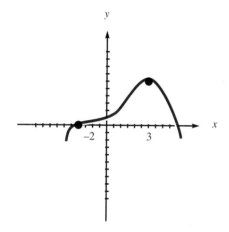

CHAPTER 2, SECTION 2 (PAGE 107)

1. $\dfrac{dy}{dx} = 2x + 2$

3. $f'(x) = 9x^8 - 40x^7 + 1$

5. $\dfrac{dy}{dx} = -\dfrac{1}{x^2} - \dfrac{2}{x^3} + \dfrac{1}{2\sqrt{x^3}}$

7. $f'(x) = \dfrac{3}{2}\sqrt{x} - \dfrac{3}{2\sqrt{x^5}}$

9. $\dfrac{dy}{dx} = -\dfrac{1}{8}x - \dfrac{2}{x^2} - \dfrac{3}{2}\sqrt{x} - \dfrac{2}{3x^3} + \dfrac{1}{3}$

11. $f'(x) = 12x - 1$

13. $\dfrac{dy}{dx} = -300x - 20$

15. $f'(x) = \dfrac{1}{3}(5x^4 - 6x^2)$

17. $\dfrac{dy}{dx} = \dfrac{-3}{(x - 2)^2}$

19. $f'(x) = \dfrac{-x^2 - 2}{(x^2 - 2)^2}$

21. $\dfrac{dy}{dx} = \dfrac{-3}{(x + 5)^2}$

23. $f'(x) = \dfrac{11x^2 - 10x - 7}{(2x^2 + 5x - 1)^2}$

25. $y = -9x + 4$

27. $y = -2x - 1$

29. $y = 4x - 14$

31. $y = 8x + 5$

33. $y' = 4x - 5$

35. If c = constant, $c' = 0$

37.

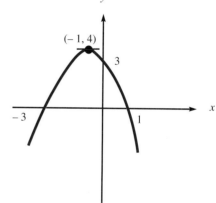

39. $f(x) = -x^2 + 5x$

41. $(0, 0), (4, 64)$

43. The rectangle is a square.

45. $\dfrac{12x^3 + 51x^2 + 70x - 33}{(3x + 5)^2}$

CHAPTER 2, SECTION 3 (PAGE 117)

1. 9

3. $\dfrac{141}{2}$

5. $\dfrac{13}{64}$

7. (a) $C'(t) = 200t + 400$
   (b) Increasing at the rate of 1,400 per year.
   (c) 1,500

9. (a) $f'(x) = -3x^2 + 12x + 15$ (b) 24 radios per hour   (c) 26

11. (a) $P'(t) = \dfrac{6}{(t + 1)^2}$ thousand per year
   (b) 1,500 per year   (c) 1,000   (d) 60 per year   (e) The rate of growth approaches zero.

13. 100 kilometers per hour

15. 66.67 percent

17. (a) 10,800 people   (b) 17.53 percent

19. (a) $\dfrac{200}{t + 100}$ percent   (b) 0

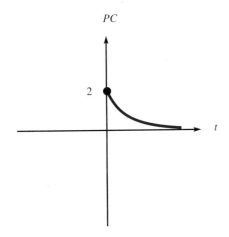

PC

2

t

21. 7.5 percent

23. (a) $V(t) = 6t + 2$ traveling to the right;
   (b) 5   (c) $a(t) = 6$ speeding up

25. (a) For $S(t) = t^4 - 4t^3 + 8t$, $V(t) = 4t^3 - 12t^2 + 8$, traveling to the right for $0 \le t < 1$ and $2.73 < t \le 4$ (b) 50   (c) $a(t) = 12t(t - 2)$, slowing down for $0 \le t < 2$

27. (a) $V(t) = 3(t - 5)(t - 1)$, traveling to the right for $0 \le t < 1$ and $5 < t \le 6$ (b) 46   (c) $a(t) = 6(t - 3)$, slowing down for $0 \le t < 3$.

29. (a) 294 feet   (b) -146 feet per second

31. (a) 32 feet per second   (b) 128 feet
   (c) 32 feet per second   (d) 96 feet per second

33. If $y = mx + b$, what is the instantaneous rate of change?

35. (a) $S(t) = -\dfrac{g}{2}t^2 + V_0 t$, so $t_1 = \dfrac{2V_0}{g}$
   (b) $V(t_1) = -V_0$

CHAPTER 2, SECTION 4 (PAGE 128)

1. 2.1

3. 20.6 percent

5. Cost will increase by approximately $50.08.

7. 200

9. 0.05 part per million

11. Daily output will increase by approximately 8 units.

13. 825

15. 27.14 cm²

17. 0.5

19. 6 percent

21. -6 percent

23. 3 percent

25. 0.5 percent

27. 0.67 percent

29. ± 2.67 percent

31. 15 percent

33. (a) 499.70   (b) 500.20

35. (a) $C'(x) = \dfrac{2}{5}x + 4$; $R'(x) = 9 - \dfrac{x}{2}$
   (b) 5.2   (c) 5.4   (d) 7.5   (e) 7.25

CHAPTER 2, SECTION 5 (PAGE 140)

1. $\dfrac{dy}{dx} = 6(3x - 2)$

3. $\dfrac{dy}{dx} = \dfrac{x + 1}{\sqrt{x^2 + 2x - 3}}$

5. $\dfrac{dy}{dx} = \dfrac{-4x}{(x^2 + 1)^3}$

7. $\dfrac{dy}{dx} = -x(x^2 - 9)^{-3/2}$

9. $\dfrac{dy}{dx} = \dfrac{-2x}{(x^2 - 1)^2}$

11. -160

13. $\dfrac{2}{3}$

15. -16

17. $f'(x) = 8(2x + 1)^3$

19. $f'(x) = 8x^2(x^5 - 4x^3 - 7)^7(5x^2 - 12)$

21. $f'(x) = -\dfrac{10x - 6}{(5x^2 - 6x + 2)^2}$

23. $f'(x) = \dfrac{-4x}{(4x^2 + 1)^{3/2}}$

25. $f'(x) = \dfrac{24x}{(1 - x^2)^5}$

27. $f'(x) = \dfrac{15}{2}\dfrac{(1 + \sqrt{3x})^4}{\sqrt{3x}}$

29. $f'(x) = (x + 2)^2(2x - 1)^4(16x + 17)$

31. $f'(x) = \dfrac{-5}{2(3x + 1)^{1/2}(2x - 1)^{3/2}}$

33. $f'(x) = \dfrac{(x + 1)^4(9 - x)}{(1 - x)^5}$

35. $f'(x) = \dfrac{5 - 6x}{(1 - 4x)^{3/2}}$

37. $y = -48x - 32$

39. $y = -12x + 13$

41. $6(3x + 5)$

43. (a) \$2,025 per year   (b) 10.125 percent per year

45. 0.31 part per million per year

47. Decreasing at the rate of 6 pounds per week

49. 5.09 percent per year

51. $\dfrac{dy}{dx} = [h(x)][h'(x)] + [h(x)][h'(x)]$

CHAPTER 2, SECTION 6 (PAGE 150)

1. $\dfrac{dy}{dx} = \dfrac{-x}{y}$

3. $\dfrac{dy}{dx} = \dfrac{y - 3x^2}{3y^2 - x}$

5. $\dfrac{dy}{dx} = \dfrac{3 - 2y^2}{2y(1 + 2x)}$

7. $\dfrac{dy}{dx} = \dfrac{1}{3(2x + y)^2} - 2$

9. $\dfrac{dy}{dx} = \dfrac{2y - 10x(x^2 + 3y^2)^4}{30y(x^2 + 3y^2)^4 - 2x}$

11. $\dfrac{1}{3}$ 13. $-\dfrac{1}{2}$ 15. $\dfrac{13}{12}$ 17. $\dfrac{8}{3}$

19. $\dfrac{dy}{dx} = \dfrac{2x - y}{x + 2} = \dfrac{x(x + 4)}{(x + 2)^2}$

21. $\dfrac{dy}{dx} = \dfrac{y - 1}{1 - x} = \dfrac{-3}{(x - 1)^2}$

23. Increase input by approximately 0.57 percent

25. Approximately a 1.28-hour increase

27. $\dfrac{x_0 x}{a^2} - \dfrac{y_0 y}{b^2} = 1$

29. (a) $x^2 + (y - 3)^2 \geq 0$ (b) $\dfrac{dy}{dx} = \dfrac{-x}{y - 3}$

31. $H: (0, 0), (0.75, \pm 1.3)$
    $V: (2, 0), (-0.25, \pm 0.433)$

CHAPTER 2, SECTION 7, (PAGE 159)

1. $f''(x) = 450x^8 - 120x^3$

3. $\dfrac{d^2y}{dx^2} = \dfrac{-5}{4x^{3/2}} + \dfrac{18}{x^4} + \dfrac{1}{4x^{5/2}}$

5. $f''(x) = 180(3x + 1)^3$

7. $\dfrac{d^2y}{dx^2} = 80(x^2 + 5)^6(3x^2 + 1)$

9. $f''(x) = \dfrac{1}{(1 + x^2)^{3/2}}$

11. $\dfrac{d^2y}{dx^2} = \dfrac{4(3x^2 - 1)}{(1 + x^2)^3}$

13. $f''(x) = 16(2x + 1)^2(5x + 1)$

15. $\dfrac{d^2y}{dx^2} = \dfrac{2(1 - 2x)}{(x + 1)^4}$

17. $\dfrac{d^2y}{dx^2} = \dfrac{15}{4y^3}$

19. $\dfrac{d^2y}{dx^2} = -\dfrac{a}{b^2y^3}$

21. (a) 28 units/hour   (b) 10 units/hour^2
    (c) 2.5   (d) 2.3125

23. (a) 75   (b) 0   (c) A decrease of 500 people   (d) A decrease of 521 people

25. (a) -32 (b) It doesn't   (c) Pulled down

27. $120x - 48$

29. $-\dfrac{15}{8\sqrt{3}}x^{-7/2} + 48x^{-5}$

CHAPTER 2, REVIEW PROBLEMS (PAGE 162)

1. (a) $f'(x) = 2x - 3$ (b) $f'(x) = \dfrac{-1}{(x - 2)^2}$

2. (a) $f'(x) = 24x^3 - 21x^2 + 2$

   (b) $f'(x) = 3x^2 + \dfrac{5}{3x^6} + \dfrac{1}{\sqrt{x}} + \dfrac{3}{x^2} - \dfrac{3}{x^4} + \dfrac{4}{x^3}$

   (c) $\dfrac{dy}{dx} = \dfrac{-14x}{(3x^2 + 1)^2}$

   (d) $\dfrac{dy}{dx} = 2(x + 1)(2x + 5)^2(5x + 8)$

   (e) $f'(x) = 20(5x^4 - 3x^2 + 2x + 1)^9(10x^3 - 3x + 1)$

   (f) $f'(x) = \dfrac{x}{\sqrt{x^2 + 1}}$

   (g) $f'(x) = 2\left(x + \dfrac{1}{x}\right)\left(1 - \dfrac{1}{x^2}\right) + \dfrac{5}{2(3x)^{3/2}}$

   (h) $\dfrac{dy}{dx} = \dfrac{4(x + 1)}{(1 - x)^3}$

   (i) $\dfrac{dy}{dx} = \dfrac{9(3x + 2)}{\sqrt{6x + 5}}$

   (j) $f'(x) = \dfrac{3(3x + 1)^2(3x + 7)}{(1 - 3x)^5}$

   (k) $f'(x) = \dfrac{-7}{2(1 - 2x)^{1/2}(3x + 2)^{3/2}}$

3. (a) $y = -x + 1$ (b) $y = -x - 1$

   (c) $y = x$ (d) $y = -\dfrac{2}{3}x + \dfrac{5}{3}$

4. (a) 31   (b) 0

5. (a) $100\left[\dfrac{2t^{3/2} - 3t^{1/2} + \frac{1}{2}}{t^{5/2} - 3t^{3/2} + t}\right]$ (b) $100\left[\dfrac{2(3 - 5t)}{t(3 - 2t)}\right]$

   (c) $100\left[\dfrac{-1}{t + 1}\right]$

6. (a) 1,652 people per week   (b) 1,514 people

7. (a) Output will increase by approximately 12,000 units.
   (b) Output will increase by 12,050 units.

8. 0.30 percent per month

9. Output will decrease by approximately 5,000 units.

10. 1.347 percent

11. 10 percent

12. 1.5 percent

13. (a) $\dfrac{dy}{dx} = 3(30x + 11)$ (b) $\dfrac{dy}{dx} = \dfrac{-4}{(2x + 3)^3}$

14. (a) 2   (b) $\dfrac{3}{2}$

15. $1,663.20$ per hour

16. (a) $V(t) = 6(t - 2)(t - 5)$, $a(t) = 6(2t - 7)$; the object moves right when $1 < t < 2$ and $5 < t < 6$, left when $2 < t < 5$; it is slowing down when $t < \dfrac{7}{2}$, speeding up when $\dfrac{7}{2} < t$.   (b) 49

17. (a) $V(t) = \dfrac{2(-t^2 - t + 12)}{(t^2 + 12)^2}$

    $a(t) = \dfrac{2(2t^3 + 3t^2 - 72t - 12)}{(t^2 + 12)^3}$
    moving left, slowing down   (b) $\dfrac{1}{4}$

18. $V(5.5) = 0$, $S(5.5) = 242$ feet

19. 16.27 percent per year

20. (a) $\dfrac{dy}{dx} = -\dfrac{5}{3}$

  (b) $\dfrac{dy}{dx} = -\dfrac{2y}{x}$

  (c) $\dfrac{dy}{dx} = \dfrac{1 - 10(2x + 3y)^4}{15(2x + 3y)^4}$

  (d) $\dfrac{dy}{dx} = -\dfrac{1 + 10y^3(1 - 2xy^3)^4}{4 + 30xy^2(1 - 2xy^3)^4}$

21. (a) $-\dfrac{2}{3}$ (b) -28

22. Decrease input y by approximately 0.172 unit.

23. (a) $f''(x) = 120x^3 - 24x^2 + 10 + \dfrac{2}{x^3}$

  (b) $\dfrac{d^2y}{dx^2} = 24(3x^2 + 2)^2(21x^2 + 2)$

  (c) $f''(x) = \dfrac{2(x - 5)}{(x + 1)^4}$

24. $\dfrac{d^2y}{dx^2} = \dfrac{-9}{2y^3}$

25. (a) 27 units per hour
  (b) 12 units per hour per hour
  (c) An increase of approximately 1.2 units per hour
  (d) An increase of 1.17 units per hour

26. (a) $\dfrac{d^4y}{dx^4} = 240x + 120 + \dfrac{24}{x^5}$

  (b) $f^{(4)}(x) = -\dfrac{15\sqrt{3}}{16x^{7/2}} + \dfrac{180}{x^6}$

CHAPTER 3, SECTION 1 (PAGE 180)

1. $f'(x) > 0$ for $-2 < x < 2$; $f'(x) < 0$ for $x < -2$ and $x > 2$.

3. $f'(x) > 0$ for $x < -4$ and $0 < x < 2$; $f'(x) < 0$ for $-4 < x < -2$, $-2 < x < 0$, and $x > 2$.

5.

7.

9.

11.

13.

15.

17.

19.

21.

23.

25.

27.

29.

31.

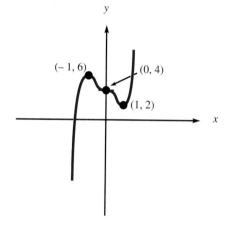

33. This answer is not unique

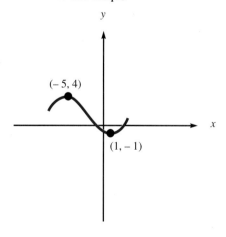

35. This curve is not unique

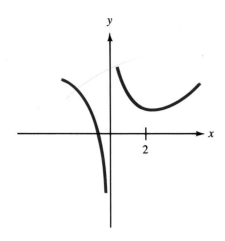

37. $a = -\dfrac{9}{25}$, $b = \dfrac{18}{5}$, $c = 3$

39. $2ax + b = 0$

CHAPTER 3, SECTION 2 (PAGE 192)

1. $f''(x) > 0$ for $x > 2$ and $f''(x) < 0$ for $x < 2$.

3.

5.

7.

9.

11.

17.

13.

19.

15.

21.

23.

25.

27.

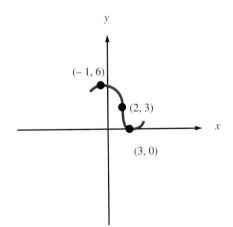

29. $x = 1$ is a vertical asymptote or a cusp exists at $x = 1$.

31.

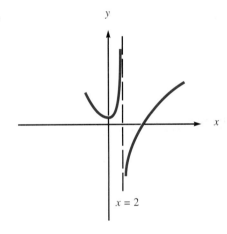

33. (a) Increasing for $x < 0$ and $x > 4$; decreasing for $0 < x < 4$.
    (b) Concave upward for $x > 2$; concave downward for $x < 2$.
    (c) Relative maximum when $x = 0$; relative minimum when $x = 4$; inflection point when $x = 2$.

35. Increasing for $x > 2$; decreasing for $x < 2$; concave upward for all x; relative minimum when $x = 2$.

37. Increasing for $x > 2$; decreasing for $x < 2$; concave upward for $x < -3$ and $x > -1$; concave downward for $-3 < x < -1$; relative minimum when $x = 2$; inflection points when $x = -3$ and $x = -1$.

39. $y'' = 2a > 0$ if $a > 0$

41. Min: $(-1, 2), (1, 2)$
    Max: $(0, 3)$

43. Min: $(1, 2)$
    Max: $(-1, -2)$

45. Min: $(4, 8)$
    Max: $(0, 0)$

CHAPTER 3 SECTION 3 (PAGE 205)

Absolute Maximum	Absolute Minimum
1. $f(1) = 10$	$f(-2) = 1$
3. $f(0) = 2$	$f(2) = -\dfrac{40}{3}$
5. $f(-1) = 2$	$f(-2) = -56$
7. $f(-3) = 3{,}125$	$f(0) = -1{,}024$
9. $f(3) = \dfrac{10}{3}$	$f(1) = 2$
11. none	$f(1) = 2$
13. none	none
15. $f(0) = 1$	none

17. (a) Membership $= 46{,}400$ in 1984
    (b) Membership $= 12{,}100$ in 1991

19. $12.50 per radio

21. At the central axis

23. $\dfrac{dR}{dP} = kB - 2kP$

25. $C = \dfrac{k_1}{V} + k_2V$, $C' = 0$ if $\dfrac{k_1}{V^2} = k_2$

27. $A' = 0$ if $C' = \dfrac{C}{q} = A$

29. $V = \left(\dfrac{w^2}{3A\rho^2 S}\right)^{1/4}$

CHAPTER 3, SECTION 4 (PAGE 221)

1. $\dfrac{1}{2}$

3. 25, 25

5. $41

7. $10.00

9. 80

11. 10 days from now

13. Field is a square.

15. $P = 2(x + y)$, $x = y = \sqrt{A}$

17. $4\sqrt{2}, 4\sqrt{2}, 8$

19. No.

21. 12 inches by 12 inches by 3 inches

23. Run the cable entirely under the water.

25. The mathematician

27. $r = 1.51$, $h = 3.02$ inches

29. 27π

31. $r = \dfrac{2h}{3}$

33. (a) 8   (b) $160   (c) $160

35. (a) 10:00 A.M.   (b) 8:00 A.M. and noon

37. Minimal slope $= \dfrac{9}{2}$ at $\left(\dfrac{1}{2}, \dfrac{5}{2}\right)$.

39. (a) 3 years from now   (b) Now

41. 17

43. 11:00 A.M.

CHAPTER 3, SECTION 5 (PAGE 240)

1. 200 cases

3. 4,000 maps

5. (a) $q = 5$
   (b) $q = 5$

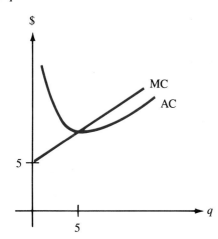

7. (a) $q = 8$

   (b) $A(q)$ is increasing for $0 < q < 8$ and decreasing for $q > 8$.

   (c)

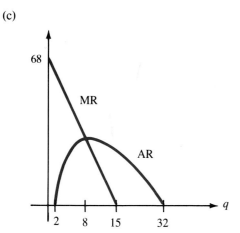

9. $C(x) = 1{,}200 + 1.20x + \dfrac{10^5}{x^2}$; $x = 55$

11. $C' = 0.4\left(2 + \dfrac{1}{\sqrt{x}}\right)$; $S' = 0.2 - \dfrac{0.4}{\sqrt{x}}$; 4

13. (a) $\eta = -\dfrac{0.1p}{60 - 0.1p}$ (b) $\eta = -0.5$ percent
    (c) $p = 300$

15. (a) Unit elasticity when $p = 125$; inelastic when $0 \le p < 125$; elastic when $125 < p \le 250$
    (b) Increasing for $0 \le p < 125$; decreasing for $125 < p \le 250$. Relative maximum at $p = 125$.
    (c) $R(p) = 500p - 2p^2$
    (d)

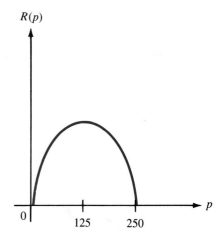

17. A 1 percent increase in price will produce a decrease in demand of m percent.

19. (a) (profit)$' = 0$ (b) $t = 5$ (c) 75 cents

21. $R = pq$

23. (a) Inelastic if $0 \le q < 10$; elastic if $10 < q \le \sqrt{300}$

(b) revenue increases if $0 \le q < 10$, decreases if $10 < q \le \sqrt{300}$, $R(10)$ is maximum

(c) $R = pq$;

(d)

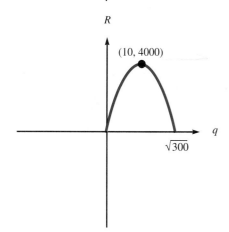

17. $x_1 < 0$, $\sqrt{x_1}$ is not defined.

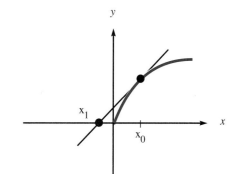

19. (a) $f'(x) = (3x^{2/3})^{-1}$

(b)

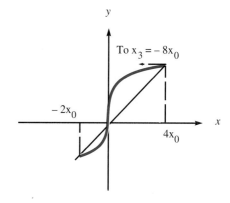

CHAPTER 3, REVIEW PROBLEMS (PAGE 252)

1. (a)

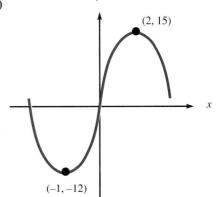

CHAPTER 3, SECTION 6 (PAGE 251)

1. 3.464   3. 0.755   5. 2.095

7. 4.791 and 0.209   9. 1.41421   11. 2.08008

13. 0.19612

15. $f(x) \ge \dfrac{26}{9}$, $f(x)$ does not intersect the x axis.

(b)

(c)

(d)

2.

3. (a)

(b)

(c)

(d)

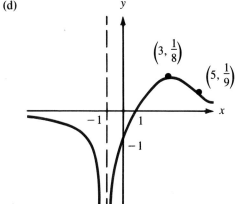

4. | Relative Maxima | Relative Minima |
    |---|---|
    | (a) (2, 15) | $(-1, -12)$ |
    | (b) $(-2, -4)$ | (0, 0) |
    | (c) $(-2, -6)$ | (2, 10) |

5. | Absolute Maximum | Absolute Minimum |
    |---|---|
    | (a) $f(-3) = 40$ | $f(-1) = -12$ |
    | (b) $f(2) = 6$ | $f(3) = -37$ |
    | (c) $f\left(-\frac{1}{2}\right) = f(1) = \frac{1}{2}$ | $f(0) = 0$ |
    | (d) None | $f(2) = 10$ |

6. The maximum speed is 52 miles per hour at 1:00 P.M. and 7:00 P.M., and the minimum speed is 20 miles per hour at 5:00 P.M.

7. $R(Q) = kQ(P - Q)$

8. $6.50

9. Each plot should be 50 meters by 37.5 meters.

10. (a) 80 feet by 80 feet   (b) 80 feet by 160 feet

11. $h = \frac{3}{2}r$

12. After 2 hours and 20 minutes on the job

13. 77 or 78   14. 12

15. Row all the way to the town.

16. $f(x) = -3x^3 + 9x^2 - 1$

17. (a) 8   (b) 13.156

18. $\frac{64}{7}$ square units

19. $x = 4$, $y = 7$

20. 5.3167 miles on shore

21. Rectangle: 3.8956 by 4.1566 feet; side of triangle: 3.8956

22. 7.41620   23. -3.104

24. -0.307, 0.459, -1.542, -2.306

CHAPTER 4, SECTION 1 (PAGE 267)

1. $e^2 \simeq 7.389$, $e^{-2} \simeq 0.135$, $e^{0.05} \simeq 1.051$, $e^{-0.05} \simeq 0.951$, $e^0 = 1$, $e \simeq 2.718$, $\sqrt{e} \simeq 1.649$, $\frac{1}{\sqrt{e}} \simeq 0.607$

3.

5.

7.

9.

11.

13.

15.

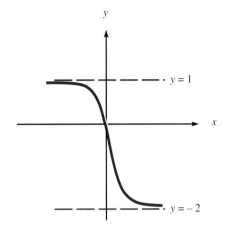

17. 400

19. 320

21. 20

23. (a) $1,967.15   (b) $2,001.60
    (c) $2,009.66   (d) $2,013.75

25. Quadruple

27. (a) $P = Be^{-rt}$ (b) $5,488.12

29. (a) 6.14 percent   (b) 6.18 percent

31.

n	1,000	10,000	25,000	50,000
$\left(1 + \dfrac{1}{n}\right)^n$	2.71692	2.71815	2.71823	2.71825

33.

n	10	100	1000	10,000	100,000
$\left(1 + \dfrac{3}{n}\right)^{2n}$	190.5	369.36	399.82	403.07	403.39

CHAPTER 4, SECTION 2 (PAGE 274)

1. (a) 50 million   (b) 91.11 million

3. 202.5 million

5. 324 billion dollars

7. (a) 12,000 people per square mile
   (b) 5,959 people per square mile

9. 204.8 grams

11. (a)

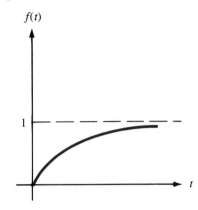

(b) 0.7408
(c) 0.0888

13. (a)

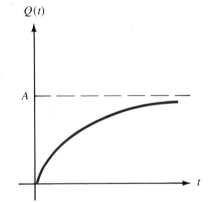

(b) The height of the graph approaches A because the number of facts recalled approaches the total number of relevant facts
in the person's memory.

15. (a)

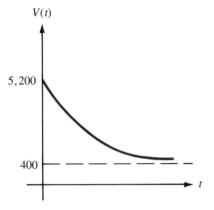

$V(t)$

5,200

400

t

(b) $5,200
(c) $1,049.61

17. 18.75° Celsius

19. (a)

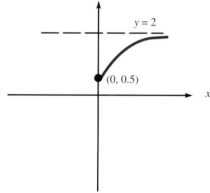

y

$y = 2$

(0, 0.5)

x

(b) 500 people   (c) 1,572   (d) 2,000.

21. (a)

$P(t)$

10

4

t

(b) 4 million
(c) 9.76 million
(d) The population will approach 10 million.

23. (a) 15   (b) 482

CHAPTER 4, SECTION 3 (PAGE 287)

1. $\ln 1 = 0$, $\ln 2 \approx 0.693$, $\ln e = 1$, $\ln 5 \approx 1.609$, $\ln \frac{1}{5} \approx -1.609$, $\ln e^2 = 2$; $\ln 0$ and $\ln -2$ are undefined.

3. 3

5. 5

7. $\frac{8}{25}$

9. 11.55

11. 0.402

13. $e^{-C-t/50}$

15. 4

17. 1.82

19. $\frac{\ln b}{\ln a} - 1$

21. -5.5

23. 11.55

25. 5.33 percent

27. 231.05 years

29. $\frac{1}{2} = e^{-k\lambda}$

31. 3,143 books

33. (a) $260 billion   (b) $324 billion

35. $P(t) = 40 \left[1 + \left(\frac{37}{3} \right) e^{-0.021t} \right]^{-1}$
    6.325 billion people

37. 5,730 years

39. $e^{-459.7} = 0$ (by calculator approximation)

41. (a) 11,090 feet   (b) 0.417

43. $e^{\ln N} = N$

45.

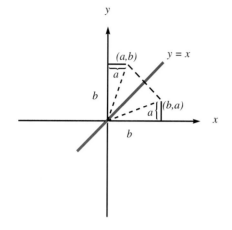

CHAPTER 4, SECTION 4 (PAGE 301)

1. $f'(x) = 5e^{5x}$

3. $f'(x) = 2(x + 1)e^{x^2+2x-1}$

5. $f'(x) = -0.5e^{-0.05x}$

7. $f'(x) = (6x^2 + 20x + 33)e^{6x}$

9. $f'(x) = (1 - x)e^{-x}$

11. $f'(x) = -6e^x(1 - 3e^x)$

13. $f'(x) = \dfrac{3}{2\sqrt{3x}} e^{\sqrt{3x}}$

15. $f'(x) = \dfrac{3}{x}$

17. $f'(x) = \dfrac{2x + 5}{x^2 + 5x - 2}$

19. $f'(x) = 2x \ln x + x$

21. $f'(x) = \dfrac{1}{x^2}(1 - \ln x)$

23. $f'(x) = \dfrac{-2}{(x + 1)(x - 1)}$

25. $f'(x) = 2$

27. $\dfrac{dy}{dx} = \dfrac{y(1 - xy)}{x(xy - 3)}$

29. $\dfrac{dy}{dx} = \dfrac{y(2x^2y^3 + 1)}{x(1 - 3x^2y^3)}$

31. (a) 1.22 million per year
    (b) Constant rate of 2 percent per year

33. (a) $1,082.68 per year
    (b) Constant rate of 40 percent per year (decreasing)

37. (a) 0.7805 billion per year
    (b) 3.375 percent per year

39. (a) Approximately 406 copies
    (b) 368 copies

41. $15 per radio

43.

45.

47.

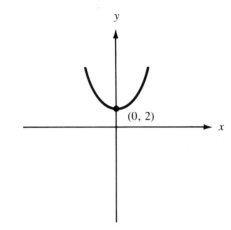

$(0, 2)$

53. $\dfrac{1}{4}f(x)\left(\dfrac{2}{2x + 1} + \dfrac{3}{1 - 3x}\right)$

55. $f(x)\left(-3 + \dfrac{1}{2x - 5} + \dfrac{20}{6 - 5x}\right)$

57. $a^x(\ln a)$

59. $f(x)(1 - \ln x)$

61. $f(x)(2 + \ln x)$

63. $f'(t) = kAC\dfrac{e^{-kt}}{(1 + Ce^{-kt})^2}$

$f\left(\dfrac{\ln C}{k}\right) = \dfrac{A}{2}$

65. $100k$

49.

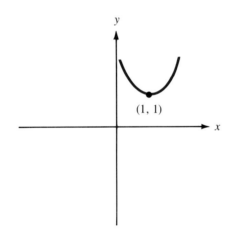

$(1, 1)$

CHAPTER 4, SECTION 5 (PAGE 314)

1. (a) $13,266.49   (b) $13,425.32
   (c) $13,591.41

3. (a) 5.86 years   (b) 5.78 years

5. Doubling time $= \dfrac{\ln 2}{k \ln(1 + r/k)}$

7. 18.58 years

9. Tripling time $= \dfrac{\ln 3}{k \ln (1 + r/k)}$

11. (a) 15.39 years   (b) 15.27 years

13. (a) 6.14 percent   (b) 6.18 percent

15. 10.51 percent and 10.74 percent

17. 10 percent

19. 5.83 percent

51.

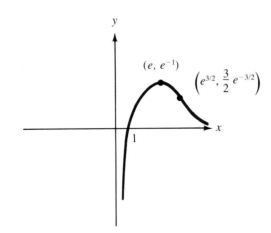

(e, e^{-1})

$\left(e^{3/2}, \dfrac{3}{2}e^{-3/2}\right)$

21. (a) $6,095.65   (b) $6,023.88

23. $6,653.15

25. $209.18

27. $1,732.55

29. 69.44 years from now

31. 6.5 years from now

CHAPTER 4, REVIEW PROBLEMS (PAGE 318)

1. (a)

(b)

(c)

(d)

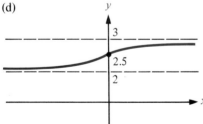

2. (a) $10\left(\dfrac{25}{10}\right)^4$ (b) $3\left(\dfrac{10}{3}\right)^2$ (c) $\dfrac{65}{2}$ (d) $\dfrac{6}{5}$

3. $8,000

4. (a) $Q(x)$

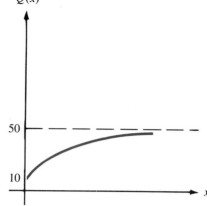

(b) 10,000
(c) 32,027
(d) $9,808
(e) 50,000

5. 60 units

6. (a) $P(t)$

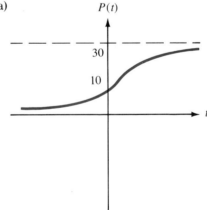

(b) 10 million
(c) 17.28 million
(d) The population will approach 30 million.

7. (a) $A = -9.0885, B = -1067.7, k = -0.00001112$
(b) 309 billion (c) If $-kBt < 0$, $P\infty = B$.
Here the model fails near $t = 186$ years and $P\infty$
is not valid.

8. (a) 5 (b) 2 (c) 32 (d) $3 \ln 3 \simeq 3.296$

9. (a) 34.66 (b) 0 (c) $e^2 \simeq 7.39$ (d) 1.86

10. 14.75 minutes

11. (a) $f'(x) = 6e^{3x+5}$

  (b) $f'(x) = x(2 - x)e^{-x}$

  (c) $f'(x) = \dfrac{x + 2}{x^2 + 4x + 1}$

  (d) $f'(x) = 2(1 + \ln x)$

  (e) $f'(x) = \dfrac{\ln 2x - 1}{(\ln 2x)^2}$

12. (a) $\dfrac{dy}{dx} = \dfrac{y(x + y)}{x(y - 2x)}$ (b) $\dfrac{dy}{dx} = \dfrac{y(1 - 3x^3y^2)}{x(2x^3y^2 + 1)}$

13. (a) 0.13 part per million per year
  (b) Constant rate of 3 percent per year

14. $140 per camera

15. (a)

(b)

(c)

(d)

(e)

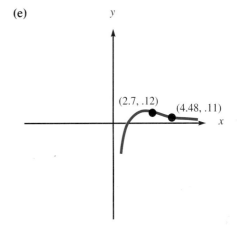

16. (a) $\dfrac{f(x)}{2}\left(\dfrac{2x}{x^2 + 1} + \dfrac{2x + 1}{x^2 + x + 2}\right)$

(b) $f(x)\left[-3 - \dfrac{1}{1 - 2x} - \dfrac{4(x + 1)}{x^2 + 2x - 3}\right]$

(c) $f(x)\left(\dfrac{2x + 1}{x} + 2 \ln x\right)$

(d) $f(x)\left(\dfrac{1}{\ln x} + \ln\sqrt{\ln x}\right)$

17. (a) 11.57 years   (b) 11.45 years

18. 8.20 percent compounded continuously

19. (a) $1,070.26   (b) $1,072.52

20. $7,433.55

21. 51.02 years from now

22. 6.52 percent

CHAPTER 5, SECTION 1 (PAGE 333)

1. $\dfrac{1}{6}x^6 + C$

3. $-\dfrac{1}{x} + C$

5. $5x + C$

7. $x^3 - \dfrac{2\sqrt{5}}{3}x^{3/2} + 2x + C$

9. $2x^{3/2} + \dfrac{1}{x^2} + \ln|x| + C$

11. $\dfrac{e^x}{2} + \dfrac{2x^{5/2}}{5} + C$

13. $\dfrac{1}{3}\ln|x| + \dfrac{3}{2x} + e^2x + \dfrac{x^{3/2}}{3} + C$

15. $x + \ln x^2 - \dfrac{1}{x} + C$

17. $-\dfrac{5x^4}{4} + \dfrac{11x^3}{3} - x^2 + C$

19. $\dfrac{2}{7}x^{7/2} - \dfrac{2}{3}x^{3/2} + C$

21. 10,128

23. 20 meters

25. $\displaystyle\int_2^4 r(t)dt$

27. $7,267,840

29. 15 units

31. $436

33. $2,300

35. $f(x) = x^3 + 3x^2 - 2x + 6$

37. $f(x) = x^3/3 - 5x^2/2 + 4x$; not unique

39. $\dfrac{3}{5}e^{5x} + C$

41. $\dfrac{2}{3}\ln|3x + 2| + C$

43. $\dfrac{1}{12}(2x + 3)^6 + C$

45. $\displaystyle\int_0^{12} n(x)p(x)\,dx$

47. 154.88 feet

49. Maybe; yes if the car's hood is positioned between the legs under the camel's belly.

CHAPTER 5, SECTION 2 (PAGE 343)

1. $\dfrac{1}{12}(2x + 6)^6 + C$

3. $\dfrac{1}{6}(4x - 1)^{3/2} + C$

5. $-e^{1-x} + C$

7. $\dfrac{1}{2}e^{x^2} + C$

9. $\dfrac{1}{12}(x^2 + 1)^6 + C$

11. $\dfrac{4}{21}(x^3 + 1)^{7/4} + C$

13. $\dfrac{2}{5}\ln|x^5 + 1| + C$

15. $\dfrac{1}{26}(x^2 + 2x + 5)^{13} + C$

17. $\frac{3}{5} \ln |x^5 + 5x^4 + 10x + 12| + C$

19. $-\frac{3}{2} \left(\frac{1}{x^2 - 2x + 6} \right) + C$

21. $\frac{1}{2} (\ln 5x)^2 + C$

23. $-\frac{1}{\ln x} + C$

25. $\frac{1}{2} [\ln(x^2 + 1)]^2 + C$

27. $x + \ln |x - 1| + C$

29. $-\frac{1}{4} (x - 5)^{-4} - (x - 5)^{-5} + C$

31. $\ln |x - 4| - \frac{7}{x - 4} + C$

33. $\frac{1}{5} (2x - 1)^{5/2} + \frac{4}{3} (2x - 1)^{3/2} + C$

35. $y = -\frac{1}{3} \ln |1 - 3x^2| + 5$

37. $849.61

39. $510.56 per acre

CHAPTER 5, SECTION 3 (PAGE 350)

1. $-(x + 1)e^{-x} + C$

3. $-5(x + 5)e^{-x/5} + C$

5. $(2 - x)e^x + C$

7. $\frac{1}{2} x^2 \left(\ln 2x - \frac{1}{2} \right) + C$

9. $\frac{2}{3} x(x - 6)^{3/2} - \frac{4}{15} (x - 6)^{5/2} + C$

11. $\frac{1}{9} x(x + 1)^9 - \frac{1}{90} (x + 1)^{10} + C$

13. $2x(x + 2)^{1/2} - \frac{4}{3} (x + 2)^{3/2} + C$

15. $-(x^2 + 2x + 2)e^{-x} + C$

17. $(x^3 - 3x^2 + 6x - 6)e^x + C$

19. $\frac{1}{3} x^3 \ln x - \frac{1}{9} x^3 + C$

21. $-\frac{1}{x} (\ln x + 1) + C$

23. $\frac{1}{2} (x^2 - 1)e^{x^2} + C$

25. $\frac{1}{36} x^4 (x^4 + 5)^9 - \frac{1}{360} (x^4 + 5)^{10} + C$

27. $f(x) = \frac{1}{4} x^2 \ln x - \frac{1}{8} x^2 - \frac{5}{2} - \ln 2$

29. 176.87

31. $666.70

33. (b) $\left(\frac{1}{5} x^3 - \frac{3}{25} x^2 + \frac{6}{125} x - \frac{6}{625} \right) e^{5x} + C$

CHAPTER 5, SECTION 4 (PAGE 356)

1. $-\frac{1}{3} \ln \left| \frac{x}{2x - 3} \right| + C$

3. $\ln |x + \sqrt{x^2 + 25}| + C$

5. $\frac{1}{4} \ln \left| \frac{2 + x}{2 - x} \right| + C$

7. $\frac{1}{2} \ln \left| \frac{x}{3x + 2} \right| + C$

9. $\left(\frac{1}{3} x^2 - \frac{2}{9} x + \frac{2}{27} \right) e^{3x} + C$

11. $-\frac{1}{2} \ln |2 - x^2| + C$

13. $x(\ln 2x)^2 - 2x \ln 2x + 2x + C$

15. $\frac{1}{3\sqrt{5}} \ln \left| \frac{\sqrt{2x + 5} - \sqrt{5}}{\sqrt{2x + 5} + \sqrt{5}} \right| + C$

17. $\frac{1}{2} x + \frac{1}{2} \ln |2 - 3e^{-x}| + C$

19. $\ln \left(\frac{a}{b} \right) = \ln a - \ln b$

21. $x(\ln x)^3 - 3x(\ln x)^2 + 6x(\ln x) - 6x + C$

CHAPTER 5, REVIEW PROBLEMS (PAGE 358)

1. $\frac{1}{6}x^6 - x^3 - \frac{1}{x} + C$

2. $\frac{3}{5}x^{5/3} - \ln|x| + 5x + \frac{2}{3}x^{3/2} + C$

3. $\frac{2}{9}(3x + 1)^{3/2} + C$

4. $\frac{1}{3}(3x^2 + 2x + 5)^{3/2} + C$

5. $\frac{1}{12}(x^2 + 4x + 2)^6 + C$

6. $\frac{1}{2}\ln|x^2 + 4x + 2| + C$

7. $-\frac{3}{4}\left(\frac{1}{2x^2 + 8x + 3}\right) + C$

8. $\frac{1}{13}(x - 5)^{13} + C$

9. $\frac{1}{14}(x - 5)^{14} + \frac{5}{13}(x - 5)^{13} + C = \frac{x}{13}(x - 5)^{13} - \frac{1}{182}(x - 5)^{14} + C$

10. $\frac{5}{3}e^{3x} + C$

11. $\left(\frac{5}{3}x - \frac{5}{9}\right)e^{3x} + C$

12. $-2(x + 2)e^{-x/2} + C$

13. $\frac{1}{3}(x^3 - 1)e^{x^3} + C$

14. $10(2x - 19)e^{0.1x} + C$

15. $\frac{1}{2}x^2 \ln 3x - \frac{1}{4}x^2 + C$

16. $x \ln 3x - x + C$

17. $\frac{1}{2}(\ln 3x)^2 + C$

18. $-\frac{1}{x}(\ln 3x + 1) + C$

19. $\frac{1}{18}x^2(x^2 + 1)^9 - \frac{1}{180}(x^2 + 1)^{10} + C$

20. $(x^2 + 1)[\ln(x^2 + 1) - 1] + C$

21. $y = \frac{1}{8}(x^2 + 1)^4 + 3$

22. 11,250

23. 10,945

24. $2,265.80

25. $1.62

26. $1,000

27. $\frac{2}{3}$ meter

28. $75

29. $\int \frac{5dx}{8 - 2x^2} = \frac{5}{8}\ln\left|\frac{2 + x}{2 - x}\right| + C$

30. $\frac{2}{3}\ln|3x + \sqrt{9x^2 + 16}| + C$ or

$\frac{2}{3}\ln\left|x + \sqrt{x^2 + \left(\frac{4}{3}\right)^2}\right| + C$

31. $e^{-x/2}(-2x^2 - 8x - 16) + C$

32. $\int \frac{4\,dx}{x(9 + 5x)} = \frac{4}{9}\ln\left|\frac{x}{9 + 5x}\right| + C$

CHAPTER 6, SECTION 1 (PAGE 366)

1. $\frac{9}{20}$

3. 144

5. $\frac{8}{3} + \ln 3$

7. $\frac{2}{9}$

9. $-\frac{16}{3}$

11. $\frac{4}{3}$

13. $\frac{7}{6}$

15. e

17. $e^2 + 1$

19. $-3e^{-2} - e^2$

21. $\frac{8}{3}$

23. 152.85

25. 98 people

27. A decrease of $1,870

29. $774

31. $75

33. $4,081,077

35. (b) 1   (c) $\frac{70}{3}$

CHAPTER 6, SECTION 2 (PAGE 378)

1. $\frac{8}{3}$ 3. 15

5. $\frac{38}{3}$ 7. $\frac{4}{3}$

9. $\frac{1}{2}$ 11. $\frac{7}{6}$

13. $\frac{128}{3}$

15. (a) 0.0577   (b) 0.4512   (c) 0.5488

17. (a) 0.1447   (b) 0.6321   (c) 0.3679

19. 12

21. 0.72

23. $\frac{32}{3}$

25. (a) $\frac{2}{3}$ (b) $\frac{14}{3}$

27. $\frac{1}{2}$

29. $\frac{40}{3}$

31. 7.99

33. $\frac{131}{4}$

CHAPTER 6, SECTION 3 (PAGE 389)

1. (a) 12 years   (b) $1,008

3. (a) 9 years   (b) $12,150

5. (a) $\frac{184}{3}$ (b) Excess production = difference between areas

7. (a) 8 weeks   (b) $15,069.26   (c) Difference between area under income and expenditure curves

9. $1,000

11. (a) 635.47

(b)

13. (a) 563.99

(b)

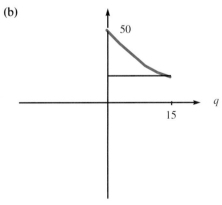

15. (a) $q = 3$; CS = \$63

(b)

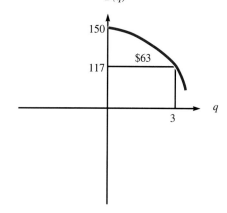

17. (a) $q = 36$, CS = \$1,122.07

(b)

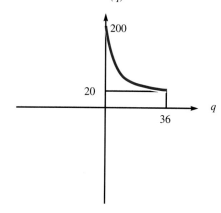

19. (a) $q = 27.47$; CS = \$225.35

(b)

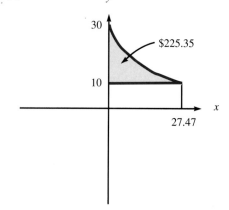

21. (a) $q = 20$ (b) CS = \$200

23. $q_E = 2$; CS = \$3.09

CHAPTER 6, SECTION 4 (PAGE 407)

1. \$13,994.35

3. \$27,124.92

5. \$63,932.98

7. \$28,591.25

9. \$76,424.11

11. 80

13. 4,207

15. 116,039

17. Velocity = quantity of blood in cubic centimeters per second through the cross section.

19. $15\pi/2$

21. $512\pi/15$

23. $0.9\,\pi$

25. 13.58

27. $\pi r^2 h/3$

29. 2/3

31. ln 2

33. \$1.57 per pound

35. 2,272

37. 30,000

39. πR^2

41. $P_0 f(N) + \displaystyle\int_0^N r(t) f(N - t)\, dt$

43. 207,360

CHAPTER 6, SECTION 5 (PAGE 420)

1. (a) 2.3438   (b) 2.3333

3. (a) 0.7828   (b) 0.7854

5. (a) 1.1515   (b) 1.1478

7. (a) 0.7430   (b) 0.7469

9. (a) 0.5090; $|E_4| \leq 0.0313$
   (b) 0.5004; $|E_4| \leq 0.0026$

11. (a) 2.7967; $|E_{10}| \leq 0.0017$
    (b) 2.7974; $|E_{10}| \leq 0.00002$

13. (a) 1.4907; $|E_4| \leq 0.0849$
    (b) 1.4637; $|E_4| \leq 0.0045$

15. (a) $n = 164$ (b) $n = 18$

17. (a) $n = 36$ (b) $n = 6$

19. (a) $n = 179$ (b) $n = 8$

21. 3.142

23. If $n = 2, A = 1.05$, if $n = 10, A = 1.107$, if $n = 20$, $A = 1.1104$, if $n = 130, A = 1.11145$

CHAPTER 6, REVIEW PROBLEMS (PAGE 423)

1. 0

2. $\dfrac{17}{3}$

3. 1,710

4. $1 - e^{-1}$

5. $\dfrac{65}{8}$

6. $\dfrac{3}{5}$

7. $2e^{-1}$

8. $\dfrac{1}{2}$

9. $\dfrac{1}{9}(2e^3 + 1)$

10. $55e^2 + 45$

11. 126 people

12. $76.80

13. 259.49 billion barrels per year

14. 36

15. $\dfrac{3}{4}$

16. $\dfrac{9}{2}$

17. $\dfrac{16}{3} + 8 \ln 2$

18. $\dfrac{3}{10}$

19. $\dfrac{13}{2}$

20. (a) 22.10 percent   (b) 55.07 percent
    (c) 44.93 percent

21. (a) 15   (b) $33,750

22. (a) 3 units   (b) $72   (c) $27

23. $7,377.37

24. $7,191.64

25. 62

26. $\dfrac{234\pi}{15}$

27. 6.32π

28. $1.32 per pound

29. (a) 1.1016; $|E_{10}| \leq 0.0133$
    (b) 1.0987; $|E_{10}| \leq 0.0004$

30. (a) 17.5651; $|E_8| \leq 10.2372$
    (b) 16.5386; $|E_8| \leq 1.0901$

31. (a) $n = 58$ (b) $n = 8$

32. (a) $n = 59$ (b) $n = 6$

CHAPTER 7, SECTION 1 (PAGE 443)

1. $\dfrac{dQ}{dt} = kQ$

3. $\dfrac{dQ}{dt} = 0.07Q$

5. $\dfrac{dP}{dt} = 500$

7. $\dfrac{dQ}{dt} = k(M - Q)$, where M is the temperature of the surrounding medium and k is a positive constant.

9. $\frac{dQ}{dt} = k(N - Q)$, where N is the total number of relevant facts in the person's memory.

11. $\frac{dQ}{dt} = kQ(N - Q)$, where N is the total number of people involved.

17. $y = x^3 + \frac{5}{2}x^2 - 6x + C$

19. $V = \ln (x + 1)^2 + C$

21. $P = 25t^2 + C_1 t + C_2$ 23. $y = Ce^{3x}$

25. $y = -\ln (C - x)$ 27. $y = \pm \sqrt{x^2 + C}$

29. $y = Ce^x - 10$ 31. $y = \frac{1}{2 + Be^{-x}}$

33. $y = x^5 - x^3 - 2x + 6$

35. $A = 4e^{-t/2} + 3t - 2$

37. $y = \frac{2}{3 - 2x^4}$

39. $y = \frac{1}{1 + 2e^{-x}}$

41. (a) $C(q) = (q - 4)^3 + C_0$ (b) $\$1,500$

43. (a) $Q(t) = 0.03(36 + 16t - t^2)^{1/2} + 0.07$
    (b) 0.37 part per million at 3:00 p.m.

45. 75

47. $C(t) = Ce^{-kt}$

49. $y' = Ce^x$

51. $S(t) = Q_0(1 - e^{-kt})$

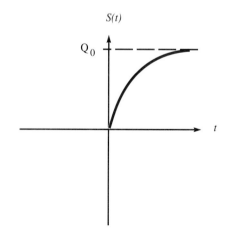

S(t)

7.1.51

53. $T = T_m + (T_0 - T_m)e^{-kt}$

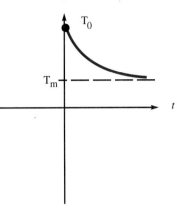

7.1.53

55. $Q(t) = 400 + 200e^{-t/40}$

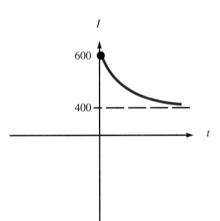

57. $Q(t) = \frac{2,000}{1 + 3e^{-0.8066t}}$

59. 45   7.1.55

61. (a) $P(t) = 3 + 3e^{-(\ln 3)t/4}$ (b) $\lim_{t \to \infty} e^{-t} = 0$

63. $S = S_0 + Ce^{-kAt}$, where A = area of cell wall

CHAPTER 7, SECTION 2 (PAGE 452)

1. $y = \frac{x^2}{5} + \frac{C}{x^3}$

3. $y = \sqrt{x} e^x - \frac{e^x}{\sqrt{x}} + \frac{C}{\sqrt{x}}$

5. $y = \dfrac{2}{x}\ln|x| + \dfrac{C}{x}$

7. $y = \dfrac{x}{2}e^{-2x} + \dfrac{C}{x}e^{-2x}$

9. $y = 2x^3$

11. $y = x - 1 + 5e^{-x}$

13. $y = \dfrac{5}{3} + Ce^{-3x}$

15. $y = -x - 1 + 2e^{x+1}$

17. \$260,578

19. \$148,043

21. (a) $S(t) = \dfrac{(60 + t)t + 300}{30 + t}$ (b) 38

23. (a) $0.03 + 0.22e^{-t/2.5}$ percent
    (b) 3.7 minutes   (c) 0.03 percent

25. (a) $x(t) = \dfrac{r}{s}(1 - e^{-st})$ (b) $\dfrac{r}{s}$

27. (b) $P(t) = \dfrac{m}{s - r} + \left(P_0 + \dfrac{m}{r - s}\right)e^{(r-s)t}$, where
    $P_0 = P(0)$ (c) 125.4613 million

CHAPTER 7, REVIEW PROBLEMS (PAGE 457)

1. $y = \dfrac{1}{4}x^4 - x^3 + 5x + C$

2. $y = Ce^{0.02x}$

3. $y = 80 - Ce^{-kx}$

4. $y = \dfrac{1}{1 + Ce^{-x}}$

5. $e^{x+y} + Ce^y = -1$ or $e^x + e^{-y} = C$

6. $\ln y = -\dfrac{1}{2}e^{-2x} + C$

7. $y = e^{-x}\left(-1 - \dfrac{4}{x} - \dfrac{12}{x^2} - \dfrac{24}{x^3} - \dfrac{24}{x^4}\right) + \dfrac{C}{x^4}$

8. $y = 2\left(1 - \dfrac{1}{x}\ln|x + 1|\right) + Cx^{-1}$

9. $y = \dfrac{1}{(x + 1)^2}(2x^2 + 4\ln|x + 2| + C)$

10. $y = 5\left(\dfrac{1}{x} + 1\right) + Ce^{1/x}$

11. $y = x^5 - x^3 - 2x + 6$

12. $y = 100e^{0.06x}$

13. $y = 3 - e^{-x}$

14. $y = x^2 + 3x + 5$

15. $y = -\dfrac{x^3}{2} - \dfrac{7x^5}{2}$

16. $y = \dfrac{x^3}{3(x + 1)} + \dfrac{x^2}{2(x + 1)} - \dfrac{27}{2(x + 1)}$

17. $y = e^{x^2/2}(x + 4)$

18. $y = \dfrac{1}{3}(1 - 7e^{-x^3})$

19. \$22,857   20. \$549,504

21. 18.75 pounds

22. $Q(t) = \dfrac{10}{1 + 1.5e^{-0.06t}}$

23. $\dfrac{dQ}{dt} = 18 - \dfrac{18Q}{5,000}$; 640 days

24. (a) $S(t) = 2(50 + t) - \dfrac{3,500}{50 + t}$ (b) 53 gallons

25. $P = Ce^{-kt/x}$ or $Ce^{-(k/x)t}$ 26. $P = C(P_0)^{\beta^t}$

CHAPTER 8, SECTION 1 (PAGE 472)

1. ∞ 3. $-\infty$ 5. $\dfrac{1}{2}$ 7. 0   9. ∞

11. $-\infty$ 13. $\dfrac{3}{2}$ 15. ∞ 17. ∞ 19. 0

21. 0   23. 0   25. 0   27. 1   29. ∞

31. 0   33. 0   35. 1   37. e^2

CHAPTER 8, SECTION 2 (PAGE 480)

1. $\dfrac{1}{2}$ 3. ∞ 5. ∞ 7. $\dfrac{1}{10}$ 9. $\dfrac{5}{2}$

11. $\dfrac{1}{9}$ 13. ∞ 15. $\dfrac{2}{e}$ 17. $\dfrac{2}{9}$ 19. $5e^{10}$

21. ∞ 23. 2   25. \$20,000

27. \$150,000   31. 200   33. 50

CHAPTER 8, SECTION 3 (PAGE 494)

1. (a) 1   (b) $\frac{1}{3}$ (c) $\frac{1}{3}$

3. (a) 1   (b) $\frac{3}{16}$ (c) $\frac{9}{16}$

5. (a) 1   (b) $\frac{7}{8}$ (c) $\frac{1}{8}$

7. (a) 1   (b) 0.3496   (c) 0.9817

9. $\frac{1}{3}$

11. $\frac{1}{6}$

13. 0.1353

15. $E(x) = \frac{7}{2}$; Var $(x) = \frac{3}{4}$

17. $E(x) = \frac{4}{3}$; Var $(x) = \frac{8}{9}$

19. $E(x) = 10$; Var $(x) = 1,900$

21. 10 minutes   23. 4 minutes

CHAPTER 8, REVIEW PROBLEMS (PAGE 499)

1. $-\infty$ 2. $\frac{-3}{2}$ 3. $-\infty$ 4. $\frac{2}{3}$

5. 0   6. ∞ 7. ∞ 8. $\frac{1}{2}$

9. ∞ 10. ∞ 11. 1   12. e^6

13. ∞ 14. 1   15. ∞ 16. $\frac{3}{5}$

17. $\frac{1}{4}$ 18. $\frac{2}{3}$ 19. $\frac{1}{4}$ 20. $\frac{1}{\ln 2}$

21. ∞ 22. $\frac{1}{3}$ 23. 10,000

24. $120,000

25. The population will increase without bound.

26. (a) 1   (b) $\frac{1}{3}$ (c) $\frac{1}{3}$

27. (a) 1   (b) $\frac{1}{3}$

28. (a) 1   (b) 0.1484   (c) 0.3679

29. $E(x) = \frac{5}{2}$; Var $(x) = \frac{3}{4}$

30. $E(x) = 1$; Var $(x) = \frac{1}{2}$

31. $E(x) = 5$; Var $(x) = 25$

32. $\frac{2}{9}$

33. (a) 0.0498   (b) 2 minutes

CHAPTER 9, SECTION 1 (PAGE 512)

1. The domain of f consists of all ordered pairs (x, y) of real numbers; $f(2, -1) = -3$ and $f(1, 2) = 16$.

3. The domain of f consists of all ordered pairs (x, y) of real numbers for which $|y| \geq |x|$; $f(4, 5) = 3$ and $f(-1, 2) = \sqrt{3} \approx 1.732$.

5. The domain of f consists of all ordered pairs (x, y) of real numbers for which $y > 0$ and $y \neq 1$; $f(-1, e^3) = -\frac{1}{3}$ and $f(\ln 9, e^2) = \ln 3 \approx 1.099$.

7.

9.

11.

13.

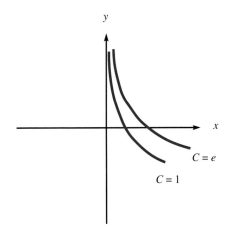

15. (a) 160,000
    (b) The level will increase by 16,400 units.
    (c) The level will increase by 4,000 units.
    (d) The level will increase by 20,810 units.

17. (a) $R(x_1, x_2) = 200x_1 - 10x_1^2 + 25x_1x_2 + 100x_2 - 10x_2^2$
    (b) $R(6, 5) = \$1,840$

19. (a) $f(x, y)$ is more than doubled (b) $f(x, y)$ is increased (if $0 < a + b$) but less than doubled (c) $f(x, y)$ is doubled

21. $R(x, y) = 60x + 50y - \dfrac{x^2}{5} - \dfrac{y^2}{10} + \dfrac{xy}{10}$; profit = $R(x, y) - $ cost

23. $c^{x+y} = c^x c^y$

25. $2x^3y^2 = 4,000$ y

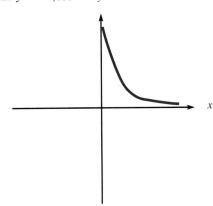

CHAPTER 9, SECTION 2 (PAGE 521)

1. $f_x = 2y^5 + 6xy + 2x; f_y = 10xy^4 + 3x^2$

3. $\dfrac{\partial z}{\partial x} = 15(3x + 2y)^4; \dfrac{\partial z}{\partial y} = 10(3x + 2y)^4$

5. $f_x = \dfrac{-3y}{2x^2}; f_y = \dfrac{3}{2x}$

7. $\dfrac{\partial z}{\partial x} = (xy + 1)e^{xy}; \dfrac{\partial z}{\partial y} = x^2e^{xy}$

9. $f_x = -\dfrac{e^{2-x}}{y^2}; f_y = -\dfrac{2e^{2-x}}{y^3}$

11. $f_x = \dfrac{5y}{(y - x)^2}; f_y = \dfrac{-5x}{(y - x)^2}$

13. $\dfrac{\partial z}{\partial x} = \ln y; \dfrac{\partial z}{\partial y} = \dfrac{x}{y}$

15. $f_x = \dfrac{1}{(x + 2y)y^2}; f_y = \dfrac{2[y - (x + 2y)\ln(x + 2y)]}{(x + 2y)y^3}$

17. Daily output will increase by approximately 10 units.

19. 12 cents

21. (a) Butter and margarine

   (b) $\dfrac{\partial Q_1}{\partial p_2} \geq 0$ and $\dfrac{\partial Q_2}{\partial p_1} \geq 0$ (c) No

23. $f_{xx} = 60x^2y^3, f_{yy} = 30x^4y, f_{xy} = f_{yx} = 60x^3y^2 + 2$

25. $f_{xx} = 2y(1 + 2x^2y)e^{x^2y}, f_{yy} = x^4e^{x^2y},$
   $f_{xy} = f_{yx} = 2x(1 + x^2y)e^{x^2y}$

27. $f_{xx} = \dfrac{y^2}{(x^2 + y^2)^{3/2}}, f_{yy} = \dfrac{x^2}{(x^2 + y^2)^{3/2}},$
   $f_{xy} = f_{yx} = \dfrac{-xy}{(x^2 + y^2)^{3/2}}$

29. $\dfrac{\partial^2 Q}{\partial K^2} \approx$ change in marginal product of capital generated by an increase in capital investment of $1,000.

31. (a) $\dfrac{\partial^2 Q}{\partial L^2} > 0$ for $L < L_0$ and $\dfrac{\partial^2 Q}{\partial L^2} < 0$ for $L > L_0$

CHAPTER 9, SECTION 3 (PAGE 533)

1. $\dfrac{dz}{dt} = 7$

3. $\dfrac{dz}{dt} = \dfrac{1}{3}$

5. $\dfrac{dz}{dt} = -3t^{-4}$

7. $\dfrac{dz}{dt} = -162t^5$

9. 23

11. $-\dfrac{1}{3}$

13. 5

15. $\dfrac{y}{x}$

17. $\dfrac{3 - 2xy - 2e^{-x}}{x^2 + 6y^2}$

19. $-\dfrac{y}{x}\ln y$

21. $-\dfrac{1}{6}$

23. 2

25. $-e$

27. -0.42

29. -0.21

31. 2.6

33. The monthly demand will be increasing at the rate of 7 per month.

35. Approximately 61.6 additional units will be produced.

37. Daily profit will increase by approximately 24 cents.

39. 112 square yards

41. 0.5 percent

43. 1 percent

45. $\dfrac{d^2z}{dt^2} = a^2f_{xx} + 2abf_{xy} + b^2f_{yy}$

CHAPTER 9, SECTION 4 (PAGE 545)

	Relative Maxima	Relative Minima	Saddle Points
1.	(0, 0)	None	None
3.	None	None	(0, 0)
5.	(−2, −1)	(1, 1)	(−2, 1) and (1, −1)
7.	None	$\left(4, \dfrac{19}{2}\right)$	$\left(2, \dfrac{7}{2}\right)$
9.	(0, ±1)	(0, 0)	(±1, 0)
11.	None	$\left(\dfrac{4}{3}, \dfrac{4}{3}\right)$	(0, 0)

13. $x = 2.7, y = 2.5$

15. $x = 4, y = 6$

17. (a) $x = 450 - 3C$ (b) $y = 500 - 5C$
   (c) $x = 450 - 3C$ and $y = 500 - 5C$
   (d) Independence of markets. Note: C = cost of manufacturing one machine

19. $x = 2, y = 0$

21. $\sqrt{3}$

23. $x = y = z = \dfrac{20}{3}$

25. Saddle point

CHAPTER 9, SECTION 5 (PAGE 559)

1. $f\left(\dfrac{1}{2}, \dfrac{1}{2}\right) = \dfrac{1}{4}$

3. $f(1, 1) = f(-1, -1) = 2$

5. $f(0, 2) = f(0, -2) = -4$

7. $\dfrac{3}{2}, -3$

9. 40 meters by 80 meters

11. 11,664 cubic inches

13. Radius = 1 inch, height = 4 inches

15. \$40,000 on labor; \$80,000 on equipment

17. Maximum output would increase by approximately 31.75 units.

19. (a) \$4,000 on development and \$6,000 on promotion   (b) $\lambda = 0$

23. $Q_x = \lambda p, Q_y = \lambda q$

25. $C_x = \lambda A \alpha x^{\alpha - 1} y^{\beta}$

CHAPTER 9, SECTION 6 (PAGE 569)

1. $y = \dfrac{1}{4}x + \dfrac{3}{2}$

3. $y = 3$

5. $y = 0.78x + 1.06$

7. $y = -\dfrac{1}{2}x + 4$

9. $y = 1.0184x + 0.8022$

11. (b) $y = 0.42x - 0.71$ (c) 1,306

13. (b) $y = 3.05x + 6.10$ (c) 42.7 percent

15. $y = 18.68x + 70.91$
   (a) 201.67 (in 1970)   (b) 239.03   (c) 257.7

CHAPTER 9, REVIEW PROBLEMS (PAGE 572)

1. (a) $f_x = 6x^2y + 3y^2 - \dfrac{y}{x^2}, f_y = 2x^3 + 6xy + \dfrac{1}{x}$

  (b) $f_x = 5y^2(xy^2 + 1)^4, f_y = 10xy(xy^2 + 1)^4$

  (c) $f_x = y(1 + xy)e^{xy}, f_y = x(1 + xy)e^{xy}$

  (d) $f_x = \dfrac{2(x^2 + xy + y^2)}{(2x + y)^2}$

     $f_y = \dfrac{-x^2 - 4xy - y^2}{(2x + y)^2}$

  (e) $f_x = \dfrac{1}{x} - \dfrac{1}{x + 3y}$

     $f_y = \dfrac{1}{y} - \dfrac{3}{x + 3y}$

2. (a) $f_{xx} = 2, f_{yy} = 6y - 4x, f_{xy} = f_{yx} = -4y$

  (b) $f_{xx} = (2 + 4x^2)e^{x^2 + y^2}, f_{yy} = (2 + 4y^2)e^{x^2 + y^2}$

     $f_{xy} = f_{yx} = 4xye^{x^2 + y^2}$

  (c) $f_{xx} = 0, f_{yy} = \dfrac{-x}{y^2}, f_{xy} = f_{yx} = \dfrac{1}{y}$

3. Daily output will increase by approximately 16 units.

4. $\dfrac{\partial^2 Q}{\partial K \partial L} > 0$

5. (a) $\dfrac{dz}{dt} = -30t^4 + 24t^2$ (b) $\dfrac{dz}{dt} = 4t$

6. The demand will drop by approximately 46 cans per week.

7. 1.33 percent

8. Decreasing at the rate of approximately 18 pies per month.

9. (a)

(b)

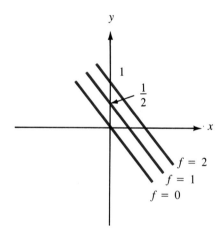

10. (a) $\frac{2}{3}$ (b) $-\frac{1}{2}$

11. The level of unskilled labor should be decreased by approximately 2 workers.

12. (a) Relative maximum at $(-2, 3)$; relative minimum at $(0, 9)$; saddle points at $(0, 3)$ and $(-2, 9)$

(b) Relative minimum at $\left(-\frac{23}{2}, 5\right)$; saddle point at $\left(\frac{1}{2}, 1\right)$

(c) Relative minimum at $(0, 9)$, relative maximum at $(-2, 3)$, saddle point $(0, 3)$ and $(-2, 9)$

13. Maximum value $= f(1, \sqrt{3}) = f(1, -\sqrt{3}) = 12$; minimum value $= f(-2, 0) = 3$

15. \$5,000 on development and \$8,000 on promotion

16. \$4,000 on development and \$7,000 on promotion

17. Maximum profit will increase by approximately \$235.

18. $D > 0, f_{xx} > 0$, relative minimum at $(2, 3)$

19. $R = -\frac{D}{2V}$

20. $Q_x = ax^{a-1} y^b$

21. 2 cubic centimeters

22. $y = \frac{4}{9}x + 1$

23. (b) $y = 11.54x + 44.45$ (c) \$102,150

CHAPTER 10, SECTION 1 (PAGE 580)

1. $\frac{7}{6}$

3. -1

5. $4 \ln 2$

7. 32

9. $\frac{1}{3}$

11. $\frac{3}{2}e^4 + \frac{1}{2}$

13. $\frac{1}{3} \ln 2$

15. $\frac{1}{2}$

17. $-\frac{1}{4}e^2 + e - \frac{1}{4}$

19. $\frac{32}{9}$

21. $\frac{e^2 - 1}{8}$

CHAPTER 10, SECTION 2 (PAGE 590)

1. $0 \le x \le 3$ and $x^2 \le y \le 3x$; $0 \le y \le 9$ and $\dfrac{y}{3} \le x \le \sqrt{y}$

3. $-1 \le x \le 2$ and $1 \le y \le 2$; $1 \le y \le 2$ and $-1 \le x \le 2$

5. $1 \le x \le e$ and $0 \le y \le \ln x$; $0 \le y \le 1$ and $e^y \le x \le e$

7. $\dfrac{3}{2}$

9. $\dfrac{1}{2}$

11. $\dfrac{44}{15}$

13. $32 \ln 2 - 12$

15. 3

17. 1

19. $\dfrac{3}{2} \ln 5$

21. $2e - 4$

23. $\displaystyle\int_0^4 \int_0^{\sqrt{4-y}} f(x, y) \, dx \, dy$

25. $\displaystyle\int_0^1 \int_{y^2}^{\sqrt[3]{y}} f(x, y) \, dx \, dy$

27. $\displaystyle\int_0^2 \int_1^{e^y} f(x, y) \, dx \, dy$

29. $\displaystyle\int_1^2 \int_{-\sqrt{y-1}}^{\sqrt{y-1}} f(x, y) \, dx \, dy$

31. $\displaystyle\int_0^1 \int_0^y f(x, y) \, dx \, dy + \int_1^2 \int_0^{2-y} f(x, y) \, dx \, dy$

33. $\displaystyle\int_0^4 \int_{-\sqrt{y}}^{\sqrt{y}} f(x, y) \, dx \, dy + \int_4^9 \int_{-\sqrt{y}}^{6-y} f(x, y) \, dx \, dy$

35. $\displaystyle\int_1^8 \int_{y^{1/3}}^y f(x, y) \, dx \, dy$

CHAPTER 10, SECTION 3 (PAGE 601)

1. 18

3. $\dfrac{16}{3}$

5. $\dfrac{4}{3}$

7. 1

9. $\dfrac{19}{6}$

11. 8

13. $e - 1$

15. $\dfrac{1}{2}e^2 - e + \dfrac{1}{2}$

17. $\dfrac{\displaystyle\iint_R f(x, y) \, dA}{\displaystyle\iint_R 1 \, dA}$

19. 188.61 per acre

21. 2

23. 0

25. $3(e - 2)$

27. 0.2285

29. 0.0803

31. 0.8452

33. 0.5269

CHAPTER 10, REVIEW PROBLEMS (PAGE 606)

1. -4

2. $\dfrac{1}{9}$

3. $\dfrac{4}{5} \ln 2$

4. $(e^{-2} - 1)(e^{-1} - 1) \simeq 0.5466$

5. $\dfrac{1}{8}(e^2 - 1)$

6. 81

7. $\dfrac{9}{2}$

8. 256

9. $e - 2$

10. 1,026

11. $\displaystyle\int_0^2 \int_{y^2}^{\sqrt{8y}} f(x, y)\, dx\, dy$

12. $\displaystyle\int_1^{e^2} \int_{\ln x}^{2} f(x, y)\, dy\, dx$

13. $\displaystyle\int_0^4 \int_0^{\sqrt{y}} f(x, y)\, dx\, dy + \int_4^8 \int_0^{4-y/2} f(x, y)\, dx\, dy$

14. $\displaystyle\int_0^1 \int_{-\sqrt{y}}^{\sqrt{y}} f(x, y)\, dx\, dy + \int_1^5 \int_{-\sqrt{y}}^{1} f(x, y)\, dx\, dy$
$\displaystyle\quad + \int_5^9 \int_{-\sqrt{y}}^{6-y} f(x, y)\, dx\, dy$

15. $\dfrac{9}{2}$

16. $e - 1$

17. $\dfrac{5}{6}$

18. $\dfrac{1}{3}$

19. $\dfrac{1}{2}(3 - e)$

20. $\dfrac{1}{2}$

21. 20

22. (a) $\displaystyle\iint_R f(x, y)\, dA$ (b) $\dfrac{\displaystyle\iint_R f(x, y)\, dA}{\displaystyle\iint_R 1\, dA}$

23. (a) 0.8625   (b) 0.9500

24. 0.4582

CHAPTER 11, SECTION 1 (PAGE 618)

1. $\displaystyle\sum_{n=1}^{\infty} \dfrac{1}{3^n}$

3. $\displaystyle\sum_{n=1}^{\infty} \dfrac{n}{n + 1}$

5. $\displaystyle\sum_{n=1}^{\infty} \dfrac{(-1)^{n+1}n^2}{n + 1}$

7. $\dfrac{15}{16}$

9. $-\dfrac{7}{12}$

11. $\dfrac{1}{4}$

13. $\dfrac{1}{2}$

15. $\dfrac{2}{3}$

17. 2

19. Diverges

21. -1

23. $-\dfrac{2}{5}$

25. Diverges

CHAPTER 11, SECTION 2 (PAGE 626)

1. 5

3. 3

5. Diverges

7. $\dfrac{3}{20}$

9. 45

11. $\dfrac{3}{16}$

13. 100

15. $\dfrac{1}{3}$

17. $\dfrac{25}{99}$

19. 575 billion dollars

21. \$12,358.32

23. $\dfrac{d}{q}(100 - q)$

25. Infinite geometric series; 2T

27. 0.8202

29. 2027

31. (a) $P(n) = 100(1 + b)^n - d\displaystyle\sum_{k=0}^{n-1}(1 + b)^k$

$= 100(1 + b)^n + \dfrac{d}{b}[1 - (1 + b)^n]$

(b) 229.8

CHAPTER 11, SECTION 3 (PAGE 640)

1. $\displaystyle\sum_{n=0}^{\infty}\dfrac{3^n}{n!}x^n$

3. $\displaystyle\sum_{n=1}^{\infty}\dfrac{(-1)^{n+1}}{n}x^n$

5. $\displaystyle\sum_{n=0}^{\infty}\dfrac{1}{(2n)!}x^{2n}$

7. $\displaystyle\sum_{n=0}^{\infty}\left(\dfrac{n+1}{n!}\right)x^n$

9. $\displaystyle\sum_{n=0}^{\infty}\dfrac{2^n e^2}{n!}(x - 1)^n$

11. $\displaystyle\sum_{n=0}^{\infty}(-1)^n(x - 1)^n$

13. $\displaystyle\sum_{n=0}^{\infty}(x - 1)^n$

15. $\displaystyle\sum_{n=0}^{\infty}5^n x^{n+1}\quad$ for $|x| < \dfrac{1}{5}$

17. $\displaystyle\sum_{n=0}^{\infty}\dfrac{x^n}{n!(-2)^n}\quad$ for all x

19. $\displaystyle\sum_{n=0}^{\infty}(1 - x)^n$

21. 1.94936

23. 0.09531

25. 1.34984

27. 0.46127

29. 0.09967

CHAPTER 11, REVIEW PROBLEMS (PAGE 642)

1. $\dfrac{5}{6}$

2. 1

3. $-\dfrac{1}{2}$

4. $-\dfrac{1}{2}$

5. Diverges

6. 1.5415

7. 72

8. $\dfrac{17}{11}$

9. 54 feet

10. \$3,921.67

11. 18.16 units

12. 0.3297

13. $\displaystyle\sum_{n=0}^{\infty} (-1)^n\left(\dfrac{1}{3}\right)^{n+1} x^n$

14. $\displaystyle\sum_{n=0}^{\infty} \dfrac{(-3)^n}{n!} x^n$

15. $\displaystyle\sum_{n=0}^{\infty} \left(\dfrac{2n+1}{n!}\right) x^n$

16. $\displaystyle\sum_{n=0}^{\infty} (n+1)(x+2)^n$

17. $\displaystyle\sum_{n=0}^{\infty} (-1)^n(4^{n+1})\, x^{2n+1}$ for $|x| < \dfrac{1}{2}$

18. 0.94869

19. 0.11919

CHAPTER 12, SECTION 1 (PAGE 658)

1. 30°

3. 120°

5. $-120°$

7.

9.

11.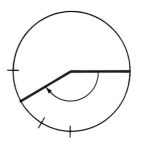

13. $\dfrac{\pi}{12}$

15. $-\dfrac{5\pi}{6}$

17. 3π

19. 150°

21. 270°

23. $-135°$

25. $\dfrac{5\pi}{6}$

27. $\dfrac{3\pi}{4}$

29. $-\dfrac{2\pi}{3}$

31.

33.

35.

37. 0 39. 0 41. 0 43. 0

45. $\cos \frac{\pi}{3} = \frac{1}{2}$, $\sin \frac{\pi}{3} = \frac{\sqrt{3}}{2}$

47.

θ	$\frac{7\pi}{6}$	$\frac{5\pi}{4}$	$\frac{4\pi}{3}$	$\frac{3\pi}{2}$	$\frac{5\pi}{3}$	$\frac{7\pi}{4}$	$\frac{11\pi}{6}$	2π
$\sin \theta$	$-\frac{1}{2}$	$-\frac{\sqrt{2}}{2}$	$-\frac{\sqrt{3}}{2}$	-1	$-\frac{\sqrt{3}}{2}$	$-\frac{\sqrt{2}}{2}$	$-\frac{1}{2}$	0
$\cos \theta$	$-\frac{\sqrt{3}}{2}$	$-\frac{\sqrt{2}}{2}$	$-\frac{1}{2}$	0	$\frac{1}{2}$	$\frac{\sqrt{2}}{2}$	$\frac{\sqrt{3}}{2}$	1

49. $-\frac{1}{2}$ 51. $\frac{1}{2}$ 53. $\frac{\sqrt{3}}{3}$ 55. 2

57. -1 59. $\frac{3}{4}$ 61. $\frac{3}{4}$ 63. $\frac{4}{5}$

65. $\theta = \frac{\pi}{2}$, $\theta = \frac{3\pi}{2}$, $\theta = \frac{\pi}{6}$, and $\theta = \frac{5\pi}{6}$

67. $\theta = \frac{\pi}{2}$, $\theta = \frac{\pi}{3}$, and $\theta = \frac{2\pi}{3}$

69. $\theta = \frac{\pi}{2}$ and $\theta = \frac{\pi}{4}$

71. $\theta = \frac{\pi}{2}$

73. $\theta = \frac{\pi}{6}$ and $\theta = \frac{5\pi}{6}$

CHAPTER 12, SECTION 2 (PAGE 668)

1. $f'(\theta) = 3 \cos 3\theta$

3. $f'(\theta) = -2 \cos (1 - 2\theta)$

5. $f'(\theta) = -3\theta^2 \sin (\theta^3 + 1)$

7. $f'(\theta) = 2 \cos \left(\frac{\pi}{2} - \theta\right) \sin \left(\frac{\pi}{2} - \theta\right) = \sin (\pi - 2\theta)$
$= \sin 2\theta$

9. $f'(\theta) = -6(1 + 3\theta) \sin (1 + 3\theta)^2$

11. $f'(\theta) = -e^{-\theta/2} \left(2\pi \sin 2\pi\theta + \frac{1}{2} \cos 2\pi\theta\right)$

13. $f'(\theta) = \frac{\cos \theta}{(1 + \sin \theta)^2}$

15. $f'(\theta) = -5\theta^4 \sec^2 (1 - \theta^5)$

17. $f'(\theta) = -4\pi \tan \left(\frac{\pi}{2} - 2\pi\theta\right) \sec^2 \left(\frac{\pi}{2} - 2\pi\theta\right)$

19. $f'(\theta) = 2 \frac{\cos \theta}{\sin \theta} = 2 \cot \theta$

21. $f'(\theta) = \tan \theta \sec \theta$

23. $f'(\theta) = -\csc^2 \theta$

CHAPTER 12, SECTION 3 (PAGE 677)

1. 60 radians per hour

3. 0.15 radian per minute

5. $\frac{\pi}{2}$ radians

7. $\frac{\pi}{3}$ radians

9. 8 feet

CHAPTER 12, REVIEW PROBLEMS (PAGE 682)

1. (a) $\frac{2\pi}{3}$ radians; 120° (b) $-\frac{5\pi}{4}$ radians; $-225°$

2. 0.8727 radian

3. 14.3239°

4. (a) $\frac{\sqrt{3}}{2}$ (b) $\frac{\sqrt{2}}{2}$ (c) 2 (d) $-\frac{1}{\sqrt{3}}$

5. $\frac{4}{3}$

6. $\frac{3}{2}$

7. $\theta = \frac{\pi}{2}$ and $\theta = \frac{3\pi}{2}$

8. $\theta = \frac{\pi}{3}$ and $\theta = \frac{2\pi}{3}$

9. $\theta = \frac{\pi}{6}$ and $\theta = \frac{5\pi}{6}$

10. $\theta = \frac{\pi}{6}$ and $\theta = \frac{5\pi}{6}$

14. $f'(\theta) = 6(3\theta + 1) \cos (3\theta + 1)^2$

15. $f'(\theta) = -6 \cos (3\theta + 1) \sin (3\theta + 1) =$
$-3 \sin (6\theta + 2)$

16. $f'(\theta) = 6\theta \sec^2 (3\theta^2 + 1)$

17. $f'(\theta) = 12\theta \tan (3\theta^2 + 1) \sec^2 (3\theta^2 + 1)$

18. $f'(\theta) = \dfrac{-1}{1 - \cos \theta}$

19. $f'(\theta) = -2 \tan \theta$

21. 0.012 radian per minute

22. $\dfrac{\pi}{3}$ radians

23. 27 feet

24. $\cos \theta = \dfrac{4}{5}$

APPENDIX (PAGE 702)

1. $1 < x \le 5$ 3. $x > -5$

5.

7.

9. 4   11. 5

13. $-3 \le x \le 3$ 15. $-6 \le x \le -2$

17. $x \ge 3$ or $x \le -7$ 19. 125

21. 4   23. 4

25. $\dfrac{1}{2}$ 27. $\dfrac{1}{2}$

29. $\dfrac{1}{4}$ 31. 2

33. $\dfrac{1}{4}$ 35. $n = 10$

37. $n = 1$ 39. $n = 4$

41. $n = \dfrac{13}{5}$ 43. $x^4(x - 4)$

45. $25(7 - x)$

47. $2(x + 1)^2(x - 2)^2(7x - 2)$

49. $2(x + 3)$

51. $\dfrac{2(x + 3)^3(5 - x)}{(1 - x)^3}$

53. $(x + 2)(x - 1)$

55. $(x - 3)(x - 4)$

57. $(x - 1)^2$

59. $(x + 2)(x - 2)$

61. $(x - 1)(x^2 + x + 1)$

63. $x^5(x + 1)(x - 1)$

65. $2x(x - 5)(x + 1)$

67. $x = 4$ and $x = -2$

69. $x = -5$

71. $x = -4$ and $x = 4$

73. $x = -\dfrac{1}{2}$ and $x = -1$

75. $x = -\dfrac{3}{2}$

77. $x = -5$ and $x = 1$

79. $x = -2$ and $x = 1$

81. $x = -\dfrac{1}{2}$ and $x = -1$

83. No real solutions

85. $x = -\dfrac{3}{2}$

87. $x = 3, y = 2$

89. $x = 4, y = 2$

91. $x = -7, y = -5$ and $x = 1, y = -1$

93. 34

95. 0

97. $\displaystyle\sum_{j=1}^{6} \dfrac{1}{j}$ 99. $\displaystyle\sum_{j=1}^{6} 2x_j$

101. $\displaystyle\sum_{j=1}^{8} (-1)^{j+1} j$

INDEX

Index